CRYSTAL CHEMISTRY AND PROPERTIES OF MATERIALS WITH QUASI-ONE-DIMENSIONAL STRUCTURES

PHYSICS AND CHEMISTRY OF MATERIALS WITH LOW-DIMENSIONAL STRUCTURES

Series B: Quasi-One-Dimensional Materials

CRYSTAL CHEMISTRY AND PROPERTIES OF MATERIALS WITH QUASI-ONE-DIMENSIONAL STRUCTURES

A Chemical and Physical Synthetic Approach

Edited by

JEAN ROUXEL

Laboratoire de Chemie des Solides,
CNRS, University of Nantes, France

D. REIDEL PUBLISHING COMPANY

A MEMBER OF THE KLUWER ACADEMIC PUBLISHERS GROUP

DORDRECHT / BOSTON / LANCASTER / TOKYO

Library of Congress of Cataloging-in-Publication Data
Main entry under title:

Crystal chemistry and properties of materials with quasi-one-dimensional structures.

(Physics and chemistry of materials with low-dimensional structures. Series B, Quasi-one-dimensional materials)
Bibliography: p.
Includes indexes.
1. Crystallography. 2. One-dimensional conductors. I. Rouxel, Jean, 1935– . II. Series.
QD906.6.C79 1986 530.4′1 85–30081
ISBN-13: 978-94-010-8516-8 e-ISBN-13: 978-94-009-4528-9
DOI: 10.1007/978-94-009-4528-9

Published by D. Reidel Publishing Company,
P.O. Box 17, 3300 AA Dordrecht, Holland

Sold and distributed in the U.S.A. and Canada
by Kluwer Academic Publishers,
190 Old Derby Street, Hingham, MA 02043, U.S.A.

In all other countries, sold and distributed
by Kluwer Academic Publishers Group,
P.O. Box 322, 3300 AH Dordrecht, Holland

CONTENTS

PREFACE

There is no doubt that in the development of the Physics and Chemistry of Solids during the last fifteen years, the very important place taken by low-dimensional compounds will be remembered as a major event. Dealing very widely at the beginning with two-dimensional structures and intercalation chemistry, this theme progressively evolved as the synthesis of one-dimensional conductors increased, along with the observation of their remarkable properties. Beyond the classical separation of the traditional disciplines, essential progress has stemmed each time from the concerted efforts of, and overlapping between, chemists, experimental physicists, and theoreticians. This book is a synthetic approach which aims to retrace these united efforts.

The observation and characterization of charge density waves in their static or dynamic aspects have been the main points to attract the interest of researchers. Two broad categories of compounds have been the material basis of these observations: transition-metal polychalcogenides and either condensed-cluster phases or bronze-type compounds. These families are referred to throughout the various chapters of this book, thus illustrating the continuous progress of concepts in this domain and, at the same time, providing the first synthetic and exhaustive view of this group of materials.

However, although written by specialists in one-dimensional compounds, this book is not meant for the sole use of specialists in this field. Not only are there very general considerations in each article, treating a given series of materials, but three important articles have been inserted at the beginning which provide a general introduction to one-dimensional crystal chemistry and phase transitions, along with orbital interaction calculations within the atomic chains. Because of their broad interest, these features go well beyond the narrower frame of one-dimensional conductors alone.

This book will also be novel for many chemists who only tend to see as one-dimensional conductors either the chain material belonging to coordination chemistry or organic conductors of various types. The book will not come as much of a surprise to physicists, who know most of the materials described in this volume well, because they are the very people who promoted the phases. However, they will find, gathered here, a great deal of chemical or structural information and, in particular, the description of a new series of materials, the potential properties of which are clearly apparent.

Today the one-dimensional conductor field remains a rapidly expanding one. Each year new questions are asked about, for example, the very nature of phase transitions, the origin of the threshold field of a charge density wave depinning, the concept of bond density wave. Already, unifying models have been proposed, placing the one-dimensional conductors and their striking properties in the wider context of the physics of low dimensionality and noncommensurate effects.

JEAN ROUXEL

AN APPROACH TO ONE-DIMENSIONAL COMPOUNDS

JEAN ROUXEL

Laboratoire de Chimie des Solides, Associé au CNRS,
2, rue de la Houssinière, F. 44072 Nantes Cédex, France

The concept of a low dimensionality for solids clearly refers to a high anisotropy in their chemical bonding. The structure of a two-dimensional (2D) or one-dimensional (1D) material results from a stacking of slabs or a juxtaposition of atomic chains. In such units there are strong ionocovalent or metallic bonds, but slabs and fibers are separated by rather large distances in agreement with weak interlayer or interchain bonding, mostly of the van der Waals type.

To this peculiar geometry there corresponds a certain number of specific properties. The first is straightforwardly found in the morphology of crystals. For example, the foliated crystallization of two-dimensional materials such as graphite or mica has been known for a long time. Easy splitting, in conjunction with weak bonding, allows us to remove thin plates in a manner parallel to the slabs. The one-dimensional materials that have been prepared in large numbers over the last few years outwardly possess a crystal shape in relation with their dimensionality. For instance, transition metal trichalcogenides are observed as very thin needles. $NbSe_3$ crystallizes as very narrow, long hair like strands.

From a chemical point of view solids with a low dimensionality provide us with most of the best host structures on which to practice intercalation chemistry. An intercalation compound is formed when ions or molecules are accepted by a host lattice. This topochemical process is reversible. It is possible to return to the initial state through appropriate electrical, thermal, or chemical actions. The definition of an intercalation compound assumes the idea of reversibility: this means that no strong bonding is broken in the host structure or formed between host and guest when intercalation takes place. Many tunnel structures are known which allow us to practice intercalation chemistry. However, in that case the host structure represents a rigid or quasi-rigid framework. The intercalation process is limited to those ions or molecules that have a suitable size to fit with the channels of the host. The ideal host structure would be deformable, thus adapting itself to each type of ion: this is precisely the case of solids with low dimensionality, and explains why intercalation chemistry developed essentially in that field. (On the other hand, a rigid framework may have some advantages in particular situations such as the search of molecular seaves or the use of size selectivity in catalytic processes by zeolites.) Among low-dimensional compounds that permit intercalation chemistry, conductors are of particular importance. Both the electronic conductivity of the host network and the ionic conductivity induced by the motion of the intercalated ions make these compounds excellent cathodic materials. In that case, the key words are electronic conductivity, ionic conductivity, and reversibility.

J. Rouxel (ed.), Crystal Chemistry and Properties of Materials with Quasi-One-Dimensional Structures, 1–26.

In an intercalation compound, the larger the degree of freedom of the intercalated species the lower the dimensionality of the host lattice. Whereas, for intercalation, a 3D solid can only offer cages or tunnels with 1D mobility, the 2D material will offer planes of mobility and the 1D compound will present a 3D space of mobility. By suggesting such a scheme, it is possible to say that the dimension of the space of mobility changes as the inverse of that of the host lattice, with a sum equal to four.

$$\text{Dimensionality of host} + \text{Dimensionality of mobility} = 4.$$

To the very strong crystal anisotropy of 2D and 1D materials there corresponds an important anisotropy of their physical properties. This, by itself, is the source of physical problems that are worthy of investigation, particularly concerning low-dimensional magnetism. However, the origin of the widespread interest born to the low-dimensional phases these last few years is to be found in the remarkable properties exhibited by the conducting phases. In agreement with a Fermi surface presenting large parallel portions, structural instabilities are observed which are related to charge density wave fluctuations. This generally leads to noncommensurate superstructures and can give rise to nonlinear electrical properties, observed above a threshold field and analyzed in terms of a Fröhlich mode, i.e. a sliding motion of the depinned charge density wave.

1. Structural Considerations

Solids of low dimensionality can be classified according to the nature of the slabs or fibers they contain and to the way these units either stack or juxtapose.

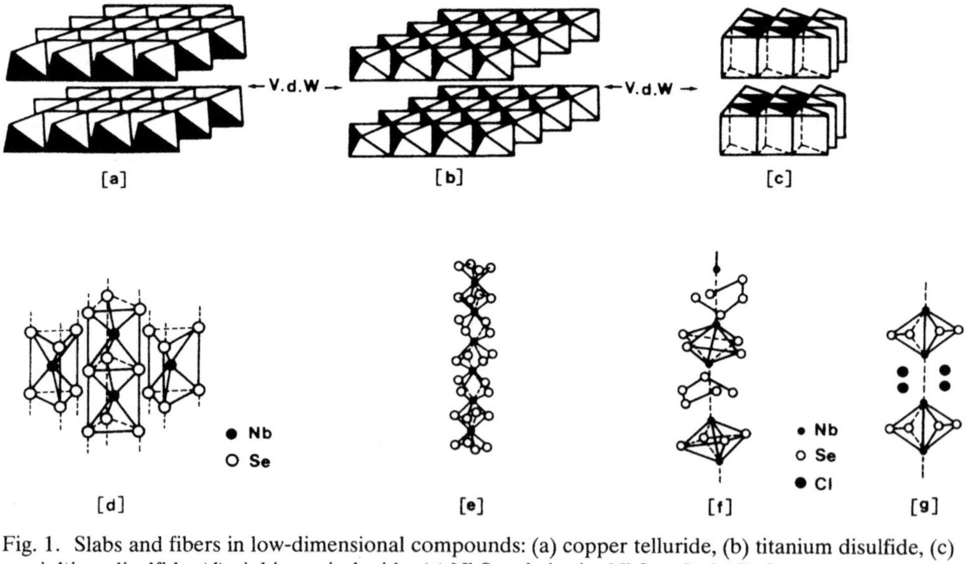

Fig. 1. Slabs and fibers in low-dimensional compounds: (a) copper telluride, (b) titanium disulfide, (c) niobium disulfide, (d) niobium triselenide, (e) $NbSe_4$ chains in $(NbSe_4)_3I$, (f) Nb_2Se_9, (g) $NbSe_2Cl_2$.

The slab is a more or less simple atomic arrangement. It may consist of an atomic layer, as in the case of graphite where carbon atoms are bound in a honeycomb lattice. However, more generally, the slab will be made of coordination polyhedra such as tetrahedra, octahedra, trigonal prisms, etc. (Figure 1). The same polyhedra will be found in 1D structures, but the symmetry conditions in that case will also permit the existence of other structural units such as rectangular or square anti-prisms [1—4].

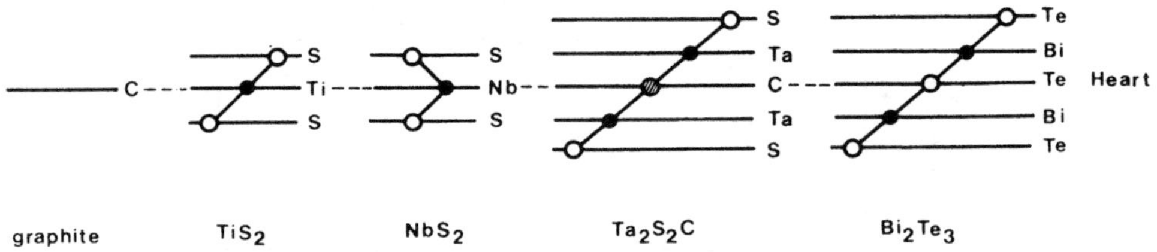

Fig. 2. Atomic layers in various types of slabs.

Thus, there are two ways of describing a slab or a fiber: one may consider the polyhedra array (which is generally done in the silicates family), or one may look at the structure as the result of a certain global atomic arrangment. The latter description is interesting in that it shows some construction rules. A slab contains atomic layers that have a symmetrical arrangement in respect of a central plane which lies at the heart of the slab (Figure 2) and which can be a simple symmetry plane or a gliding plane. The fiber sets itself around an axis which is often a rotation axis or a screw axis.

The van der Waals gap, between slabs or fibers, is bounded on each side by atomic arrangements of the same nature. Besides geometrical factors, its width will depend on the electronegativity of these atoms. For a given structural type to be stable the bonding through the van der Waals gap must stabilize the structure in front of the repulsion between the same atomic layers situated on each side. In the case of oxides this repulsion is quite strong and very destabilizing. The rutile structure is observed in most of the MO_2 transition metal oxides, and not the layered forms of the parent MS_2 chalcogenides. However, Na_xMO_2 derivatives have been prepared and are often isostructural with Na_xMS_2 intercalation compounds [5]; the sodium ions located in the van der Waals gap stabilize the 2D arrangement. If one draws an ionicity—structure diagram, layered compounds cannot be found in the regions corresponding to the highest ionicity of the bonds (or highest electronegativity difference Δx of the components). Figure 3 shows the classical ionicity—structure diagram of the Mooser—Pearson type for AB_2 compounds [6]. Layered oxides can exist only for the highest oxidation states of transition metals (less ionic character) or if $(OH)^-$ groups are present allowing hydrogen bonds to stabilize the structures.

From these general frames various possibilities exist. In particular, partial or total substitutions can be performed in the anionic as well as in the cationic planes. In TiS_2, titanium can be substituted 2/3 by a M^{2+} cation and 1/3 by a (P—P)

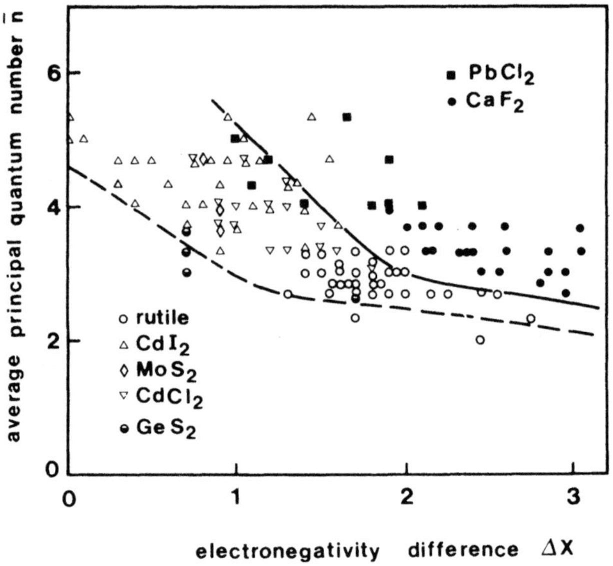

Fig. 3. Mooser—Pearson ionicity—structure diagram for AB_2 derivatives.

pair leading, for example, to a $NiPS_3$ slab [7]. In the same octahedral-type slab the occurrence of vacancies illustrate the structure of many MX_3 halogenides [$XM_{2/3}\square_{1/3}X$].

The slabs exhibit a large relative freedom because of the weak van der Waals bonding that ties them together. This is related to the way they stack and can be illustrated by the numerous polytypes that are found, for example, for tantalum dichalcogenides. In that case one can even find for $TaSe_2$ a particular structure (4Hb in the Ramsdell notation [8]) which consists of alternate stackings of octahedral and trigonal prismatic slabs. Other complications may still arise from folding or distortion of the slabs or from the occurrence of clusters related to the setting of metal—metal bondings: this is the case of the high-temperature form of $MoTe_2$ and of rhenium chalcogenides.

In these materials of low dimensionality, one can find finally a few of the best examples of modulated structures. It is possible to imagine atom density waves, as suggested by F. Jellinek, for titanium sulfides [9], or even more simply, distribution waves among the slabs.

All these considerations, which are essentially based on 2D materials because those are the types of phases in which the more numerous examples have been found until now, can be entirely applied to 1D phases.

What, then, are the factors which allow us to go from a 2D structure to a 1D model?

From a purely geometric point of view, it is convenient to describe a structure as being made of coordination polyhedra constituted by the anions around the cations.

We may point out at once that a 3D or even a 2D construction implies that the apex of the polyhedra can be shared and that this is possible only beyond a minimum coordination number. With a coordination number of 2, only chains or finite molecules can form. But we are interested in discovering the situations in which the usual polyhedra — as tetrahedra, octahedra and trigonal prisms — can lead to chains rather than sheets or 3D structures.

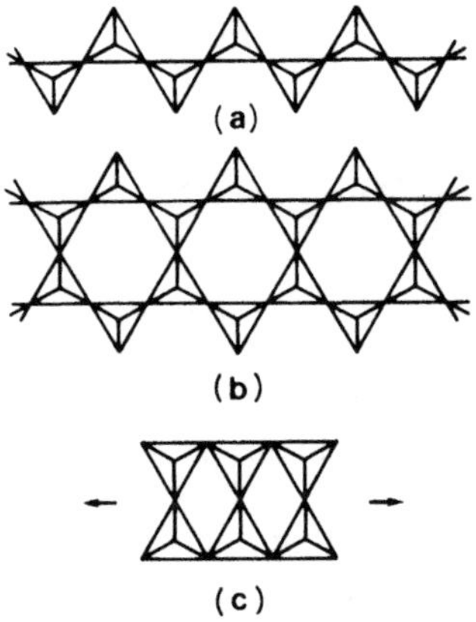

Fig. 4. Chains made of sharing corners tetrahedra in (a) pyroxenes, (b) amphiboles, (c) asbestos.

With tetrahedra, chains can be made by joining them together through apex sharing. This very simple type of array is well illustrated in some silicates (e.g. Li_2SiO_3) and pyroxenes (e.g. diopside $CaMg(SiO_3)_2$; see Figure 4(a)). The chain can degenerate in ribbons if three apexes are shared, as in the mineral amphibole with chains made of hexagonal rings of tetrahedra (Figure 4(b)). Ribbons made of pairs of tetrahedra are found in asbestos (Figure 4(c)). Similar arrangements may be found in other derivatives such as metaphosphates or CrO_3. Those sharing corners structures minimize cationic repulsions in structures which remain largely ionic and often need the presence of interchain cations to be stabilized. When covalency increases, chains of tetrahedra sharing edges can be found, as in the classical models of SiS_2 or $KFeS_2$ (Figure 5). $GaPS_4$ shows an interesting intermediate situation of tetrahedra sharing one edge and two corners. $CsCu_2Cl_3$ also represents a particular situation with SiS_2-type chains associated in ribbons of two chains.

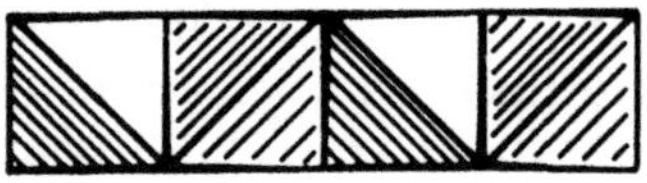

Fig. 5. Sharing-edge tetrahedra in SiS_2 or $K[FeS_2]$ chain structures.

The formation of chains from octahedra could result in the sharing of two opposite apexes (Figure 6(a)), of two opposite edges (Figure 6(b)), or of two opposite

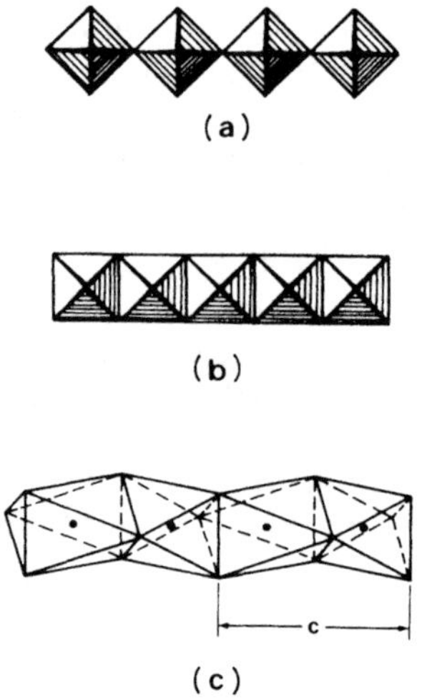

Fig. 6. Building chains from octahedra. (a) BiF_5 type, (b) $CoCl_2$, $2H_2O$, (c) $CsNiF_3$.

faces (Figure 6(c)). The examples that can be given to illustrate these possibilities are very few and most of the time are related either to chains that are here again stabilized by the occurrence of cations or to chains with reduced anionic charges as some free apexes are occupied by water molecules or $(OH)^-$ groups. In Figure 6, BiF_5 and UF_5 correspond to (a), $CoCl_2 2H_2O$ to (b), and $CsNiF_3$ to (c). Pentavalent oxyhalides MOX_3 of transition metals such as Nb, Mo, or Ta exhibit double chains of octahedra sharing corners.

Although the trigonal prism corresponds to a coordination of six, like the octahedron, it is found in much less ionic systems. The octahedron represents the best balance between similar charges and it is indeed found in oxides and in compounds with strong anionic charges. The trigonal prism appears in less ionic chalcogenides. Considering the layered chalcogenides MS_2, the prism, with D_{3h} symmetry, results in the splitting of the wide t_{2g} band observed in *Oh* mode for TiS_2 for instance, with a resulting strong stabilization for d^1 or d^2 systems [10—11]. Let us then consider $NbSe_2$, which is a layered selenide made from sheets of prisms sharing edges. The most straightforward 1D construction we can imagine consists of stacking prisms above each other in order to form a trigonal prismatic frame in which the cross section is constituted by the common basis of all the prisms. This is the situation encountered in $NbSe_3$, and corresponds to Nb^{4+} with a d^1 configuration, one Se^{2-} ion, and a $(Se_2)^{2-}$ pair. Depending on the relative positions of *d* metal orbitals and selenium orbitals, a hole is created on selenium leading to a $(Se_2)^{2-}$ pairing and causing a deformation of the structure [12]. This situation, commonly found in numerous chalcogenides, occurs each time the (maximum —1)+ oxidation state of the metal (here 4+ for niobium) lies low enough below the top of the *sp* anionic valence band. A cation is reduced and holes are created in the valence band [13].

Starting with the use of tetrahedra, octahedra, and trigonal prisms as modules in a construction game, we have imagined various ways of building 1D structures.

A much more exhaustive description of all the geometric possibities is given by Hulliger [14], whereas Day [15] classified the 1D compounds in respect of the co-ordination of both the metal atoms and the number of bridging groups. Then the obtention of a metallic conduction implies in a given structure direct cation—cation interactions. This excludes the chain structures formed from polyhedra sharing corners. The sharing of edges allows, in principle, direct interaction but actually, to date, 1D conducting systems are concerned with the most favorable case — that is, that of chains formed through a face-sharing arrangement.

Classical examples are now those of $NbSe_3$, of the tetrachalcogenides of the $(NbSe_4)n$-halogen type, and of Krogmann salts that show still more open structures. The metallic chain hence spread amid a frame made by the ligands. This frame is relatively covalent and rather open so as to lessen the potential variations. The models are then related to direct interaction of the metal *d* orbitals and form a conduction band with a mainly dz^2 character in the direction of the chain axis (Figure 7).

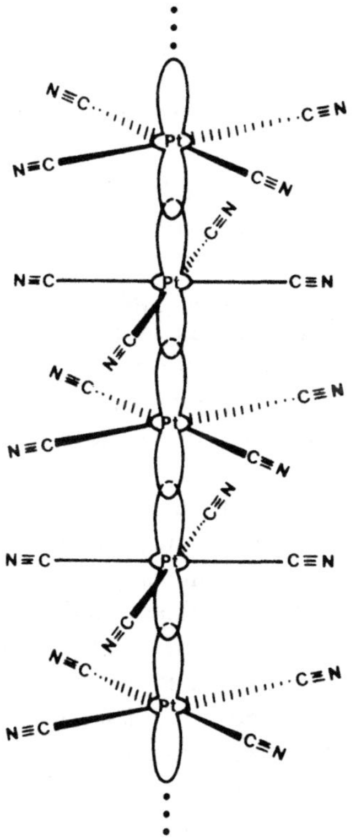

Fig. 7. dz^2 interactions in linear platinum chains of Krogmann's salts.

The occurrence of such metallic chains with possible direct interactions does not mean that there will necessarily be metallic conduction. As we shall see in Section 3, distortions associated with atomic pairing along the chain may lead to a semicon-

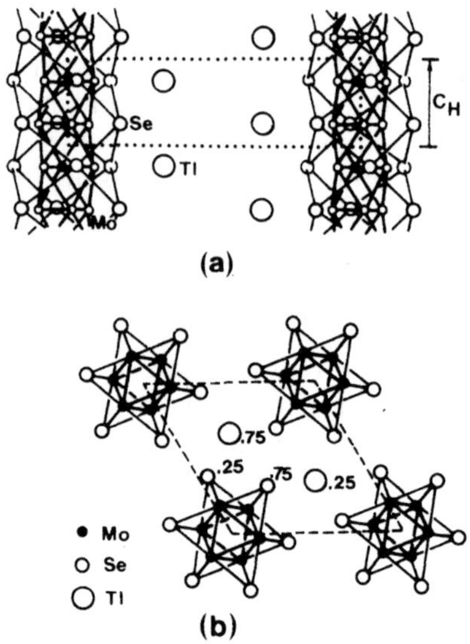

Fig. 8. Condensation of Mo_6 clusters in $Tl_2Mo_6X_6$. Projection of the structure on (a) the $11\bar{2}0$ plane and on (b) the 0 0 1 plane.

ducting state. Examples thus abound of such insulating compounds of transition elements corresponding to a d^1—d^1 pairing. Remaining with Nb^{4+}, let us mention the case of $NbCl_4$.

During the last few years a particular type of chemistry based on a condensation of metallic clusters has been developed by Simon [16], Corbett [17], and Sergent and Chevrel [18]. Layers or chains of closely spaced metal atoms may be found in these compounds (Figure 8) which are essentially metal-rich halides, oxides, or chalcogenides. Finally, we must not forget the properties of polymeric structures such as $(SN)_x$.

All the types of 1D compounds discussed above will be successively considered in this book: linear chain compounds, in particular tri- and tetrachalcogenides of niobium and tantalum, pentachalcogenides, phases with extended condensed clusters (oxides, chlorides, and chalcogenides), and molybdenum bronze type compounds which represent a promising new category of materials with typical 1D effects. Hence, it is no longer necessary to consider them in this introduction. Rather we are going to envisage the two essential sources of interest that make the 1D materials so attractive: their use in intercalation chemistry, and the phase transitions and remarkable physical effects associated with a 1D conduction. This last point is the best justification for this book.

2. 1D Structures for Intercalation Chemistry

We shall only briefly mention this aspect, not because it is unimportant but because the theme of intercalation chemistry has already been widely treated these last few years [19—20]. 1D materials show, however, a few interesting specificities that are useful to note in the general frame of modern concepts in intercalation chemistry.

If we look for host structures in order to do intercalation chemistry, two conditions will have to be fulfilled: (i) there must be a skeleton of 1D chains with ions moving easily between them; (ii) there must be, within the chains, redox centers able to accept electrons donated by the intercalated species. It is in fact the second condition which generally determines the intercalation occurrence. In other words, the structure must possess an empty level with an associate energy low enough to avoid any destabilization. This can be a discrete atomic level, or it can be a molecular level of a polyatomic entity existing in the structure, and it can also be situated in the conduction band.

These three situations also correspond to three different levels of delocalization for the transferred electron. Let us illustrate them briefly with a few examples. The first case corresponds to a reduction on an ion in the host network; for instance, Fe^{3+} to Fe^{2+} in Li_xFeOCl intercalates. Actually, the localization is only complete at low temperatures, otherwise a hopping mechanism between Fe^{3+} and Fe^{2+} ions induces an electronic conductivity in the slabs. The second situation occurs when the electron is partially delocalized among a finite number of directly bonded metal atoms. This is the case of Chevrel phases built up from $(Mo)_6$ octahedral clusters enclosed in a pseudo-cube of eight chalcogen atoms [21]. Extra metal atoms can be seated between the Mo_6X_8 units. Schöllhorn *et al.* [22] and Steele *et al.* [23] developed a chemistry of exchange and intercalation reactions in Chevrel phases. the Mo_6 cluster controls the process. It can accommodate four electrons more than it has. It will be possible to introduce 4 lithium, 2 zinc, etc., in the structure. The third case is that of a delocalization in the conduction band. This is encountered in the layered chalcogenides such as TiS_2 or NbS_2. If the band is wide — for example, a t_{2g} band with six electrons to be accepted in TiS_2 — there is no electronic limitation, only the available geometrical sites can be limiting. On the other hand, if the band is narrow and partially filled, things are different. This is clearly illustrated by TaS_2 which can be prepared in two forms: a 1T form with octahedral slabs and a broad t_{2g} band, a 2H form with trigonal prismatic slabs and a narrow half-filled d_{z^2} band. In both cases similar sites are available in the van der Waals gap. They can be easily filled in the 1T structure, but not in the 2H structure [24].

To be able to separate the particularities of a 1D intercalation system, it is necessary to recall a basic concept of intercalation chemistry. We think that the driving force for intercalation chemistry is to be found in a favorable competition between the two main energy terms in the system, i.e. the cost in elastic energy associated to the distortion of the host when intercalation takes place, and the gain in electronic energy due to the difference between the initial and final levels of the tansferred electrons (Figure 9(a)). Of course, other terms such as configurational entropy and ion—electron interactions are to be considered, but they are quite 'corrective' terms compared to the two others. The MPS_3 series of phases clearly illustrate that problem. They are semiconducting host materials for lithium intercalation. When the band gap increases, i.e. the accepting levels get higher and higher, intercalation becomes more and more difficult and even stops in $MnPS_3$ (Figure 9(b)). On the

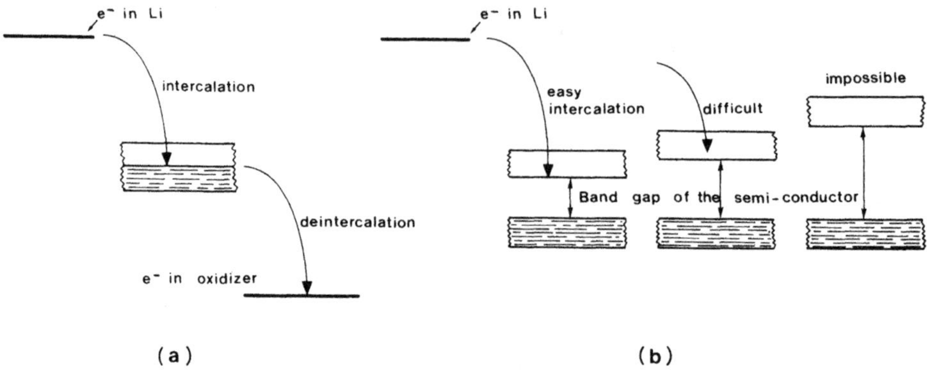

Fig. 9. Electronic transfers and intercalation—deintercalation processes.

other hand, a deintercalation process may occur if an oxidizing reagent presents accepting levels lower than the last occupied levels of the host. This competition between elastic and electronic energies also explains the phase transitions in a given intercalation system, such as $NaTiS_2$ [25]. It also sheds some light on the ionicity—structure diagram of intercalates. The ratio of ionic radius indirectly expresses in this case the strain effect, whereas the functions of ionicities are related to electronic transfer [26].

What are the advantages and disadvantages of using 1D structures to intercalate? As has already been said, the fiber interspace is much more favorable to a good diffusion of intercalated ions. It is also to be noted that diffusion will not be limited to a definite direction by a tunnel or a flat van der Waals gap as in 3D or 2D structures. This will limit the preferred orientation unfavorable effects that appear on packing of powdered samples.

In a 1D compound the fibers have a higher relative freedom and they are also more flexible than in a 2D structure. This will minimize the cost in elastic distortion energy.

On the other hand, in 1D conductors the conduction band is always narrow and is essentially made from dz^2 orbitals along the chain. This is an unfavorable factor. To face this last drawback while keeping the low-dimensional advantages, we have searched for new low-dimensional materials possessing discrete redox centers to be reduced by accepting transferred electrons. These redox centers are no longer reducible metallic cations, but are anionic groups and, in particular, S_2^{-2} pairs. In chalcogenide chemistry, S_2^{-2} or Se_2^{-2} pairs begin to appear in the MS_n (or MSe_n) formula when n becomes larger than 2. One leaves the layered models (MS_2 or MSe_2) to obtain, for example, $NbSe_3$ with one (Se_2^{-2}) pair, $NbSe_4$ with two pairs, or Nb_2Se_9 with a (Se_5) complex group.

As an other example, by increasing the sulfur content in M—P—S systems one leaves the layered MPS_3 phases to get compounds that are new, low-dimensional, and in particular 1D, presenting $(S—S)^{2-}$ pairs. These latter phases are noteworthy new systems for intercalation. The numerous S—S pairs constitute as many redox

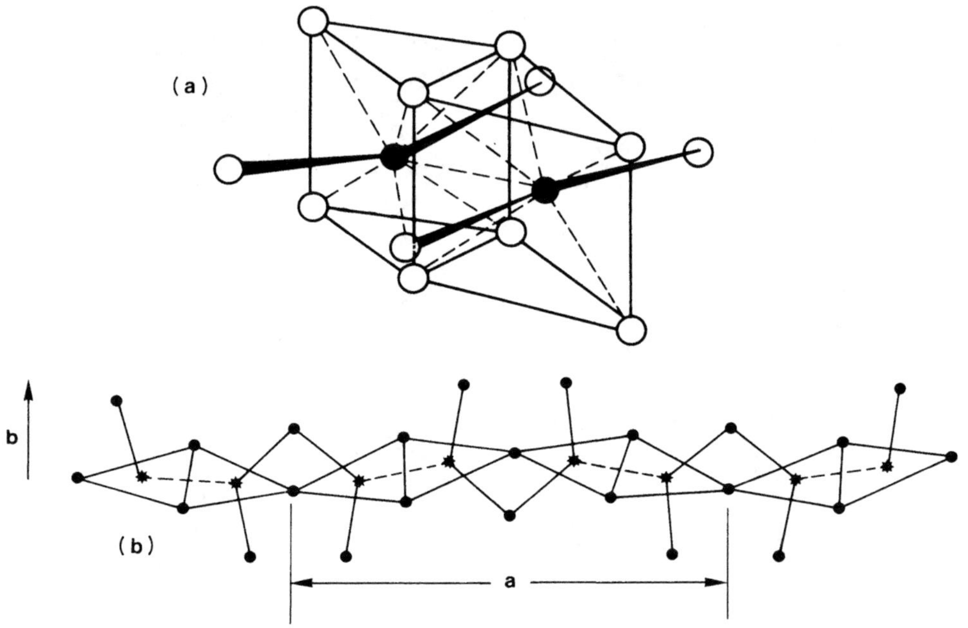

Fig. 10. (a) Bicapped M_2S_{12} biprisms as building units of (b) $(M_2S_9)_\infty$ chains.

centers located at the chain edges, and because of an expectedly high level of elasticity, should lead to fair reversibility.

Thus low-dimensional phases (2D and 1D) have been found in the P—V—S and P—Nb—S systems, with the general formula $MP_2S_{10+\varepsilon}$ ($\varepsilon = 0$ or 0.5). To date three phases were characterized [27, 28]; they are (1D) PV_2S_{10}, (2D) PNb_2S_{10}, and (2D) $PNb_2S_{10.5}$ ($P_2Nb_4S_{21}$). Although presenting either a dimensionality or a chemical composition different from each other, the compounds derive from the same basic building units. A structural comparison between the three phases show in effect that:

(1) all the phases utilize, as building units, bicapped biprisms $[M_2S_{12}]$ (Figure 10(a));

(2) all these biprisms are bonded together to form infinite puckered $[M_2S_9]_\infty$ chains (Figure 10(b));

(3) the bonding between these chains takes place through $[PS_4]$ units, and the way such bondings are made determines the dimensionality of the phases (see Figure 11(a)—(e)) which is:

(i) 1D (PV_2S_{10}) when the $[M_2S_9]$ lines are linked two by two (front linking);

(ii) 2D (PNb_2S_{10}) when PS_4 groups unite the chains on both their sides (side linking);

(iii) 2D ($PNb_2S_{10.5}$) when the phosphorus sulfide groups link the chains above and sideways to each other (step linking).

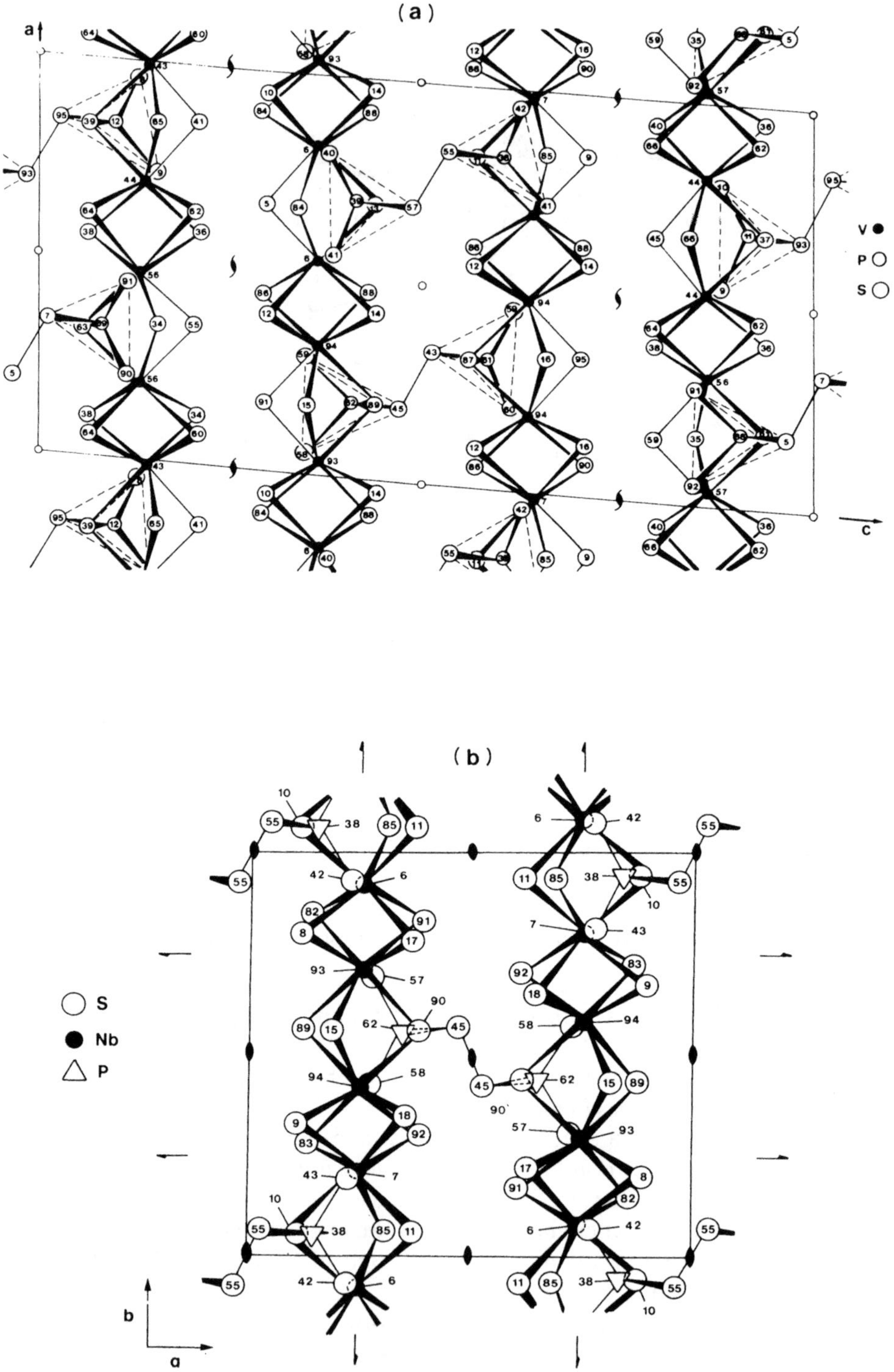
(a)
V
P
S
a
c
(b)
S
Nb
P
b
a

(c)

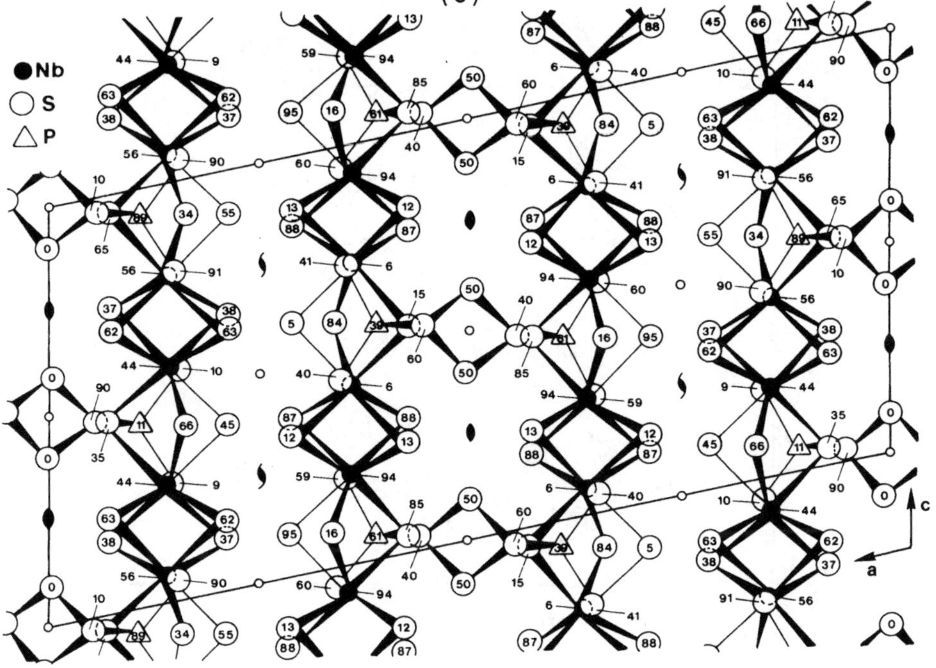
Nb
S
P
c
a

(d)

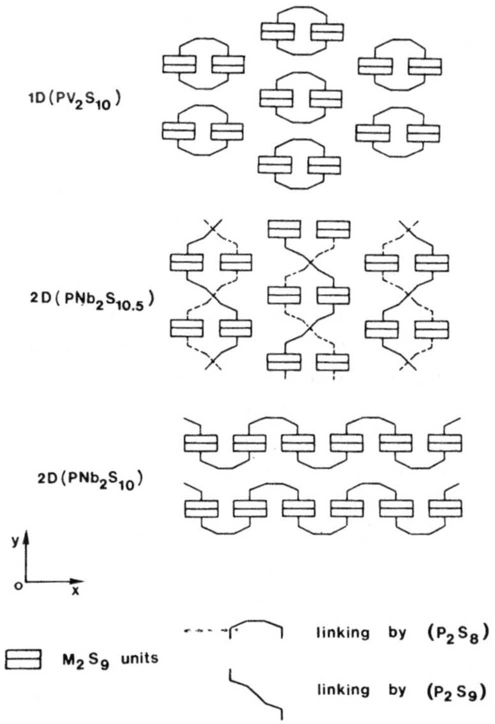
1D (PV_2S_{10})
2D ($PNb_2S_{10.5}$)
2D (PNb_2S_{10})
y
o
x
M_2S_9 units
linking by (P_2S_8)
linking by (P_2S_9)

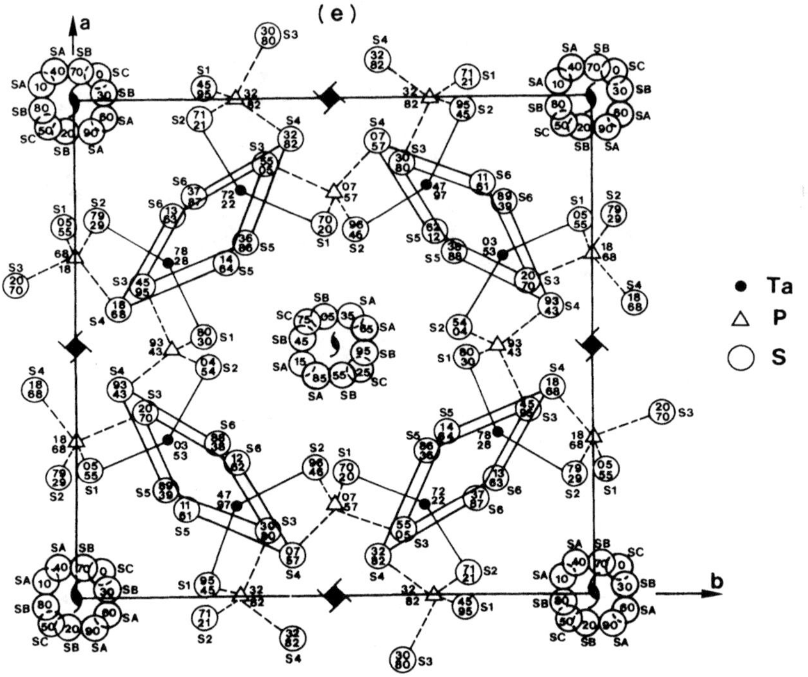

Fig. 11. (a) 1D structure from M_2S_9 chains: PV_2S_{10} with ribbons of two chains linked by PS_4 units. Fig. 11(b). 2D structure from M_2S_9 chains: PNb_2S_{10} with chains linked on each side. Fig. 11(c). 2D structure from M_2S_9 chains: $PNb_2S_{10.5}(P_2Nb_4S_{21})$ with a 'step linking' of the chains. Fig. 11(d). 1D and 2D structures from M_2S_9 chains: schematic representation. If the figure plane is referred as $x0y$ (z being the axis of the M_2S_9 chains) twinned fibers develop along z in the case of 1D PV_2S_{10}, slabs develop parallel to $y0z$ for $PNb_2S_{10.5}$ and to $x0z$ for PNb_2S_{10}, respectively. Fig. 11(e). 1D and 2D structures from M_2S_9 chains: $P_4Ta_4S_{29}$ with large channels containing polymeric sulfur.

The occurrence of the last structural type (step linking) is made possible by the extra half sulfur atom which constitutes longer P_2S_9 (S_3P(—S—S—S—)PS_3 linking groups, whereas only P_2S_8 (S_3P(—S—S—)PS_3 linking units are found in PV_2S_{10} and PNb_2S_{10}.

In respect of such a structural arrangement, higher sulfur content results in anionic catenation, as shown in $PNb_2S_{10.5}$ with the occurrence of S_3^{-2} anions in addition to the S_2^{-2} and S^{-2} species already present in the structures.

This last aspect reveals another originality in respect to the classical systems of intercalation. It constitutes a first approach towards the separation of functions, towards composite structures with some elements determining the frame in which intercalation takes places whereas some others constitute redox centers. This point is fully exemplified by $P_4Ta_4S_{29}$ in which M_2S_{12} biprisms and PS_4 enclose a large channel containing long chains of polymeric sulfur [28].

The preparation of new 1D materials presenting anionic pairs will emphasize the use of these compounds as host structures in intercalation chemistry. Anionic pairs may function as redox centers, depending on their relative position towards the empty cationic levels. This approach is the reciprocal of the oxidation of sulfides to polysulfide anions preferentially to an oxidation of Fe^{2+} to Fe^{3+} in iron sulfides.

Finally, we must take notice that intercalation chemistry can be used to modulate the properties of 1D conductors (next section). Through intercalation the Fermi level is modified, this one defining the distortion vectors to be expected during phase transitions of the charge density wave type.

3. 1D Materials as Conductors

In low-dimensional conductors structural instabilities may appear in connection with the occurrence of charge density wave (CDW) fluctuations. This results in some very important consequences that are related at the same time to the very structure of matter and to the experimental manifestation of new physical phenomena. Indeed, CDW usually lead to noncommensurate structures. Pretransitional effects illustrate lateral correlations between chains. Depinning of CDW emphasizes the role of impurities, the gap to commensurability, and may suggest the existence of discommensuration domains separating commensurable regions. 1D conductors strongly depart from the classical model of metals in which single particle scattering processes determine the electrical resistivity and are illustrated in the Ohm's law and a frequency-independent conductivity. In 1D conductors nonlinear effects have been observed at moderate fields and the transport properties may also be frequency dependent.

3.1. Nature and Origin of Instabilities

Charge density waves (CDWs) are modulations of the electronic density which may occur in low-dimensional conductors, as predicted by Overhauser [29]. In this state, the electronic density at a point represented by a vector position **r** along a metallic chain can be written as

$$\rho(\mathbf{r}) = \rho_0[1 + \alpha \cos(\mathbf{q} \cdot \mathbf{r} + \phi)],$$

where ρ_0 is the average electronic density as before the oscillation, **q** the vector of the charge oscillation, α the relative amplitude modulation, and ϕ the phase. The phase ϕ states precisely the position of the CDW in respect to the ions of the lattice.

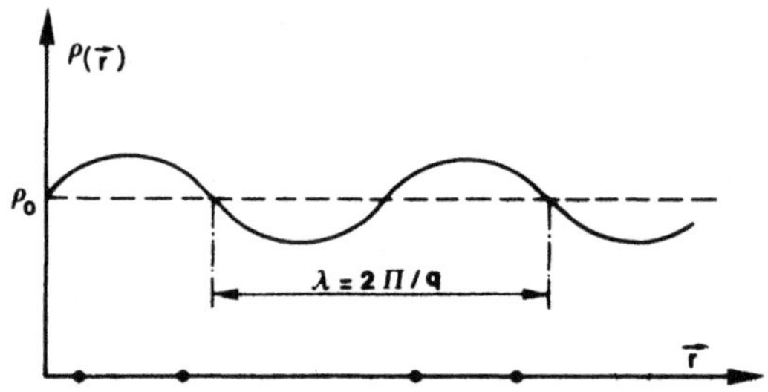

Fig. 12. Charge density wave and periodic lattice distortion along a metallic chain.

An atomic periodic rearrangement appears at the same time. Positive ions are drifted towards each maximum of electronic density (Figure 12): the amplitude of the displacement being small in front of the periodicity a of the network.

Finally, a charge density wave state is a coupled distortion of both the electron density (CDW) and the lattice (periodic lattice distortion PLD). The distortion opens a gap in the conduction band which, in a metallic case with an incompletely filled band, will lead to a lowering of the electronic energy. If the gain in electronic energy exceeds the cost in elastic energy necessary to distort the structure, the coupled distortion may appear. As the temperature is increased electrons are excited across the energy gap and a distortion, stable at 0 K, will disappear above a critical temperature.

The reason why 1D and 2D systems are concerned is to be found in the expression of the generalized electronic susceptibility $\varkappa_q$ of an electron gas (Lindhard function):

$$\varkappa_q = \sum_k \frac{f_{(k)} - f_{(k+q)}}{\varepsilon_{k+q} - \varepsilon_k},$$

where $f_{(k)}$ is the Fermi—Dirac distribution function of the state k with energy ε_k. Indeed, the response $\Delta\rho(\mathbf{r})$ of a free electron gas to a periodic external potential $V(q, r)$ is expressed through the Lindhard function

$$\Delta(\rho) = \varkappa_q \cdot V(\mathbf{q}, \mathbf{r}).$$

The Lindhard function is represented on Figure 13 for 3D, 2D, and 1D systems at 0 K. In the case of an ideal 1D system this function, which is calculated to be equal to $\ln|(2k_F + q)/(2k_F - q)|$, diverges for $q_c = 2k_F$. The ideal case is represented

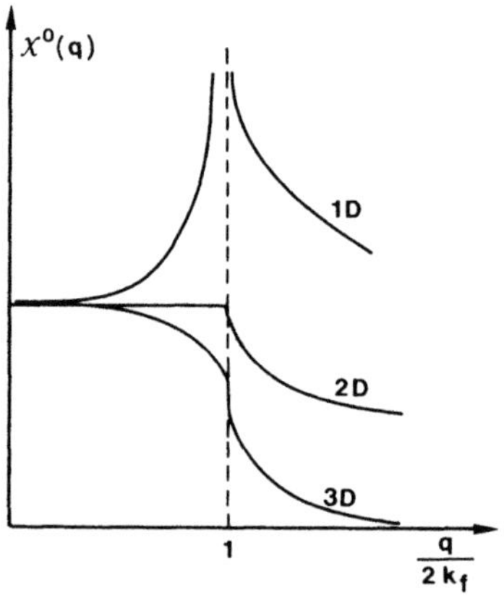

Fig. 13. The divergence of the Lindhard function at $q = 2k_F$.

by 1D conductors in which all k states are occupied up to $|k_x| = k_f$. For a particular value of q the states connected are not on the Fermi surface (FS). But for the critical value $q_c = 2k_F$ both sides of the FS, at $\pm k_f$, will be brought into coincidence (Figure 14). We have a plane of degeneracy (with $\varepsilon_k = \varepsilon_{k+q}$), which is at the origin of the divergence.

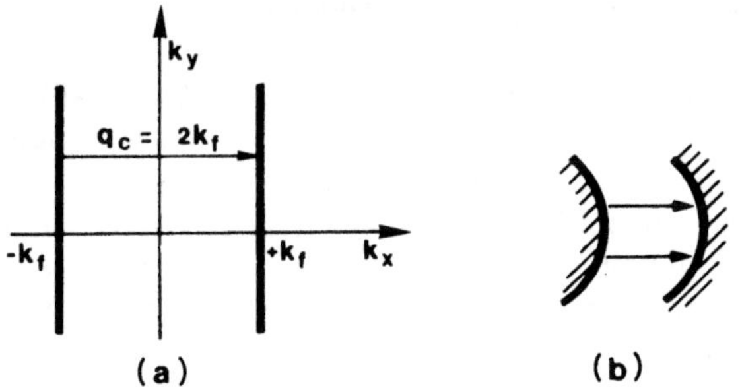

Fig. 14. (a) Perfect nesting in 1D materials and (b) partial nesting in cylindrical portions of the Fermi surface of a 2D material.

In pseudo-1D systems, the greater the number of states of the same energy which are coupled by q_c (limited nesting condition), the higher the instability taking place.

Because the q vector of the distortion comes from the FS, there is no reason why the distortion should be commensurate with the underlying sublattice, and generally it is not.

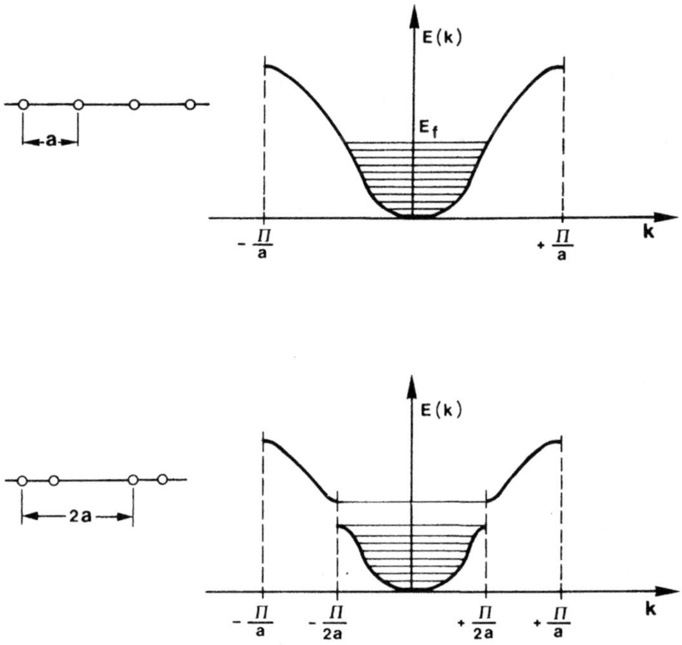

Fig. 15. Peierls distortion.

3.2. From Krogmann's Salts to $NbSe_3$

Such phase transitions leading to a coupled system CDW—PLD are called Peierls distortions [30]. If the compound is purely 1D the FS is completely destroyed at the transition (perfect nesting of a FS consisting of two parallel plans) and the compound becomes a semiconductor. Let us consider, for example, a 1D conductor with periodicity a and presenting a half-filled conduction band (Figure 15). The opening of a gap at the Fermi level leads to a $2a$ periodicity, illustrated by a pairing of the atoms along the chain.

This type of transition may happen whatever the position of the Fermi level. The Krogmann's salts have played an essential role in the development of that type of research. Indeed the 1D conductor $K_2Pt(CN)_4Br_{0.30}$, xH_2O provided in 1973 the first experimental evidence of a Peierls instability in a real system [31]. Metallic conductivity is related to a 5/6 filled dz^2 band and the transition leads to a superstructure involving six Pt—Pt distances.

The studies on niobium triselenide started immediately after its characterization in 1975 [32]. $NbSe_3$ departs largely from a purely 1D model. The transition does not lead to an insulating state. When the temperature is decreased, the metallic-like resistivity curve shows two strong anomalies, typical of the manifestation of CDWs. An increase in resistivity, observed at each transition, corresponds to the diminution of the area of the Fermi surface in relation with the opening of gaps at those portions which satisfy the nesting condition. Three different $NbSe_3$ chains are present in the structure. The two CDWs are sited along two of them, the third probably being insulating.

Such a phenomena gives, through its nature and its physical manifestations, the means for its own characterization. Electron diffraction studies, neutron or X-rays, allow us to visualize the superstructure which gives the q vector of the distortion. The application of a pressure, which increases the strain energy, will impede the CDW formation [33]. Hall effect measurements represent a convenient way to observe a change in carrier concentration at the transitions. R_H increases, indicating a loss of carriers. This is well in connection with the opening of gaps at the FS when CDW takes place. But the measured values does not correspond only to a loss of carriers. A change in the mobility anisotropies occurs simultaneously in relation with the structural distortion. A better approach to the real loss of carriers is given by specific heat measurements [34]. The change in entropy at the transition can be estimated from the areas under the C_p versus temperature peaks at the transition, and compared to the total value calculated from density of states provided by the band structure.

A more 1D structure is achieved when changing niobium for tantalum (more electropositive) and selenium for sulfur (more electronegative). This leads to a full metal-to-insulator transition observed, for example, at 240 K in monoclinic TaS_3 [35]. In addition, this material emphasizes an other important point of the behavior of 1D conductors, i.e. the existence of pretransitional effects well above the upper temperature. This is manifested by diffuse lines in the diffraction experiments and corresponds to 1D fluctuations of the CDWs. Above the transition their modulations exist dynamically with phase incoherence between chains. Below the transition the phases of each individual CDW become coherent with each other and a long-range 3D order is established. The diffuse lines condense into superlattice rows of spots. These different points will be emphasized in the article devoted to tri-chalcogenides.

3.3. CHARGE DENSITY WAVE DEPINNING

$NbSe_3$ and TaS_3 exhibit a very striking nonlinear resistivity behavior below the CDW transitions. The voltage—intensity characteristic, $V(I)$, becomes nonlinear beyond a threshold current I_c (Figure 16 for TaS_3). For a better accuracy, one usually measures the differential resistance $\mathrm{d}V/\mathrm{d}I$ as a function of the direct current I that is passing through the sample. In a linear regime, $\mathrm{d}V/\mathrm{d}I$ represents the ohmic resistance of the sample. The difference from that value defines afterwards the non-linear regime. The occurrence of a critical current, related to a threshold field, is shown clearly on Figure 17, which represents an example given by $NbSe_3$.

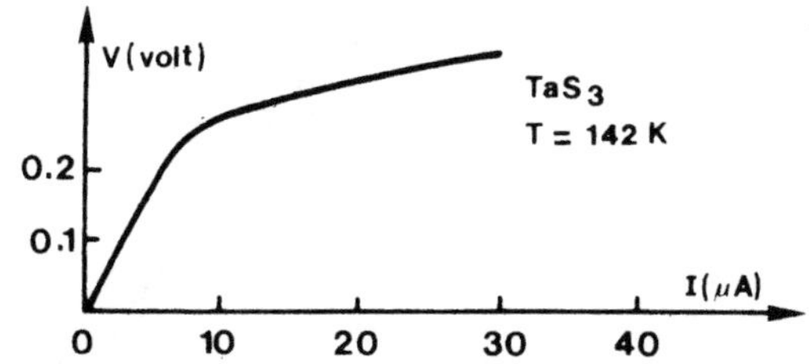

Fig. 16. Voltage—intensity characteristic showing the appearance of a nonlinear regime.

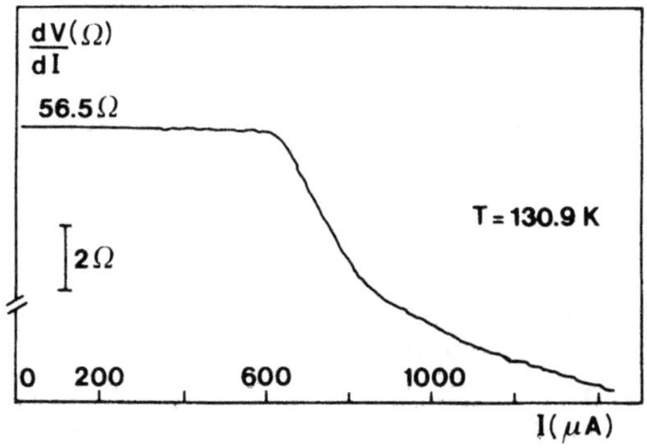

Fig. 17. Differential resistance $\mathrm{d}V/\mathrm{d}I$ as a function of I in the case of $NbSe_3$

The nonlinear properties are to be associated with a motion of the CDW. Below the threshold field the CDW was pinned to the lattice. A sufficient electric field is necessary to overcome this pinning energy. Various theories have been proposed to better understand this phenomenon (see, for example, [36—38]). These theories will be discussed elsewhere in this series of books, but the nonlinear effects are strongly dependent on the crystal chemistry of the materials and, consequently, must be dealt with here.

It appears that pinning depends essentially on three factors: interchain coupling, pinning by impurities, and pinning caused by the gap to commensurability of the CDW with the underlying lattice. These remarks justify comparisons between TaS_3 and $NbSe_3$, measurements on samples doped by different types of impurities [39] or presenting defects created by irradiation [40, 41], preparation of polytypes such as monoclinic TaS_3 which has an incommensurate transition whereas orthorhombic TaS_3 presents a commensurate transition [42].

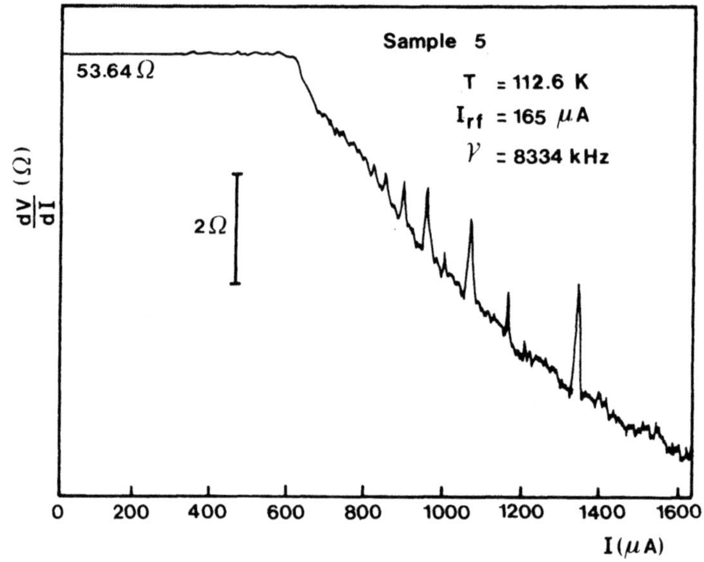

Fig. 18. Noise is generated above the threshold field ($NbSe_3$). Peaks are emphasized by superimposing a radio frequency (I_{rF}).

Above the threshold field, noise is generated in the crystal [43, 44]. This noise is the superposition of a broad band noise and a periodic noise (Figure 18). The Fourier spectra of the periodic noise can be perfectly described with the help of three fundamental frequencies and their harmonics.

The nonlinear effects in $NbSe_3$ are an illustration of a Fröhlich mode. They correspond to a sliding motion of the CDW as a whole. A perfect Fröhlich mode supposes a perfect invariance through the translation of the CDW energy, this one being independent of the ϕ phase in respect of the host lattice. Displacement would occur without dissipation and would lead to a superconducting system (Fröhlich superconductivity). In fact, the invariance by translation is broken because the phase is trapped by the lattice — in particular, by the impurities, the potential variation of the host itself, etc. However, the application of an electric field can bring to the wave an energy higher than that of the trapping and, beyond a threshold field, the wave can move and carry an electric current. The mechanisms of the trapping draws along a friction of the wave with the lattice resulting in a finite conductivity and an energy of the Fröhlich mode dependent on ϕ.

In the nonlinear regime the current is the sum of the ohmic contribution and a contribution due to the movement of the CDW with velocity v:

$$I = I_{\text{ohm}} + nev$$

where n is the number of carriers in the CDW mode.

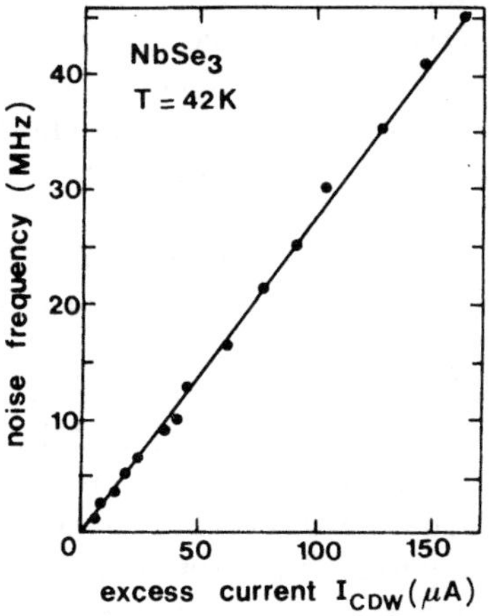

Fig. 19. Linear relationship between the current carried by the CDW and the fundamental frequency of the noise (in the case of $NbSe_3$). (After [45].)

In the noise spectra the observed frequencies are due to the modulation of the velocity of the CDW. The peaks are displaced towards higher frequencies if the applied field increases. The current carried by the CDW is directly proportional to the frequency ν_0 of the fundamental. The corresponding linear relation is represented on Figure 19 in the case of $NbSe_3$ [45].

We have

$$\nu_0 = \frac{v_{\text{CDW}}}{\lambda};$$

but

$$I_{\text{CDW}} = nev_{\text{CDW}},$$

leading to

$$I_{\text{CDW}} = ne\lambda \cdot \nu_0.$$

I_{CDW} represents the excess current which can be measured in the non-ohmic region. Determining ν_0, and using the values of ne obtained from other experiments (specific heat variations for example), one gets $\lambda = 2\pi/q$, in good agreement with the observed value in the case of $NbSe_3$. Measuring q by electron diffraction techniques gives *ne*. This is a success of that model. The inverse of the slope of ν versus J is a measure of the electron density in bands disturbed by the Peierls distortion. These electrons, situated beneath the forbidden band created by the distortion, contribute to the conductivity when the CDW moves.

The amplitude of the alternative current measured at the edges of the sample is about 1% of the direct current transported by the wave. Such a very weak amplitude can be explained by a nonsynchronization of the CDW speeds in various domains presenting different critical fields. Only the in-phase domains in the whole sample would be concerned, the summation over all the other frequencies leads to a broad band noise. However, if the CDW moves in the crystal at a speed $v = (2\pi\nu_0)/q$, it must be possible to synchronize this displacement by applying on the sample a frequency equal to ν. The previously described experiences show that by applying a d.c. electric field an alternative tension can be detected. The experience of synchronization is the conjugated experience: an alternative tension is applied and an effect is expected in the continuous characteristic $V(I)$. The change in differential resistance dV/dI is recorded in respect of the direct current applied and by superposing a fixed frequency having a constant amplitude.

Figure 18 shows that dV/dI presents some peaks that correspond to the interference between the imposed frequency (8.3 MHz) and the displacement frequency of the CDW. When both frequencies become close, there is a hooking in phase on the outer frequency of the frequency corresponding to the movement of the wave.

The strong point of the suggested interpretation lies in the good agreement between the electronic densities thus deduced and the densities obtained otherwise. Everything is not totally understood, however. Deeper discussions are given in references [45, 46] and in the references already mentioned [36—38].

Charge density waves may be expressed with in classical chemical concepts. When the electronic density increases between ions, we have a true picture of a chemical bond. On the other hand, the diminution of electronic density, related to a lengthening of the ion—ion distance, can go as far as an absence of direct interaction. This chemical approach has led us to illustrate the phenomena through the distribution of waves of bonds.

The extreme case is well illustrated by Peierls distortion occurring along chains of ions with a d^1 electronic configuration. This may be exemplified by NbS_3 [47]. A doubling of the periodicity along the chains corresponds to a d^1—d^1 pairing of niobium atoms. NbS_3 is a diamagnetic semiconductor and not a metal.

A more flexible and richer situation is given by the derivatives of the $(MY_4)_nX$ series, with M = Nb, Ta, X = Cl, Br, I, Y = S = Se and n = 2, 3, 4, . . . [48]. In these materials, described in detail in this book, the halogens form columns between the MY_4 chains. According to the value of n, long and short metal—metal distances alternate along the chains with a more or less complicated sequence. We have the bond density wave, which represents the fixed form of a charge density wave. To check this hypothesis, depinning experiences have thus been tempted. They have been successful in the case of $(TaSe_4)_2I$ and $(NbSe_4)_{10/3}I$ [48, 49]. The former com-

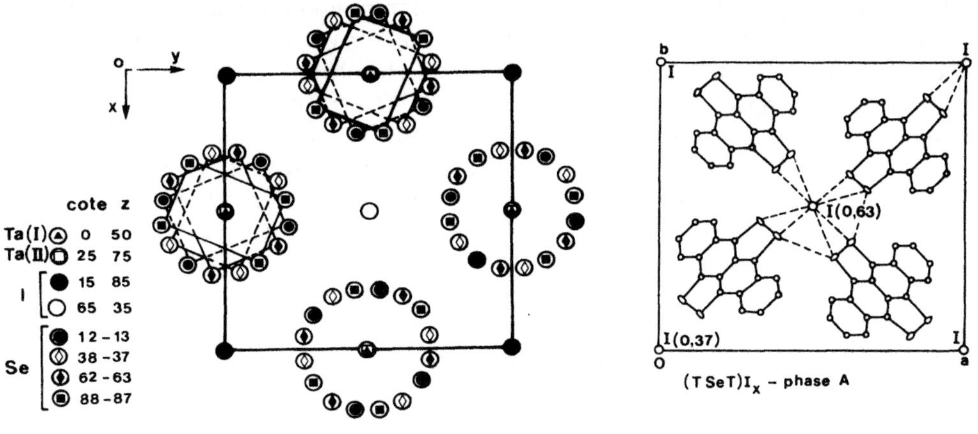

Fig. 20. $(NbSe_4)_3I$ reminds us of some organic conductors such as $(TSeT)I_x$ (tetraselenotetracene-iodine) phase A with a stacking of molecules to built columns that are separated by iodine counter ions.

pound even is an ideal material with a perfect verification of the linear relation between ν and I_{CDW}.

The study of the $(MY_4)_nX$ phases presents many other centers of interest. In particular, these phases recall the Krogmann salts and some organic derivatives. They remind us of Krogmann salts because of their metallic chains (along which are found d_{z^2} interactions) and because of the X^- counter ions. They also remind us of some 1D organic conductors with columns of molecules stacked in such a way that they satisfy strong π interactions and are separated by iodine counter ions. Figure 20 illustrates projections on planes, perpendicular to the column direction, of the $(NbSe_4)_3I$ and tetraselenotetracene-iodine (phase A) structures [50].

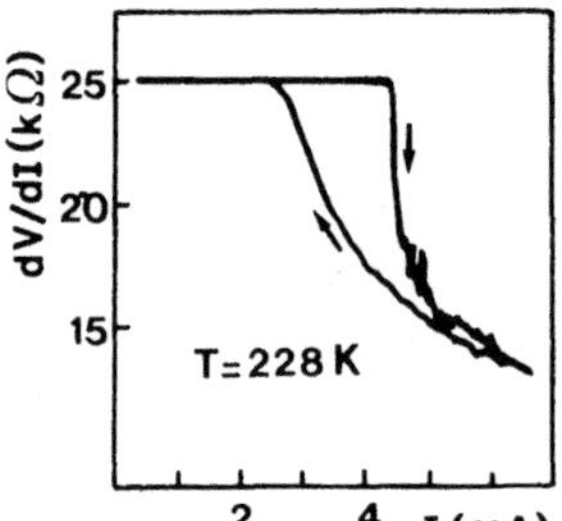

Fig. 21. Memory effect in $(NbSe_4)_{10/3}I$ at 228 K.

A new striking effect was recently observed in relation with charge density waves depinning. We may have a threshold field when the field is increased and a different one when it is decreased. This type of hysteresis phenomenon takes place in orthorhombic TaS_3 [51] in the blue bronzes $K_{0.30}MoO_3$ [52] and in $(NbSe_4)_{10/3}I$ [53]. It is illustrated in Figure 21 in the case of the latter compound for which it is observed at each temperature betwen 220 and 285 K. Such a memory effect depends on the electrical history of the sample. It implies that closely related metastable states of comparable energies are concerned in pinning effects. Its origin is probably to be found in an interaction between CDW and impurities. The impurity, depending on its position towards the maximum of the CDW, can induce different types of local deformation of the CDW [54]. The movement of the CDW, when depinned, can also determine slight displacements of the impurities which have probably been observed in the case of bronzes [55]. Memory effects will be presented in more detail by C. Schlenker and J. Dumas in this book. A unified discussion will probably be possible in the future in connection with the idea of direct density waves in modulated systems proposed by P. Lederer and coworkers [56].

4. Conclusion

In a few years, research about 1D conductors have undoubtedly enriched the chemistry and physics of solids. Even in a now classical field as that of intercalation chemistry, new prospects have been uncovered. However, it is indeed at the level of new mechanisms of the phase transitions that our knowledge has most increased.

Now questions are asked about more accurate and intimate knowledge of matter as:

(i) the role of impurities that could be better labelled as local variations of potential, including a change in the anionic pair length as a true chemical impurity or as a cationic vacancy;

(ii) the effect on pinning of the gap to commensurability and the existence of discommensuration walls.

In effect, a first period corresponding to a static study of the charge density waves, has been followed from dynamic studies which have considerably developed just after the discovery of possible depinning. Today, memory effects attract interest. They might lead to a unified theory of the role of defects in modulated systems.

This book gives a description of the main 1D inorganic compound families. Two articles also introduce the difficult problem of the structural characterization of transitions and give a new general approach of the the electronic structures from orbital interactions.

References

1. A. Meerschaut and J. Rouxel, *J. Less Common Met.* **39**, 498 (1975); J. L. Hodeau *et al., J. Phys. C*, **91**, 4117 (1978).
2. A. Meerschaut, P. Palvadeau, and J. Rouxel, *J. Solid State chem.* **20**, 21 (1977).
3. M. Meerschaut, L. Guemas, R. Berger, and J. Rouxel, *Acta Cryst.* **B35**, 1747 (1979).
4. H. G. von Schnering and W. Beckman, *Z. Anorg. allg. Chem.* **347**, 231 (1966); J. Rijnsdorp, G. J. de Langes, and G. A. Weigers, *J. Solid State Chem.* **30** (3), 365 (1979).

5. C. Fouassier, C. Delmas, and P. Hagenmuller, *Mat. Sci. Engl.* **31**, 97 (1977).
6. E. Mooser and W. B. Pearson, *Acta Cryst.* **12**, 1015 (1979).
7. W. Klingen, G. Eulenberger, and H. Hahn, *Narturwiss.* **55**, 229 (1968).
8. L. S. Ramsdell, *Am. Mineralogist* **32**, 64 (1967).
9. G. A. Wiegers and F. Jellinek, *J. Solid State Chem.* **1**, 519 (1970).
10. r. Huisman, R. de Jonge, C. Haas, and F. Jellinek, *J. Solid State Chem.* **3**, 56 (1971).
11. J. A. Wilson and A. D. Yoffee: *Adv. in Physics* **18**, 193 (1969).
12. R. Hoffmann, S. Shaik, J. C. Scott, M. H. Whangbo, and M. Foshee, *J. Solid State Chem.* **34**, 263 (1980).
13. F. Jellinek, *Inorg. Sulfur Chemistry* (ed. G. Nickless), Elsevier (1968).
14. F. Hulliger, *Structural Chemistry of Layer Phases* (ed. E. Lévy) Reidel, Dordrecht (1976).
15. P. Day, *Educational Modules in Mat. Science EMMSE* **3**, **4**, 594 (1981); see also P. Day, *Ann. N.Y. Acad. Sci.* **313**, 9 (1978).
16. A. Simon, *Structure and Bonding* **36**, 81 (1979); and *Angewandte Chemie.*
17. J. Corbett, *Acc. Chem. Res.* **14**, 239 (1981); and this volume, p. 000.
18. M. Sergent and R. Chevrel (this volume, p. 000, and references therein).
19. *Physics and Chemistry of Materials with Layered Structures*, Vol. 6, Reidel, Dordrecht (1979).
20. *Intercalation Chemistry* (eds M. S. Whittingham and A. Jacobson), Academic Press (1982).
21. R. Chevrel and M. Sergent, *J. Solid State Chem.* **3**, 8807 (1971).
22. R. Schöllhorn, M. Kumpers, and D. Plorin, *J. Less Common Met.* **58**, 55 (1978).
23. B. C. H. Steele, G. J. Dudley, and K. Y. Cheung, *Second Int. Meeting on Solid Electrolytes, St Andrews, U.K.* 1978, Abstract 7—3.
24. J. Rouxel, P. Molinie, and L. H. Top, *J. Power Sources* **9**, 345 (1983).
25. C. Berthier, Y. Chabre, P. Segransan, L. Trichet, and J. Rouxel, *J. Phys. Chem. Solids* **45** (1), 105 (1984).
26. J. Rouxel, *J. Solid State Chem.* **17**, 223 (1976).
27. R. Brec, M. Evain, P. Grenouilleau, and J. Rouxel, *Rev. Chim. Min.* **20**, 283 (1983); R. Brec, G. Ouvrard, M. Evain, P. Grenouilleau, and J. Rouxel, *J. Solid State Chem.* **47**, 174 (1983).
28. M. Evain, M. Queignec, R. Brec, and J. Rouxel, *J. Solid State Chem.* (in press).
29. A. W. Overhauser, *Phys. Rev.* **167**, 691 (1968).
30. R. E. Peierls, *Quantum Theory of Solids*, Clarendon Press, Oxford (1955).
31. R. Comès, M. Lambert, H. Launois, and H. R. Zeller, *Phys. Rev.* **B8**, 571 (1973).
32. A. Meerschaut and J. Rouxel, *J. Less Common Met.* **39**, 197 (1975); M. Marezio, C. Roucau, R. Ayroles, A. Meerschaut, J. Rouxel, and P. Monceau, *J. Phys. C,* **11**, 4117 (1978).
33. A. Briggs, P. Monceau, M. Nunez-Regueiro, J. Peyrard, M. Ribault, and J. Richard, *J. Phys. C,* **13**, 2117 (1980).
34. S. Tomic, K. Biljaksvić, D. Djurek, J. R. Cooper, P. Monceau, and A. Meerschaut, *Solid State commun.* **38**, 109 (1981).
35. A. Meerschaut, J. Rouxel, P. Haen, P. Monceau, and M. Nunez-Regueiro, *J. Phys. Lett.* **40**, L157 (1970).
36. P. A. Lee and T. M. Rice, *Phys. Rev.* **B19**, 3970 (1979).
37. J. Bardeen, *Phys. Rev. Lett.* **45**, 1978 (1980).
38. P. Monceau, J. Richard, and M. Renard, *Phys. Rev.* **B25**, 931 (1982).
39. J. W. Brill, N. P. Ong, J. C. Eckert, J. W. Savage, S. K. Khanna, and R. B. Somoano, *Phys. Rev.* **B23**, 1517 (1981).
40. L. Zuppiroli, *Radiation Effects* **62**, 53 (1982).
41. P. Monceau, J. Richard, and R. Lagnier, *J. Phys. C,* **14**, 2995 (1981).
42. T. Sambongi, K. Tsutsumi, Y. Shiozaki, M. Yamamoto, K. Yamaya, and Y. Abe, *Solid State Commun.* **22**, 729 (1977).
43. R. Fleming and C. C. Grimes, *Phys. Rev. Lett.* **42**, 1423 (1979).
44. J. Richard, Thesis, Grenoble (1981).
45. J. Bardeen, E. Ben Jacobs, G. Grüner, and A. Zettl, *Phys. Rev. Lett.* **49**, 493 (1982).
46. G. Grüner, *Physica D,* 1—34, (1983).
47. J. Rinjsdorp and F. Jellinek, *J. Solid State Chem.* **25**, 325 (1978).

48. P. Gressier, L. Guemas, J. Rouxel, and P. Monceau, *J. Solid State Chem.* **51**, 141 (1984).
49 G. Mozurkewich, M. Maki, and G. Grüner, *Solid State Commun.* **48** (5), 453 (1983); M. Maki, M. Kaiser, A. Zettl, and G. Grüner, *Solid State Commun.* **46** (7), 497 (1983).
50. P. Delhaes, C. Coulon, S. Flandrois, B. Hilte, C. W. Mayer, G. Riho, and J. Rivory, *J. Chem. Phys.* **73**, 1452 (1980).
51. L. Mihaly and G. Grüner, *Solid State Commun.* **48**, 449 (1983).
52. J. Dumas and C. Schlenker (this volume, p. 000, and references therein).
53. Z. Z. Wang, P. Monceau, M. Renard, P. Gressier, L. Guemas, and A. Meerschaut, *Solid State Commun.* **47** (6), 439 (1983).
54. R. Fleming and L. Schneemeyer (personal communication).
55. C. Berthier, Grenoble (personal communication).
56. P. Lederer, G. Montambaux, J. P. Jamet, and M. Chauvin, *J. Phys. Lett.* **45**, L627 (1984).

ORBITAL INTERACTION ANALYSIS FOR THE ELECTRONIC STRUCTURES OF LOW-DIMENSIONAL SOLIDS

MYUNG-HWAN WHANGBO*

Department of Chemistry, North Carolina State University, Raleigh, NC 27695—8204, U.S.A.

1. Introduction

The concept of orbital interaction has been extremely fruitful in understanding various molecular structure and reactivity problems [1]. In recent years, structural and electronic properties of numerous crystalline solids have also been examined by orbital interaction analysis [2—6]. This approach has been invaluable in identifying analogies between discrete molecules and extended solids in their structural and electronic properties [2(l), 3(d), 3(e)]. Orbital interaction analysis provides a qualitative conceptual framework of understanding both molecular and solid-state problems, and therefore is indispensable in building true bridges of comprehension between chemistry and physics associated with solid-state materials.

The major objective of the present article is to describe in some detail how the concept of orbital interaction is employed in analyzing the structural and electronic properties of various low-dimensional solids. In Section 2 we examine how band orbitals are constructed for any given crystal structure. The various structural instabilities associated with low-dimensional solids are examined in Section 3 by analyzing how band orbitals are modified upon structural distortion. In Section 4 we briefly survey the electronic structures of some low-dimensional solids derived from transition metal ML_6 octahedra, ML_6 trigonal prisms, and ML_8 tetragonal antiprisms.

CHART I

Crystal structure
Direct lattice

↓

Reciprocal lattice
Brillouin zone

↓

Wave vector
Band orbitals

* Camille and Henry Dreyfus Teacher—Scholar (1980—1985).

J. Rouxel (ed.), Crystal Chemistry and Properties of Materials with Quasi-One-Dimensional Structures, 27—85.

2. Crystal Structure and Band Orbital

Electronic structures of crystalline solids are described by band orbitals and their energies, which are invariably expressed as a function of wave vector **k**. From the viewpoint of chemical bonding in solids, it is essential to know how wave vector is related to a crystal structure under consideration. Chart I shows a brief summary of how band orbitals are related to crystal structures, which will be examined in this section.

2.1. Direct and Reciprocal Lattice [7]

2.1.1. *One-Dimensional Systems*

An example of a one-dimensional (1D) chain, *trans*-polyacetylene, is shown in **1**. The primitive (repeat) vector **a** of this chain is given by

$$\mathbf{a} = a\hat{\mathbf{a}} \tag{1}$$

in terms of its length a and unit vector $\hat{\mathbf{a}}$. This primitive vector leads to a set of special points (or lattice sites) in the coordinate (or direct) space defined by

$$\mathbf{R} = n\mathbf{a} = na\hat{\mathbf{a}} \quad (n = \text{integer}), \tag{2}$$

which is known as the direct lattice. A schematic representation of the direct lattice **R** is given by **2**, where one might think of each lattice site as representing a unit cell.

The reciprocal lattice **K** corresponding to **R** is defined by the requirement

$$\exp(i\mathbf{K} \cdot \mathbf{R}) = 1 \tag{3}$$

which is satisfied when $\mathbf{K} \cdot \mathbf{R}$ is an integral multiple of 2π. In terms of the reciprocal vector $\mathbf{a}^*$ with length a^* and unit vector $\hat{\mathbf{a}}^*$, **K** can be written as

$$\mathbf{K} = m\mathbf{a}^* = ma^*\hat{\mathbf{a}}^* \quad (m = \text{integer}). \tag{4}$$

For any 1D direct lattice **R**, it is found that

$$a^* = 2\pi/a, \qquad \hat{\mathbf{a}}^* = \hat{\mathbf{a}}. \tag{5}$$

A general position in the direct space and that in the reciprocal space may be written as **r** and **k**, respectively:

$$\mathbf{r} = r\hat{\mathbf{a}}, \qquad \mathbf{k} = k\hat{\mathbf{a}}^*. \tag{6}$$

1

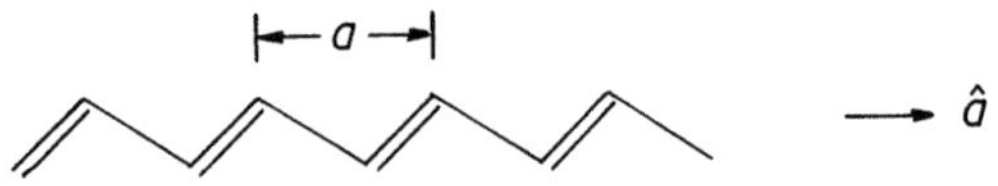

2

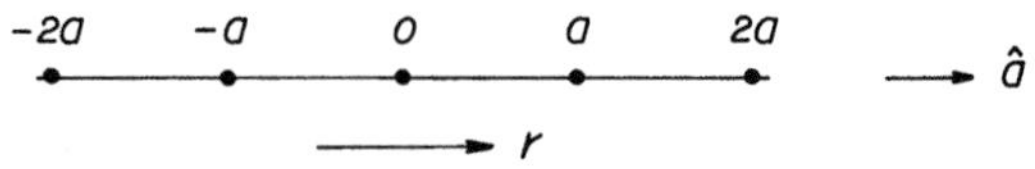

3 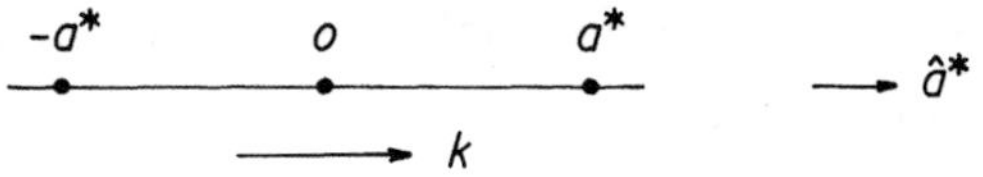

4 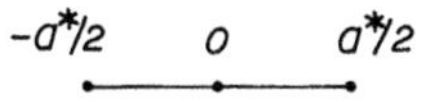

As far as chemical bonding is concerned, important values of **k** (i.e. wave vector) are those within the first Brillouin zone (BZ). This is defined as the smallest region around the reference site (i.e. $m = 0$ in Equation (4)) enclosed by those planes bisecting every straight line that connects the reference site to its neighbor sites. Thus, the first BZ of **3** is given by **4**.

2.1.2. *Two- and Three-Dimensional Systems*

It is straightforward to generalize the preceding discussion to two-dimensional (2D) and three-dimensional (3D) systems. A 3D direct lattice **R** is shown in **5**, where the lattice sites are generated by three primitive (repeat) vectors **a**, **b**, and **c** as follows:

$$\mathbf{R} = n_a\mathbf{a} + n_b\mathbf{b} + n_c\mathbf{c} \quad (n_a, n_b, n_c = \text{integer}), \tag{7}$$

where the primitive vectors are expressed as

$$\mathbf{a} = a\hat{\mathbf{a}}, \qquad \mathbf{b} = b\hat{\mathbf{b}}, \qquad \mathbf{c} = c\hat{\mathbf{c}} \tag{8}$$

in terms of their lengths a, b, c and unit vectors $\hat{\mathbf{a}}$, $\hat{\mathbf{b}}$, $\hat{\mathbf{c}}$. The reciprocal lattice **K** corresponding to **R** is given in terms of the three reciprocal vectors $\mathbf{a}^*$, $\mathbf{b}^*$, $\mathbf{c}^*$ as follows

$$\mathbf{K} = m_a\mathbf{a}^* + m_b\mathbf{b}^* + m_c\mathbf{c}^* \quad (m_a, m_b, m_c = \text{integer}). \tag{9}$$

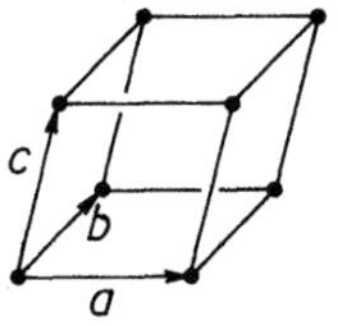

5

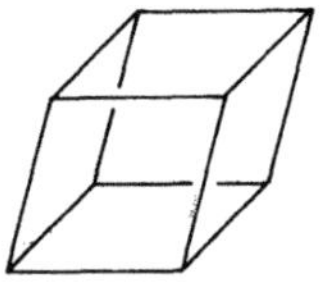

6

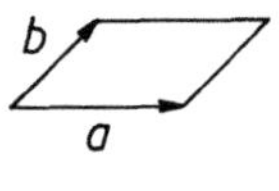

7

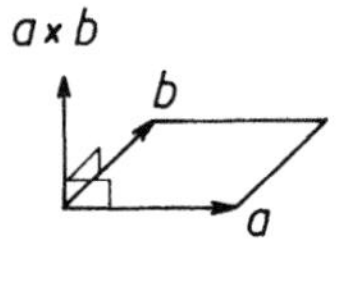

8

The reciprocal vectors are related to the primitive vectors as

$$\begin{aligned} \mathbf{a}^* &= (2\pi/V)\mathbf{b}\times\mathbf{c}, \\ \mathbf{b}^* &= (2\pi/V)\mathbf{c}\times\mathbf{a}, \\ \mathbf{c}^* &= (2\pi/V)\mathbf{a}\times\mathbf{b} \end{aligned} \tag{10}$$

where V is the volume of the parallelepiped **6** defined by **a, b**, and **c**. The lengths and directions of the vectors $\mathbf{b}\times\mathbf{c}$, $\mathbf{c}\times\mathbf{a}$, and $\mathbf{a}\times\mathbf{b}$ are simple to determine. For instance, $\mathbf{a}\times\mathbf{b}$ has the length equal to the area of the parallelogram **7** defined by **a** and **b**, and is perpendicular to both **a** and **b** as shown in **8**. The positive direction of $\mathbf{a}\times\mathbf{b}$ is obtained by adopting the right-handed screw convention: rotate **a** around the axis of $\mathbf{a}\times\mathbf{b}$ towards **b** as if to advance a right-handed screw. Then the screw-advancing direction is the positive direction of $\mathbf{a}\times\mathbf{b}$. It is convenient to write the reciprocal vectors as

$$\mathbf{a}^* = a^*\hat{\mathbf{a}}^*, \qquad \mathbf{b}^* = b^*\hat{\mathbf{b}}^*, \qquad \mathbf{c}^* = c^*\hat{\mathbf{c}}^* \tag{11}$$

in terms of their lengths a^*, b^*, c^* and unit vectors $\hat{\mathbf{a}}^*$, $\hat{\mathbf{b}}^*$, $\hat{\mathbf{c}}^*$. A general position in the direct space and that in the reciprocal space may be written as **r** and **k**, respectively:

$$\begin{aligned} \mathbf{r} &= r_a\hat{\mathbf{a}} + r_b\hat{\mathbf{b}} + r_c\hat{\mathbf{c}} \equiv (r_a, r_b, r_c) \\ \mathbf{k} &= k_a\hat{\mathbf{a}}^* + k_b\hat{\mathbf{b}}^* + k_c\hat{\mathbf{c}}^* \equiv (k_a, k_b, k_c). \end{aligned} \tag{12}$$

Important values of **k**, wave vector, are those of the first BZ. For a 3D crystal, it is somewhat complicated to show the first BZ, so that the reader is referred to the literature [8] for the first BZs of various crystal systems. However, the essence of how to derive the first BZ can be shown by considering the reciprocal lattice **K** corresponding to a 2D direct lattice **R**,

$$\mathbf{R} = n_a\mathbf{a} + n_b\mathbf{b}, \qquad \mathbf{K} = m_a\mathbf{a}^* + m_b\mathbf{b}^*. \tag{13}$$

The reciprocal vectors $\mathbf{a}^*$ and $\mathbf{b}^*$ can be determined by using Equation (10) once a unit vector $\hat{\mathbf{c}}$ is added along the direction of $\mathbf{a}\times\mathbf{b}$ as shown in **9**. Then the volume

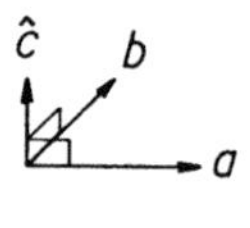

9

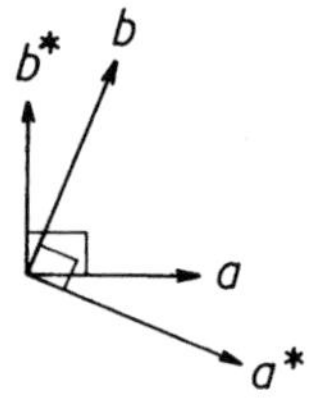

10

of the parallelepiped defined by **a, b**, and $\hat{\mathbf{c}}$ becomes identical to the area of the parallelogram defined by **a** and **b**. The reciprocal vectors $\mathbf{a}^*$ and $\mathbf{b}^*$ lie in the plane defined by **a** and **b** as indicated in **10**, and lead to the 2D reciprocal lattice **K** shown in **11**. The first BZ is the smallest region around the reference site (i.e. $m_a = m_b = 0$ for **K** in Equation (13)) enclosed by those planes bisecting every straight line that

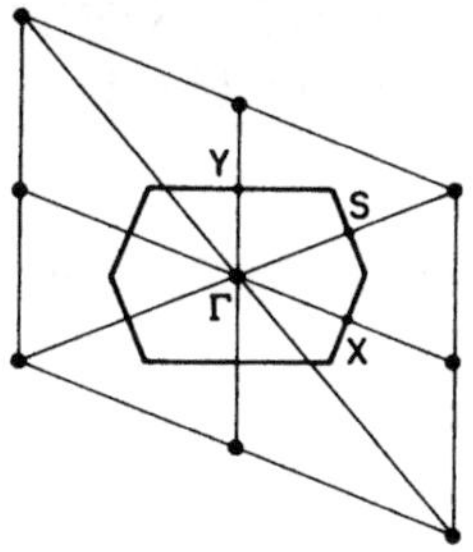

11

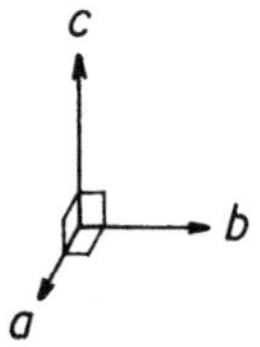

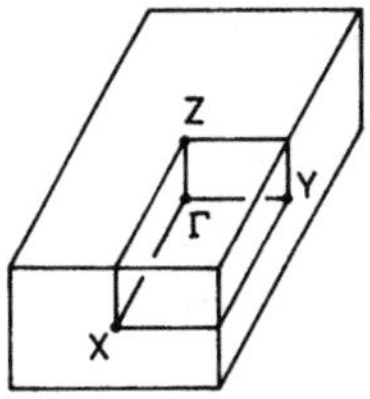

12 13

connects the reference site to its neighbor sites. This is shown in **11**. The special points Γ, *X*, *Y*, and *S* of the first BZ are given by

$$\begin{aligned} \Gamma &= (0, 0), \qquad X = (a^*/2, 0), \\ Y &= (0, b^*/2), \qquad S = (a^*/2, b^*/2). \end{aligned} \tag{14}$$

As a simple example of 3D systems, let us consider the direct lattice generated by three orthogonal vectors **a, b**, and **c** shown in **12**. For this lattice, the reciprocal vectors **a***, **b***, and **c*** satisfy the following:

$$a^* = 2\pi/a, \qquad b^* = 2\pi/b, \qquad c^* = 2\pi/c \tag{15}$$

and $\hat{\mathbf{a}}^*$, $\hat{\mathbf{b}}^*$, $\hat{\mathbf{c}}^*$ are directed along $\hat{\mathbf{a}}$, $\hat{\mathbf{b}}$, $\hat{\mathbf{c}}$, respectively. The first BZ of the resulting reciprocal lattice is shown in **13**, where the special points Γ, *X*, *Y*, and *Z* are given by

$$\begin{aligned} \Gamma &= (0, 0, 0), \qquad X = (a^*/2, 0, 0), \\ Y &= (0, b^*/2, 0), \qquad Z = (0, 0, c^*/2). \end{aligned} \tag{16}$$

2.2. Band Orbital Formation [2(c), 7]

2.2.1. *Bloch Orbital*

In the previous section we considered wave vector **k** as representing a general position in the reciprocal space. From the viewpoint of chemical bonding, important **k** values are those enclosed within the first BZ. For simplicity, let us first consider a lattice with one orbital, χ, per site. The orbital χ located at the site $\mathbf{R}$ is written as $\chi(\mathbf{r}-\mathbf{R})$. Then all possible bonding combinations of those site orbitals $\chi(\mathbf{r}-\mathbf{R})$ are given by the Bloch orbital $\phi(\mathbf{k})$

$$\phi(\mathbf{k}) = (1/\sqrt{N}) \sum_{\mathbf{R}} \exp(i\mathbf{k}\cdot\mathbf{R})\chi(\mathbf{r}-\mathbf{R}), \tag{17}$$

where N is the number of lattice sites in a crystal under consideration (i.e. $N \to \infty$). Each site orbital $\chi(\mathbf{r}-\mathbf{R})$ has the coefficient $\exp(i\mathbf{k}\cdot\mathbf{R})$, which is rewritten as

$$\exp(i\mathbf{k}\cdot\mathbf{R}) = \exp(in_a k_a a)\exp(in_b k_b b)\exp(in_c k_c c) \tag{18}$$

for a 3D system.

For a 1D chain, Equation (17) is simplified as

$$\phi(k) = (1/\sqrt{N}) \sum_{n} \exp(inka)\chi(r-na) \tag{19}$$

At the BZ center (k = 0) and edge (k = $\pm a^*/2 = \pm \pi/a$), the coefficient $\exp(inka)$ has the following value:

$$\exp(inka) = \begin{cases} (1)^n & \text{at } k=0 \\ (-1)^n & \text{at } k=\pm a^*/2. \end{cases} \tag{20}$$

With the site orbital χ taken as an s orbital, for instance, the Bloch orbitals $\phi(k)$ at the BZ center and edge are represented by **14a** and **14b**, respectively. Thus **14a** and **14b** are the maximum and minimum bonding combinations of the given site

14a

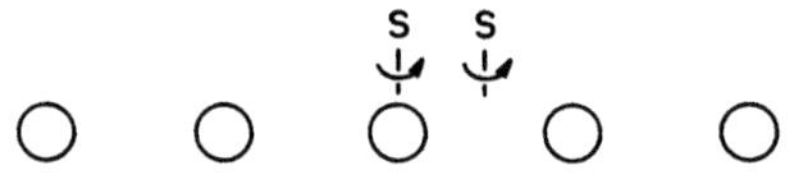

14b

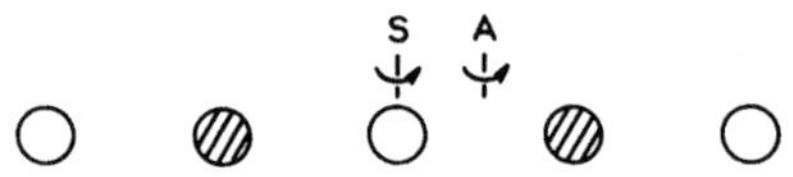

15a

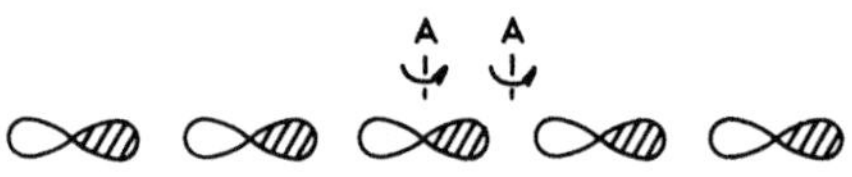

15b

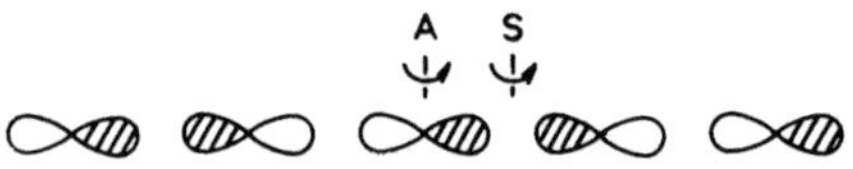

orbitals. With respect to the two two-fold rotations shown, **14a** and **14b** have SS and SA symmetry, respectively. In general, the energy and symmetry of a Bloch orbital $\phi(k)$ is dependent upon the value of k. At any other values of k within the first BZ, the energy of $\phi(k)$ lies in between the two levels **14a** and **14b**. Specifically, the energy $e(k)$ of a Bloch orbital $\phi(k)$ can be given by

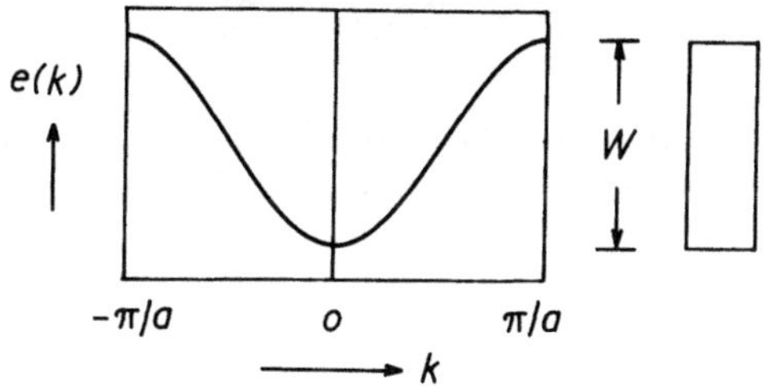

Fig. 1. Variation of the band orbital energy $e(k)$ as a function of wave vector k for the case when $t < 0$.

$$e(k) = \varepsilon + 2t\cos(ka), \tag{21}$$

where the mean band energy ε and the resonance (or hopping) integral t are expressed as

$$\begin{aligned} \varepsilon &= \langle \chi_n | H | \chi_n \rangle \\ t &= \langle \chi_n | H | \chi_{n+1} \rangle \end{aligned} \tag{22}$$

where χ_n refers to the orbital $\chi(r - na)$. Thus for $t < 0$, the energy $e(k)$ varies with k as shown in Figure 1. The Bloch orbitals $\phi(k)$ have the energy value from $\varepsilon + 2t$ to $\varepsilon - 2t$ so that the bandwidth W is given by

$$W = |4t|. \tag{23}$$

When the resonance integral t between nearest neighbors is positive, the minimum and maximum energy of $e(k)$ occur at $k = \pm a^*/2$ and 0, respectively. Suppose, for example, that each site has a p orbital pointed along the chain axis. Then the Bloch orbitals $\phi(k)$ at $k = 0$ and $\pm a^*/2$ are represented by **15a** and **15b**, respectively, and the $e(k)$ versus k plot is given by Figure 2.

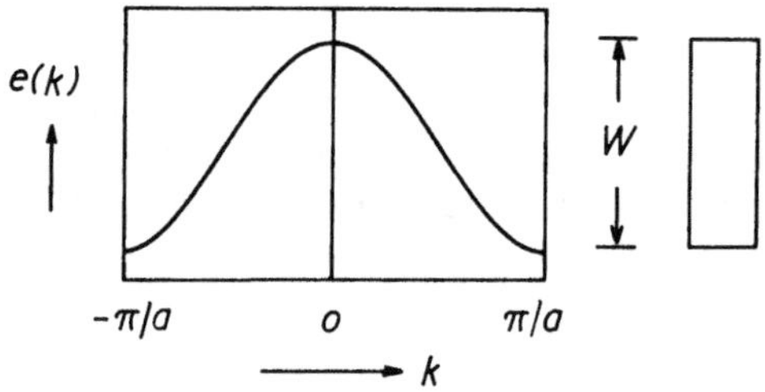

Fig. 2. Variation of the band orbital energy $e(k)$ as a function of wave vector k for the case when $t > 0$.

2.2.2. *Band Orbital*

Let us now consider a lattice in which each site has a set of orbitals $\{\chi_1, \chi_2, \ldots \chi_\mu, \ldots\}$. For each site orbital χ_μ, the Bloch orbital $\phi_\mu(\mathbf{k})$ is defined as

$$\phi_\mu(\mathbf{k}) = (1/\sqrt{N}) \sum_{\mathbf{R}} \exp(i\mathbf{k} \cdot \mathbf{R}) \chi_\mu(\mathbf{r} - \mathbf{R}) \tag{24}$$

at various **k** values of the first BZ. Then the band orbitals $\psi_i(\mathbf{k})$ describing such a system can be expressed as a linear combination of Bloch orbitals

$$\psi_i(\mathbf{k}) = \sum_{\mu} c_{\mu i}(\mathbf{k}) \phi_\mu(\mathbf{k}), \tag{25}$$

where $c_{\mu i}(\mathbf{k})$ is the coefficient for $\phi_\mu(\mathbf{k})$. As representative examples of how to derive band orbitals from a set of Bloch orbitals, we will examine the band orbitals of two chain structures **16** and **17**. These chain structures are obtained from ML_6 octahedra upon sharing their opposite corners and edges, respectively. Throughout our discussion it is important to note that *d* orbitals of transition metal atoms M lie higher in energy than *s, p* orbitals of ligand L (**18**). Therefore, the '*d*-block' orbitals of transition metal complexes have M—L antibonding character, while their '*s, p*-block' orbitals have M—L bonding character [1].

16

17

18

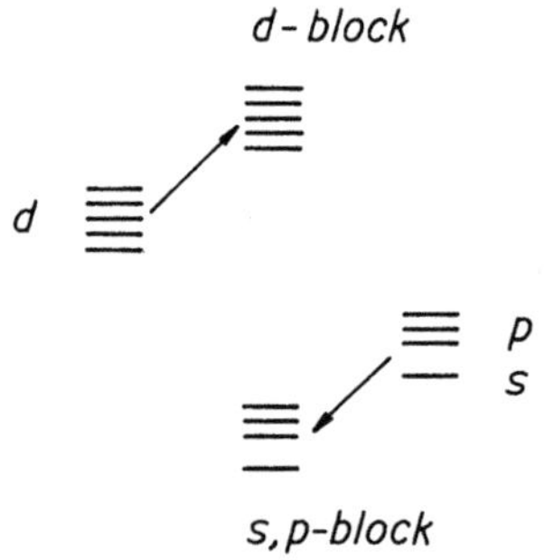

(a) *Chains with single atom bridge* [2(d)]. It is convenient to consider the chain structure **16** in terms of square planar ML_4 units and bridging ligand atoms L. In **16** the bridging ligands are at the midpoints between ML_4 units, so that each

19

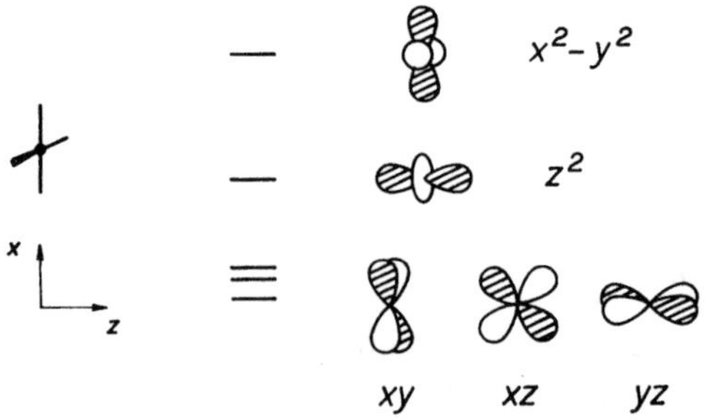

M—L—M bridge is symmetrical. The *d*-block orbitals of ML_4 are shown in **19**, where small ligand contribution is not shown for clarity.

σ band. The z^2 orbital of ML_4 has σ-type overlap with the *s*, *z* orbitals of bridging ligands. (For simplicity, p_x, p_y, p_z, $d_{x^2-y^2}$, d_{z^2}, d_{xy}, d_{yz} orbitals will be referred to as *x*, *y*, *z*, $x^2 - y^2$, z^2, *xy*, *xz*, *yz* orbitals, respectively.) The Bloch orbitals of z^2, *z*, and *s* at Γ ($k = 0$) are given by **20a, 21a**, and **22a**, respectively, while those at X ($k = \pm\pi/a$) are given by **20b, 21b**, and **22b**, respectively. The symmetry of these Bloch orbitals may be classified as symmetric (S) or antisymmetric (A) with respect to the two-fold rotations centered at an ML_4 unit and at a bridging ligand atom.

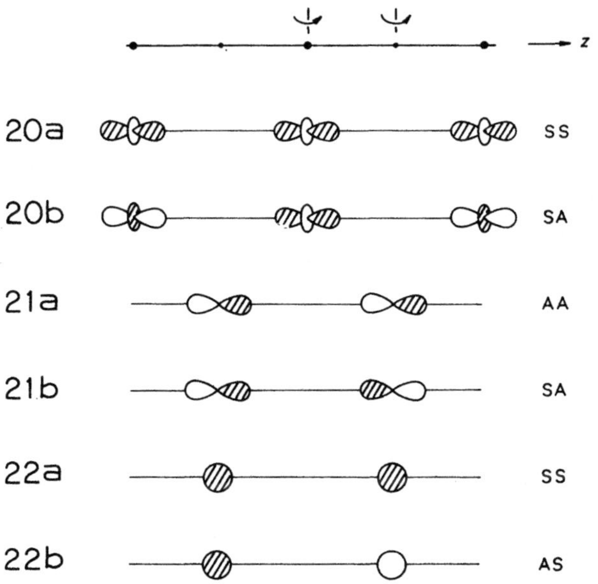

Direct interaction between ML_4 units and that between bridging ligands are weak. Consequently, the $e(k)$ versus k plots for the z^2, *z*, and *s* Bloch orbitals are nearly flat (dispersionless) as indicated by the dashed lines in Figure 3. However, the z^2 Bloch orbitals can interact with the *s* and *z* Bloch orbitals when allowed by symmetry. For example, the z^2 and *z* Bloch orbitals mix at *X* while the z^2 and *s* Block orbitals do at Γ. Since the overlap between z^2 and *z* is more effective than that between z^2 and *s*, the z^2 band is raised more at *X* than at Γ. As *k* varies from Γ to *X*, the magnitude of interaction between the z^2 and *s* Bloch orbitals decreases gradually but that between the z^2 and *z* Bloch orbitals increases gradually. Therefore the z^2, *z*, and *s* band levels vary smoothly as shown by the full lines in Figure 3. The z^2 and *s* bands are raised in energy as *k* varies Γ to *X*, while the opposite is the case for the *z* band.

π band. The *xz* and *yz* orbitals of ML_4 make π-type overlap with the *x* and *y* orbitals of bridging ligand L, respectively. The band orbitals arising from *xz* and *x*

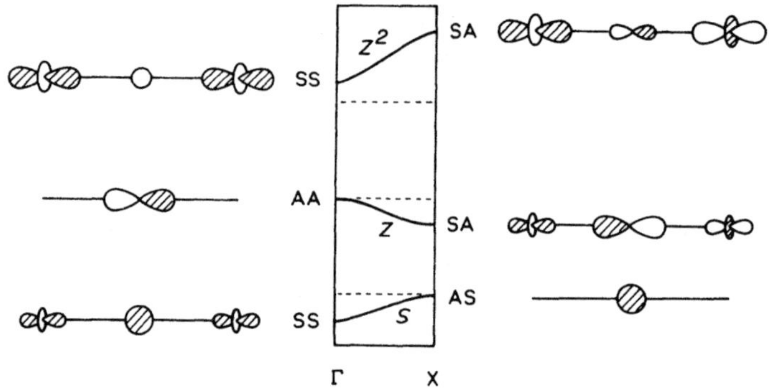

Fig. 3. Formation of the z^2, z, and s bands (full lines) in terms of the z^2, z, and s Bloch orbitals (dashed lines) for the ML_5 chain **16**.

orbitals are similar in nature to those from yz and y orbitals. Thus we will examine only those band orbitals associated with yz and y orbitals.

The yz and y Bloch orbitals at Γ are given by **23a** and **24a**, respectively, and those at X are given by **23b** and **24b**, respectively. By symmetry, **23a** and **24a** do

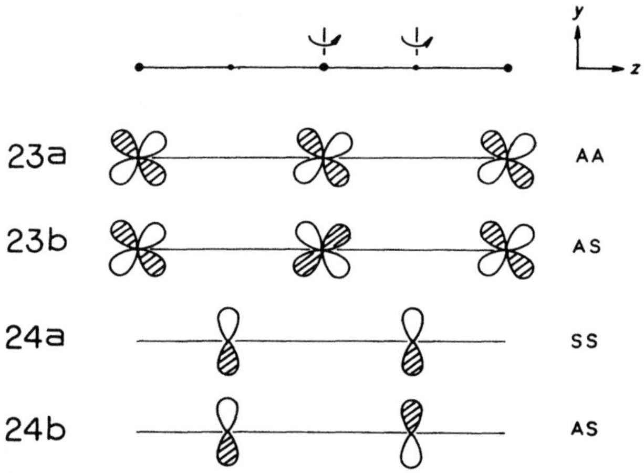

not interact, but **23b** and **24b** do. The yz and y Bloch orbitals are dispersionless as shown by the dashed lines in Figure 4. However, on going from Γ to X, the magnitude of interaction between the yz and y Bloch orbitals increases gradually. Thus the yz and y bands are smoothly raised and lowered, respectively, as shown by the full lines in Figure 4.

The yz and y Bloch orbitals do not mix at Γ because the M—L—M bridge is symmetrical. When this bridge is made asymmetrical (i.e. M—L · · · M), the two can mix at Γ, thereby lowering the y band and raising the yz band in the vicinity of Γ.

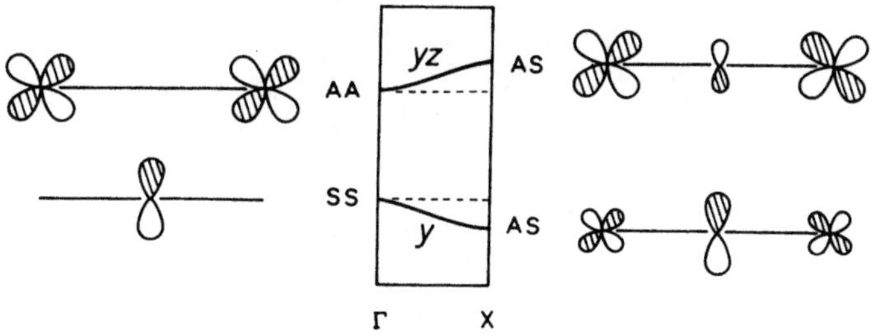

Fig. 4. Formation of the *yz* and *y* bands (full lines) in terms of the *yz* and *y* Bloch orbitals (dashed lines) for the ML_5 chain **16**.

(b) *Chains with double atom bridge* [2(d)]. Each metal atom of **17** is approximately under an octahedral environment, so that its *d* orbitals are split as shown in **25**. In the following only those **d**-block bands resulting from the three low-lying levels of **25** will be examined.

25

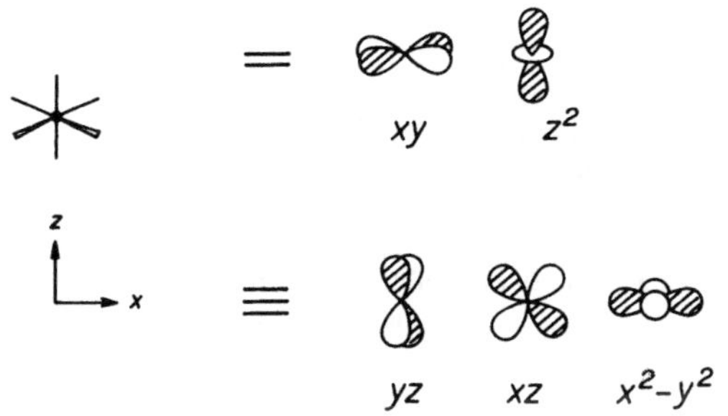

σ band. The $x^2 - y^2$ orbitals make *σ*-type overlap along the chain. The $x^2 - y^2$ Bloch orbitals at Γ and *X* are given by **26a** and **26b**, respectively. Since the metal—metal distance is relatively short, the energy of the $x^2 - y^2$ Bloch orbital is raised quite substantially on going from Γ to *X*, as shown by the dashed line in Figure 5.

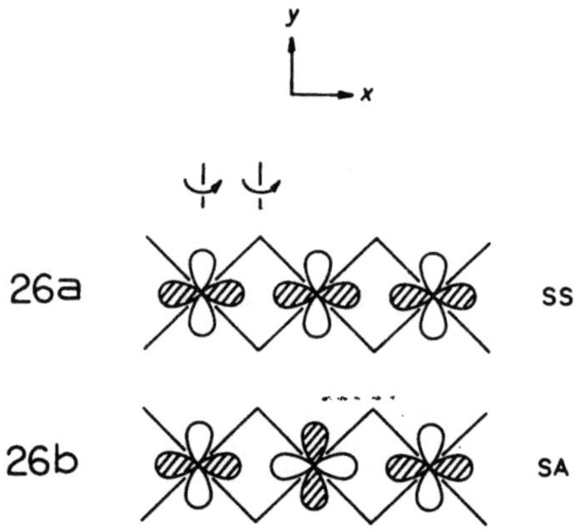

27a
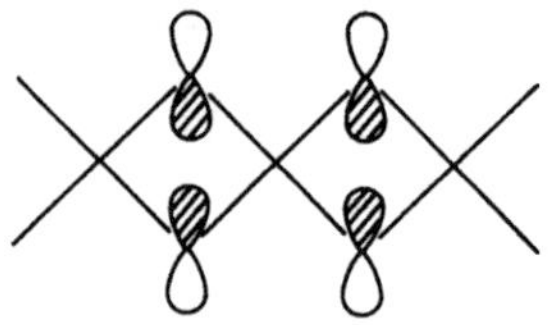
SS

27b
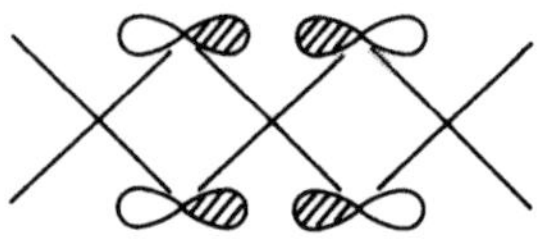
SA

The symmetry adapted orbitals of bridging ligands **27a** and **27b** can interact with **26a** and **26b**, respectively. Such ligand orbitals, when allowed to mix, raise the $x^2 - y^2$ Bloch levels as shown by the full line in Figure 5.

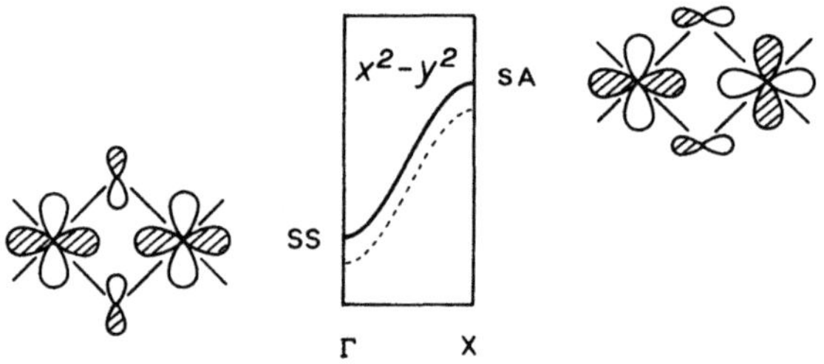

Fig. 5. Formation of the $x^2 - y^2$ band (full line) in terms of the $x^2 - y^2$ Bloch orbital (dashed line) for the ML_4 chain **17**.

π band. The *xz* orbitals make π-type overlap along the chain. The *xz* Bloch orbital at Γ and *X* are given by **28a** and **28b**, respectively. Thus the *xz* Bloch orbital is stabilized gradually on going from Γ to *X*, as shown by the dashed line in Figure 6. No orbitals of bridging ligands can interact with **28a** at Γ. However, **28b** can interact with the symmetry adapted orbital **29** at *X*. Thus the *xz* Bloch orbital is

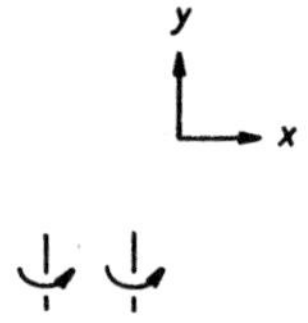

28a

SS

28b

SA

29

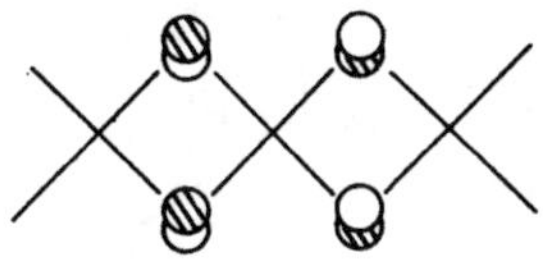

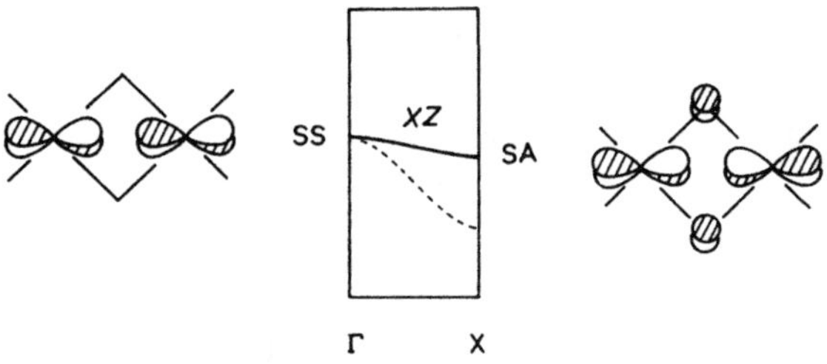

Fig. 6. Formation of the *xz* band (full line) in terms of the *xz* Bloch orbital (dashed line) for the ML_4 chain **17**.

gradually raised by bridging ligand orbitals on going from Γ to *X*. As a result of the opposing effects of the metal—metal and metal-bridging ligand interactions, the resulting *xz* band becomes nearly flat.

δ band. The *yz* orbitals make δ-type overlap along the chain. The *yz* Bloch orbitals at Γ and *X* are given by **30a** and **30b**, respectively. Since δ-type overlap is weak, the *yz* Bloch orbital is nearly flat as indicated by the dashed line in Figure 7. No symmetry adapted orbitals of bridging ligands can interact with **30b** at *X*. However, **30a** can interact effectively with **31**. Consequently, the *yz* Bloch orbital is raised at Γ, and the resulting *yz* band varies as indicated by a full line in Figure 7.

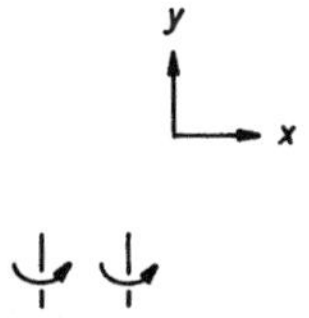

30a

30b

31

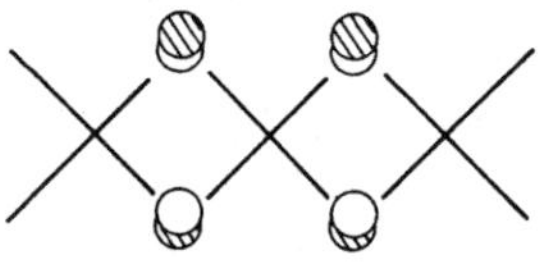

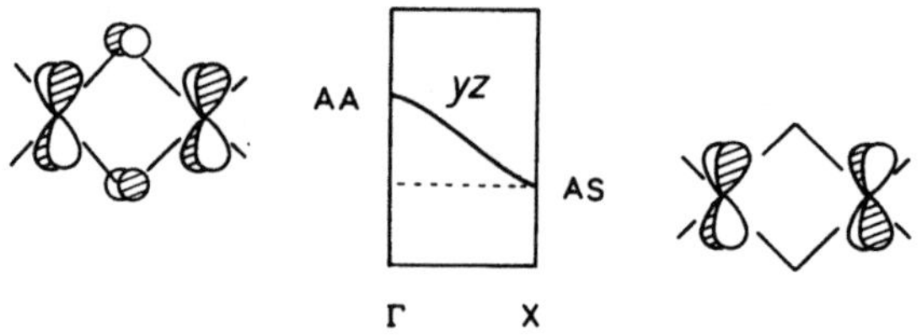

Fig. 7. Formation of the *yz* band (full line) in terms of the *yz* Bloch orbital (dashed line) for the ML_4 chain **17**.

32

With respect to the two mirror planes of symmetry, shown in **32**, the $x^2 - y^2$, xz and yz band orbitals are SS, AS, and AA, respectively. Therefore the three bands cross as indicated in Figure 8. The $x^2 - y^2$ band is wide (dispersive) because metal—metal interaction is strong. The xz band is almost dispersionless because metal—metal interaction is opposed by metal-bridging ligand interaction. It is metal-bridging ligand interaction that makes the yz band somewhat dispersive.

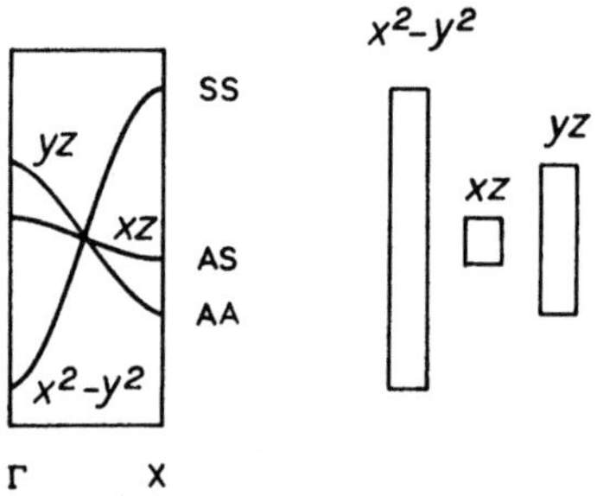

Fig. 8. Three low-lying *d*-block bands of the ML_4 chain **17**.

So far we have examined only those Bloch orbitals whose energies are well separated so that the band orbitals resulting from them become well separated as well. Figure 9 shows an example in which two interacting Bloch orbitals (dashed lines) overlap in energy. When these orbitals interact, the intended crossing is avoided by orbital mixing so that band orbital character switches from primarily one Block orbital to another as k varies. Consequently, there result in two bands (full lines) separated by an energy gap. In such a case, the widths of the resulting bands are narrower than expected from the Bloch orbitals alone.

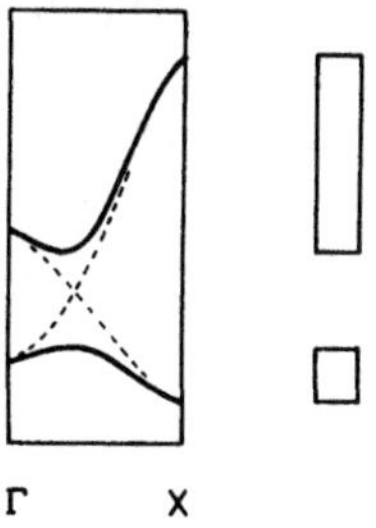

Fig. 9. Bands (full line) derived from two interacting Bloch orbitals (dashed lines) for the case when the Bloch orbitals overlap in energy.

2.2.3. *Interchain Interaction*

(a) *One-dimensional systems.* It is often convenient to consider a 2D layer structure as formed from 1D chains. Likewise a 3D crystal may be considered as made up of 2D layers. A 1D system consisting of two chains is shown in **33**, where the rectangular box represents a repeat unit. It is typical that interaction between nearest-neighbor sites is stronger within a chain than between chains. For simplicity, we again represent each site by an s orbital. Thus two s orbitals of each unit cell lead to s_+ and s_- orbitals shown in **34**. These orbitals differ in their symmetry with respect

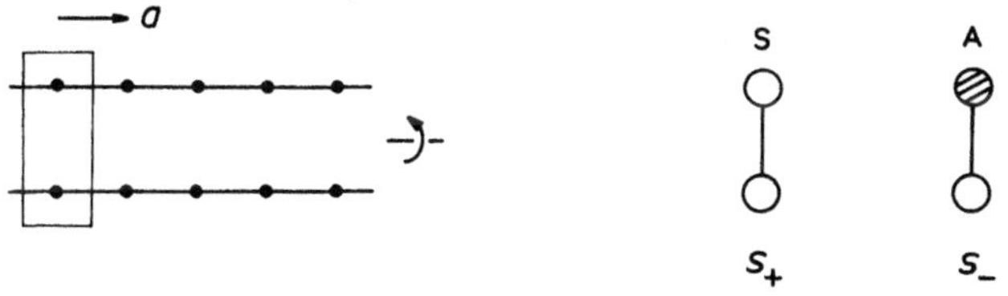

33 34

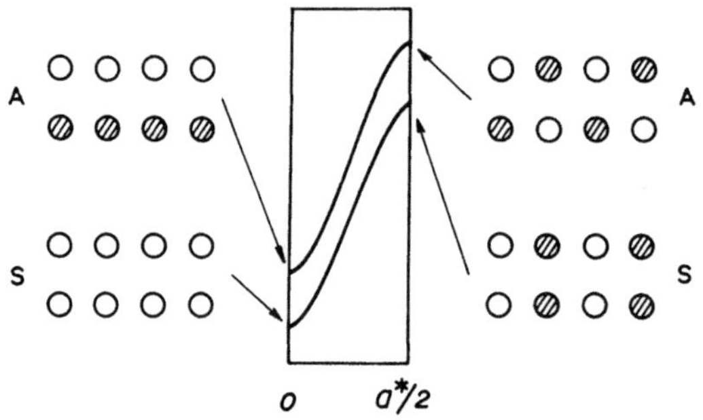

Fig. 10. Band structure of a one-dimensional system **33** with two chains per unit cell.

to the two-fold rotation of **33**, so that the Bloch orbitals constructed from s_+ and s_- will not mix throughout the region $\Gamma \to X$. Thus **33** has two bands, as shown in Figure 10. In general, a 1D system with n chains per unit cell will have n band levels, which is shown in Figure 11. The energy difference between any two successive bands is small compared with their bandwidths, when interchain interaction is weaker than intrachain interaction.

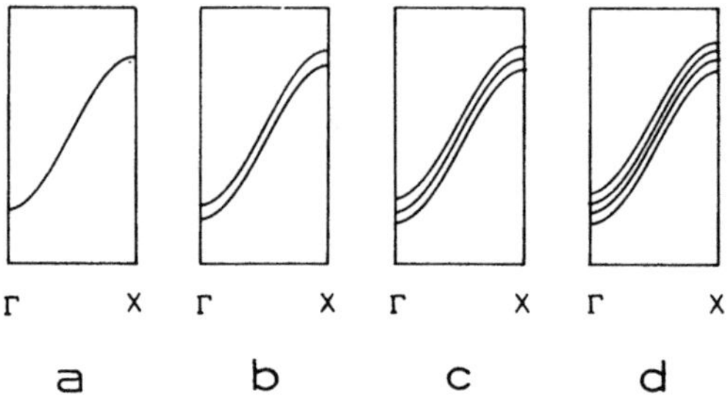

Fig. 11. Band structures of one-dimensional systems with n chains per unit cell. (a), (b), (c), and (d) refer to $n = 1, 2, 3$, and 4, respectively.

(b) *Two-dimensional system.* Shown in **35** is a 2D layer constructed by repeating 1D chains along **b**. For convenience, the primitive vectors **a** and **b** are assumed to be orthogonal. The first BZ for **35** is given by **36**, where the special points Γ, X, Y, and S are expressed as in Equation (14). Again each site may be represented by an s orbital. Then the coefficient $\exp(i\mathbf{k} \cdot \mathbf{R})$ of the Bloch orbital $\phi(\mathbf{k})$ for this lattice is given by

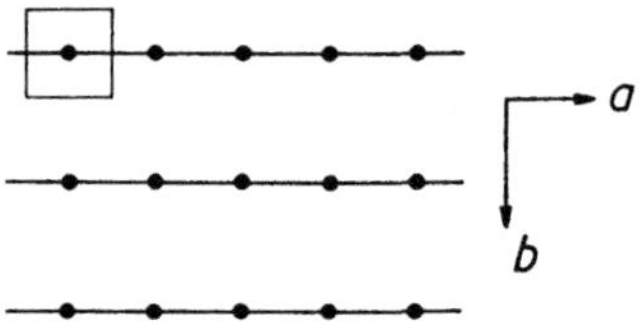

35

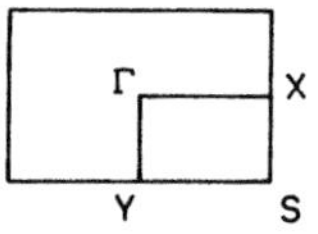

36

$$\exp(i\mathbf{k} \cdot \mathbf{R}) = \exp(in_a k_a a) \cdot \exp(in_b k_b b) = \begin{cases} (1)^{na}(1)^{nb} & \text{at } \Gamma \\ (-1)^{na}(1)^{nb} & \text{at } X \\ (1)^{na}(-1)^{nb} & \text{at } Y \\ (-1)^{na}(-1)^{nb} & \text{at } S. \end{cases} \tag{26}$$

Therefore, the site orbitals are combined as in **37a, 37b, 37c**, and **37d** at Γ, X, Y, and S, respectively. If intrachain interaction is stronger than interchain interaction, the 2D band structure for **35** can be sketched as in Figure 12. The band structure is more dispersive along the chain directions $\Gamma \to X$ and $Y \to S$ than along interchain directions $\Gamma \to Y$ and $X \to S$.

37

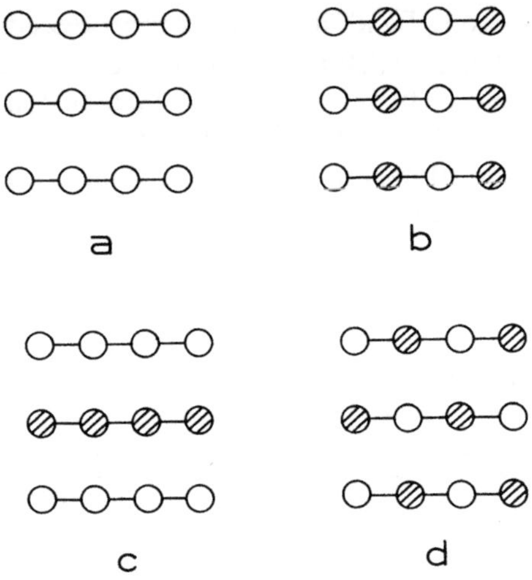

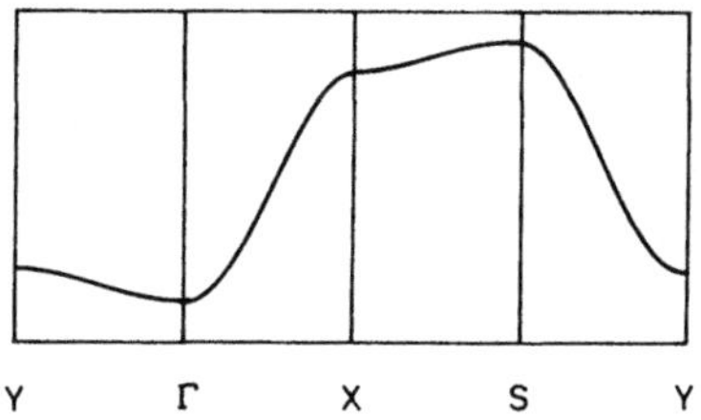

Fig. 12. Band structure of a two-dimensional layer **35**.

3. Peierls Distortion

It is often worthwhile to consider a crystal structure under consideration as distorted from a related structure of higher symmetry. In understanding the structural and electronic properties of low-dimensional solids, the concept of Peierls distortion [9] has been indispensable. In essence, Peierls distortions are solid-state counterparts of Jahn—Teller distortions [10(c)] in molecules. Molecular distortions, depending upon their electronic origin, may be classified as first- or second-order Jahn—Teller distortions and also as low- or high-spin Jahn—Teller distortions. Similarly, distortions of low-dimensional solids may be classified as first- or second-order Peierls distortions [4(d), 6(b)] and also as low- or high-spin Peierls distortions [2(l)].

3.1. Band Folding and Splitting

The effect of distortion in a low-dimensional solid upon its band electronic structure can be examined by considering the 1D chain **2**. In Section 2.2.1, the band structure of **2** was derived in terms of the unit cell comprising one site and one orbital. The corresponding band structure is reproduced in Figure 13(a). Without loss of generality, the band structure of **2** may also be calculated by adopting a unit cell of n consecutive sites. Then the repeat distance becomes na, so that the first BZ is given by $-a^*/2n \leqslant k \leqslant a^*/2n$. Within this zone, there are n band orbitals since a unit cell consists of n orbitals in such a case. Those n bands are derived by subdividing the band structure of Figure 13(a) into n sections of equal zone length and then folding those sections onto the first (i.e. $0 \leqslant k \leqslant a^*/2n$) [2(j)]. This is illustrated in Figures 13(b), 13(c), and 13(d) for $n = 2$, 3, and 4, respectively.

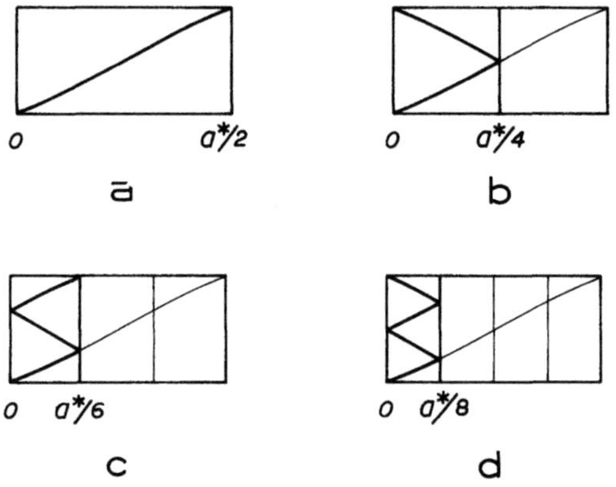

Fig. 13. Effect of increasing unit cell size upon band folding. An increase of a unit cell size by n-fold leads to n subbands in the region $0 \leqslant k \leqslant a^*/2n$. (a), (b), (c), and (d) refer to $n = 1$, 2, 3, and 4, respectively.

38

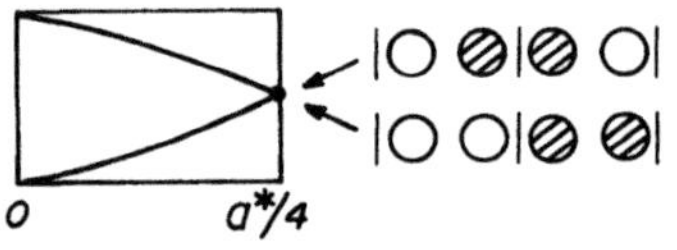

39

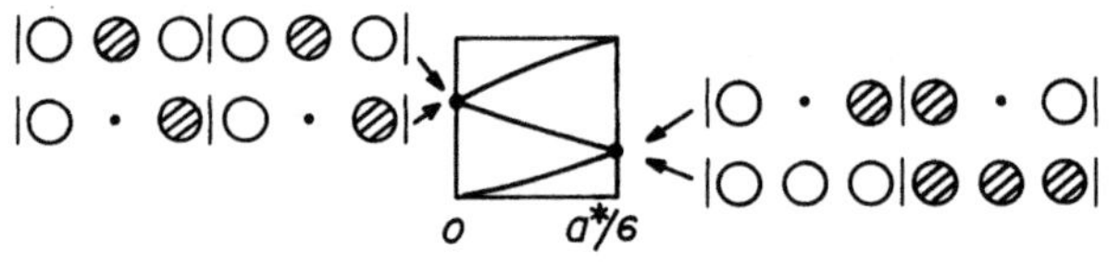

40

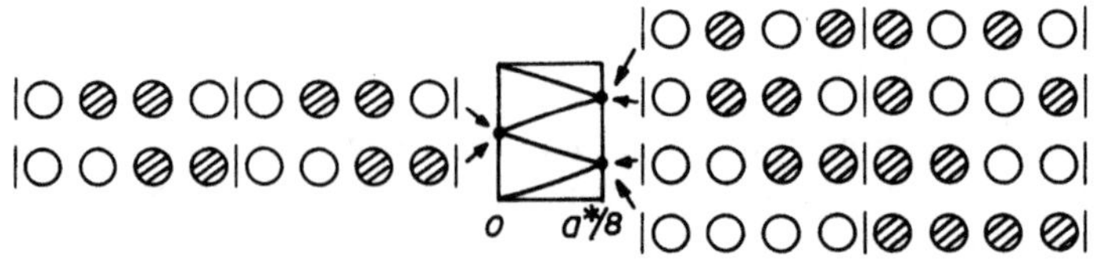

A particularly important aspect of Figures 13(b)—(d) is the degeneracy that occurs at the zone center and/or edge. The repeat orbital patterns of those degenerate band levels are shown in **38, 39**, and **40** for $n = 2$, 3, and 4, respectively. These degeneracies, artificially induced by an n-fold increase of the unit cell size, are removed by a geometry distortion that really increases the unit cell size by n-fold. For example, the dimerization **41a** → **41b** makes the degenerate levels of **38** different in energy (**42**). Consequently, the dimerization leads to the band splitting shown in Figure 14. It can be easily deduced from **39** that the trimerization **43a** → **43b** or **44a** → **44b** leads to the band splitting of Figure 15. Likewise, the tetramerization **45a** → **45b** or **46a** → **46b** splits the band as shown in Figure 16.

41

42

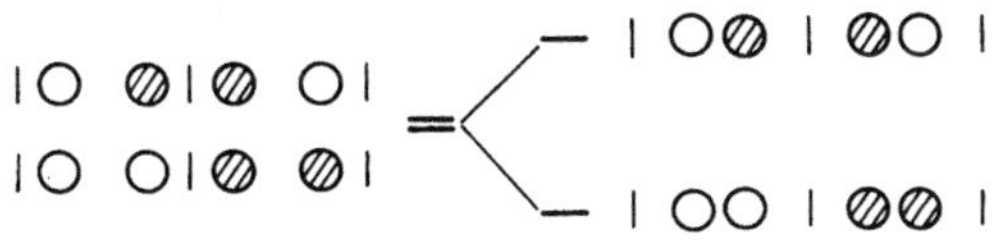

43

44

45

46

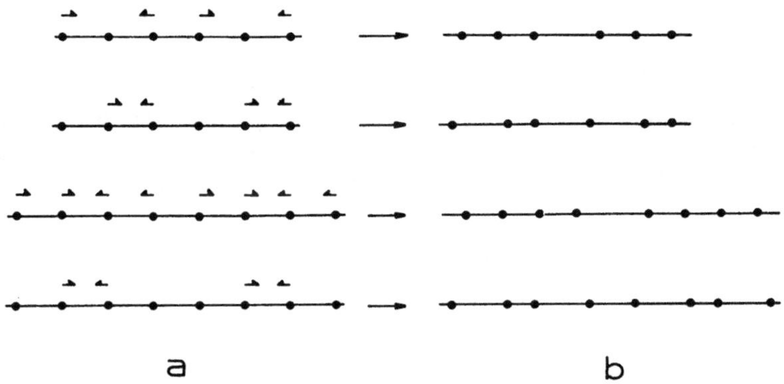

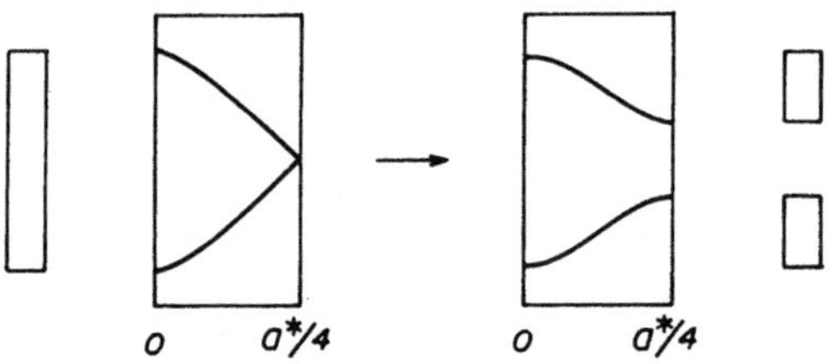

Fig. 14. Band splitting upon the dimerization **41a** → **41b**.

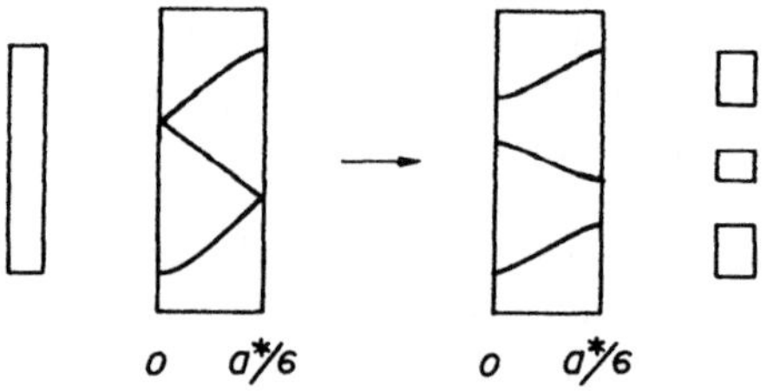

Fig. 15. Band splitting upon the trimerization **43a** → **43b** or **44a** → **44b**.

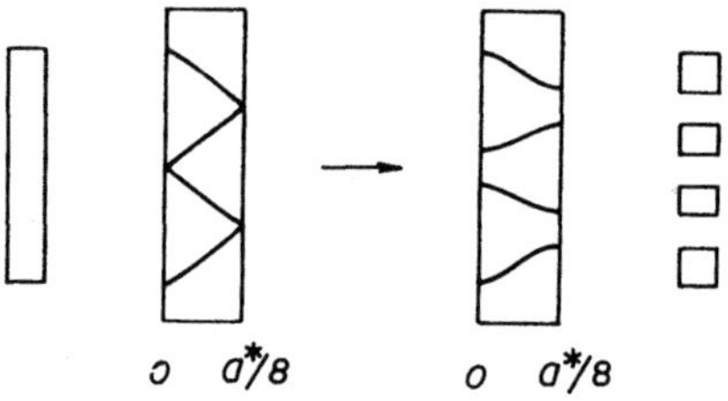

Fig. 16. Band splitting upon the tetramerization **45a** → **45b** or **46a** → **46b**.

3.2. First-Order Peierls Distortion

3.2.1. $2k_f$ *Distortion*

The band resulting from a chain with one orbital per site becomes completely filled (i.e. the band occupancy $f = 1$) when each site contributes two electrons to the band (**47**), and $\frac{1}{2}$-filled ($f = \frac{1}{2}$) with one electron per site (**48**). A general situation of band

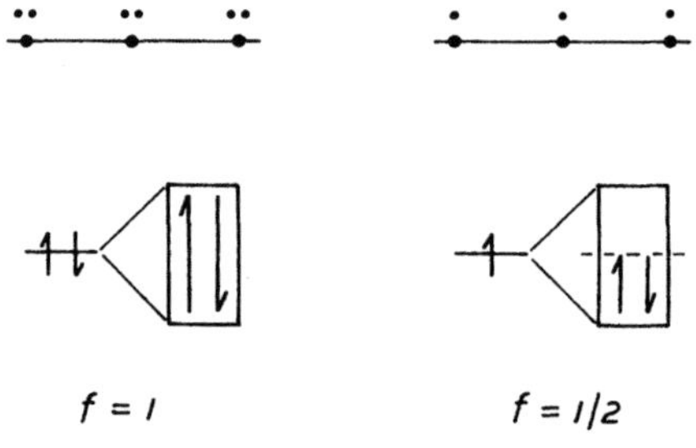

47 48

filling is shown in Figure 17, where the highest occupied band level is the Fermi level e_f, and the wave vector corresponding to this energy level is the Fermi wave vector k_f. In Figure 17 the band orbitals are occupied in the region $-k_f \leqslant k \leqslant k_f$, so that the length of the occupied region is $2k_f$ out of the total first BZ length of a^*.

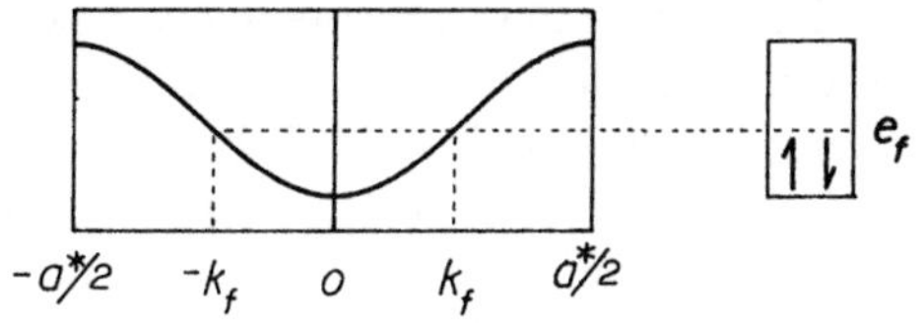

Fig. 17. Relationship between the Fermi level e_f and the Fermi wave vector k_f for the low-spin arrangement of a partially filled band.

For a tight-binding band given by Equation (20), the band occupancy f is related to $2k_f$ as follows

$$f = 2k_f/a^*. \quad (27)$$

When $f = 1/n$ (n = integer), the wave vector $2k_f$ is commensurate in that the value of $a^*/2k_f$ (= n) is an integer. In such a case, the Fermi wave vector k_f is given by $k_f = a^*/2n$. Thus, according to our discussion of Section 3.1, any distortion that increases the unit cell size by n-fold will introduce a band gap at the Fermi level. Introduction of a band gap at the Fermi level lowers the energies of the occupied levels in the vicinity of the Fermi level, and therefore acts as a driving force towards distortion for a chain with a partially filled band. Shown in **49** is band splitting of a $\frac{1}{2}$-filled band under dimerization. **50a** and **50b** refer to band splitting of $\frac{1}{3}$-filled and

49

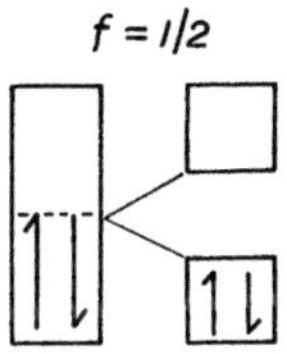

50

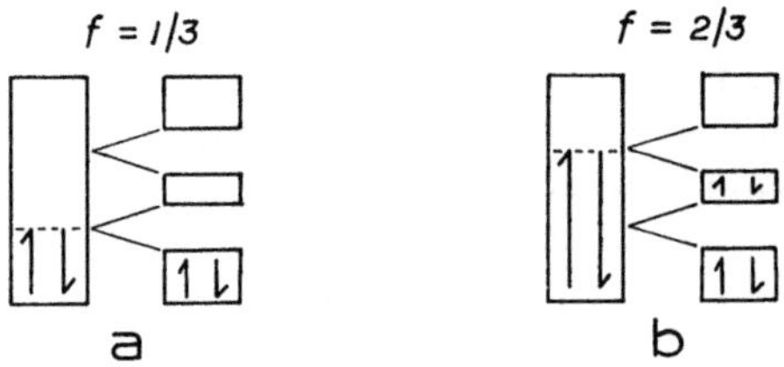

51

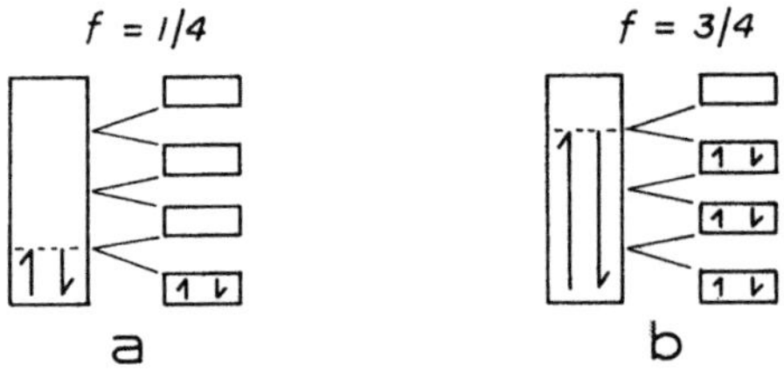

and $\frac{1}{3}$-empty bands under trimerization, respectively. Likewise, **51a** and **51b** show band splitting under tetramerization for $\frac{1}{4}$-filled and $\frac{1}{4}$-empty bands, respectively. It is noted from these examples that a chain with band occupancy f ($0 < f < 1$) is identical to that with the band occupancy $1 - f$ in its structural and electronic properties. When $f = \frac{1}{2}$, the Fermi level occurs right in the middle of the band. This level is nonbonding in that bonding and antibonding interactions between nearest-neighbor sites alternate along the chain. The dimerization (pairing distortion) **41a** → **41b** enhances every bonding interaction between sites and reduces every anti-bonding interaction between sites (**42**). Thus the driving force for distortion is strong when $f = \frac{1}{2}$. As f decreases from $\frac{1}{2}$, the orbitals around the Fermi level become more bonding and less antibonding between sites. Therefore the amount of electronic energy lowering that may be gained by distortion becomes smaller when f decreases from $\frac{1}{2}$ (or when f increases from $\frac{1}{2}$ to 1, since bands with the occupancies f and $1 - f$ have the same structural properties).

In general, 1D systems with a partially filled band are susceptible to a distortion that creates a band gap at the Fermi level, which is known as a Peierls distortion. The number of sites involved in a Peierls distortion is given by $a^*/2k_f$, so that a Peierls distortion is also termed a $2k_f$ distortion. Before distortion, there is no energy gap between the highest occupied and the lowest unoccupied band levels since the Fermi level occurs within a band; that is, the undistorted system is in a metallic state. After distortion, there is an energy gap between the highest occupied and lowest unoccupied band levels due to the band gap created at the Fermi level. Thus the distorted system is in an insulating state (or semiconducting state if the band gap is nonzero but small). Thus $2k_f$ distortion causes a metal-to-insulator transition.

3.2.2. $4k_f$ *Distortion*

(a) *Electron localization* [11]. In our discussion so far a partially filled band was assumed to be metallic in which each occupied level has two electrons. This kind of band filling is energetically favorable in terms of occupying low-lying orbitals but unfavorable in terms of electron—electron repulsion. When electron—electron repulsion dominates over orbital energy contribution, a magnetic insulating state of a partially filled band becomes more stable than its metallic state [2(c), 2(l), 12].

52

53

54

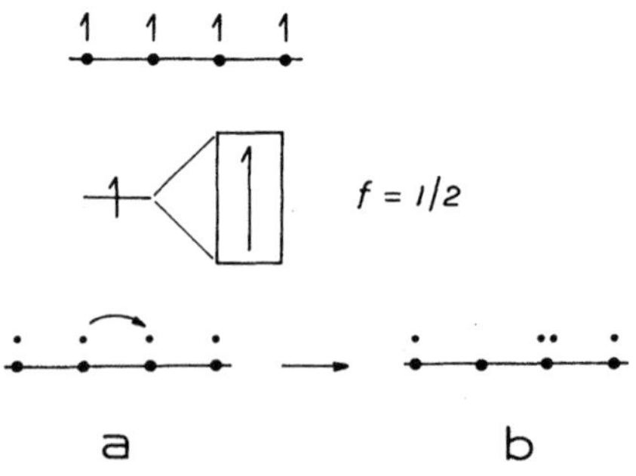

Let us examine how to represent a magnetic insulating state within the framework of a band orbital description. For simplicity, we will consider a $\frac{1}{2}$-filled band obtained from a chain with one orbital and one electron per site. Electrons of a magnetic insulating state are regarded as localized, which implies 'no interaction' between electrons of neighboring sites and 'vanishing bond order' between nearest-neighbor sites. From the viewpoint of a delocalized orbital picture, of which our band orbital description is a part, this situation is achieved when all the bonding and antibonding levels are equally occupied. For instance, the ferromagnetic state **52** can be represented by **53** which has every band orbital singly occupied by an up-spin electron [12].

The meaning of electron localization may be examined further in terms of **54**, which shows electron-hopping from one lattice site to its nearest neighbor. For a hopping electron the mean time of stay at a lattice site may be given by τ. Then for a localized electron, $\tau \rightarrow \infty$. In principle, a hopping electron may assume any energy value within the allowed energy band of width W. Thus the uncertainty in energy for a hopping electron is equal to the bandwidth W [13]. According to the uncertainty principle, τ is related to W as follows:

$$\tau \propto 1/W \tag{28}$$

so that $\tau \rightarrow \infty$ when $W \rightarrow 0$. Namely, electron-localization occurs when the bandwidth W $(= |4t|)$ is narrow [2(l)]. Electron-hopping is enhanced when the resonance integral t (and hence W) increases in magnitude. As shown in **54b**, electron-hopping creates two electrons at a lattice site thereby causing on-site Coulomb repulsion U, which opposes electron-hopping. It is found that electron-hopping is not favorable energetically when W is small compared with U. Under this condition, the ferromagnetic state **53** (an example of localized states) is more stable than the metallic state **48** (i.e. delocalized state). The relative stability problem of the metallic and magnetic insulating states is analogous to that of the low-spin and high-spin states of a molecular dimer **55** [2(c), 2(l), 12(a)]. With one electron and one orbital per site, the dimer leads to two molecular orbitals which are separated in energy by $|2t| = W/2$. The high-spin state **56b** becomes more stable than the low-spin state **56a** when W is small compared with U.

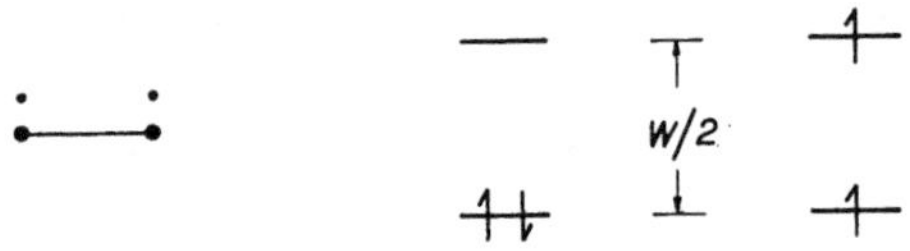

55 56a 56b

The spin arrangement of localized electrons may have a long-range order as in the ferromagnetic state **52** or in the antiferromagnetic state **57**. Alternatively, localized electrons may not have any long-range order (e.g. **58**). The energy difference between any two localized states with different spin arrangement is very small. In

57 58

fact, when only on-site Coulomb repulsion U and intersite Coulomb repulsion V are included in evaluating the total electron—electron repulsion, the electronic energy of the ferromagnetic state **52** becomes identical to that of any other localized state with different spin arrangement. Consequently, any localized state can be approximated by a ferromagnetic band filling (or high-spin band filling) within a band orbital description. Certainly, small energy differences between various spin arrangements of localized electrons are important for magnetic properties, which are analyzed in terms of spin-Hamiltonians [11, 14].

(b) *High-spin distortion.* Let us examine one important consequence of electron localization upon Peierls distortions in terms of a chain with a $\frac{1}{4}$-filled band, which is obtained when every pair of sites has one electron (**59a**). Low-spin and high-spin band-filling situations are shown in Figure 18. For the low-spin arrangement the

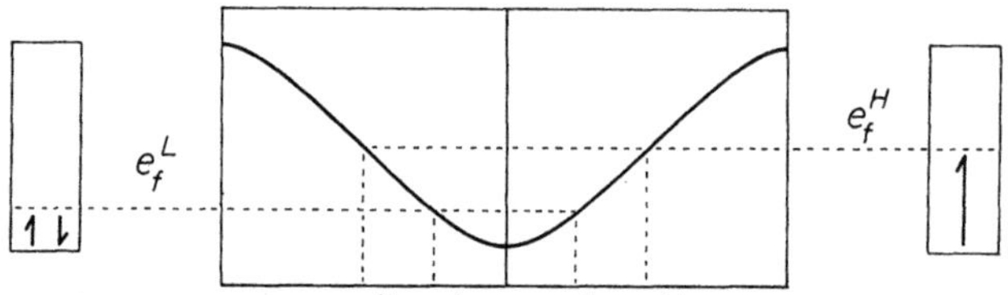

Fig. 18. Relationship between the Fermi level and the Fermi wave vector for the low-spin and high-spin arrangements of a partially filled band. The superscripts L and H signify low-spin and high-spin arrangements, respectively.

Fermi wave vector is given by $k_f^L = a^*/8$, which is therefore susceptible to a tetramerization (i.e. $a^*/2k_f^L = 4$). The Fermi wave vector of the high-spin arrangement is twice that of the low-spin arrangement, i.e. $k_f^H = 2k_f^L$. That there is no band gap in the high-spin arrangement reflects the fact that, as long as the lattice sites are equi-

59a 

59b 

60 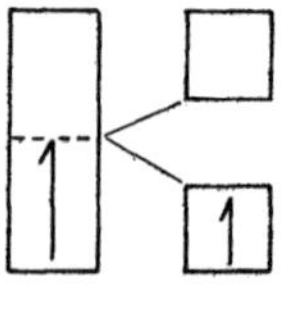

distant, the electron arrangement **59a** is equivalent to its alternative **59b**. Now the distortion that will introduce a band gap at the Fermi level e_f^H is dimerization, since $a^*/2k_f^H = 2$. This lowers the electronic energy of the resulting localized state (**60**). Since $2k_f^H = 4k_f^L$, the kind of Peierls distortion that occurs in a localized state is called a $4k_f$ distortion [2(l), 15].

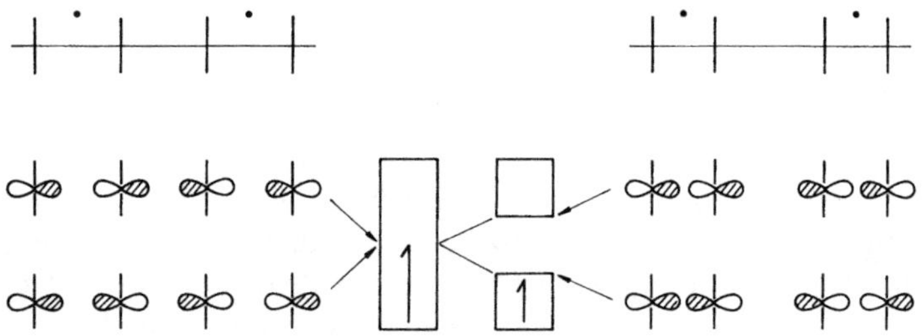

Fig. 19. Band orbital character for the highest occupied and lowest unoccupied levels of the high-spin band electronic states of **62a** and **62b**.

An example of a $4k_f$ distortion is found in MEM$^+$ (TCNQ)$_2^-$ (see **61** for the structures of MEM$^+$ and TCNQ) [15(e)—(g)] which consists of stacks of planar TCNQ units. In each TCNQ stack, every pair of TCNQ units gain an extra electron (**62a**). These electrons are accommodated by the band constructed from the lowest unoccupied molecular orbital (LUMO) of each neutral TCNQ, which becomes $\frac{1}{4}$-filled. Figure 19 shows the band splitting for the high-spin $\frac{1}{4}$-filled band under

61

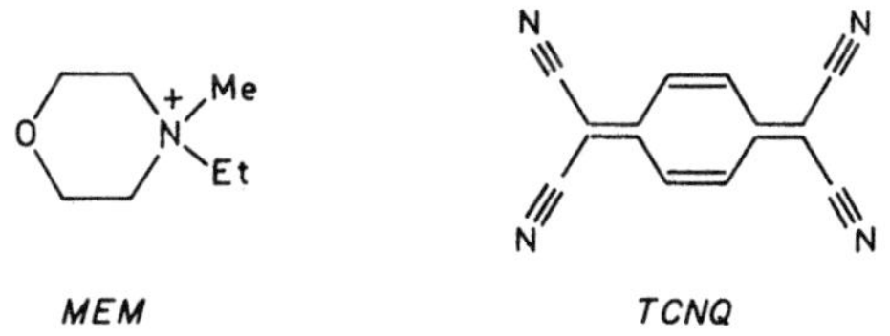

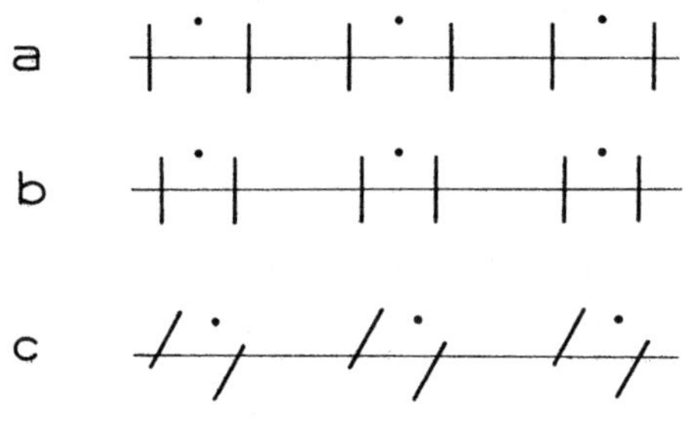

62

dimerization **62a** → **62b**, where the LUMO of each TCNQ is represented, for simplicity, by a *p* orbital perpendicular to the molecular plane. In **62b** the top portion of the high-spin $\frac{1}{4}$-filled band is primarily antibonding between TNCQ dimeric units. This antibonding interaction can be further reduced by tilting those dimeric units away from the chain axis as shown in **62c**, which is experimentally found.

3.3. Second-order Peierls Distortion

The primary cause of a first-order Peierls distortion is degeneracy lifting at the Fermi level of a partially filled band, regardless of whether the electronic state involved is high or low spin. Thus $2k_f$ and $4k_f$ distortions of extended systems are analogous to low- and high-spin Jahn—Teller distortions of molecules, respectively. Molecules with small energy gap between the occupied and unoccupied levels are susceptible to a structural distortion that allows orbital mixing between them, which is known as a second-order Jahn—Teller distortion. An analogous phenomenon also occurs in extended systems, and may be termed a second-order Peierls distortion.

Let us now re-examine the electronic structure of the double chain **33**, whose electronic structure is shown in Figure 10. When the size of the repeat unit is doubled in this chain, the resulting band structure has the folded structure shown in Figure 20. With one electron per site, the Fermi level occurs where the middle two

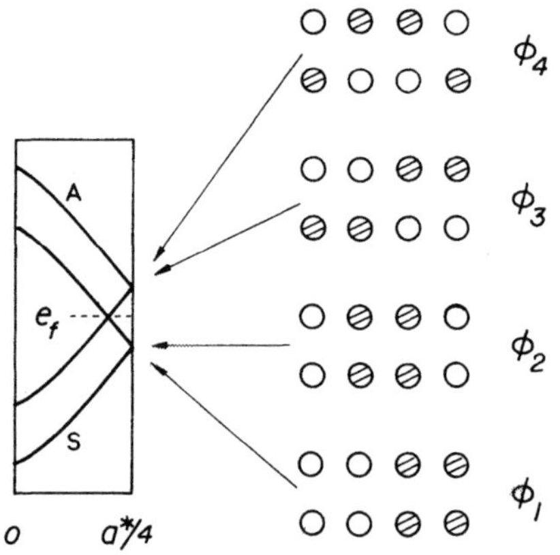

Fig. 20. Band structure of the double chain **33** for the case when the size of a unit cell is doubled.

bands cross. The degenerate levels ϕ_1 and ϕ_2 at the zone edge are symmetric with respect to the two-fold rotation of **33**. The degenerate levels ϕ_3 and ϕ_4 are antisymmetric for the same rotation. For the dimerization **63a** → **63b**, the band structure changes as depicted in Figure 21. The S and A bands each split upon the distortion, but the middle two bands of **63b** remain overlapping so that the distorted structure **63b** remains metallic.

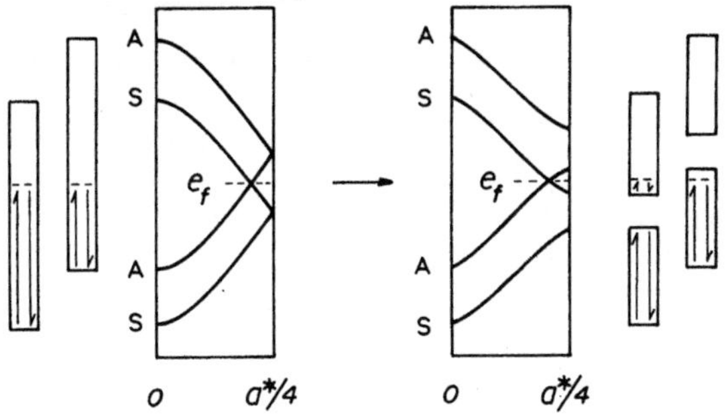

Fig. 21. Band splitting for the dimerization **63a** → **63b**. (a) and (b) show the band structures of **63a** and **63b**, respectively.

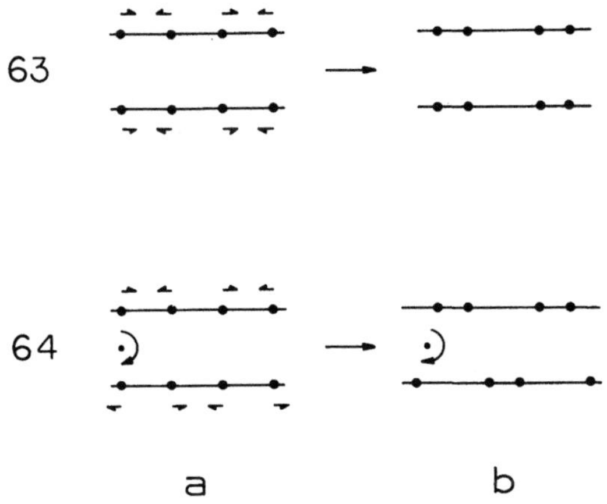

A distortion that makes the middle two bands of Figure 20 different in symmetry will introduce a band gap at the Fermi level and thus lower the energy of the resulting chain. An example of such a distortion is **64a** → **64b**. To examine this distortion, it is convenient to have an alternative representation for ϕ_1 and ϕ_2 and for ϕ_3 and ϕ_4. They are shown in **65**, where

$$\begin{aligned} \psi_1 &\propto \phi_1 + \phi_2 \\ \psi_2 &\propto \phi_1 - \phi_2 \\ \psi_3 &\propto \phi_3 + \phi_4 \\ \psi_4 &\propto \phi_3 - \phi_4. \end{aligned} \tag{29}$$

The symmetry of the new orbitals in **65** was determined with respect to the two-fold rotation shown in **64a**. Under the **64a** → **64b** distortion, the higher lying orbitals ψ_3 and ψ_4 mix into the lower lying orbitals ψ_2 and ψ_1, respectively. Namely, the **64a** → **64b** distortion allows mixing between orbitals of different energies. Figure 22 shows the band structure change for the **64a** → **64b** distortion, which introduces a band gap at the Fermi level of **64a**. Since the band gap

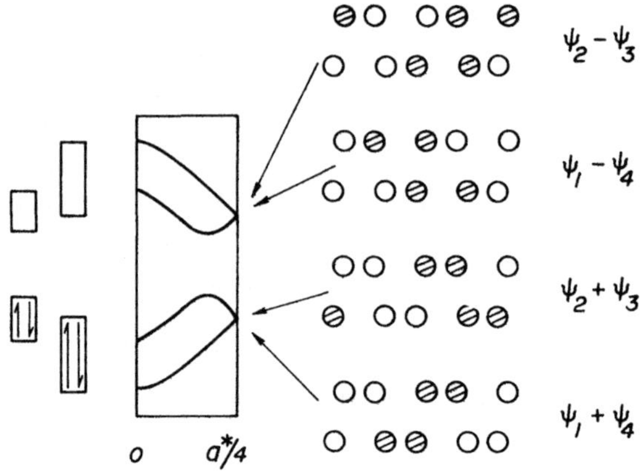

Fig. 22. Band structure of **64b**.

65

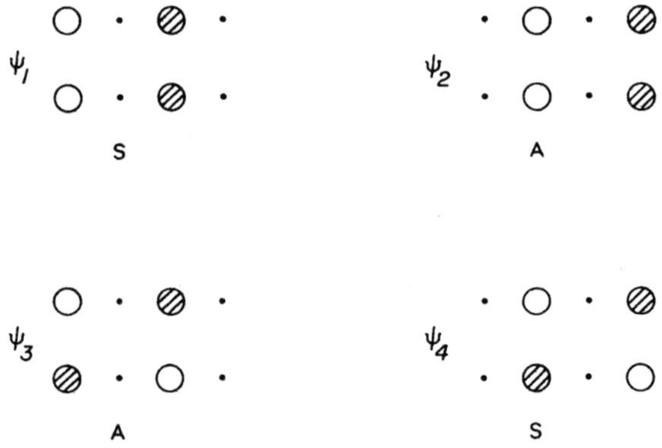

introduction is a result of mixing between orbitals of different energies, **64a** → **64b** is a second-order Peierls distortion. The degeneracies of the band structure of **64b** at the zone edge result from the fact that the two repeat units of **64b, 66a**, and **66b** are identical except for their orientation.

66

An interesting example of a second-order Peierls distoriton is found in $Sc_{3/4}Zn_{5/4}Mo_4O_7$ and $Ti_{1/2}Zn_{3/4}Mo_4O_7$ [16], which contain metal cluster chains that are derived from Mo_6 octahedra by sharing their opposite edges as indicated in **67** [4(d)]. The $x^2 - y^2$ orbitals of two apical Mo atoms in each Mo_6 octahedron leads to

67

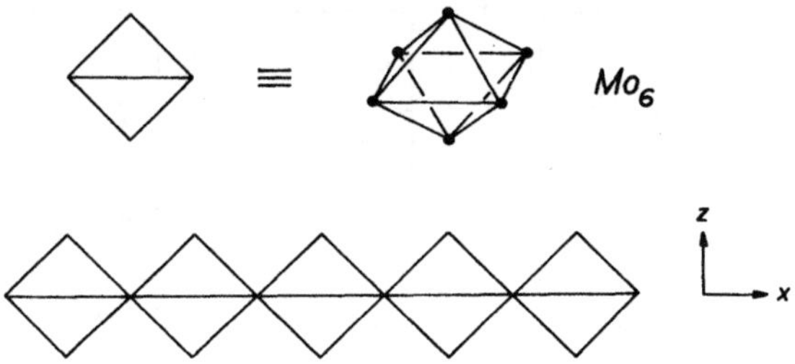

68

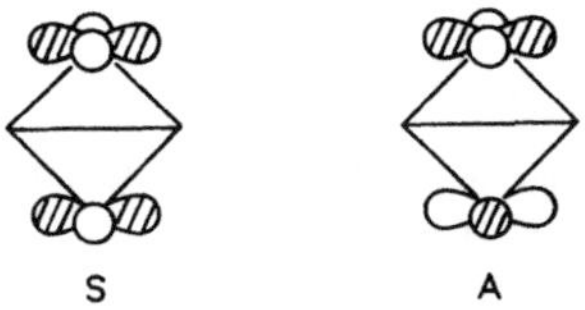

69

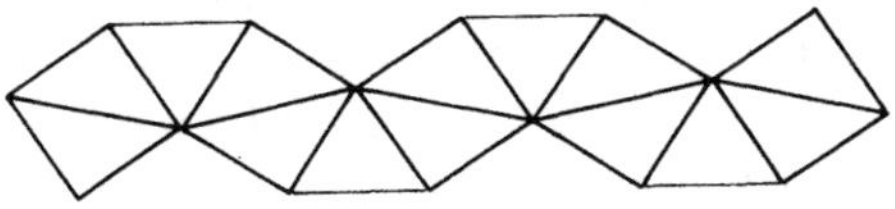

70

a

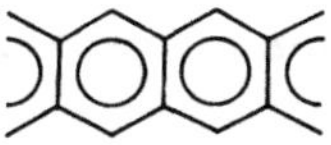

b

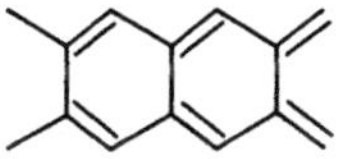

c

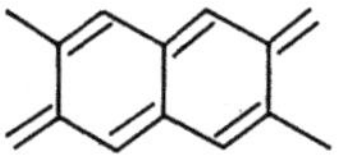

71

a b

the bonding and antibonding levels in **68**. The $x^2 - y^2$ bands resulting from these orbitals are similar in nature to the *s* bands of Figure 20. Thus, when the $x^2 - y^2$ bands are $\frac{1}{2}$-filled, the metal cluster chain **67** undergoes a second-order Peierls distortion **67** → **69**, which leads to the alternation of short and long apical Mo—Mo contacts along the chain [4(d)]. Another example of a second-order Peierls distortion is found in polyacene [2(b), 6(b)] which may adopt the structures **70a, 70b**, or

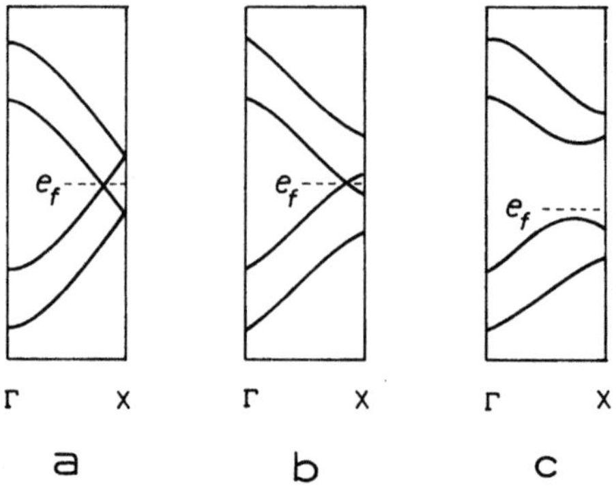

Fig. 23. π-band structures of polyacene. (a), (b), and (c) refer to the band structures of **70a**, **70b**, and **70c**, respectively.

70c. The π-band structures of **70a, 70b**, and **70c** are depicted in Figures 23(a), (b), and (c), respectively. The distortion **70a** → **70b** does not introduce a band gap at the Fermi level, but the distortion **70a** → **70c** does. The band structure of **70c** does not exhibit degeneracy at the zone end, unlike in Figure 22. This is due to the fact that the two repeat units of **70c**, **71a** and **71b**, are not chemically equivalent.

4. Low-Dimensional Solids

In this section we will briefly survey the electronic structures of some low-dimensional solids derived from transition metal ML_6 octahedra, ML_6 triangular prisms, and ML_8 tetragonal antiprisms. The prime goal of this survey is to describe the qualitative band structures of those systems and thereby rationalize their structural and electronic properties.

4.1. Structures Derived from ML_6 Octahedra

4.1.1. *NbX_4 Chain* [2(d), 17(a)—(c)]

The ideal NbX_4 chain (X = halogen) **72a** is derived from NbX_6 octahedra upon sharing their opposite edges, which is identical to the chain structure **17**. Thus the *d*-block bands of Figure 8, discussed in detail in Section 2.2.2(b), are relevant for **72a**. With the oxidation formalism $Nb^{4+}(X^-)_4$, there is one *d* electron per NbX_4 unit so that the $x^2 - y^2$ and *yz* bands, which overlap significantly, become partially filled (**73a**).

The **72a** → **72b** distortion doubles the size of the repeat unit cell and also splits each *d*-block band in two. The **72a** → **72b** distortion can be studied in terms of the Nb atom displacement δ defined in **74**. Figure 24 shows how the band structure of NbX_4 changes as a function of δ, where the *xz* band was omitted for simplicity. Figure 24(a) is simply the band structure of **72a** with $(NbX_4)_2$ as a repeat unit. For

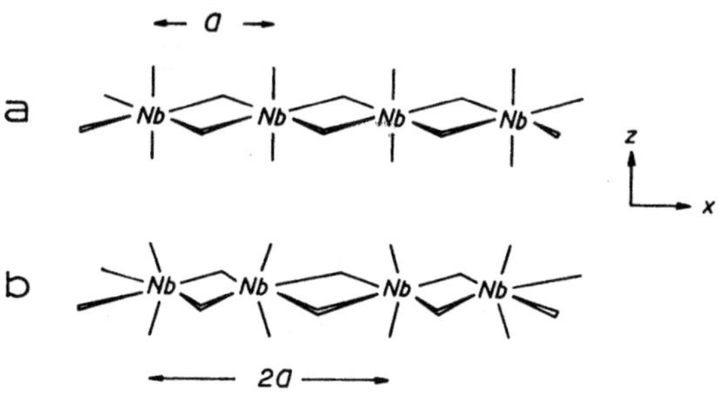

72

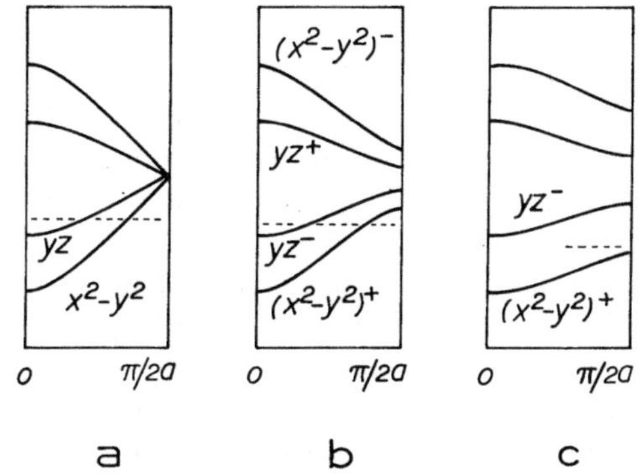

Fig. 24. Low-lying d-block bands of NbX_4 as a function of the Nb displacement δ defined in **74**. The band structure of NbX_4 with a unit cell $(NbX_4)_2$ and $\delta = 0$ is given in (a). The band splittings for small and large values of δ are shown in (b) and (c), respectively.

nonzero δ, the $x^2 - y^2$ and yz bands each split in two as shown in Figures 24(b) and 24(c), where the superscripts '+' and '−' signify that two metal d orbitals of a dimer repeat unit combine in-phase and out-of-phase, respectively. When δ is small, the splitting of the $x^2 - y^2$ band is small so that the $(x^2 - y^2)^+$ and yz^- bands remain overlapping (Figure 24(b)). For large δ the splitting of the $x^2 - y^2$ band is large, so

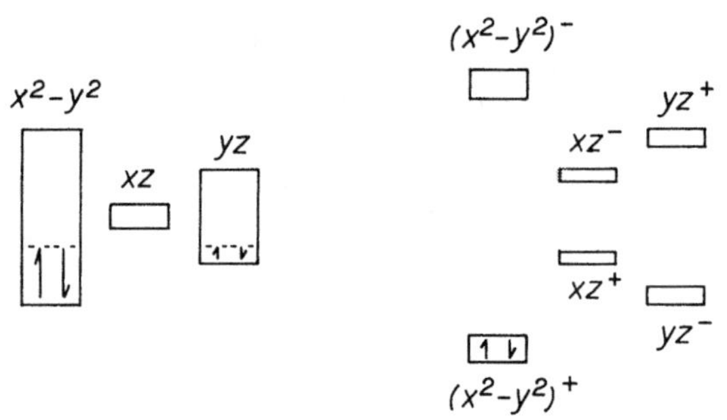

73a 73b

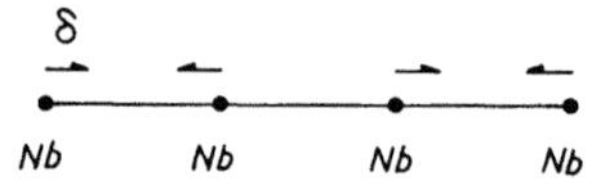

74

75 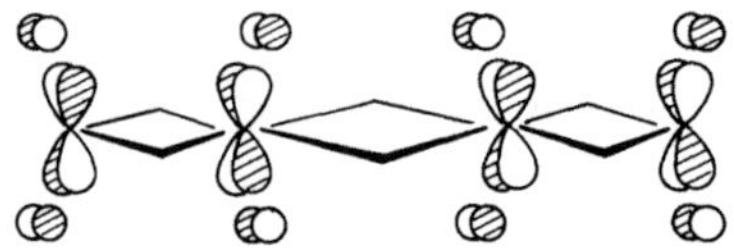

76 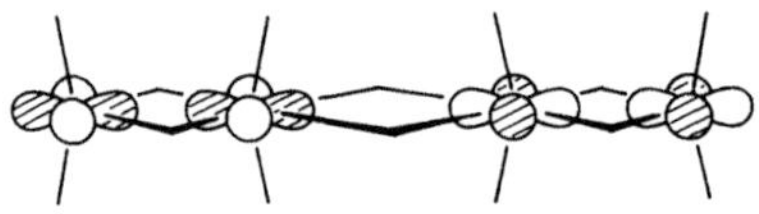

that the $(x^2 - y^2)^+$ and yz^- band do not overlap any more (Figure 24(c)). There are two d electrons per unit cell $(NbX_4)_2$ to fill the d-block bands of Figure 24. Therefore a band gap is introduced when the extent of the **72a** → **72b** distortion is beyond a certain value of δ. The d-block bands of **72b** are occupied as shown in **73b**, thereby providing semiconducting properties for NbX_4.

Whether or not the $(x^2 - y^2)^+$ and yz^- bands overlap depends upon the relative energies of the top of the $(x^2 - y^2)^+$ band (**76**) and the bottom of the yz^- band (**75**). **75** has antibonding character between Nb and the apical halogen atoms. The overlap between yz orbitals is weak so that the bottom of the yz^- band is not very sensitive to δ. On the other hand, the orbital character of **76** is such that the top of the $(x^2 - y^2)^+$ band becomes lower in energy very sharply as δ increases.

4.1.2. *$NbOX_2$ Net* [2(e), 17(c)–(f)]

The ideal 2D net structure of $NbOX_2$ (**77**) is derived from NbO_2X_4 octahedra (**78**) by sharing the opposite halogen edges along **a** and by sharing the opposite oxygen corners along **b**. With the oxidation formalism $Nb^{4+}O^{2-}(X^-)_2$, each $NbOX_2$ unit

77 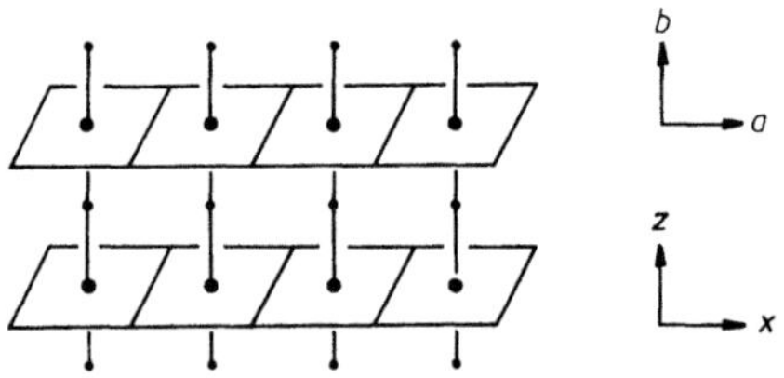

78 

contributes one *d* electron to the net structure. As expected from our discussion of the NbX_4 chain, the real structure of the $NbOX_2$ net shows an alternation of short and long Nb—Nb separations along **a** (i.e. Nb—Nb···Nb alternation). In addition, the $NbOX_2$ net also shows an alternation of short and long Nb—O distances (i.e. Nb—O···Nb alternation) along **b**. The Nb—O···Nb alternation may be achieved by the oxygen displacement **79a** → **79b** or by the Nb displacement from the X_4 plane **79a** → **79c**. It is the latter that is experimentally found.

79

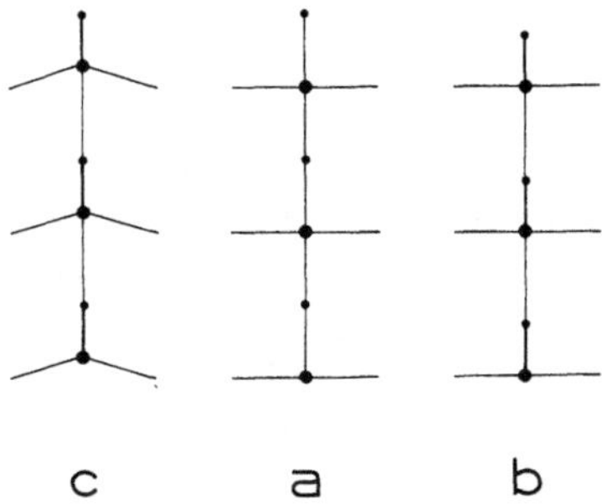

The reason why $NbOX_2$ net adopts the distortion mode **79a** → **79c** is important in connection with its structural stability [17(f)] and semiconducting property [17(c)]. Figure 25 shows the $(x^2 - y^2)^+$ and yz^- bands of $NbOX_2$ net with the Nb—Nb···Nb alternation as found in the real structure but without the

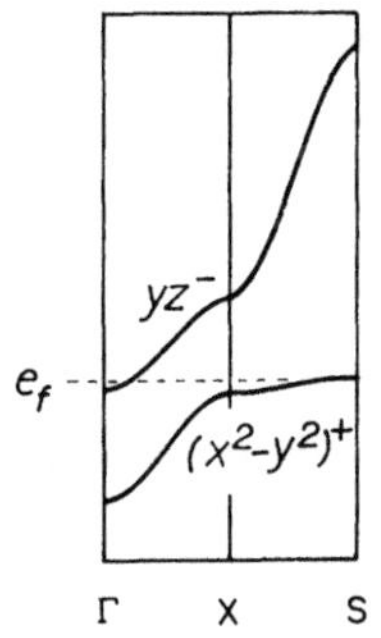

Fig. 25. Low-lying *d*-block bands of $NbOX_2$ net with the Nb—Nb···Nb alternation but without the Nb—O···Nb alternation.

Nb—O···Nb alternation [2(e)]. There are two *d* electrons to fill these bands, which are found to overlap. Thus, unlike the case of NbX_4 chain, the Nb—Nb···Nb alternation alone does not lead to a semiconducting state for $NbOX_2$. Let us examine why this is so. The essence of how the yz^- band varies along $\Gamma \rightarrow Y$ (i.e. along the Nb—O···Nb direction) can be shown by considering the *yz* band resulting from the niobium *yz* orbitals and oxygen *y* orbitals shown in **80** for the symmetrical structure **79a**. In **80**, *yz* orbitals are considered to be empty since one *d* electron from each

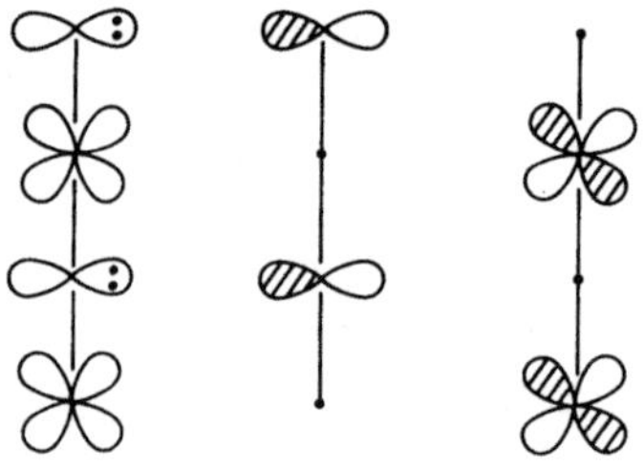

80 81a 81b

Nb is accommodated by the $(x^2 - y^2)^+$ band (i.e. into the σ bonds of niobium atoms in the dimeric units along **a**). The *y* and *yz* bands resulting from **80** were already examined in Section 2.2.2(a), which are reproduced in Figure 26(a). At $\Gamma = (0, 0)$ the *y* and *yz* bands have the orbital patterns **81a** and **81b**, respectively. Namely, for the symmetrical Nb—O—Nb bridge (**79a**), the *y* orbitals of oxygen cannot interact with the *yz* orbitals of Nb at Γ by symmetry. Thus the oxygen *y* band is not stabilized by the niobium *yz* orbitals around Γ, nor is the niobium *yz* band destabilized by the oxygen *y* orbitals.

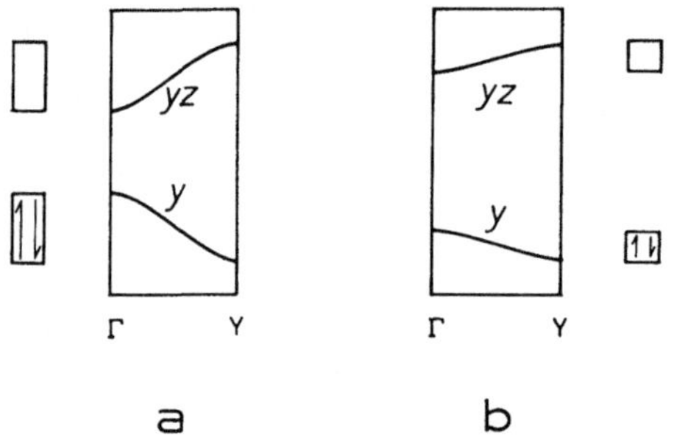

Fig. 26. Change in the *yz* and *y* bands of $NbOX_2$ upon the Nb—O···Nb alternation. (a) and (b) refer to **79a** and **79c**, respectively.

In contrast, the yz^- band of the NbX_4 chain has an antibonding contribution from the apical halogen atoms as shown in **75**. Therefore, the bottom of the yz^- band is lower in energy for the $NbOX_2$ net with symmetrical Nb—O—Nb bridge, which is why the $(x^2 - y^2)^+$ and yz^- bands overlap.

The *y* band of Figure 26(a) may be stabilized by introducing an asymmetrical Nb—O···Nb bridge since this will allow orbital mixing between *y* and *yz* in the region of Γ. For the Nb—O···Nb bridge **79b**, the *y* and *yz* orbital arrangement is given by **82**. The *y* and *yz* band at Γ may be represented by **83a** and **83b**, respectively. In **83a**, bonding combination between *yz* and *y* for an Nb—O bond leads to antibonding interaction between *y* and *yz* for the adjacent O···Nb bonds. Similarly, in **83b**, antibonding combination between *yz* and *y* for an Nb—O bond leads to

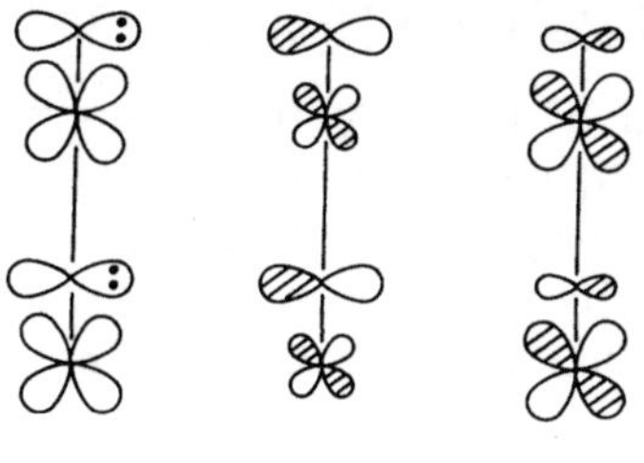

82 83a 83b

bonding interaction between *y* and *yz* for the adjacent O· · ·Nb bonds. Thus **79a** → **79b** distortion is not effective in lowering the *y* band nor in introducing a band gap between the yz^- and $(x^2 - y^2)^+$ bands. This situation can be rectified by strengthening the interaction of Nb—O bonds and weakening that of O· · ·Nb bonds, which is achieved by the distortion **79a** → **79c**. The out-of-plane displacement of Nb leads to the hybridization of *yz* shown in **84**. Thus for **79c** the orbital arrangement is given by **85. 86a** and **86b** show the *y* and *yz* bands at Γ, respectively. Now the

84

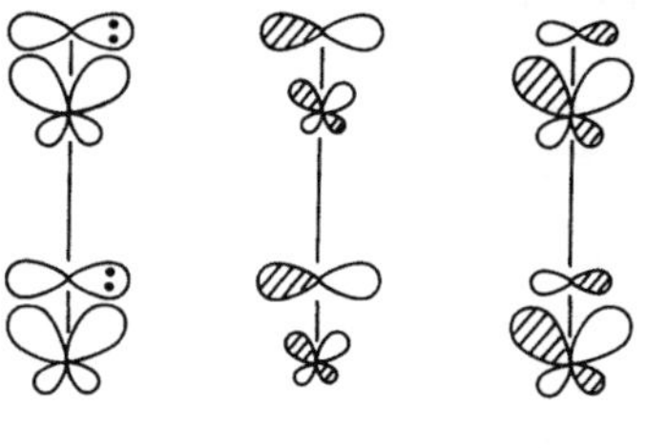

85 86a 86b

Nb—O bonding is more effective than the O· · ·Nb antibonding in **86a**, while the Nb—O antibonding is more effective than the O· · ·Nb bonding in **86b**. Consequently, as shown in Figure 26(b), the **79a** → **79c** distortion is effective in lowering the *y* band and raising the *yz* around Γ. This explains why $NbOX_2$ adopts the Nb—O· · ·Nb alternation **79c** and why $NbOX_2$ net is semiconducting as a result.

In essence, the **79a** → **79c** distortion allows mixing between the *yz* and *y* bands, thereby stabilizing the latter. Therefore this distortion may be classified as a second-order Peierls distortion, although the two bands involved are not close in energy compared with the examples discussed in Section 3.3. The Nb—Nb· · ·Nb alternation along **a** is, as in the case of NbX_4 chain, an example of a first-order Peierls distortion. Thus $NbOX_2$ net is an example which exhibit both first and second-order Peierls distortions.

4.1.3. $[Pt(II)L_4 \cdot Pt(IV)L_4X_2]$ *Chain* [2(d), 2(l), 18]

The ideal Pt(III)L_4X chain **87a** has Pt(III)L_4 units bridged by halide ions with a symmetrical Pt(III)—X—Pt(III) bridge. The band structure of this chain structure was derived in Section 2.2.2(a). With a Pt(III) ion, the *d* orbitals of each Pt(III)L_4 unit (see **19**) are occupied by seven electrons. This leads to a singly filled z^2 level in each Pt(III)L_4 unit, so that the z^2 band of **87a** becomes $\frac{1}{2}$-filled. Thus **87a** is susceptile to dimerization.

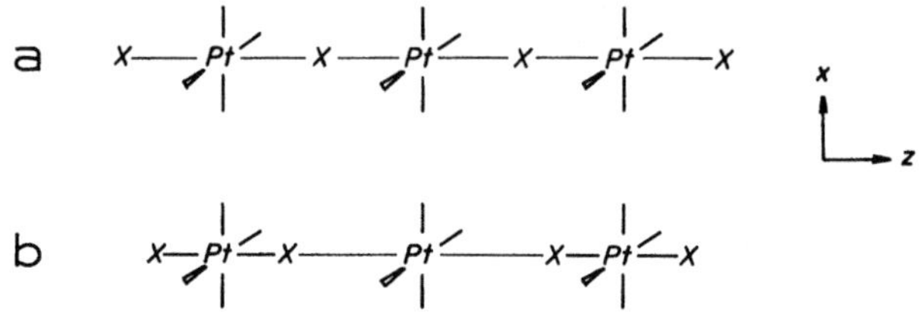

87

To examine the structural distortion that **87a** adopts, we represent the z^2 band of **87a** in Figure 27(a) for the unit cell of $(PtL_4X)_2$. This is obtained by folding the z^2 band of Figure 3. Let us now examine the band orbital degeneracy at the zone edge of Figure 27(a). The z^2 Bloch orbitals **88a** and **88b** are degenerate. The two symmetry-adapted *z* orbitals of bridging halides **89a** and **89b** interact with

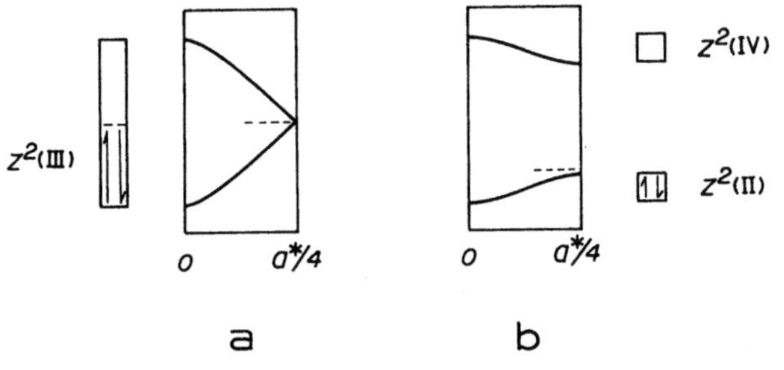

Fig. 27. z^2 bands of PtL_4X chains. (a) and (b) show the band structures of the monovalent chain **87a** and the mixed-valent chain **87b**, respectively.

88a and **88b**, respectively. The two symmetry-adapted *s* orbitals **90a** and **90b** interact with **88a** and **88b**, respectively. Both **89a** and **90a** mix into **88a** in an antibonding manner to give **91a**. Likewise, **89b** and **90b** mix into **88b** in an antibonding manner to give **91b**. **91a** and **91b** are responsible for the degeneracy at the zone end of Figure 27(a). This degeneracy is lifted, and hence a band gap is introduced at the Fermi level, by the pairing distortion of bridging halides **87a** → **87b**; that is, this distortion reduces metal—halogen antibonding in **91a** but

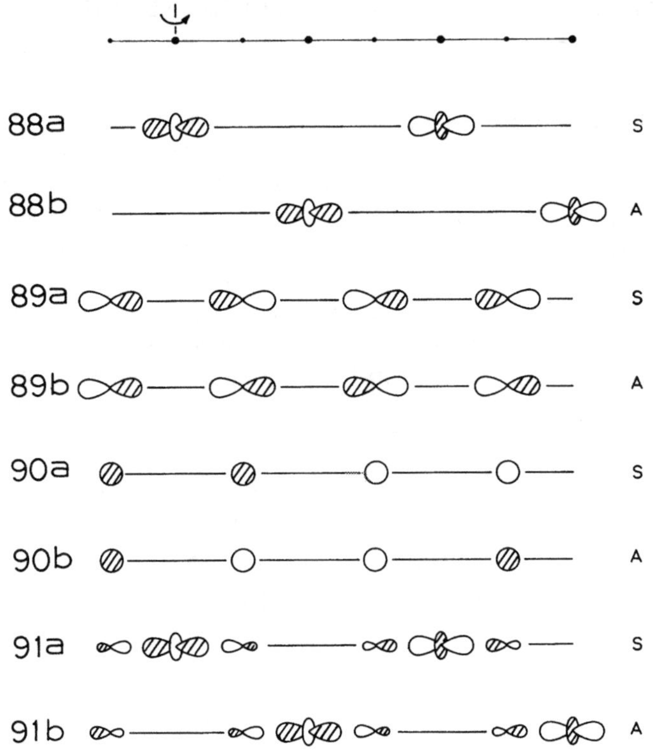

enhances that in **91b**. The z^2 band structure of **87b** is shown in Figure 27(b). The filled z^2 band of **87b** has nonbonding character in that it is primarily composed of the z^2 orbital of each Pt(II)L_4 unit. The empty z^2 band of **87b** has the z^2 orbital character of each Pt(IV)L_4X_2 unit which has strong Pt—X antibonding contribution.

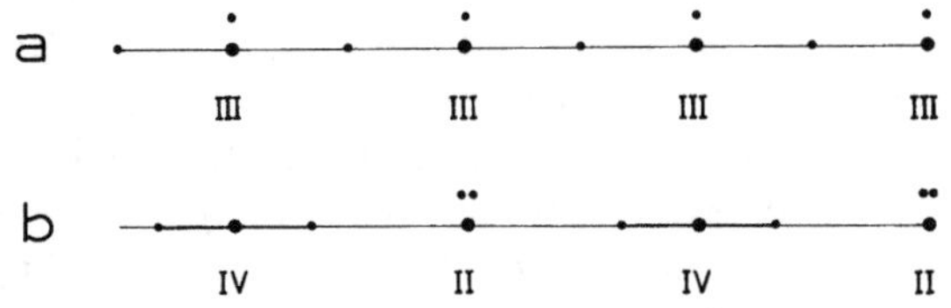

92

As far as the z^2 band is concerned, the pairing distortion **87a** → **87b** creates an electron pair at each Pt(II) center (**92a** → **92b**) and gives rise to on-site Coulomb repulsion. This is energetically unfavorable. Nevertheless, the mixed-valent structure [Pt(II)L_4 · Pt(IV)L_4X_2] is observed because its alternative, monovalent structure Pt(III)L_4X, is forced to accommodate the electron pair into the band with strong Pt(III)—X—Pt(III) antibonding character [2(l)].

4.2. Structures Derived from ML_6 Trigonal Prisms

4.2.1. *Structural Pattern*

A number of transition metal chalcogenides MX_3 (X = S, Se, Te) have structures that are made up of trigonal prismatic chains (**93**) parallel to one crystallographic direction (usually taken as **b**) [19—21]. In each prismatic chain, every metal atom is sandwiched by two X_3 triangles. A prismatic chain may be termed equilateral-like when all three sides of the X_3 triangle are substantially longer than the X—X bond length of an $(X_2)^{2-}$ dimer. Likewise, a prismatic chain may be termed isosceles-like if one side of the X_3 triangle is close in value to the X—X bond length of $(X_2)^{2-}$ while the other two sides are substantially longer. In equilateral- and isosceles-like chains, the appropriate oxidation formalisms of X_3 are $(X^{2-})_3$ and $X^{2-}(X_2)^{2-}$, respectively.

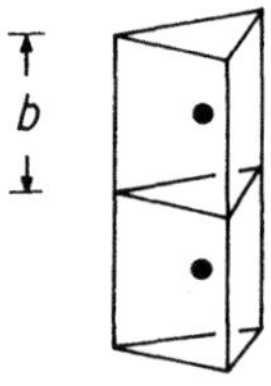

93

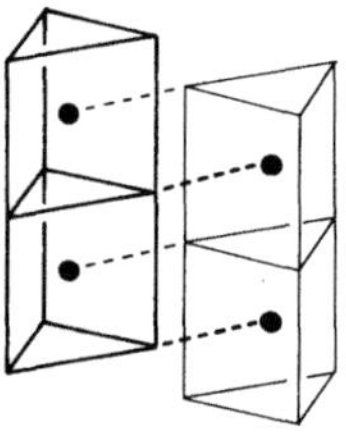

94

95

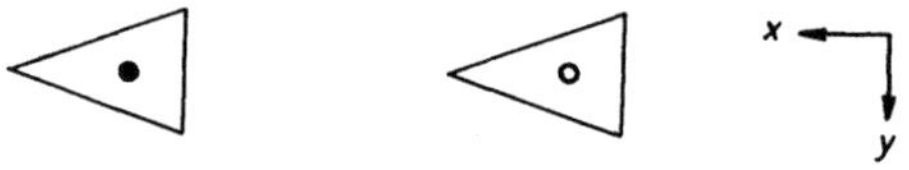

a b

Equilateral- and isosceles-like prismatic chains crosslink together to form a 2D layer structure, in which every chain interacts with its two nearest-neighbor chains shifted along the chain axis by $b/2$. As illustrated in **94**, a 2D layer structure results when two equivalent rectangular faces of each MX_6 prisms of one chain are capped by the X atoms of its nearest-neighbor chains thereby leading to interchain M···X interactions. An isosceles-like chain viewed along **b** may be represented by **95a**. When this chain is shifted along **b** by $b/2$, the corresponding projected view may be represented by **95b**. The three known 2D MX_3 layer structures may be described as in Figure 28 [22]. The $ZrSe_3$ structure [19] is characterized by one type of isosceles-like chain. The $TaSe_3$ structure [20] has two types of chains, in which pairs of equilateral-like chains are separated by those of 'isosceles-like' chains. Here the shortest Se—Se distance of the 'isosceles-like' chains (2.58 Å) is too large to represent Se—Se single bonds but too short for there to be negligible bonding interaction. The $NbSe_3$ structure [21] is characterized by three types (I, II, III) of chains, in which type I and III chains are isosceles-like and type II chains are equilateral-like.

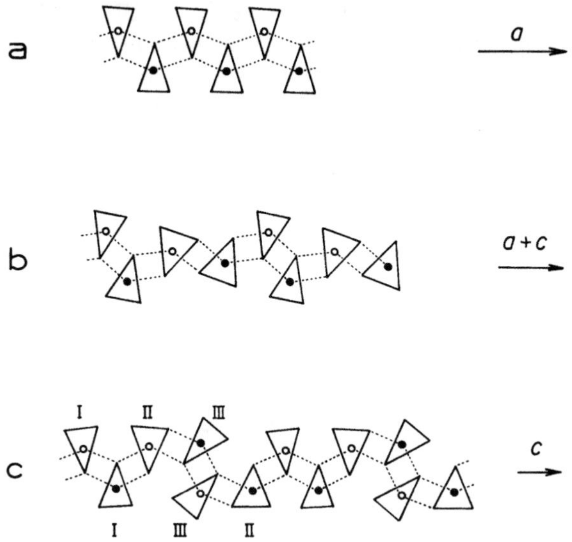

Fig. 28. Three types of two-dimensional MX_3 layers. (a), (b), and (c) refer to the $ZrSe_3$, $TaSe_3$, and $NbSe_3$ structures, respectively.

A 3D MX_3 is formed when 2D MX_3 layers stack together. Interaction between those 2D MX_3 layers is weak, because they are typically separated in the order of van der Waals radii. Thus in the following we will not discuss the interaction between 2D MX_3 layers.

In understanding the structures derived from prismatic MX_3 chains, it is important to know the valence orbitals of isosceles and equilateral X_3 triangles. Figure 29 shows the three sets of group orbitals arising from the *p* orbitals of each

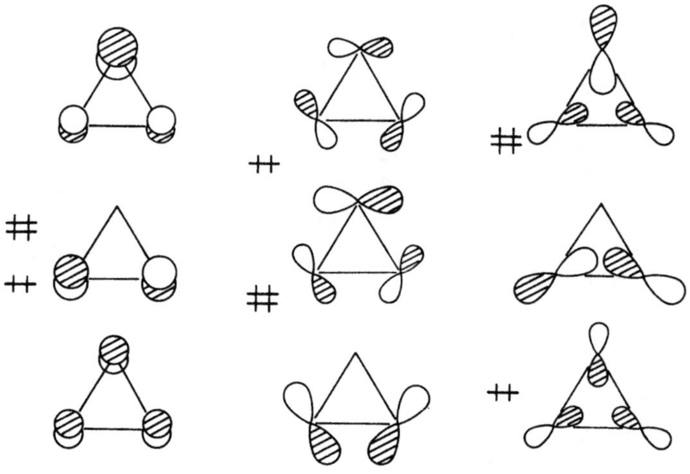

Fig. 29. *p* orbitals of an equilateral X_3 triangle.

X in an equilateral triangle. For the convenience of simple electron counting, further orbital mixing between the two sets of in-plane *p* orbitals is omitted. With the oxidation $(X^{2-})_3$, all the orbitals of Figure 29 are completely occupied. Note that we have not shown the three lower-lying orbitals of X_3 resulting from an *s* orbital of each X.

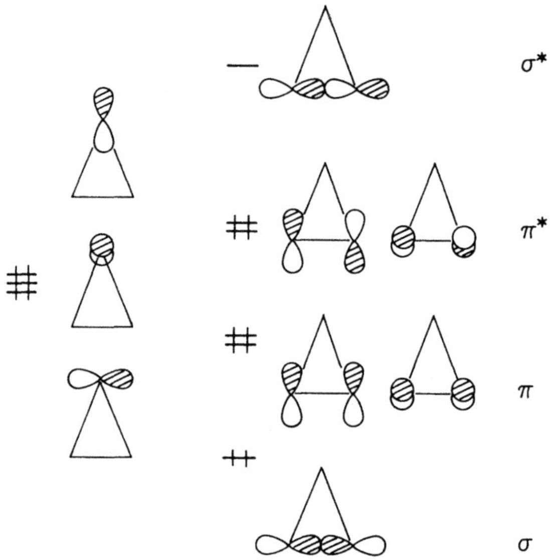

Fig. 30. *p* orbitals of an isosceles X_3 triangle.

Figure 30 shows the orbitals of an isosceles X_3 triangle derived from the *p* orbitals of each X. For the convenience of simple electron counting, an isosceles triangle is considered as $X^{2-} + (X_2)^{2-}$ and mixing between the orbitals of X^{2-} and $(X_2)^{2-}$ is neglected in Figure 30. With an isosceles X_3, all the levels are occupied except for the σ^* level of its $(X_2)^{2-}$ dimer unit.

The in-plane *p* orbitals of an equilateral X_3 (Figure 29) and those of an isosceles X_3 (Figure 30) are important in connection with the interchain M···X interactions in a 2D MX_3 layer. These in-plane orbitals interact effectively with the metal $x^2 - y^2$ and xy orbitals in each plane containing the zig-zag ···X···M···X···M··· linkage. This has a very important consequence upon the electronic structure of a 2D MX_3 layer, as will be discussed later.

4.2.2. *Trigonal Prismatic Chain*

Since MX_3 chains are essential building blocks, it is important to examine the band structure of a single MX_3 chain and subsequently to study how this band structure is modified by the interchain M···X interactions in a 2D MX_3 layer. For an isosceles-like MX_3 chain with the oxidation formalism $M^{4+}X^{2-}(X_2)^{2-}$, M^{4+} ions are

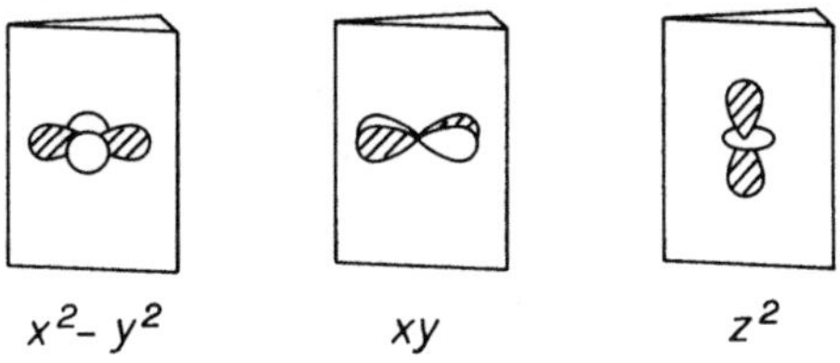

96

typically d^0 or d^1. Therefore, it is sufficient to study only the low-lying d-block bands, which are derived from the z^2, $x^2 - y^2$, and xy orbitals (**96**) since they avoid strong overlap with the orbitals of X_3.

The z^2 Bloch orbitals at the zone center and edge are shown in **97a** and **97b**, respectively. Interaction of z^2 orbitals with the orbitals of X_3 is not effective. As shown in **98** and **99**, however, the $x^2 - y^2$ and xy Bloch orbitals combine more effectively with the orbitals of X_3. Direct metal—metal interaction along the chain

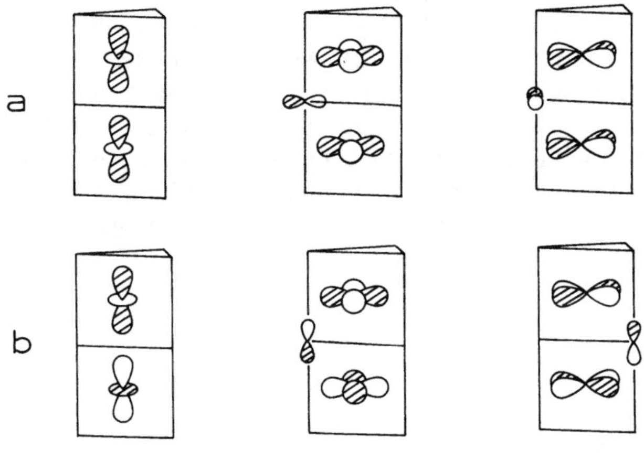

97 98 99

axis is much stronger with z^2 orbitals than with $x^2 - y^2$ or xy orbitals so that the resulting z^2 Bloch orbital is more dispersive than the $x^2 - y^2$ and xy Bloch orbitals. The band structures resulting from those Bloch orbitals are sketched in Figure 31 [4(f)]. The z^2 and $x^2 - y^2$ Bloch orbitals are the same in their symmetry, so that their mixing around the intended crossing region (dashed lines) splits the two bands thereby leading to the band character switching. The xy Bloch orbitals are different in symmetry from the z^2 Bloch orbitals throughout the BZ, so that the xy band crosses the z^2 band. With a d^1 metal ion per MX_3 as in a single $NbSe_3$ chain, there is one electron per unit to fill the bottom band of Figure 31. This band is narrow due to the interaction between the z^2 and $x^2 - y^2$ Bloch orbitals.

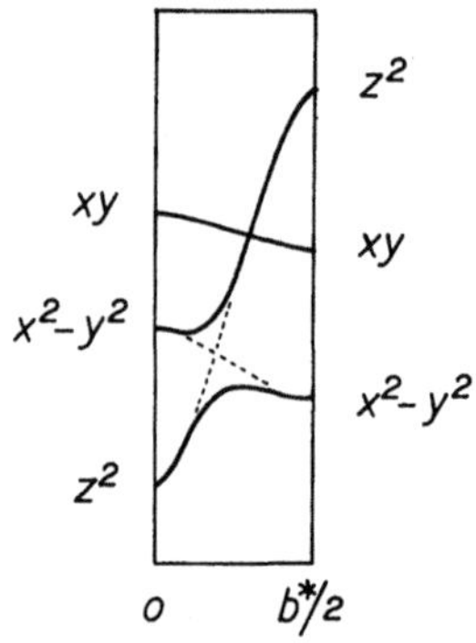

Fig. 31. Low-lying d-block bands of a single, isosceles-like MX_3 chain.

4.2.3. *Two-Dimensional MX_3 Layer*

(a) *Interchain M$\cdots$X interaction.* NbS_3 has the $ZrSe_3$ structure shown in Figure 28(a) [19], and the oxidation formalism $S^{2-}(S_2)^{2-}$ leads to $Nb^{4+}(d^1)$ ions. Thus the bottom band of Figure 31 would be $\frac{1}{2}$-filled for a single NbS_3 chain. Let us now examine how the band structure of a 2D MX_3 layer is derived by considering the interchain M$\cdots$X interaction. The $x^2 - y^2$ and xy orbitals of each metal in a chain are extended towards the capping X atoms of the nearest-neighbor chains, while this is not the case with the z^2 orbitals. Therefore, the $x^2 - y^2$ and xy orbitals of one chain interact more effectively with the X_3 orbitals of the nearest-neighbor chains so that the $x^2 - y^2$ and xy band are raised in energy with respect to the z^2

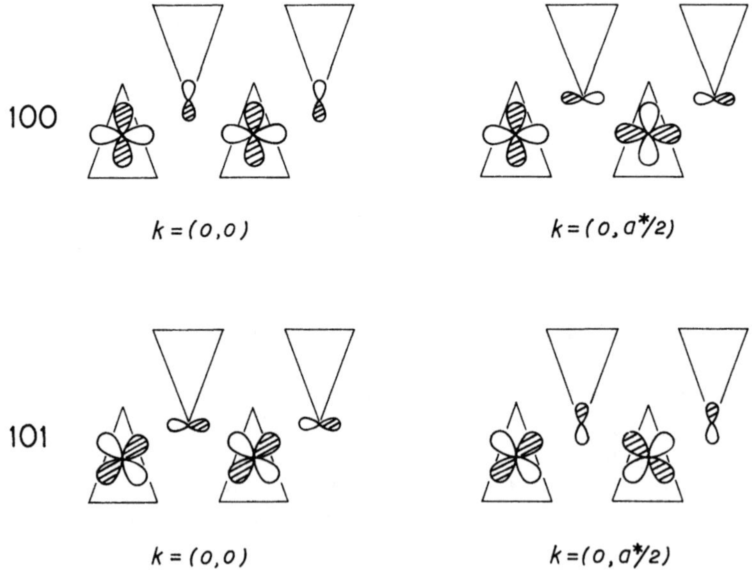

bands. This effect of the capping orbitals is illustrated in **100** and **101** for the $x^2 - y^2$ and xy bands, respectively, where the in-plane p orbitals of X_3 units combine out-of-phase with the $x^2 - y^2$ and xy orbitals.

The in-plane p orbitals of X_3, regardless of whether equilateral- or isosceles-like, interact effectively with the $x^2 - y^2$ and xy orbitals at all values of wave vector as long as the interchain M···X contacts are present. As a consequence, for a 2D MX_3 layer, the $x^2 - y^2$ and xy bands are raised beyond the z^2 bands shown in **102** [4(f)]. The presence of the two z^2 bands is due to the fact that there are two MX_3 chains per unit cell in NbS_3. It should be noted that the p-block bands resulting

102

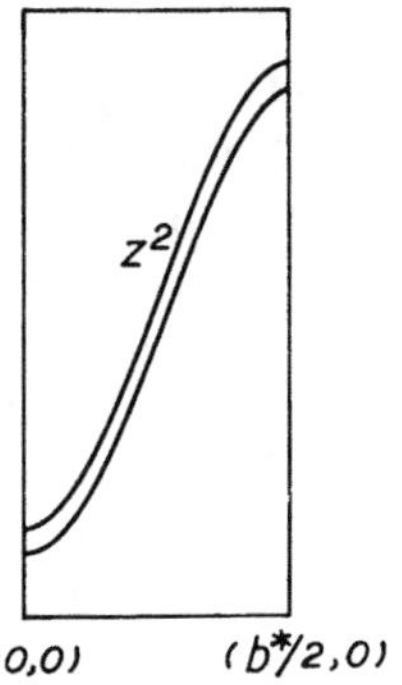

from the in-plane p orbitals of X_3 units are stabilized by the interchain M···X interactions because the $x^2 - y^2$ and xy orbitals combine in-phase with the in-plane p orbitals of X_3 units. It is the stabilization of these p-block bands that allows 1D MX_3 chains to crosslink together to form a 2D MX_3 layer. The most important consequence of the interchain M···X interaction is to remove the $x^2 - y^2$ and xy bands from the region of the z^2 band, which is dispersive primarily along the chain direction (i.e. one-dimensional-like in nature).

With two d electrons per unit to occupy, the z^2 bands of **102** are $\frac{1}{2}$-filled for a 2D NbS_3 layer. Thus, NbS_3 undergoes a pairing distortion in which Nb atoms along the chain dimerize to give Ne—Nb···Nb alternation [19(d), 23].

(b) *$NbSe_3$ and TaS_3*. Monoclinic $NbSe_3$ and TaS_3 have a 2D layer structure consisting of three types of chains as shown in Figure 28(c), where type II chains are equilateral-like while type I and III chains are isosceles-like. The X—X length of a $(X_2)^{2-}$ dimer is slightly larger in type I than in type III chains (2.105 versus 2.068 Å in TaS_3 and 2.485 versus 2.374 Å in $NbSe_3$) [21]. An important structural feature of the $NbSe_3$-type structure is that every pair of type I chains is separated from that of type III chains by a type II chain.

As in the case of a 2D NbS_3 layer examined in the previous section, the interchain M···X contacts in $NbSe_3$ and TaS_3 allow interaction between the $x^2 - y^2$, xy orbitals of one chain and the in-plane p orbitals of X_3 in the nearest-neighbor chains. Thus, for 2D $NbSe_3$ and TaS_3 layers, the $x^2 - y^2$ and xy bands are raised with respect to the z^2 bands. The raising of these two bands is somewhat smaller in the $NbSe_3$ structure than in the $ZrSe_3$ structure [4(f), 24]. The former is lower in symmetry so that its interchain M···X interactions are expected to be less cooperative in raising the $x^2 - y^2$ and xy bands.

Metal z^2 orbitals are not well extended towards the interchain M···X interaction directions, and type I chains do not have direct interaction with type III chains. Thus, qualitative aspects of the z^2 bands of a 2D $NbSe_3$ or TaS_3 layer may be approximated by assuming, at least to a first approximation, no interaction between pairs of type I and type III chains. Each pair of these chains is expected to have two z^2 bands, as shown in **102**. The extent of the z^2 band separation in **102** reflects how strongly two nearest-neighbor chains interact via the interchain M···X contacts. The interchain M···X distance is slightly shorter within a pair of type I chains than within a pair of type III chains (2.564 versus 2.589 Å in TaS_3 and 2.73 versus 2.76 Å in $NbSe_3$) [21]. Thus it might be suggested that the interchain M···X interaction is greater between type I chains. However, the magnitude of this interaction would also be affected by other geometrical factors such as the X···M···X angle and the X—X length of $(X_2)^{2-}$. It is not intuitively obvious how these factors are related to the magnitude of the M···X interaction. *Provided* that the interchain M···X interaction is stronger between type I chains than between type III chains, the z^2 bands of type I and type III chains may be given as in Figures 32(a) and (b), respectively. Superposition of these two leads to the four z^2 bands shown in Figure 32(c), which is consistent with the z^2 band structure of $NbSe_3$ recently calculated [22(c), 24].

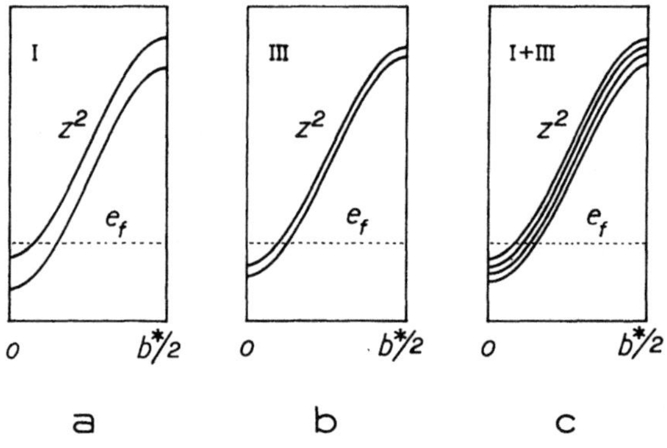

Fig. 32. Qualitative z^2 band structure of $NbSe_3$. Two z^2 bands for a pair of type I chains are shown in (a), and that for a pair of type III chains in (b). Superposition of these two leads to (c), which approximates the essence of the z^2 band structure of $NbSe_3$.

With the oxidation formalism of $(X^{2-})_3$ for type II chains and that of X^{2-} $(X_2)^{2-}$ for type I and type III chains, there remain two d electrons to fill the four z^2 bands of Figure 32(c). Thus $NbSe_3$ and TaS_3 are expected to have four approximately $\frac{1}{4}$-filled z^2 bands, i.e. the Fermi wave vectors k_f^b along the chain direction **b** are anticipated to be ~ 0.25 $(b^*/2)$ for the four z^2 bands [22(b), (c), 24]. This is in good agreement with the k_f^b values derived from the two CDWs observed from

$NbSe_3$ ($2k_f^b = 0.243b^*$ and $0.263b^*$ for the CDWs at 144 and 59 K, respectively [25]) and TaS_3 ($2k_f^b = 0.254b^*$ and $0.245b^*$ for the CDWs at 240 and 160 K, respectively [22(a), 26]). These CDWs, which arise from type I and type III chains [22(a), 24] are independent and lead to spectacular resistivity anomalies [27]. The type II chains of a unit cell lead to two z^2 bands, which lie above the Fermi level [24] in agreement with the oxidation formalism of Nb^{5+} for the Nb atoms in those chains.

(c) *$FeNb_3Se_{10}$*. A recent attempt to dope $NbSe_3$ with Fe led to a new compound of nominal composition $FeNb_3Se_{10}$ [28], which has a 2D layer structure shown in **103**. The unit cell has a pair of isosceles-like chains made up of $NbSe_6$ prisms and a double chain of edge-shared MSe_6 (M = Fe, Nb) octahedra which share corners with the $NbSe_6$ prisms. The Fe atoms occur only in the octahedral chains, and the

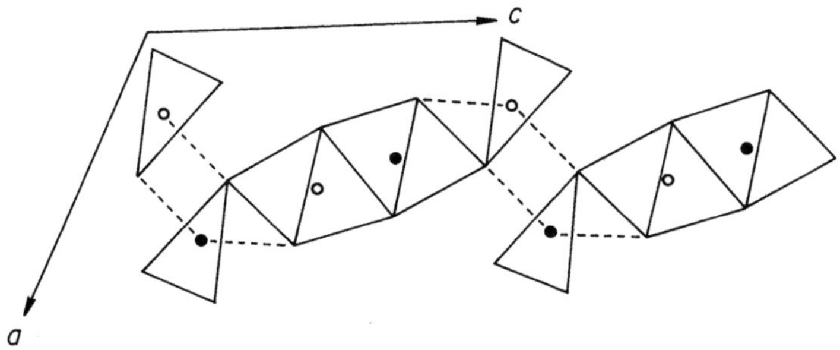

103

Fe and Nb atoms in the octahedral chains are randomly distributed. To emphasize the presence of prismatic and octahedral chains, $FeNb_3Se_{10}$ may be written as $[NbSe_3]_2$ $(FeNbSe_4)$. The square brackets refer to those atoms in the prismatic chains, and the parentheses to those in the octahedral chains. For convenience, one Se atom shared by the two types of chains is included in the square brackets.

The prismatic chains of $[NbSe_3]_2$ $(FeNbSe_4)$ have a short Se—Se distance of 2.348 Å in the triangular base, and thus are similar in structure to type III chains of $NbSe_3$ [28(a)]. In $[NbSe_3]_2$ $(FeNbSe_4)$, each Nb atom of a prismatic chain has the interchain Nb$\cdots$Se interactions with the Se atoms of the neighboring prismatic and octahedral chains. Therefore, the z^2 band structures of $[NbSe_3]_2$ $(FeNbSe_4)$ that originate from the prismatic chains are expected to be similar to those shown in **102**.

With the oxidation formalism $Se^{2-}(Se_2)^{2-}$ and $(Se^{2-})_4$ for the Se atoms in the prismatic and octahedral chains, respectively, the oxidation states of the metal atoms may be written as $[Nb^{4+}]_2$ $(Fe^{3+}Nb^{5+})$. Each Fe atom, under a local octahedral environment, would have three low-lying *d* levels derived from the t_{2g} level. Thus six *d* electrons can be accommodated by each Fe, so that the oxidation formalism for the metal atoms may be rewritten as $[2Nb]^{9+}$ $(Fe^{2+}Nb^{5+})$; that is, there is one *d* electron to fill the two z^2 bands associated with a pair of prismatic chains, so that $[NbSe_3]_2$ $(FeNbSe_4)$ is expected to have two approximately $\frac{1}{4}$-filled z^2 bands. This analysis is consistent with the CDW wavevector observed in this compound $(2k_f^b = 0.26b^*)$ [29].

In principle, a band structure calculation on a disordered material would require an infinite unit cell. Nevertheless, the effect of disorder may be estimated by studying a number of differently ordered arrangements. The two simplest ordered arrangements of the Fe and Nb atoms in the octahedral chains of $[NbSe_3]_2$ $(FeNbSe_4)$ are shown in **104**, where the rectangles and squares refer to the $NbSe_6$ prisms and MSe_6 (M = Fe, Nb) octahedra, respectively. The essential features of the z^2 band structure [30] for the arrangement **104a** may be summarized as in Figure 33. The folded nature of the band reflects the fact that the repeat unit cell of **104a** is $(FeNb_3Se_{10})_2$. With two electrons to fill the two z^2 bands, the lower two z^2 bands are $\frac{1}{2}$-filled so that $k_b^f \simeq 0.5(b^*/4) = 0.25(b^*/2)$; that is, the z^2 bands are approximately $\frac{1}{4}$-filled with the unit cell size of $(FeNb_3Se_{10})$.

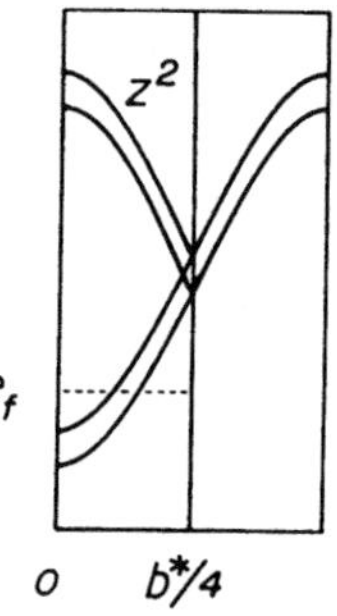

Fig. 33. z^2 band structure of $FeNb_3Se_{10}$ for the ordered arrangement **104a** of Fe and Nb atoms in octahedral chains.

In general, $[NbSe_3]_2$ $(FeNbSe_4)$ is a special case of $[NbSe_3]_2$ $(Fe_{1+x}Nb_{1-x}Se_4)$. Thus the oxidation formalism for the metal atoms can be written as $[2Nb]^{9+3x}$ $(Fe_{1+x}^{2+}Nb_{1-x}^{5+})$. For nonzero x, therefore, the prismatic chains are expected to lose $3x$ electrons per formula unit to the octahedral chains. Therefore the z^2 bands of

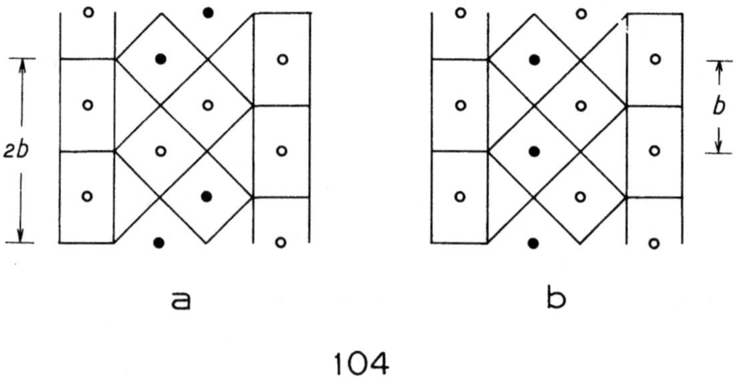

104

$[NbSe_3]_2$ ($Fe_{1+x}Nb_{1-x}Se_4$) would be less than $\frac{1}{4}$-filled. Apparently, this is not consistent with the observed CDW wave vector. It is necessary to examine the dependence of the CDW wave vector upon *x*.

The oxidation formalism $[2Nb]^{9+}$ ($Fe^{2+}Nb^{5+}$) for **104a** implies the presence of six *d*-block bands of largely Fe character below the z^2 bands of prismatic Nb atoms shown in Figure 33. The calculated band structure [24] differs from this picture somewhat in that five *d*-block bands of primarily Fe character lie below the z^2 bands but one flat *d*-block band of mainly octahedral Nb and Fe character overlaps with the z^2 bands below the Fermi level. Another *d*-block band of largely octahedral Nb and Fe character overlaps with the z^2 bands slightly above the Fermi level.

4.3. STRUCTURES DERIVED FROM ML_8 TETRAGONAL ANTIPRISMS

4.3.1. *Structural Pattern*

Chains of formula MX_4 (M = V, Nb, Ta; X = S, Se, Te) are found in binary compounds MX_4 and in ternary compounds $(MX_4)_nY$ (Y = halogen) [31–33]. Each metal atom M of an MX_4 chain, sandwiched by two planar X_4 units, is located at the center of a tetragonal antiprism ML_8. The X_4 units of MSe_4 and MS_4 chains are rectangular in shape, so that every two adjacent X_4 units form

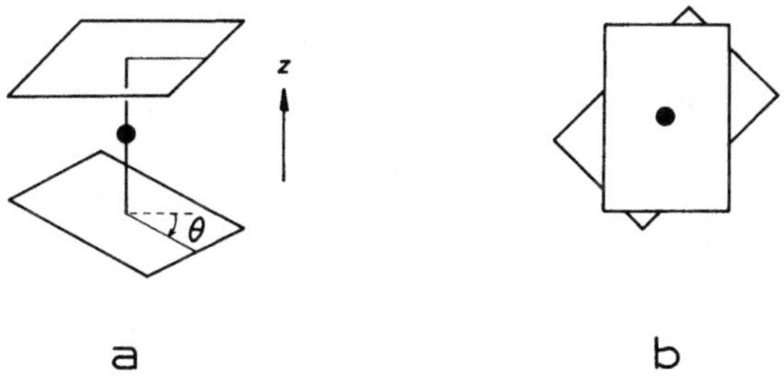

105

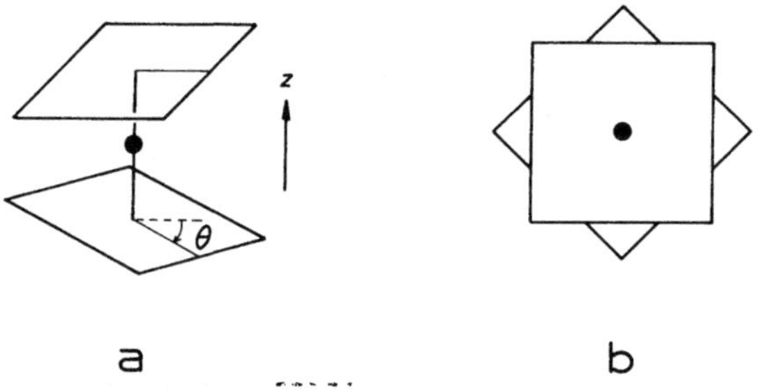

106

a rectangular antiprism **105a**. The X_4 units of MTe_4 chains are square planar, so that every two neighboring X_4 units form a square antiprism **106a**. The projected views of **105a** and **106a** along the chain axis are shown in **105b** and **106b**, respectively. The dihedral angle of Θ between two adjacent X_4 units is $\sim 45°$ both in rectangular and in square antiprisms. MX_4 chains, whether in a binary or in a ternary compound, are arranged parallel to the chain axis **c** and form a tetragonal structure.

4.3.2. $(MSe_4)_nI$

(a) *MSe_4 chain.* In each Se_4 rectangle of $(MSe_4)_nI$, the shorter Se—Se side has a length typical of an $(Se_2)^{2-}$ dimer. The longer Se—Se side is comparable to the shortest interchain Se···Se distance. Iodine atoms in $(MSe_4)_nI$, present between MSe_4 chains, are isolated from one another [33]. Therefore, the average number of d electrons on each metal is $(n - 1)/n$ in $(MSe_4)_nI$. With the coordinate z axis taken along the chain direction, significant interaction between metal ions occur through overlap between metal z^2 orbitals. If all the metal ion d electrons are accommodated into the resulting z^2 band, the extent of its band occupancy f is given by $f = (n - 1)/2n$ for $(MSe_4)_nI$.

Every two adjacent Se_4^{4-} units in an MSe_4 chain have a dihedral angle Θ of $\sim 45°$. Two electronic factors are responsible for this. The overlap of the occupied π, π^* orbitals of one Se_4^{4-} unit (resulting from the π, π^* orbitals of two Se_2^{2-} units shown in Figure 30) with those of the adjacent Se_4^{4-} unit is minimized at $\Theta = 45°$, which therefore minimizes the extent of two-orbital four-electron repulsive interactions [1(c)]. In addition, at $\Theta = 45°$, the π, π^* orbitals of two adjacent Se_4^{4-} units provide various symmetry-adapted orbitals that interact effectively with all the d orbitals except z^2. Thus at $\Theta = 45°$, the p-block orbitals of Se_4^{4-} units have an energy reduction while all the d-block levels except z^2 have an energy increase. This is shown in Figure 34, which plots the d-block levels of a monomeric unit $M^{4+}(Se_4^{4-})_2$ (**105a**) as a function of Θ. With a d^1 ion, the z^2 level is singly oc-

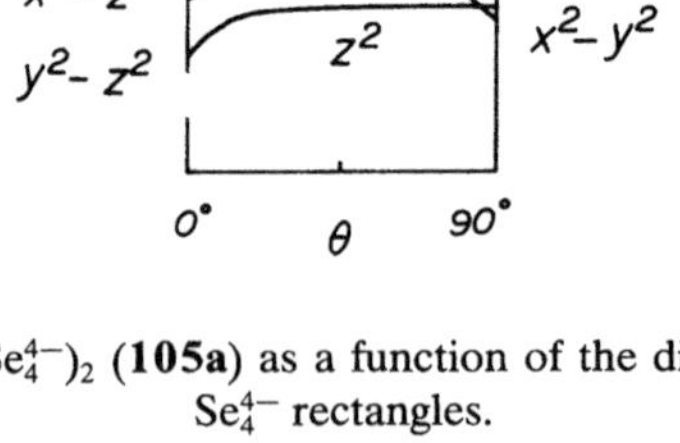

Fig. 34. d-block levels of $M^{4+}(Se_4^{4-})_2$ (**105a**) as a function of the dihedral angle Θ between the two Se_4^{4-} rectangles.

cupied. Thus, in each rectangular antiprism of an MSe_4 chain, the z^2 level is well separated from the p-block levels lying below and from the other d-block levels lying above. As a result, the electronic structures of an MSe_4 chain can be approximated by considering only the z^2 bands.

With $\Theta = 45°$ for any two adjacent Se_4^{4-} units, a number of relative orientations of Se_4^{4-} units are possible for MSe_4 chains. To specify such orientations, simple vector notations may be introduced [2(j)]. Let us represent each Se_4^{4-} rectangle by an arrow starting from its center to the mid-point of a shorter Se—Se side. For example, two Se_4^{4-} units with $\Theta = 90°$ (**107a**) can be represented by **107b**. Since $\Theta = 45°$ for any adjacent Se_4^{4-} units, the projected view of all possible orientations of the arrows along the chain is given by **108**. The arrow heads may be

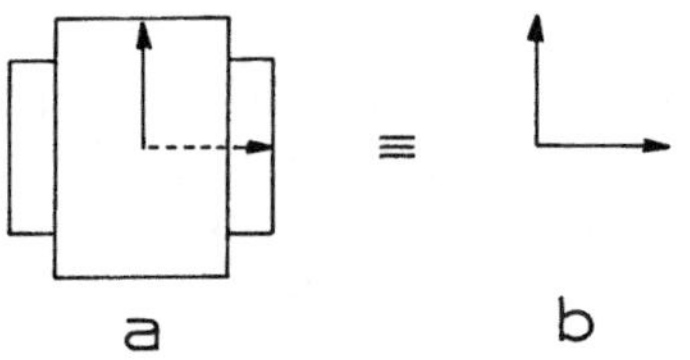

107

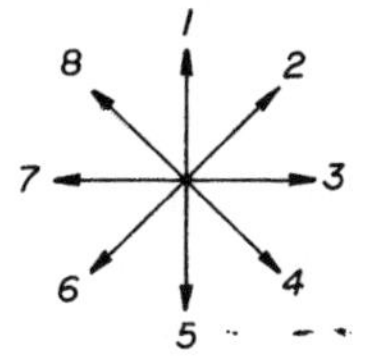

108

109 

110 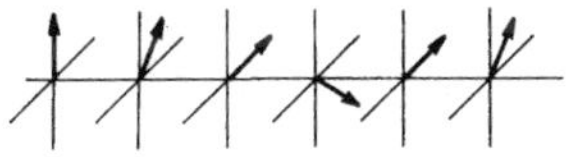

111 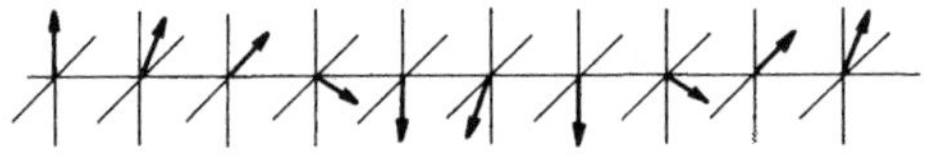

numbered successively starting from an arbitrary starting point, as shown in **108**. Then, using such numbers, the relative arrangement of n consecutive Se_4^{4-} units can be specified by a set of n integers, $(\nu_1 \nu_2 \ldots \nu_n)$, where ν_i refers to the ith Se_4^{4-} unit. Then the dihedral angle between two Se_4^{4-} units with the sequence numbers ν_i and ν_j is given by $\Theta = 45° \times (\nu_j - \nu_i)$. What matters in specifying the relative orientation of any two Se_4^{4-} units with ν_i and ν_j is the difference $(\nu_j - \nu_i)$, so that ν_1 may be taken to be 1. In **108** the sequence numbers 1, 2, 3, and 4 may refer to the same Se_4^{4-} units as given by numbers 5, 6, 7, and 8, respectively. Complication that may arise from this equivalence can be eliminated by requiring that any two consecutive numbers ν_i and ν_j be chosen to give minimum possible difference in magnitude. VS_4 and $(TaSe_4)_2I$ are found to have the (1234) arrangement of

Se_4^{4-} units (**109**) [31—33(d)]. In $(NbSe_4)_3I$ and $(NbSe_4)_{10/3}I$, the $NbSe_4$ chains have the (123432) and (1234565432) arrangements of Se_4^{4-} units shown in **110** and **111**, respectively [33(a), (e)].

An MSe_4 chain with the $(\nu_1\nu_2 \ldots \nu_n)$ arrangement of Se_4^{4-} units may simply be written as $(\nu_1\nu_2 \ldots \nu_n)$—$MSe_4$. With the M—M distance of d, the repeat distance of a $(\nu_1\nu_2 \ldots \nu_n)$—$MSe_4$ chain is given by nd. The z^2 band structure of such a chain has n subbands in the region $0 \leqslant k \leqslant d^*/2n$, as shown in Figure 13 for $n = 1-4$.

The (1234)—VS_4 chain of VS_4 has a $\frac{1}{2}$-filled z^2 band, and thus the V atoms undergo a pairing distortion (**41a** → **41b**) along the chain axis [31]. The (123432)—$NbSe_4$ chain of $(NbSe_4)_3I$ has a $\frac{1}{3}$-filled z^2 band, which is susceptible to a trimerization such as **43a** → **43b** or **44a** → **44b**. It is the latter that is experimentally observed [33(a)].

(b) *Interchain interaction.* The (1234)—$TaSe_4$ chain of $(TaSe_4)_2I$ has a $\frac{1}{4}$-filled z^2 band, whose band structure can be represented by Figure 13(d) [2(j)]. This chain does not undergo a tetramerization such as **45a** → **45b** or **46a** → **46b**, but exhibits an incommensurate CDW [34]. Two factors may be responsible for this observation [2(j)]: first, when a band occupancy f is small compared with $\frac{1}{2}$, the gain in electronic energy that results from a Peierls distortion is small in magnitude; second, interchain interaction leads to incommensurate CDW wave vectors, which is described below.

112

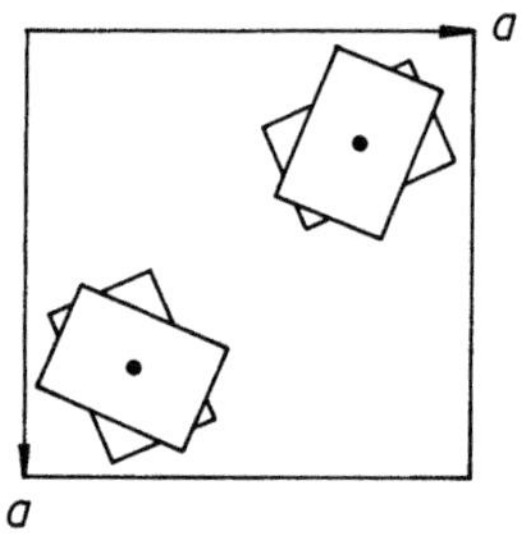

113

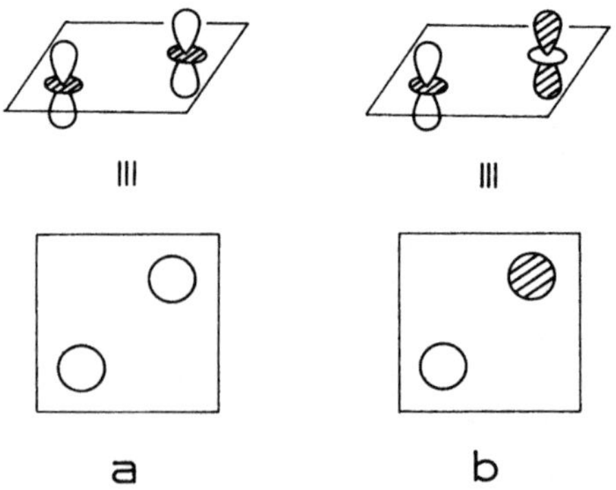

As shown in **112**, there are two chains per unit cell for 3D $(TaSe_4)_2I$ so that there are two z^2 bands for 3D $(TaSe_4)_2I$. To examine how these two bands are affected by the interchain interaction, we note that 3D $(TaSe_4)_2I$ has planes of metal ions sandwiched by those of Se_4^{4-} units, perpendicular to the chain axis **c**. A unit cell **112** contributes two metal ions at any metal ion plane. The z^2 orbitals of such metal ions lead to bonding and antibonding levels, shown in **113a** and **113b**, respectively, where z^2 orbitals perpendicular to the plane are represented by *s* orbitals for simplicity. The first BZ of 3D $(TaSe_4)_2I$ is shown in **114**, where wave vectors describing the chain direction are found along the $\Gamma \rightarrow Z$, $X \rightarrow R$, and $M \rightarrow A$ lines.

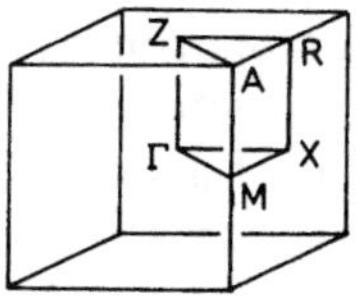

114

$\Gamma = (0,0,0)$

$Z = (0,0,c^*/2)$

$R = (a^*/2,0,c^*/2)$

$A = (a^*/2,a^*/2,c^*/2)$

115

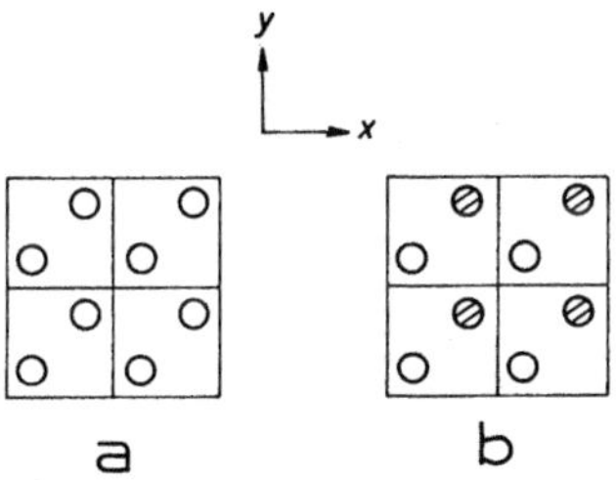

116

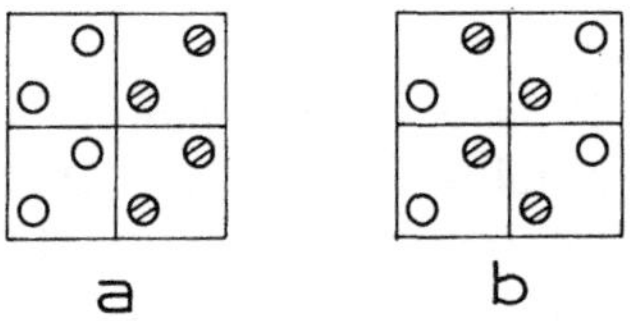

117

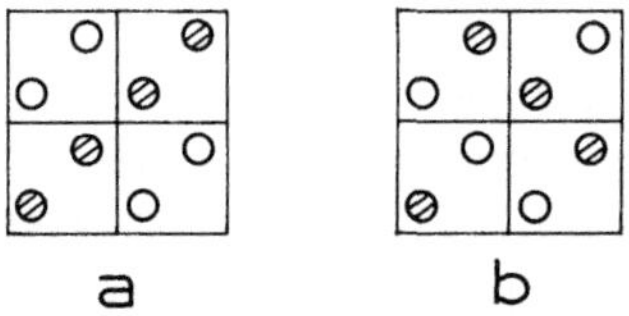

Along $\Gamma \rightarrow Z$, $k_x = k_y = 0$ so that the unit cell orbitals **113a** and **113b** are each repeated in the x and y directions without sign change to give the plane orbitals **115a** and **115b**, respectively. These two plane orbitals are different in energy, so that the two z^2 bands resulting from them are split in energy at all values of wave vector along $\Gamma \rightarrow Z$. This z^2 band splitting is substantial only along $\Gamma \rightarrow Z$. Along $X \rightarrow R$, $k_x = a^*/2$ and $k_y = 0$. Thus the unit cell orbitals **113a** and **113b** are each repeated without sign change in the y direction but with sign alternation in the x direction. This leads to the plane orbitals **116a** and **116b**, which are degenerate in energy along $X \rightarrow R$. Therefore the z^2 bands are degenerate along $X \rightarrow R$. Similarly, along $M \rightarrow A$ for which $k_x = a^*/2$ and $k_y = a^*/2$, the unit cell orbitals **113a** and **113b** lead to the plane orbitals **117a** and **117b**, respectively, which are degenerate in energy.

Consequently, the z^2 band structure of 3D $(TaSe_4)_2I$ can be estimated by simply doubling the z^2 band structure of a single $TaSe_4$ chain with an appropriate energy split, as illustrated in Figure 35. This approach [2(j)] led to the Fermi wave vector

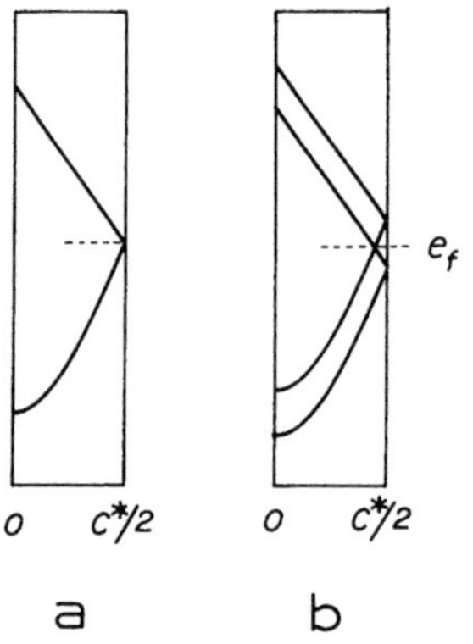

Fig. 35. Effect of the interchain interaction on the z^2 band structure of $(MSe_4)_2I$ (M = Nb, Ta). The bottom two subbands of a (1234)—MSe_4 chain (e.g. see Figure 13(d)) are shown in (a), and the corresponding band structures expected for 3D MSe_4 are shown in (b).

along the chain direction $k_f^c \simeq 0.88(c^*/2) = 0.22(d^*/2)$, where $c = 4d$ since 3D $(TaSe_4)_2I$ consists of (1234)—$TaSe_4$ chains. This gives rise to the CDW wave vector $2k_f^c \simeq 0.22d^*$, which is in good agreement with the experimental value of $2k_f^c = 0.915c^* = 0.229d^*$ obtained by X-ray diffraction [34(c)] and that of $2k_f^c = 0.94c^* = 0.235d^*$ obtained by electron diffraction [34(d)].

$(MSe_4)_nI$ systems are remarkable in that they lead to z^2 bands of varying degrees of band occupancies. The distortion pattern of the metal ions in $(NbSe_4)_{10/3}I$ ($f = 0.350$) is found to be somewhat different from that in $(NbSe_4)_3I$ ($f = 0.333$) [33(b), (e)], although the band occupancies of the two compounds are similar. In addition, $(NbSe_4)_{10/3}I$ exhibits an incommensurate CDW despite the metal ion distortion along the chain [2(j), 34(d), 35]. Thus $(MSe_4)_nI$ systems, with varying degrees of z^2 band filling, provide an excellent series of compounds by which to investigate how structural and electronic properties are affected by small differences in band occupancies.

4.3.2. MTe_4 ($M = Nb, Ta$)

(a) *Ideal structure.* As shown in **106**, each metal M of an MTe_4 chain is located at the center of a square antiprism. The arrangement of MTe_4 chains, parallel to the chain axis **c**, is shown in **118a**. In this ideal structure there is one MTe_4 chain per unit cell. As shown in **118b**, however, there are four MTe_4 chains per unit cell in real MTe_4 since it has a superstructure with a doubled *a* axis and a tripled *c* axis [32].

In **118** the side of a Te_4 square (~3.3 Å) is considerably longer than the shortest interchain Te—Te distance (~2.90 Å), which is comparable to the Te—Te distance of Te_2 dimeric units in the isosceles-like chains of $ZrTe_5$ and $HfTe_5$ [36]. Consequently, it is more reasonable to represent the crystal structure of MTe_4 in **118a** as in **119** [2(k)]. With each Te_2 dimer counted as a $(Te_2)^{2-}$ unit, each metal atom of MTe_4 exists as an M^{4+} (d^1) ion. Thus each metal ion M^{4+} of MTe_4 is surrounded by eight $(Te_2)^{2-}$ units, the π, π^* orbitals of which provide various symmetry-adapted orbitals that interact effectively with all the *d* orbitals of M^{4+} except for z^2. Therefore, the z^2 level of M^{4+} is well separated from other *d* levels lying above, and the z^2 band of ideal MTe_4 is $\frac{1}{2}$-filled [2(k)].

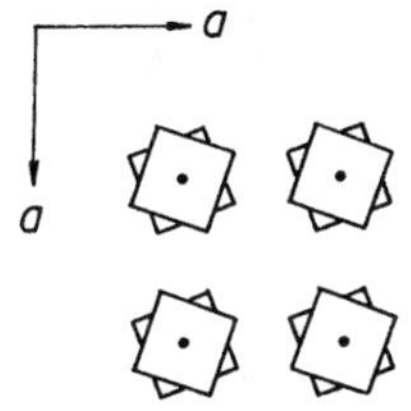

118a

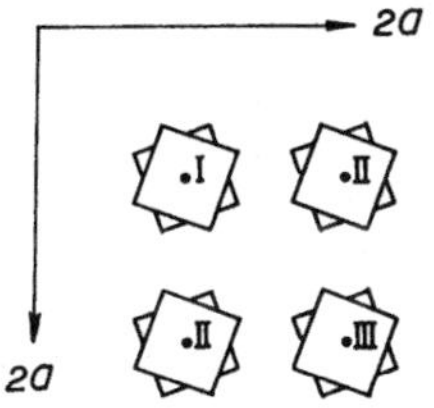

118b

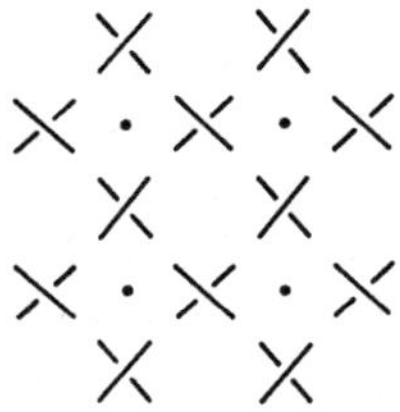

119

Any two adjacent Te_4 units of a MTe_4 chain (**118a**) are staggered so that a unit cell of ideal MTe_4 has the formula $(MTe_4)_2$. Consequently, the z^2 band of ideal MTe_4 has the folded shape in Figure 13(b) along the chain directions $\Gamma \rightarrow Z$, $X \rightarrow K$, and $X \rightarrow A$ of its first BZ (**114**). The effect of interchain interaction

120

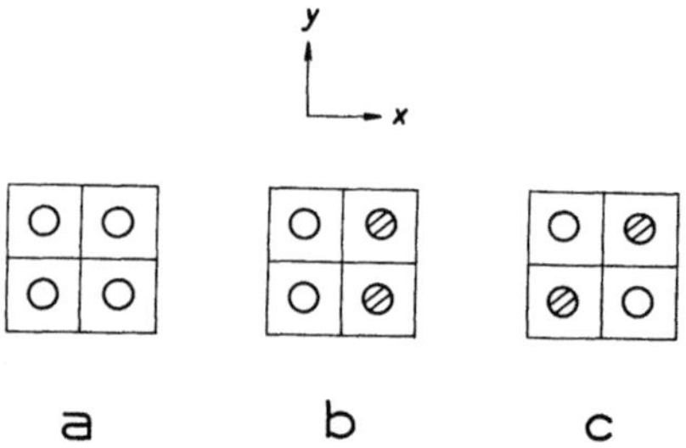

upon the z^2 band structure may be estimated as follows: in each metal ion plane of 3D MTe_4, there is only one z^2 orbital per unit cell. Therefore the z^2 plane orbitals of metal ions are given by **120a**, **120b**, and **120c**, respectively, for the wave vectors along the chain directions $\Gamma \to Z$, $X \to R$, and $M \to A$ of **114**. The stability of these plane orbitals increases in the order **120a** > **120b** > **120c**. Therefore, the z^2 bands arising from these plane orbitals are summarized as shown in Figure 36. Namely, the z^2 bands of the $\Gamma \to Z$ region are somewhat lower in

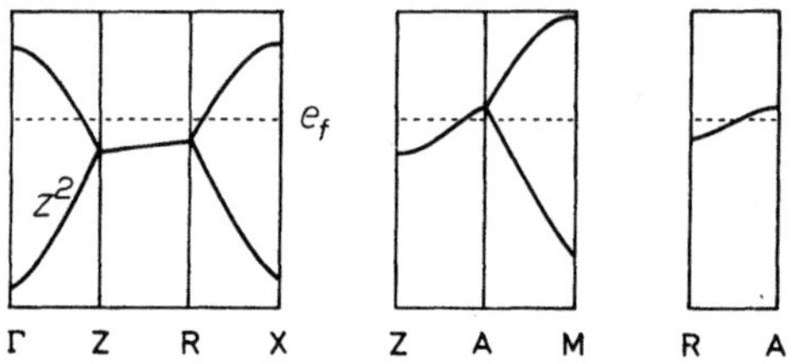

Fig. 36. z^2 band structure of ideal 3D MTe_4 (M = Nb, Ta) (**118a**) along some cross sections of the Brillouin zone.

energy than those of the $X \to R$ region, which in turn are somewhat lower in energy than those of the $M \to A$ region. With two d electrons per $(MTe_4)_2$, the z^2 bands are now $\frac{1}{2}$-filled. Thus the Fermi level cuts the z^2 bands as indicated by the dashed lines in Figure 36.

(b) *Real structure.* For a 3D system the Fermi wave vectors $\mathbf{k}_f$ of a partially filled band form a surface, which separates those wave vectors leading to the energy above e_f from those leading to the energy below e_f. Such a wave vector surface of the first BZ is called a Fermi surface. For a 2D system, the Fermi wave vectors $\mathbf{k}_f$ form a line of the first BZ that separates occupied wave vectors from unoccupied wave vectors. The upper subband of the z^2 bands of Figure 36 leads to the Fermi surface shown in Figure 37, which consists of two nearly flat pieces. The centers of these two are estimated to be apart by $\sim 0.38d^*$ ($d^* = 2\pi/d$, and d is the M—M distance along the chain) [2(k)].

Experimentally, three independent CDWs are observed for both $NbTe_4$ and $TaTe_4$ [32(c), 37]. The components, $2k_f^c$, of these CDWs along the chain direction **c** are exactly $d^*/3$ for $TaTe_4$ but $0.311d^*$, $0.344d^*$, and $d^*/3$ for $NbTe_4$. As

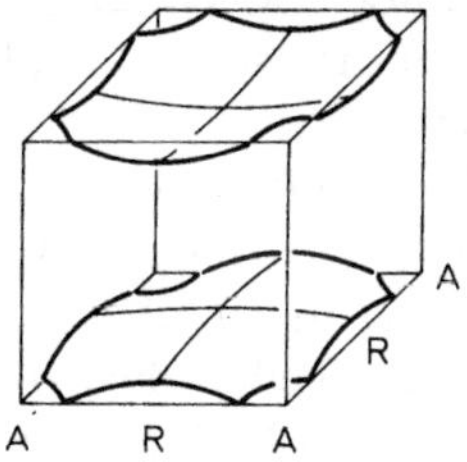

Fig. 37. Fermi surface resulting from the upper subband of the z^2 band in ideal 3D MTe_4 (**118a**).

indicated in **118b**, the presence of three independent CDWs led to the suggestion that real 3D MTe_4 consists of three types (I, II, III) of slightly different MTe_4 chains, each of which is considered to be about $\frac{1}{3}$-filled since $f = 2k_f^c/d^* \simeq \frac{1}{3}$ [32(c)]. We note that the Fermi surface of ideal MTe_4 (**118a**) shown in Figure 37 is derived from the z^2 band which is $\frac{1}{2}$-filled. The Fermi surface of real MTe_4 (**118b**) would have four sets of two nearly flat pieces such as shown in Figure 37 and hence would give rise to three independent CDWs [2(k)]. However, this raises a question of whether or not such calculations would lead to the $2k_f^c$ values very close $d^*/3$ as experimentally found for both $NbTe_4$ and $TaTe_4$ [38].

Alternatively, one might simply assume the presence of $\frac{1}{3}$-filled z^2 bands in the MTe_4 chains to be the real cause for the observed $2k_f^c$ values [32(c)]. This brings up a question of how such a band occupancy is possible for MTe_4. It is recalled that the existence of $\frac{1}{2}$-filled bands in **118** arises from the oxidation formalism $M^{4+}(Te_2^{2-})_2$. The z^2 band occupance in MTe_4 may be decreased from $\frac{1}{2}$ to $\frac{1}{3}$ by introducing a particular geometry distortion that lengthens the Te—Te distance of some Te_2 dimer units (in the representation of **119**) so as to change the oxidation formalism from $(Te_2)^{2-}$ to $(Te^{2-})_2$. The superstructure of MTe_4, with a doubled a axis and a tripled c axis, has the formula $M_{24}(Te_2)_{48}$ as a repeat unit. Thus if 4 out of these 48 Te_2 units undergo the aforementioned Te—Te bond lengthening, there remain 16 d electrons to share among 24 M atoms. This will lead to $\frac{1}{3}$-filled z^2 bands for the MTe_4 chains. The Te—Te bond lengthening in the representation of **119** may lead to square Te_4 units of unequal size [39] or even to nonsquare Te_4 units in the representation of **118**. It is an experimental challenge to pin down the details of the superstructures in MTe_4.

5. Concluding Remarks

In the present article, we have examined in some detail how to employ the concept of orbital interaction in understanding the electronic and structural properties of low-dimensional solids. As far as qualitative aspects are concerned, the band electronic structure of any given system can be readily estimated by first constructing Bloch orbitals and then analyzing their orbital interactions. In most practical cases this analysis is necessary only at a few high-symmetry k points of the first BZ (e.g. the BZ center and edge), as illustrated for the various low-dimensional solids derived from transition-metal ML_6 octahedra, ML_6 trigonal prisms, and ML_8 tetragonal antiprisms.

Examination of how the band electronic structure of a given structure is modified upon geometrical distortion leads to the concept of Peierls distortion, which is simply a solid-state counterpart of a Jahn—Teller distortion in discrete molecules. By analogy with Jahn—Teller distortions, distortions associated with low-dimensional solids may be classified as first- or second-order Peierls distortions and also as low- or high-spin Peierls distortions based upon their band electronic origin.

A Peierls distortion introduces a band gap at the Fermi level and therefore lowers the electronic energy for 1D systems with low-spin partially filled bands. However, not all such systems exhibit permanent lattice distortions. The magnitude of the electronic-energy lowering decreases as the band occupancy f decreases from $\frac{1}{2}$. Low-dimensional systems with several chains per repeat unit often lead to the Fermi wave vectors $\mathbf{k}_f$ which are not commensurate due to the interchain interactions. When such systems have a low band occupancy (typically, $f \simeq 0.25$), the driving force for permanent lattice distortion diminishes and an interesting resistivity anomaly associated with sliding CDWs is observed [9(b), 40].

Acknowledgements

This work is in part supported by the Camille and Henry Dreyfus Foundation through a Teacher—Scholar Award to the author. I would like to thank Professor T. A. Albright, Professor F. W. Boswell, Professor J. K. Brandon, and Dr F. J. Di Salvo for valuable discussions. I am deeply indebted to Professor J. Rouxel (Laboratoire de Chimie des Solides, Université de Nantes) for providing me with an opportunity to visit his laboratory for an extended period (May—June, 1984). The present article was revised during this visit after a number of stimulating discussions with Professor J. Rouxel, Dr A. Meerschaut, and Dr P. Gressier, for which I am very grateful.

References

1. (a) R. Hoffmann, *Acc. Chem. Res.* **4**, 1 (1971); (b) R. Hoffmann, *Angew. Chem. Int. Ed. Engl.* **21**, 711 (1982); (c) T. A. Albright, J. K. Burdett, and M.-H. Whangbo, *Orbital Interactions in Chemistry*, Wiley, New York (1985).
2. (a) M.-H. Whangbo and R. Hoffmann, *J. Am. Chem. Soc.* **100**, 6093 (1978); (b) M.-H. Whangbo, R. Hoffmann, and R. B. Woodward, *Proc. Roy. Soc.* **A366**, 23 (1979); (c) M.-H. Whangbo, *Extended Linear Chain Compounds* (ed. J. S. Miller), Plenum, New York, Vol. 2, p. 127 (1982); (d) M.-H. Whangbo and M. J. Foshee, *Inorg. Chem.* **20**, 113 (1981); (e) M.-H. Whangbo, *Inorg, Chem.* **21**, 1721 (1982); (f) M.-H. Whangbo, M. J. Foshee, and R. Hoffmann, *Inorg. Chem.* **19**, 1723 (1980); (g) M.-H. Whangbo and K. R. Stewart, *Israel J. Chem.* **23**, 133 (1983); (h) M.-H. Whangbo, J. M. Williams, M. A. Beno, and J. R. Dorfman, *J. Am. Chem. Soc.*

105, 645 (1983); (i) M.-H. Whangbo, F. J. Di Salvo, and R. M. Fleming, *Phys. Rev.* **B26**, 687 (1982); (j) P. Gressier, M.-H. Whangbo, A. Meerschaut, and J. Rouxel, *Inorg. Chem.* **23**, 1221 (1984); (k) M.-H. Whangbo and P. Gressier, *Inorg. Chem.* **23**, 1228; (l) M.-H. Whangbo, *Acc. Chem. Res.* **16**, 95 (1983).

3. (a) J. K. Burdett, *Structure and Bonding in Crystals* (eds M. O'Keefe and A. Navrotsky), Academic Press, New York, Vol. 1, p. 255 (1981); (b) J. K. Burdett, *Prog. Solid State Chem.* **15**, 173 (1984); (c) J. K. Burdett, *J. Am. Chem. Soc.* **102**, 5458 (1980); (d) J. K. Burdett and S. Lee, *J. Solid State Chem.* **44**, 415 (1982); (e) J. K. Burdett and S. L. Price, *Phys. Rev.* **B25**, 5778 (1982).
4. (a) T. Hughbands and R. Hoffmann, *J. Am. Chem. Soc.* **105**, 1150 (1983); (b) T. Hughbands, R. Hoffmann, and M.-H. Whangbo, K. R. Stewart, O. Eistenstein, and E. Canadell, *J. Am. Chem. Soc.* **104**, 3876 (1982); (c) T. Hughbands and R. Hoffmann, *Inorg. Chem.* **21**, 3578 (1982); (d) T. Hughbanks and R. Hoffmann *J. Am. Chem. Soc.* **105**, 3528 (1983); (e) S. D. Wijeyesekra and R. Hoffmann, *Inorg. Chem.* **22**, 2387 (1983); (f) R. Hoffmann, S. Shaik, J. C. Scott, M.-H. Whangbo, and M. J. Foshee, *J. Solid State Chem.* **34**, 263 (1980).
5. (a) E. Canadell and O. Eisenstein, *Inorg. Chem.* **22**, 3856 (1983); (b) E. Canadell and S. Alvarez, *Inorg. Chem.* **23**, 575 (1984); (c) E. Canadell and O. Eisenstein, *Inorg. Chem.* **22**, 2398 (1983).
6. (a) M. Kertész and F. Vonderviszt, *J. Am. Chem. Soc.* **104**, 5889 (1982); (b) M. Kertész and R. Hoffmann *Solid State Commun.* **47**, 97 (1983); (c) R. Biswas and M. Kertész, *Phys. Rev.* **B29**, 1791 (1984).
7. N. W. Ashcroft and N. D. Mermin, *Solid State Physics*, Holt, Rinehart and Winston, New York (1976).
8. M. Lax, *Symmetry Principles in Solid State and Molecular Physics*, Wiley, New York (1974).
9. (a) R. E. Peierls, *Quantum Theory of Solids*, Oxford Univ. Press, London, p. 108 (1955); (b) A. Berlinsky, *J. Contemp. Phys.* **17**, 331 (1976).
10. (a) L. Salem, *The Molecular Orbital Theory of Conjugated Systems*, Benjamin, New York (1966); (b) R. G. Pearson, *Symmetry Rules for Chemical Reactions*, Wiley, New York (1976); (c) H. A. Jahn and E. Teller, *Proc. Roy. Soc.* **A161**, 220 (1937); (d) U. Öpik and M. H. L. Pryce, *Proc. Roy. Soc.* **A238**, 425 (1957); (e) R. F. W. Bader, *Canad. J. Chem.* **40**, 1164 (1962).
11. (a) B. H. Brandow, *Adv. Phys.* **26**, 651 (1977); (b) N. F. Mott, *Metal-Insulator Transitions*, Barnes and Noble, New York (1977); (c) C. Haas, *Current Topics in Materials Science* (ed. E. Kaldis), North-Holland, Amsterdam, chap. 1 (1979); (d) S. Doniach, *Adv. Phys.* **18**, 819 (1969); (e) C. Herring, *Magnetism*, (eds G. T. Rado and H. Suhl), Academic Press, New York, Vol. 4 (1966); (f) P. W. Anderson *Solid State Phys.* **14**, 99 (1963); (g) J. Kanamori, *Prog. Theor. Phys.* **30**, 275 (1963); (h) J. Hubbard, *Proc. Roy. Soc.* **A276**, 238 (1963).
12. (a) M.-H. Whangbo, *J. Chem. Phys.* **70**, 4963 (1979); (b) M.-H. Whangbo, *J. Chem. Phys.* **73**, 3854 (1980); (c) M.-H. Whangbo, *J. Chem. Phys.* **75**, 4983 (1981); (d) M.-H. Whangbo, *Inorg. Chem.* **19**, 1728 (1980).
13. E. Krüger, *Phys. Stat. Sol.* **B85**, 261 (1978).
14. Effective spin-Hamiltonians have been employed to examine nonmagnetic problems as well. For example, see: (a) A. O. Ovchinikov, *Theoret, Chim. Acta* **47**, 297 (1978); (b) D. J. Klein, *Int. J. Quantum Chem.* **12**, 255 (1966); (c) D. J. Klein and M. A. Garcia Bach, *Phys. Rev.* **B19**, 877 (1979); (d) M. Saïd, D. Maynau, J.-P. Malrieu, and M. A. Garcia Bach, *J. Am. Chem. Soc.* **106**, 571 (1984); (e) M. Saïd, D. Maynau, and J.-P. Malrieu, *J. Am. Chem. Soc.* **106**, 580 (1984); (f) J.-P. Malrieu, D. Maynau, and J. P. Daudley, *Phys. Rev.* **B30**, 1817 (1984).
15. (a) J. B. Torrance, *Phys. Rev.* **B17**, 3099 (1978); (b) M. Weger and H. Gutfreund, *Solid State Commun.* **32**, 1259 (1979); (c) J. Hubbard, *Phys. Rev.* **B17**, 494 (1978); (d) V. J. Emery, *Phys. Rev. Lett.* **37**, 107 (1976); (e) S. Huizinga, J. Kommandeur, G. A. Sawatzky, B. T. Thole, K. Kopinga, W. J. M. deJonge, and J. Roos, *Phys. Rev.* **B19**, 4723 (1979); (f) H. Kobayashi and A. Kobayashi, *Extended Linear Chain Compounds* (ed. J. S. Miller), Plenum, New York, Vol. 2, p. 259 (1982); (g) S. Kagoshima, *Extended Linear Chain Compounds* (ed. J. S. Miller), Plenum, New York, Vol. 2, p. 303 (1982).
16. McCarley, R. E. *Inorganic Chemistry: Toward 21st Century* (ed. M. Chisholm), American Chemical Society, New York, p. 221 (1983).

17. (a) D. R. Taylor, J. C. Calabrese, and E. M. Larsen, *Inorg. Chem.* **16**, 721 (1977); (b) D. G. Bright and D. L. Kerpert, *Phys. Rev. Lett.* **27**, 504 (1971); (c) D. L. Kerpert and R. E. Marshall, *J. Less Common Met.* **34**, 153 (1974); (d) H. G. Schnering and H. Wohrle, *Angew. Chem.* **75**, 684 (1963); (e) H. Schafer and H. G. von Schnering, *Angew. Chem.* **76**, 833 (1964); (f) J. Rijnsdorp and F. Jellinek, *J. Less Common Met.* **61**, 79 (1978).
18. (a) O. Berkaroglu, H. Beer, H. Endres, H. J. Keller, and H. Nam Gung, *Inorg. Chim. Acta* **21**, 183 (1977); (b) P. Day, *Chemistry and Physics of One-Dimensional Metals*, Plenum, New York, p. 197 (1977); (c) K. L. Brown and D. Hall, *Acta Crystallogr.* **B32**, 279 (1976); (d) R. J. H. Clark, M. L. Franks, and W. R. Trumble, *Chem. Phys. Lett.* **41**, 287 (1976); (e) M. B. Robin and P. Day, *Adv. Inorg. Radiochem.* **10**, 247 (1967).
19. (a) W. Krönert and V. Plieth, *Z. Anorg. allg. Chem.* **336**, 207 (1965); (b) J. Rijnsdorp and F. Jellink, *J. Solid State Chem.* **25**, 325 (1978); (c) F. W. Boswell and A. Prodan, *Physica* **B99**, 361 (1980); (d) T. Cornelissens, G. Van Tendeloo, J. Van Landuyt, and S. Amelinckx, *Phys. Stat. Sol.* **48**, K5 (1978).
20. E. Bjerklund, J. H. Fermor, and A. Kjekshus, *Acta Chem. Scand.* **20**, 1836 (1966).
21. (a) A. Meerschaut and J. Rouxel, *J. Less Common Met.* **39**, 197 (1975); (b) A. Meerschaut, J. Rouxel, P. Haen, P. Monceau, and M. Nunez-Regueiro, *J. Phys.* **40**, L157 (1979); (c) S. Kikkawa, M. Koizumi, S. Yamanaka, Y. Onuki, R. Inada, and S. Tanuma, *J. Solid State Chem.* **40**, 28 (1981); (d) A. Meerschaut, L. Guemas, and J. Rouxel, *C.R. Acad. sci. Paris, Serie C,* **290**, 215 (1980).
22. (a) J. Rouxel, A. Meerschaut, L. Guemas, and P. Gressier, *Ann. Chim. Fr.* **7**, 445 (1982); (b) J. A. Wilson, *Phys. Rev.* **B19**, 6456 (1979); (c) D. W. Bullett, *J. Phys. C,* **12**, 277 (1979).
23. D. W. Bullett, *J. Solid State Chem.* **33**, 13 (1980).
24. M.-H. Whangbo and P. Gressier, *Inorg. Chem.* **23**, 1305 (1984).
25. (a) J. L. Hodeau, M. Marezio, C. Roucau, R. Ayroles, A. Meerschaut, J. Rouxel, and P. Monceau, *J. Phys. C,* **11**, 4117 (1978); (b) K. Tsutsumi, T. Tugagaki, M. Yamamoto, Y. Shiozaki, M. Ido, T. Sambongi, K. Yamaya, and Y. Abe, *Phys. Rev. Lett.* **39**, 1675 (1977); (c) R. M. Fleming, D. E. Moncton, and D. B. McWhan, *Phys. Rev.* **B18**, 5560 (1978).
26. C. Roucau, R. Ayroles, P. Monceau, L. Guemas, A. Meerschaut, and J. Rouxel, *Phys. Stat. sol.* **A62**, 483.
27. (a) J. Chaussy, P. Haen, J. C. Lasjaunias, P. Monceau, G. Waysand, A. Waintal, A. Meerschaut, P. Molinié, and J. Rouxel, *Solid State Commun.* **20**, 759 (1976); (b) P. Monceau, N. P. Ong, A. M. Portis, A. Meerschaut, and J. Rouxel, *Phys. Rev. Lett.* **37**, 602 (1976); (c) N. P. Ong and P. Monceau, *Phys. Rev.* **B16**, 3443 (1977); (d) G. Grüner, L. C. Tippie, J. Sanny, W. C. Clark, and N. P. Ong, *Phys. Rev. Lett.* **45**, 935 (1980); (e) R. M. Fleming and C. C. Grimes, *Phys. Rev. Lett.* **42**, 1923 (1979); (f) J. C. Gill, *J. Phys. F,* **10**, 281 (1980); (g) P. Monceau, J. Richard, and M. Renard, *Phys. Rev.* **B25**, 931 (1982); (h) J. Richard, P. Monceau, and M. Renard, *Phys. Rev.* **B25**, 948 (1982); (i) P. Monceau, J. Richard, and M. Renard, *Phys. Rev. Lett.* **45**, 43 (1980); (j) M. Weger, G. Grüner, and W. C. Clark, *Solid State Commun.* **35**, 243 (1980).
28. (a) A. Meerschaut, P. Gressier, L. Guemas, and J. Rouxel, *Mat. Res. Bull.* **16**, 1035 (1981); (b) R. J. Cava, V. L. Himes, A. D. Mighell, and R. S. Roth, *Phys. Rev.* **B24**, 3634 (1981).
29. S. J. Hillenius, R. V. Coleman, R. M. Fleming, and R. J. Cava, *Phys. Rev.* **B23**, 1567 (1981).
30. (a) M.-H. Whangbo, R. J. Cava, F. J. Di Salvo, and R. M. Fleming, *Solid State Commun.* **43**, 277 (1982); (b) D. W. Bullett, *J. Phys. C,* **15**, 3069 (1982).
31. (a) R. Allmann, L. Baumann, A. Kutoglu, H. Rösch, and E. Hellner, *Naturwiss.* **51**, 263 (1964); (b) A. Kutoglu and R. Allmann, *Neues Jahrb. Miner. Monat.* **8**, 339 (1972).
32. (a) K. Selte and A. Kjekshus, *Acta Chem. Scand.* **18**, 690 (1964); (b) E. Bjerkelund and A. Kjekshus, *J. Less Common Met.* **7**, 231 (1964); (c) F. W. Boswell, A. Prodan, and J. K. Brandon, *J. Phys. C,* **16**, 1067 (1983).
33. (a) P. Gressier, A. Meerschaut, L. Guemas, J. Rouxel, and P. Monceau, *J. Solid State Chem.* **51**, 141 (1984); (b) A. Meerschaut, P. Palvadeau, and J. Rouxel, *J. Solid State Chem.* **20**, 21 (1977); (c) L. Guemas, P. Gressier, A. Meerschaut, D. Louër, and D. Grandjean, *Rev. Chim. Miner.* **18**, 91 (1981); (d) P. Gressier, L. Guemas, and A. Meerschaut, *Acta Crystallogr.* **B38**, 2877 (1982); (e) A. Meerschaut, P. Gressier, L. Guemas, and J. Rouxel, *J. Solid State Chem.* **51**, 307 (1984).

34. (a) Z. Z. Wang, M. C. Saint-Lager, P. Monceau, M. Renard, P. Gressier, A. Meerschaut, L. Guemas, and J. Rouxel, *Solid State Commun.* **46**, 325 (1983); (b) M. Maki, M. Kaiser, A. Zettl, and G. Grüner, *Solid State Commun.* **46**, 497 (1983); (c) H. Fujishita, M. Sato, and S. Hoshino, *Solid State Commun.* **49**, 313 (1984); (d) C. Roucau, R. Ayroles, P. Gressier, and A. Meerschaut, *J. Phys. C*, **17**, 2993 (1984).
35. Z. Z. Wang, P. Monceau, M. Renard, P. Gressier, L. Guemas, and A. Meerschaut, *Solid State Commun.* **47**, 439 (1983).
36. (a) S. Furuseth, L. Brattås, and A. Kjekshus, *Acta Chem. Scand.* **27**, 2367 (1973); (b) F. J. Di Salvo, R. M. Fleming, and J. V. Waszczak, *Phys. Rev.* **B24**, 2935 (1981).
37 F. W. Boswell and A. Prodan, *Mat. Res. Bull.* **19**, 93 (1984).
38. F. W. Boswell, (private communication).
39. H. Böhm and H. G. von Schnering, *Z. Kristallogr.* **162**, 26 (1983).
40. (a) F. J. Di Salvo, *Electron—Phonon Interactions* (ed. T. Riste), Plenum, New York, p. 107 (1977); (b) R. M. White, and T. H. Geballe, *Long Range Order in Solids*, Academic Press, New York (1979).

STRUCTURAL STUDIES OF PHASE TRANSITIONS IN ONE-DIMENSIONAL CONDUCTORS

R. MORET and J. P. POUGET

Laboratoire de Physique des Solides, Associé au CNRS, Université de Paris-Sud, 91405 Orsay, France

This paper attempts to survey the important role played by structural studies in the field of one-dimensional conductors over a little more than ten years. Our aim is by no means to provide a comprehensive review of the vast amount of research which has been devoted to these materials. Neither is it to give an account of the present state of knowledge resulting from the most recent structural results. Instead, we shall try to illustrate *the type of information* that can be obtained from X-ray, electron, and neutron studies in this class of materials. This requires, first, to recall some basic ideas and to give some information about the structural methods which are used. Then we shall present a collection of representative examples of the structural results obtained on phase transitions, in both organic and inorganic materials. It turns out that, for various reasons, a wider spectrum of quantitative data is available in the class of organic compounds so that our selection is necessarily slightly unbalanced. However, we believe that the properties which are illustrated are, in most cases, sufficiently general to apply to both classes of materials.

The subject matter of this article is organized as follows. Section 1 recalls basic concepts about phase transitions [1] with some results of the Landau theory. In Section 2 we introduce the Peierls transition [2] and the associated charge density wave phenomenon. The main part (Section 3) is devoted to a description of selected examples of structural studies, mainly on phenomena associated with the Peierls transition but also on the orientational order—disorder transition encountered in the TMTS(T)F_2X series of organic superconductors.

1. General Aspects of Phase Transitions in Condensed Matter

1.1. Thermodynamic Classification

Through a small change of an external parameter (usually temperature or pressure) a physical system may undergo a transition between two phases, I and II, which display different properties and often different structures. Each phase is characterized by its Gibbs free energy $G = U + PV - TS$ and the most stable phase is the one that presents the lowest free energy. The Ehrenfest classification allows us to distinguish the different situations which arise at a transition (in the following we shall consider only temperature-induced transitions and we neglect small volume variations):

J. Rouxel (ed.), Crystal Chemistry and Properties of Materials with Quasi-One-Dimensional Structures, 87—134.

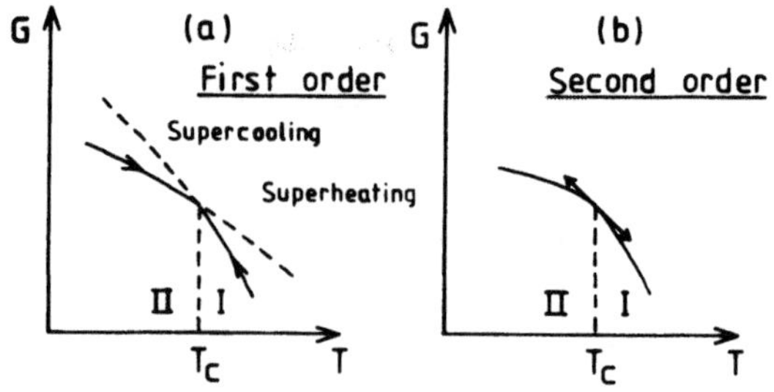

Fig. 1. Behavior of the free energy with temperature for (a) a first-order phase transition with a change of slope at T_c, (b) a second-order transition in which the curvature changes at T_c but not the slope.

(1) If the first-order derivatives of G are discontinuous at the transition (example: $\partial G/\partial T = -S$) the transition is called first-order (Figure 1(a)). It has a latent heat $T_c(S_{\mathrm{I}} - S_{\mathrm{II}})$, where T_c is the transition temperature and S_{I} and S_{II} are the entropies of the two phases I and II at T_c. Hysteresis effects can also be observed.

(2) If the first-order derivatives of G are continuous but its second (or nth) derivatives are discontinuous, the phase transition is said to be second order (or nth order) (Figure 1(b)). The entropy is then continuous but the specific heat is discontinuous at T_c. The two phases are in equilibrium only at the transition.

The type of singularity of the free energy and, therefore, the order of a phase transition can be determined from the study of physical properties associated with the transition.

1.2. Symmetry and Order Parameter

The above thermodynamic classification does not consider symmetry, which is an important notion especially in solid state where a phase transition is most often accompanied by a break of symmetry. The two phases also correspond to different 'types or degrees' of order which can be characterized quantitatively introducing an order paramater η. This quantity can be a scalar, a vector, or a tensor. Its choice is sometimes clear (for instance, in the case where there is a group—subgroup relation, see below): the magnetization vector in paramagnetic—ferromagnetic transitions, the amplitude of a lattice modulation wave in commensurate displacive transitions, an element of the deformation tensor in ferroelastic transitions, or the difference between the probabilities of occupancy of sites in order—disorder transitions.

The symmetry is best taken into account of considering the symmetry groups $\mathcal{G}_{\mathrm{I}}$ and $\mathcal{G}_{\mathrm{II}}$ of the two phases (i.e. the sets of orientational and translational symmetry operations for structural phase transitions). Then one can distinguish several cases:

(1) $\mathscr{G}_{I} \equiv \mathscr{G}_{II}$. No change of symmetry, the transition is usually first order.
(2) $\mathscr{G}_{I} \not\subset \mathscr{G}_{II}$ and $\mathscr{G}_{II} \not\subset \mathscr{G}_{I}$. There is no group—subgroup relation between the two phases. This is the case of reconstructive transitions which involve long-range diffusion of atoms. There is no continuity at the transition, it is first order (ex: graphite—diamond transition).
(3) $\mathscr{G}_{II} \subset \mathscr{G}_{I}$ (or $\mathscr{G}_{I} \subset \mathscr{G}_{II}$). Phase I is usually the high-temperature 'disordered' phase while phase II is the low-temperature 'ordered' phase (by analogy with order—disorder transitions in alloys). The order parameter is then 0 in phase I and $\neq 0$ in phase II. Its behavior near T_c reflects the order of the transition:
   (a) if η is discontinuous at T_c the transition is first order (e.g. the orientational order—disorder transition of NH_4Cl)
   (b) if η is continuous the transition is second order (e.g. the displacive transition of $SrTiO_3$ where there is rotation of oxygen tetrahedra, the Jahn—Teller transition of $PrAlO_3$ at 151 K).

1.3. Elements of the Phenomenological Landau Theory

The Landau theory links the above considerations based on thermodynamics and symmetry. It makes simple assumptions about the type of singularity of the free energy at T_c using an analytical expansion which allows the prediction of general properties of the transition. The theory also relates the order parameter to symmetry and establishes symmetry conditions for second-order transitions. Our purpose here is to give only some of its basic results.

Near T_c it is assumed that the free energy can be expanded in powers of η (η being small and pressure effects being left aside):

$$G(\eta, T) = G_0(T) + \sum_i G_i(T)\eta^i, \tag{1}$$

where G_0 is the free energy in the disordered phase ($\eta = 0$).

The order parameter in the low-temperature ordered phase corresponds to a minimum of the free energy determined by:

$$\frac{\partial G}{\partial \eta} = 0; \qquad \frac{\partial^2 G}{\partial \eta^2} > 0. \tag{2}$$

Solving these equations, and using simple arguments about the relative stability of the two phases, we get conditions for the existence of first- and second-order transitions. It is also possible to predict the variation of fundamental quantities like the order parameter, the entropy $S = -(\partial G/\partial T)$ or the specific heat $C = T(\partial S/\partial T)$. Another important quantity is the susceptibility $\varkappa = \partial\eta/\partial H$ which is the linear response of the system to an external field H conjugated to the order parameter (for instance, in a ferroelectric transition H is the electric field E, η is the polarization P, and therefore $\varkappa = \partial P/\partial E = \varepsilon$, the dielectric constant).

Depending on the value of the coefficients $G_i(T)$ in (1), different types of transition can occur. This is summarized in Figure 2 where the variations of $G(\eta,$

T), $\eta(T)$, $C(T)$, and $\varkappa^{-1}(T)$ are sketched for the three main cases, assuming a free energy of the form:

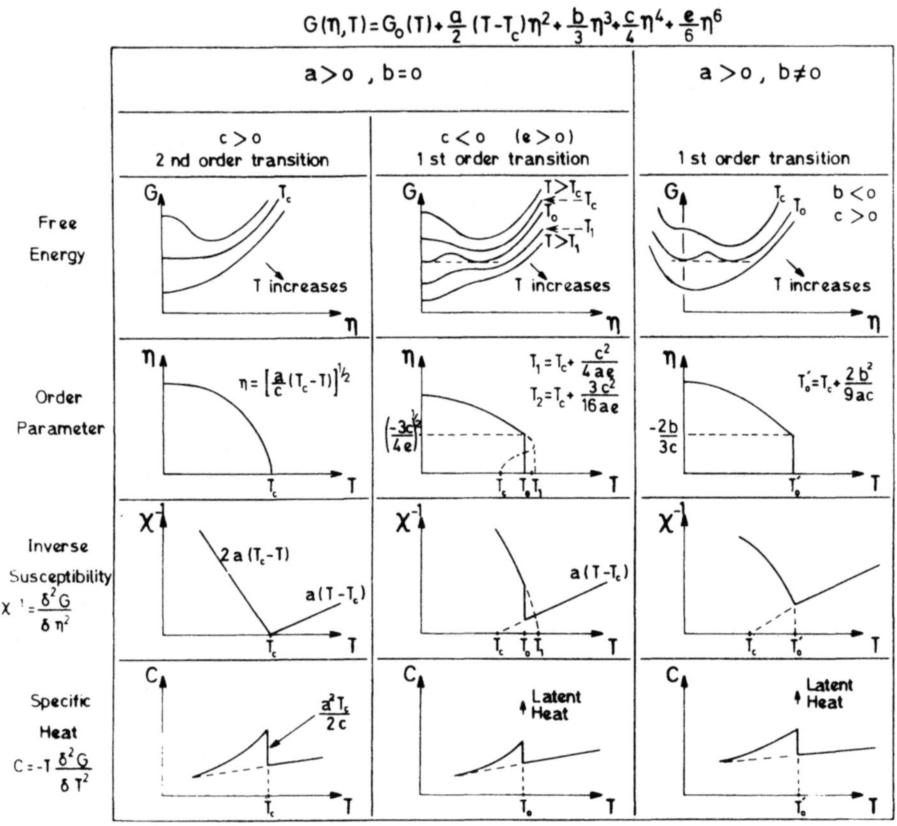

Fig. 2. Temperature dependence of the free energy $G(\eta, T)$, the order parameter η, the specific heat C, and the inverse susceptibility $\varkappa^{-1}$ in the Landau model for first- and second-order transitions.

$$G(\eta, T) = G_0(T) + \frac{a}{2}(T - T_c)\eta^2 + \frac{b}{3}\eta^3 + \frac{c}{4}\eta^4 + \frac{d}{5}\eta^5 + \frac{e}{6}\eta^6. \tag{3}$$

In the Landau model it is found that a second-order phase transition is obtained only when $b = 0$ (no third-order term in G) and $c > 0$. In this case the singularities of the thermodynamic quantities η, C, and $\varkappa$ near T_c can be characterized by critical exponents which are the exponents of the leading terms in the expansion of η, C, and $\varkappa$ in the reduced temperature $t = (T - T_c)/T_c$.

From Figure 2 one obtains:

$$T = T_c \pm \varepsilon \begin{cases} C \propto (\pm t)^{-\alpha} & \alpha = 0 \\ \chi \propto (\pm t)^{-\gamma} & \gamma = 1 \end{cases}$$

$$T = T_c - \varepsilon \to \eta \propto (-t)^{-\beta} \qquad \beta = \tfrac{1}{2}$$

$$(\eta \equiv 0 \text{ above } T_c).$$

The Landau theory also makes use of the group theory to establish symmetry conditions for the ordered state (in particular, its symmetry is defined by only one irreducible representation of the high-temperature group $\mathscr{G}_1$) but this goes beyond the scope of this section.

1.4. Fluctuations and Correlations

A second-order phase transition is announced in the high-temperature phase by precursor effects corresponding to the appearance of dynamic short-range order in the sample. Simplifying, one can considered that ordered domains of average size ξ (the correlation length) are created and destroyed as a function of time. This involves both spatial and dynamical fluctuations of the order parameter which increases as T_c is approached, leading to the divergence of the susceptibility (see Figure 2).

The spatial fluctuations can be characterized by the correlation function of the local order parameter for two points separated by **r**. Above T_c where $\langle\eta\rangle = 0$ this quantity takes the form

$$\langle\eta(0)\eta(\mathbf{r})\rangle = \langle\eta^2\rangle g(\mathbf{r}).$$

The spatial dependence of the correlation function is given by $g(\mathbf{r})$.

1.4.1. *Isotropic Correlations*

In the case of isotropic systems and for r large enough, the Ornstein—Zernike form of the correlation function is a reasonable approximation

$$g(r) = \frac{1}{r}\, e^{-r/\xi}. \tag{4}$$

By Fourier transform of $g(\mathbf{r})$ one gets a function $g(\mathbf{q})$ describing the fluctuations of the wave vector around the critical wave vector $\mathbf{q}_c$ which defines the low-temperature ordered state:

$$g(\mathbf{q}) = \int d^3\mathbf{r}\, g(\mathbf{r})\, e^{i\mathbf{q}\cdot\mathbf{r}}. \tag{5}$$

Using (4) we obtain

$$g(\mathbf{q}) \propto \frac{g(\mathbf{q}_c)}{1 + \delta q^2 \xi^2}, \tag{6}$$

where $\delta\mathbf{q} = \mathbf{q} - \mathbf{q}_c$.

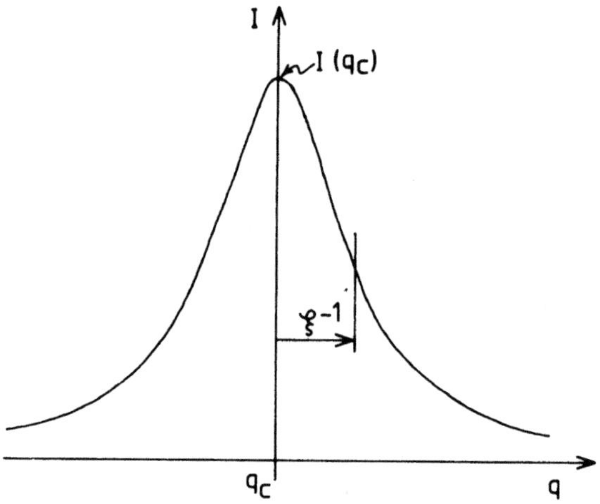

Fig. 3. Lorentzian intensity profile $I(\mathbf{q})$ corresponding to an Ornstein—Zernike form of the correlation function. The HWHM is the inverse correlation length ξ^{-1}.

This function $g(\mathbf{q})$ corresponds to the diffuse intensity $I(\mathbf{q})$ which is scattered around the wave vector $\mathbf{q}_c$, above the transition. It is a Lorentzian whose half width at half maximum (HWHM), is the inverse of the correlation length (ξ) in direct space so that the measure of $I(\mathbf{q})$ as a function of temperature allows us to determine $\xi(T)$ as the transition is approached from above (Figure 3).

$g(\mathbf{q})$ is also related to the generalized susceptibility $\chi(\mathbf{q})$ by:

$$g(\mathbf{q}) = kT\chi(\mathbf{q}) \tag{7}$$

(this comes from the classical limit of the fluctuation—dissipation theorem, valid if kT is larger than the energy $h\omega_{q_c}$, which characterizes the dynamics of the phase transition). In the mean-field approximation (or Landau theory) $\chi(\mathbf{q}_c)$ varies as $(T - T_c)^{-1}$ and assuming an Ornstein—Zernike law one obtains:

$$\xi(T) \propto (T - T_c)^{-1/2} \tag{8}$$

The above treatment did not consider the time dependence of the fluctuations. This applies, for example, to X-ray measurements which give information on the instantaneous spatial fluctuations of the order parameter. The dynamics of these fluctuations can be studied using inelastic neutron scattering which gives $g(\mathbf{q}, \omega)$, where $\hbar\omega$ is the energy transfered to the sample (g is also noted S by neutron people). $g(\mathbf{q}, \omega)$ is the Fourier transform of the time-dependent correlation function

$$\langle \eta(0, 0)\eta(\mathbf{r}, t)\rangle.$$

Let us just give the main results obtained in the case of displacive transitions arising from the decrease with temperature of the frequency ω_{q_c} of a particular phonon branch (so-called soft mode) at the wave vector $\mathbf{q}_c$. It can be shown that, in the high-temperature limit and for the critical wave vector $\mathbf{q}_c$:

$$S(\mathbf{q}_c, \omega) \equiv g(\mathbf{q}_c, \omega) \propto kT \frac{2\Gamma}{(\omega_{q_c}^2 - \omega^2)^2 + 4\Gamma^2\omega^2}, \tag{9}$$

where Γ is a damping parameter traducing anharmonic contributions. Therefore, the frequency dependence of $S(\mathbf{q}, \omega)$ shows a maximum near ω_{q_c} for Γ, as shown in Figure 4. Finally, let us note that the X-ray diffuse scattering associated with the soft mode is obtained by integration of $S(\mathbf{q}, \omega)$ against ω

$$I(\mathbf{q}_c) = \int S(\mathbf{q}_c, \omega)\, d\omega \propto \frac{kT}{(\hbar\omega_{q_c})^2}. \tag{10}$$

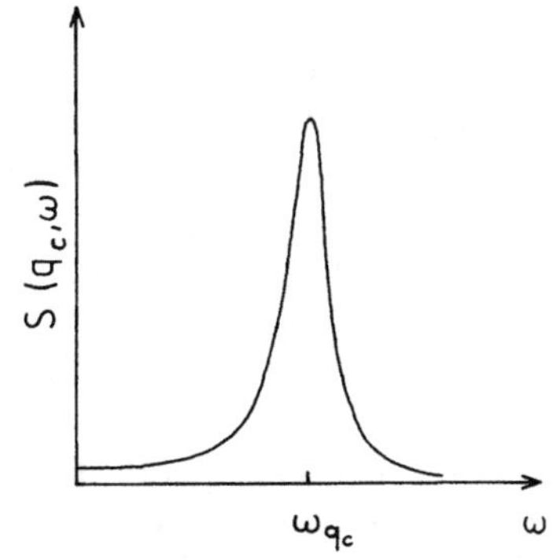

Fig. 4. Behavior of $S(q_c, \omega)$ with ω in the case of weak anharmonicity. A peak of half width Γ occurs near ω_{q_c}.

Thus the diffuse intensity is proportional to the inverse of the square of the phonon frequency, leading to the relation: $\chi(q_c)^{-1} \propto (\hbar\omega_{q_c})^2$ (i.e. $\hbar^2\omega_{q_c}^2 \sim (T - T_c)$ in the mean-field approximation).

1.4.2. *Anisotropic Correlations and X-ray Scattering*

In the case of quasi-1D conductors the correlations usually reflect the structural anisotropy, as will be shown in Section 3, and it is interesting to see how the previous results are modified. Thus, Equation (6) becomes:

$$g(\mathbf{q}) \propto \frac{g(\mathbf{q}_c)}{1 + \delta q_a^2 \xi_a^2 + \delta q_b^2 \xi_b^2 + \delta q_c^2 \xi_c^2} \tag{11}$$

and the diffuse intensity scattered around the wave vector $\mathbf{q}_c(I(\mathbf{q}) \equiv g(\mathbf{q}))$ presents a Lorentzian shape along the a^*, b^*, and c^* directions of the reciprocal space with HWHMs equal to ξ_a^{-1}, ξ_b^{-1}, and ξ_c^{-1}, respectively. If we consider first a set of (*a*,

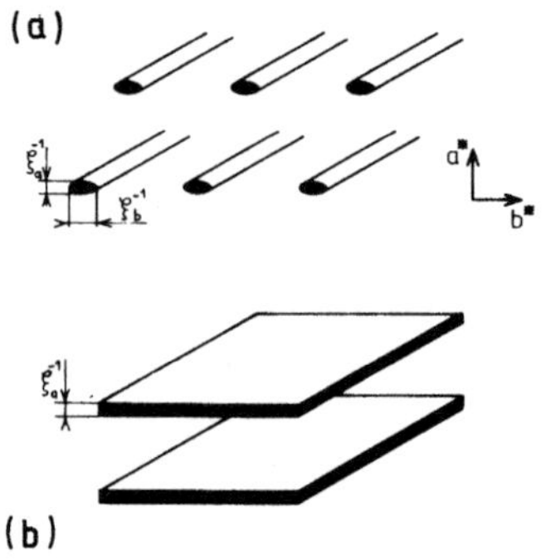

Fig. 5. Schematic diffuse intensity distribution in reciprocal space for (a) planar correlations, (b) linear correlations.

b) planes with no correlation between the planes in the c direction ($\xi_c = 0$), the scattered intensity will be localized on diffuse rods along c^*. The in-plane transverse correlations (ξ_a and ξ_b) will be given by the corresponding inverse HWHMs (Figure 5(a)).

On the other hand, for a system of uncorrelated chains along a one obtains diffuse sheets parallel to b^* and c^* ($\xi_b = \xi_c = 0$). From the Lorentzian intensity profile across the sheets one can therefore determine the correlation length ξ_a (Figure 5(b)).

Similar anisotropic effects arise for diffuse scattering due to soft phonon modes through the dependence of the corresponding phonon frequency with **q**.

Such effects are very useful in the field of 1D conductors since they allow us to analyze and follow the dimensionality of the fluctuations of the system, as will be shown in Section 3.

2. Peierls Instability of a One-Dimensional Metal

Quasi-one-dimensional (quasi-1D) conductors consist of parallel chains of atoms or molecules of which the highest occupied electronic states overlap strongly in the chain direction to form energy bands. The interaction between the chains being much weaker, these materials exhibit a marked anisotropy in most of their properties. This is the origin of the metallic conduction along the chains and of the conductivity anisotropy.

2.1. $2k_F$ INSTABILITY OF A ONE-DIMENSIONAL ELECTRON GAS

Because of the one-dimensional (1D) nature of these conductors it is appropriate to consider the instability of a 1D electron gas coupled to a lattice through electron–phonon interaction. Peierls pointed out that such a system is unstable and undergoes an energetically favorable distortion. This is best explained using an

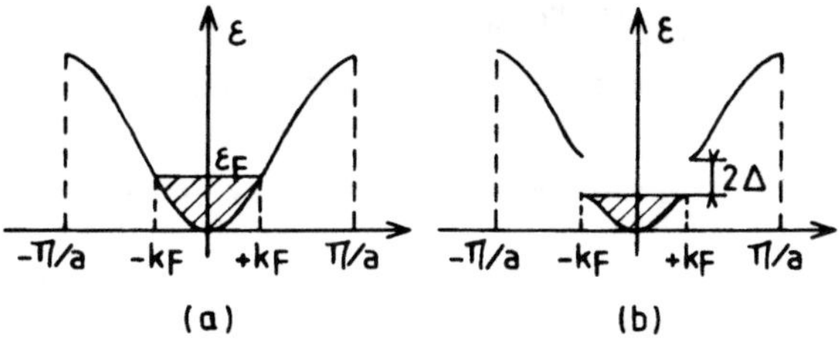

Fig. 6. Electronic energy dispersion curves for a one-dimensional metal. The band is filled for $|\mathbf{q}| < k_F$. In (b) the opening of a gap 2Δ at $\pm\mathbf{k}_F$ lowers the electronic energy.

electronic energy band scheme such as shown in Figure 6. In the conduction band (Figure 6(a)) the Fermi surface is restricted to two points at $\pm\mathbf{k}_F$ ($\mathbf{k}_F$ being the Fermi wave vector). A lattice modulation of wave vector $\mathbf{q} = 2\mathbf{k}_F$ will open an energy gap at the Fermi level just in the same way as the periodic lattice potential opens a gap at the Brillouin zone boundary in the general case (Figure 6(b)). The energy of the occupied states will be lowered and it is shown that for small amplitudes (u) of distortion, the electronic energy gain (proportional to $-u^2 \ln u$) is larger than the cost in elastic energy (proportional to u^2) which demonstrates the instability of the system against the distortion. The ground state induced by this distortion is either a semiconductor or an insulator, depending on the size of the gap 2Δ which is proportional to u. This is the situation at $T = 0$, but as the temperature increases, electrons are thermally excited across the gap. Its size is reduced and it disappears at some finite temperature T_p. The electronic gap Δ, as well as the amplitude of the lattice modulation can be chosen as the order parameter of the Peierls transition.

2.2. Charge Density Waves, Incommensurability

The modulation of wave vector $2\mathbf{k}_F$ consists in a periodic lattice distortion (PLD) involving small atomic or molecular displacements. There is, moreover, a modulation of the conduction electron density with the same wave vector $2\mathbf{k}_F$, which is coupled to the lattice distortion. Strictly speaking this electronic modulation is the so-called charge density wave (CDW), although in practice this term often covers both effects. Since the $2\mathbf{k}_F$ wave vector only depends on the shape and filling of the electronic energy bands it has, in principle, no relation with the reciprocal vectors of the average or undistorted lattice so that the charge density wave can be incommensurate with this lattice. This occurs frequently in real systems because the band filling is often determined by a partial charge transfer between two chains of atoms or molecules, as will be shown in Section 3.

The incommensurability of the CDW with respect to the lattice has interesting consequences. It implies that a translation of the CDW — that is, a change of its phase — does not modify the energy of the system (Figure 7). As a result of the translational invariance of the CDW energy, the charge modulation could move

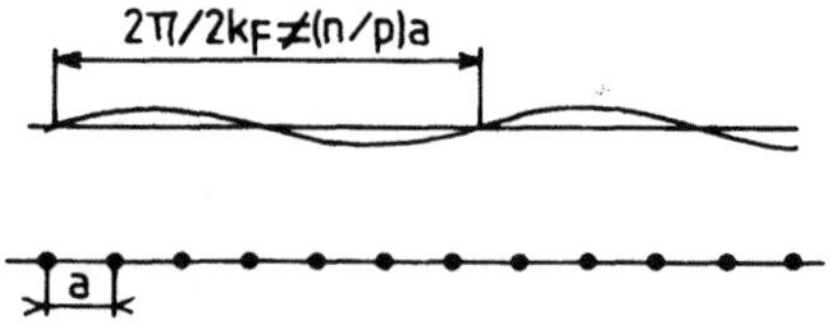

Fig. 7. Schematic picture of a periodic row of atoms with an incommensurate modulation of wavelength $2\pi/2k_F \neq (n/p)a$ (n and p integers).

through the crystal (together with the lattice distortion) which corresponds to a dc current. Neglecting all damping effects this would give rise to 'superconductivity' as suggested originally by Fröhlich. However, in real materials the situation is not so favorable and several mechanisms actually limit the motion of the CDW, like commensurability locking effects, pinning by impurities, or Coulomb interactions between CDWs of opposite charges on neighboring chains. Still, significant contributions of CDW motion to the electrical conductivity have been observed in TTF—TCNQ and in several inorganic conductors where it is associated with nonlinear and frequency-dependent conductivity effects.

2.3. Precursor Effects – Kohn Anomaly

Because of the electron—phonon coupling the Peierls transition is signalled in the high-temperature phase by dynamic fluctuations. They are characterized by the softening of a particular phonon mode at the wave vector $\mathbf{q} = 2\mathbf{k}_F$. This can be seen if one considers that the phonon frequencies, $\omega_0(\mathbf{q})$, are renormalized by screening of the conduction electrons as expressed by

$$\omega^2(\mathbf{q}) = \omega_0^2(\mathbf{q})[1 + \lambda\varkappa(\mathbf{q})], \tag{12}$$

where λ is the electron—phonon coupling and $\varkappa(\mathbf{q})$ is the electronic susceptibility of the electron gas

$$\varkappa(\mathbf{q}) = \sum_k \frac{f(\varepsilon_{k+q}) - f(\varepsilon_k)}{\varepsilon_{k+q} - \varepsilon_k}\,, \tag{13}$$

k is the wave vector of electrons with energy ε_k and $f(\varepsilon_k)$ is the Fermi distribution function ($f(\varepsilon_k) = [\exp[(\varepsilon_k - \varepsilon_F)/kT] + 1]^{-1}$). $\varkappa(\mathbf{q})$ is the response of the electron gas to a periodic field of wave vector $\mathbf{q}$. It can be shown, in the 1D case, that $\varkappa_{1D}(\mathbf{q})$ presents a logarithmic divergence at $\mathbf{q} = 2\mathbf{k}_F$ (the denominator is equal to 0 for all states at the Fermi level because of the planar Fermi surface)

$$\varkappa_{1D} = N(\varepsilon_F) \ln \frac{kT}{1.14\varepsilon_F} \tag{14}$$

with $N(\varepsilon_F)$ the density of states at the Fermi level. From (12) one finds that the frequency goes to 0 (the distortion becomes static) at the Peierls temperature $kT_p = 1.14\varepsilon_F \exp(-1/N(\varepsilon_F)\lambda)$.

This is the Kohn anomaly, which can be studied by inelastic neutron scattering (Figure 8). The Kohn anomaly also manifests itself by X-ray scattering since the

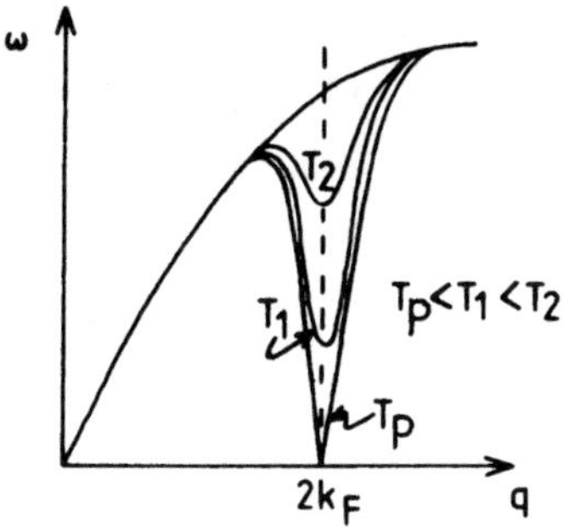

Fig. 8. Kohn anomaly in a phonon branch as the temperature decreases towards the Peierls transition temperature.

diffuse intensity is proportional to $\omega^{-2}(2\mathbf{k}_F)$ (Equation (10)). An important feature of the Kohn anomaly in 1D conductors is its 1D nature and the precursor scattering above the Peierls transition takes the form of diffuse sheets normal to the chain axis. As will be shown in Section 3, the broadening of these sheets gives the inverse of the correlation length of the CDW fluctuations.

2.4. Interchain Couplings

In real materials there are, of course, significant (although weak by definition) couplings between chains which lead to transverse ordering between the $2\mathbf{k}_F$ CDWs below T_p. First, one has to take into account Coulomb forces between charge modulations on neighboring chains. This effect will generally favor a sort of antiphase ordering of the CDWs on adjacent chains, positive charges tend to face negative ones. Another important transverse coupling is due to interchain tunneling of electrons related to deviation from the ideal planar character of the Fermi surface. This is best described in two dimensions using an energy dispersion:

$$\varepsilon(\mathbf{k}) = \varepsilon_{\|}(k_{\|}) + \varepsilon_{\perp}(k_{\perp}),$$

where $\varepsilon_{\|}$ and $\varepsilon_{\perp}$ are the longitudinal (in chain direction) and transverse energies. $\varepsilon_{\perp}$ can be of the form $\varepsilon_{\perp}(k) = -2t_{\perp}\cos k_b b$, where $t_{\perp}$ is the transverse transfer integral (supposed to be much smaller than the longitudinal one $t_{\|}$). This corresponds to a Fermi surface which is warped, as shown in Figure 9(b). Consequently a wave vector $(2\mathbf{k}_F, q_b)$ which can open a gap for all states of the Fermi surface if $t_{\perp} = 0$ (whatever the q_b value), would connect only a limited number of states if $t_{\perp} \neq 0$ (Figure 9(a)). Instead, a wave vector such as $(2\mathbf{k}_F, \pi/b)$, called a nesting vector, is able to open a complete gap on the Fermi surface and the corresponding distortion will be stabilized. This causes a phase shift of π from chain to chain in the transverse b direction which implies that the CDWs will be in antiphase on neighboring chains in this direction (Figure 9(b)).

In this simple picture Coulomb interactions and interchain tunneling of electrons both lead to the same type of CDW ordering in the transverse direction. In

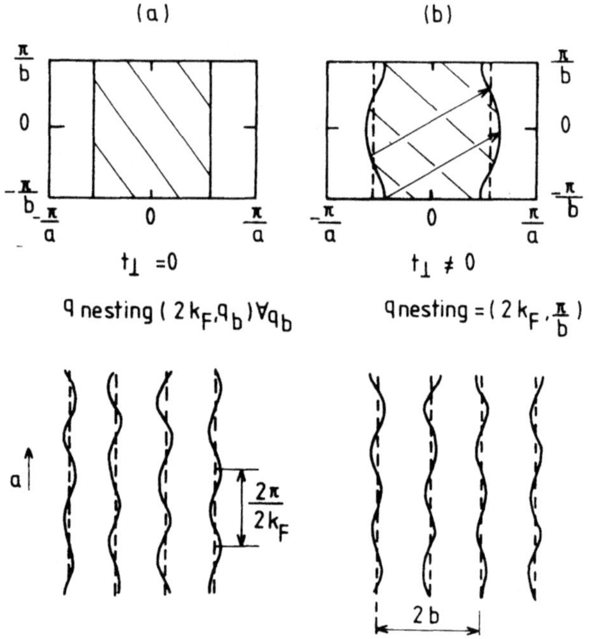

Fig. 9. Effect of transverse coupling (along b) on the Fermi surface, the nesting properties and the charge density waves ordering: (a) no coupling ($t_\perp = 0$), any $\mathbf{q}$ vector of the form $(2k_F, q_b)$ is a nesting vector so that the CDWs on neighboring chains are independent; (b) finite coupling $t_\perp$ with a sinusoidal warping of the Fermi surface, only $\mathbf{q} = (2k_F, \pi/b)$ is a good nesting vector (there is a phase shift of π from chain to chain).

real systems the situation is far more complex since several types of chains often have to be considered, as will be shown in Section 3.

2.5. Beyond the Independent Electron Picture and the Mean-Field Treatment

The Peierls—Fröhlich approach summarized above is quite useful; however, it is oversimplified and incomplete. For instance it is known that, in a strictly 1D system, long-range order is destroyed by fluctuations even at the lowest temperatures while the Peierls—Fröhlich mean field theory, which neglects the fluctuations, predicts a finite critical temperature of long-range ordering. By including the fluctuations and the weak interchain coupling it results that the Peierls transition occurs at temperatures much lower than those predicted by the mean-field theory. It follows also that the Peierls transition is announced in a wide temperature range by a regime of 1D or quasi-1D structural fluctuations which will be described in Section 3.1.4(c).

Another problem concerns electron—electron interactions which have been neglected although they are crucial in many cases. For instance, if one assumes that the intrasite Coulomb repulsion is so large that two electrons of opposite spins cannot occupy the same site, then the band is filled from $-2k_F$ to $2k_F$ (instead of $-k_F$ to k_F in case of spin degeneracy), which gives rise to a $4k_F$ instability actually observed in some organic conductors (Section 3.1.3). Other types of instability are important and can eventually prevail in 1D conductors to generate contrasted ground states. Let us just mention $2k_F$ spin density waves (where the spin density is modulated), $2k_F$ spin-Peierls distortions (where the spin density modulation is coupled to a lattice distortion by a modulation of the exchange integrals) or the formation of Cooper pairs (superconductivity).

Interaction between electrons also affects the nature of the divergence of the response function of the 1D gas. In particular, instead of a logarithmic response (Equation (14)), one shows that the CDW response function χ_{1D} presents a power law divergence in temperature. Experimental evidence will be given in Section 3.1.4(b) for organic conductors.

3. Representative Structural Studies of One-Dimensional Conductors

3.1. PEIERLS TRANSITIONS

3.1.1. *Introduction*

In Section 2 we introduced the general features of the Peierls transition and in particular we defined the transition temperature T_p above which 1D structural fluctuations are observed. However, the situation is a little more complicated and it is appropriate to begin this section by showing that three temperature regimes have to be considered to describe the structural features of a Peierls transition, as illustrated in Figure 10.

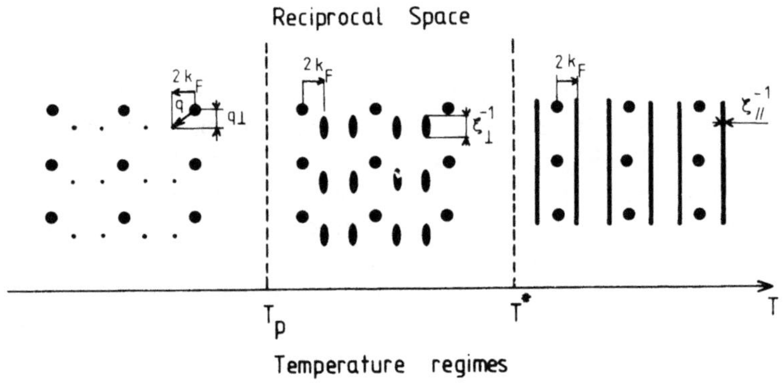

Fig. 10. Schematic picture of the three temperature ranges which can be considered to analyze the dimensionality of quasi-one-dimensional conductors presenting Peierls transitions.

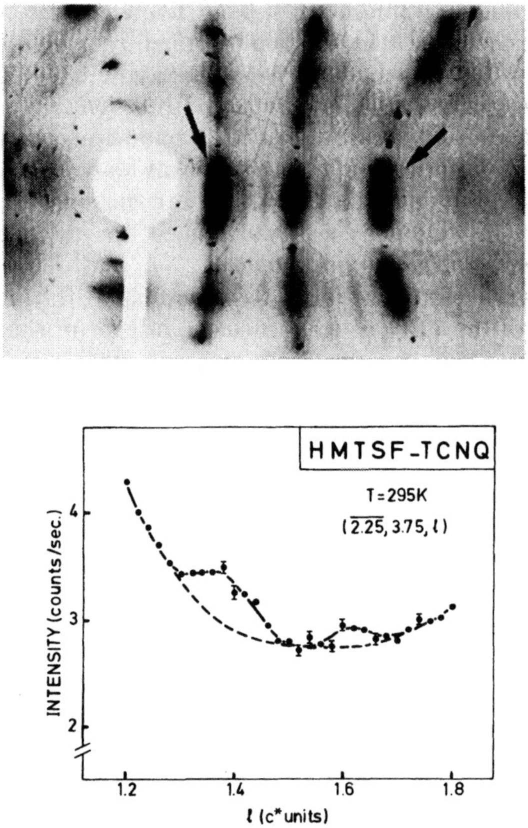

Fig. 11. $2k_F$ diffuse scattering in HMTSF—TCNQ. The X-ray monochromatic Laue photograph (a) shows diffuse lines indicated by white arrows. They are the intersections of diffuse sheets in reciprocal space with the Ewald sphere. Note that the Bragg reflections (black arrows) are overexposed indicating that the intensity of this scattering is extremely weak. This is clear on the diffractometer scan (b) across two of the diffuse lines.

(a) Above T^*, the structural fluctuations are decoupled from chain to chain. They appear as X-ray diffuse sheets in the reciprocal space. Because of the linear electron—phonon coupling, structural fluctuations reflect the CDW 1D instability of the 1D electron gas. The position of the diffuse scattering, with respect to the layers of main Bragg reflections in the chain direction, gives the wave vector of instability $2k_F$. Its HWHM in the chain direction is directly related to the inverse correlation length of the CDW, ξ_{1D}^{-1}, while its intensity at maximum value is proportional to the mean square fluctuation of the CDW amplitude (i.e. of the order parameter $I_{q_c = 2k_F} \propto \langle|\eta|^2\rangle$). A typical illustration of such a regime is given by the X-ray pattern of Figure 11(a) showing at room temperature '$2k_F$' diffuse lines (intersection of the above-mentioned diffuse sheets with the Ewald sphere) in the 1D organic conductor HMTSF—TCNQ. Figure 11(b) gives a diffractometric scan across the $1 + 2k_F$ and $2 - 2k_F$ diffuse lines of the same compound at the same temperature, showing that they are of extremely weak intensity (10^{-5}—10^{-7} times the intensity of a Bragg reflection) [3].

(b) Below T^* the intensity of the diffuse sheets begins to be sharply modulated, which means that lateral coupling between structural (and CDW) fluctuations occurs. The extent of the coupling in a given direction, $\xi_\perp$, can be obtained from the measure of the inverse HWHM of the modulation of the diffuse scattering in

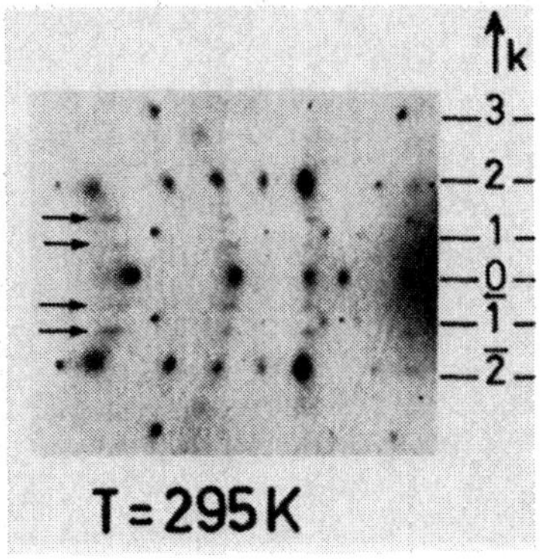

Fig. 12. Room temperature X-ray photograph of $K_{0.3}MoO_3$ showing diffuse segments (arrows) corresponding to two-dimensional correlations (After Pouget *et al.* [4].)

the same direction. T^* is defined as the crossover temperature for which $\xi_\perp \sim d_\perp$, where $d_\perp$ is the corresponding interchain distance. Below T^*, the structural fluctuations are 3D (or sometimes 2D between a T^*_{2D} and a T^*_{3D}, for very anisotropic transverse couplings; at T^*_{2D}: $\varepsilon^1_\perp \sim d^1_\perp$ but $\varepsilon^2_\perp < d^2_\perp$ and at T^*_{3D} $\varepsilon^2_\perp \sim d^2_\perp$, where 1 and 2 denote different transverse directions). The X-ray pattern of Figure 12 gives an example of 2D coupling between CDWs at room temperature in the blue bronze $K_{0.3}MoO_3$ (the diffuse segments shown by the arrow are the intersection of diffuse streaks with the Ewald sphere) [4].

(c) at T_p the diffuse scattering 'condenses' into sharp satellite reflections. Thus, a long-range modulation of the high-temperature lattice occurs, and below T_p the CDW distortions form a well-defined 3D ordered structure. A typical illustration of such a CDW lattice is provided by the electron diffraction patterns of Figure 13, showing in $NbSe_3$ the formation below 144 K of a long-range modulation of wave vector (0, 0.241, 0) and an additional long-range modulation of wave vector (0.5, 0.260, 0.5) below 59 K [5].

If, for simplicity, one assumes a plane wave modulation, for which the spatial dependence of the order parameter is described by a phase factor only, the order parameter for the chain (n, p) can be written as:

$$\eta_{n,p}(x) = \eta_0 \, e^{-i(2\pi \cdot 2k_F \cdot x + \phi np)},$$

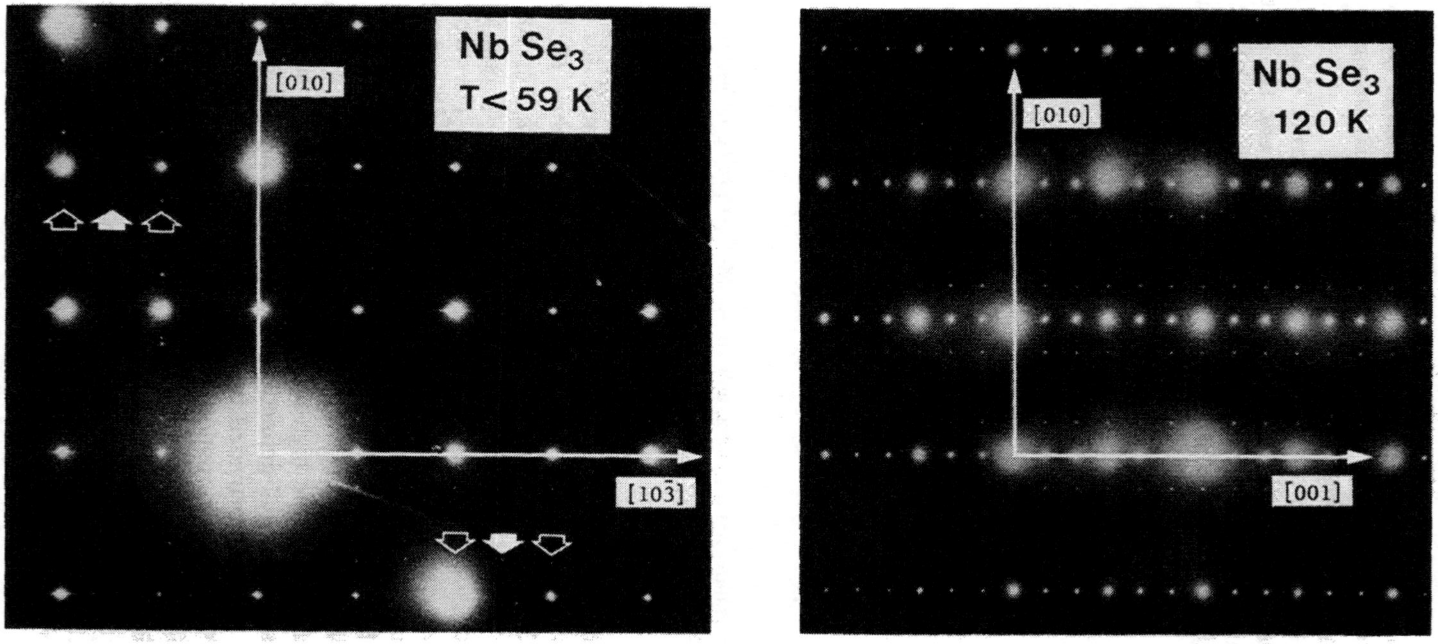

Fig. 13. Electron diffraction photographs of $NbSe_3$ showing (a) CDW satellites with wave vectors (0, 0.241, 0) at 120 K. In (b) an extra set of CDW satellites at (0.5, 0.26, 0.5) (solid arrows) corresponding to the low temperature ($T < 59$ K) distortion are visible. (After Roucau, Thesis, Toulouse (1981) and [5].)

where we have used a continuous notation (x) in the chain direction. The spatial dependence of the phase with the chain index is given by

$$\phi_{n,p} = 2\pi q_\perp^1 n + 2\pi q_\perp^2 p + \phi_0$$

with ϕ_0 the phase of the chain at the origin. In this description η_0, the amplitude of the order parameter, is taken independent of the chain index (no modulation of amplitude).

In this introduction we only considered the spatial anisotropy of the structural fluctuations above the Peierls transition which can be easily obtained by X-ray (and sometimes electron) scattering experiments. Very important information concerning the dynamics of these fluctuations is given by inelastic neutron scattering experiments. Below T_p, X-ray, neutron and electron diffraction experiments have been used to characterize the structural distortions. Before describing some of these results (Sections 3.1.4 and 3.1.5) we shall introduce the relevant series of compounds showing $2k_F$ (3.1.2) and $4k_F$ structural instabilities (3.1.3).

3.1.2. *Examples of $2k_F$ Peierls Instabilities*

(a) *$K_2Pt(CN)_4Br_{0.3}$, xH_2O.* Perhaps the simplest example of a 1D conductor is provided by the Krogmann salt $K_2Pt(CN)_4Br_{0.3}$, xH_2O which is composed of infinite chains of the planar complex $Pt(CN)_4$, with direct overlap between the d_{z^2}

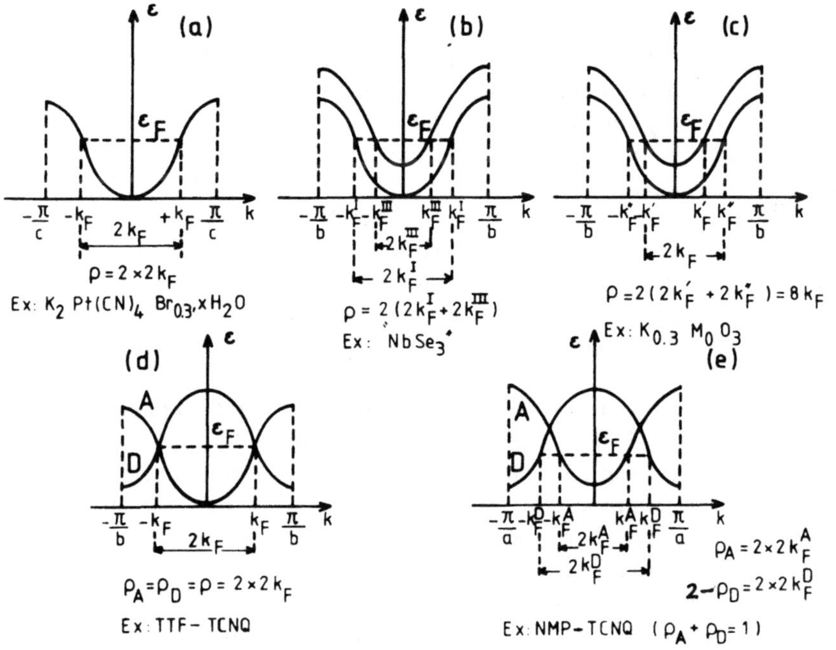

Fig. 14. Schematic electron energy dispersion for 1D conduction bands showing the different situations that can arise: (a) one single band ($K_2Pt(CN)_4Br_{0.3}$, xH_2O); (b, c) two bands ($NbSe_3$, $K_{0.3}MoO_3$); (d, e) acceptor and donor bands in the charge transfer salts (TTF—TCNQ, NMP—TCNQ).

Pt orbitals forming the conduction band. The Br takes 0.3 electron per $Pt(CN)_4$. With two $Pt(CN)_4$ units per repeat distance along the chain the conduction band is emptied by $\rho = 0.6$ electrons, leading to $2k_F = \rho/2 = 0.3$. The resulting wave vector of instability is schematically shown in Figure 14(a). At room temperature well-defined 1D structural fluctuations at the wave vector $2\mathbf{k}_F = 0.3\mathbf{c}^*$ are observed [6]. However, because of an intrinsic disorder due mainly to the Br atoms, the short-range lateral coupling between CDWs does not increase appreciably below 100 K and no real 3D phase transition is observed. The local ordering occurs with the $(\frac{1}{2}, \frac{1}{2}, 2k_F)$ reduced wave vector [7—9].

(b) *$NbSe_3$*. The structure of transition metal trichalcogenides is described in detail on pages 205—234 of this book. For the present purpose let us briefly mention that each unit cell contains six chains running along the *b* direction and belonging to three different types. Along the *c* direction there is a group of two chains of type I separated from a group of two chains of type III by a chain of type II. In this compound the conductivity is usually associated with chains of type I and III. Band calculations [10, 11] show that four bands, built with *d* orbitals originating mainly from these two groups of two chains, cross the Fermi level and accommodate two electrons per unit cell.

Two structural phase transitions characterized by the reduced wave vectors (0, 0.241, 0) and (0.5, 0.260, 0.5) occur at $T_{p_1} = 144$ K and $T_{p_2} = 59$ K, respectively (see Figure 13) [5, 12]. NMR study [13] and the observation of a single phase transition at about 140 K, characterized by the wave vector (0, 0.27, 0) in $FeNb_3Se_{10}$ [14, 15], which does not contain chains of type I, suggest that one can associate the T_{p_1} transition with type III chains and the T_{p_2} transition with type I chains. Thus the two phase transitions can be viewed as those originating in two structurally decoupled electronic systems having their own $2k_F$ instability (as schematically shown in Figure 14(b)). However, in a given electronic system containing two chains of the same type, and thus two bands, the unstable $2k_F$ wave vector may join the Fermi wave vector of one band, $-k_F'$ to the Fermi wave vector of the other band, $+k_F''$, as proposed for $K_{0.3}MoO_3$ (see Section (c)). Quasi-1D structural fluctuations have been observed at and below room temperature in $NbSe_3$ [16], but a systematic study of their temperature dependence has not yet been performed.

(c) *$K_{0.3}MoO_3$*. The structure of the blue bronzes is detailed in pages 137—142 of this book. Essentially it is built with clusters containing 10 MoO_6 octahedra. These clusters are stacked on top of each other in the *b* direction forming infinite chains, while they share corners along the [102] direction. Along the third direction, layers of octahedra are separated by alkaline ions. In this compound the 1D electronic behavior is more likely related to the presence of infinite chains of clusters than to infinite chains of individual MoO_6 octahedra. Band calculations [17] suggest that each cluster gives rise to two low-lying conduction bands which are partly filled by the three electrons provided by alkaline ions. The blue bronzes undergo a Peierls transition at 180 K which stabilizes a modulation characterized by the reduced wave vector (0, q_b, 0.5) with $q_b \sim 0.25$ [4]. It has been suggested [18] that in this two-band system the metal-insulator phase transition accompanies the structural transition because the unstable $2k_F$ wave vector connects the Fermi wave vector of one band, $-k'_F$, to the Fermi wave vector of the other band, $+k'_F$ (as schematically shown by Figure 14(c)). In that case one has

$$2k_F = k'_F + k''_F.$$

Thus, $8k_F$ represents the total amount of electrons to share between the two bands. With about three electrons to share, the unstable wave vector is $2\mathbf{k}_F \simeq 0.75\mathbf{b}^*$ which corresponds to the measured value of $1 - q_b$. Figure 15 shows more

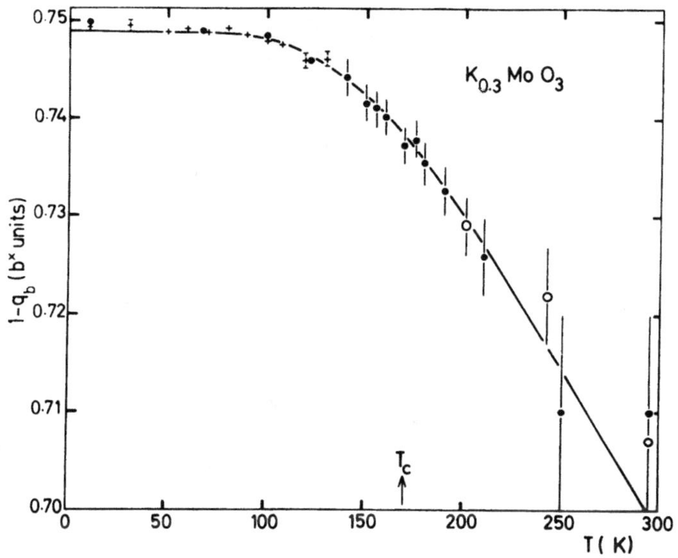

Fig. 15. Temperature dependence of $2k_F = 1 - q_b$ in $K_{0.3}MoO_3$. This continuous variation can be explained by thermal excitation of electrons to upper levels, thus reducing the number of electrons in the two bands. (After Pouget *et al.* [18].)

precisely that the $2k_F$ value decreases slightly and continuously when the temperature increases. A quantitative interpretation [18] of this variation is that by thermal excitation in the upper electronic levels, the total number of electrons available in the two conduction bands decreases. Another interesting peculiarity of the blue bronzes is that the crossover temperature T^* above which 1D fluctuations are observed, is situated well above the room temperature. At room temperature (Figure 12), anisotropic 2D structural fluctuations in the layers of MoO_6 octahedra are observed in agreement with the structural anisotropy. The coupling between layers occurs around $T^*_{3D} \sim 220$ K.

(d) *Organic charge transfer salts.* These salts are composed of segregated stacks of donor (D) and acceptor (A) molecules, as schematically drawn in Figure 16 for the

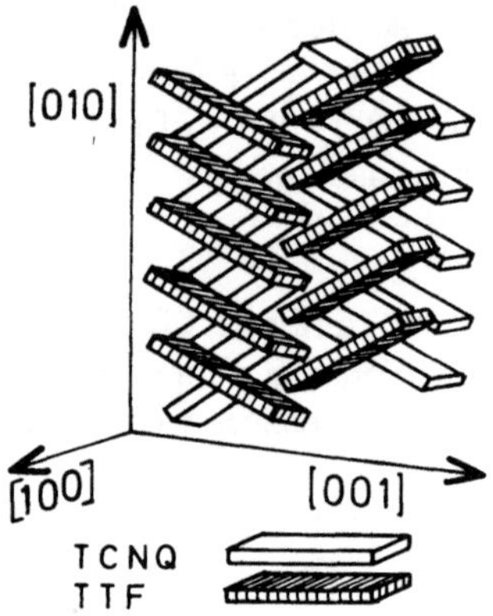

Fig. 16. Schematic herringbone stacking of TTF and TCNQ molecules in TTF—TCNQ (after R. Comès in *Chemistry and Physics of One-dimensional Metals* (ed. H. J. Keller), Plenum, New York, p. 315 (1977).

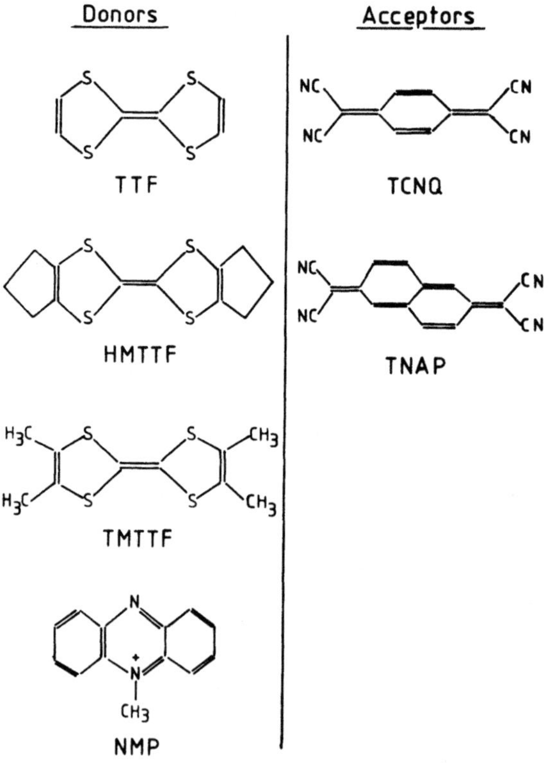

Fig. 17. Molecular structure of typical donor and acceptor molecules.

most studied compound TTF—TCNQ. Typical molecules are shown in Figure 17. There is an overlap between the $p\pi$ molecular orbitals in the chain direction, thus two systems of bands originating from donor and acceptor molecules are formed. If the band spectrum overlaps there is a partial charge transfer from donor to acceptor molecules in order to establish a common Fermi level. It is extremely difficult to perform an *ab initio* calculation of the degree of charge transfer, by lack of knowledge of the electronic parameter of the electronic gas [82]. Two cases, corresponding to different amounts of electrons to share between the donor and acceptor, will be considered here:

TABLE I

Material	$\rho (= 2 \times 2k_F)$	Occurrence of instabilities $2k_F$	$4k_F$	T_p (K)	Modulation wave vector
TMTSF—DMTCNQ	0.50	Yes	Yes	42	$(2k_F, (\pm)?\frac{1}{3}, 0)$
TMTSF—TCNQ	0.57	Yes	No	57	$(2k_F, \frac{1}{2}, ?)$
TMTTF—TCNQ	0.57	Yes	Yes	36	$(0, 2k_F, 0)$
TMTSF—TNAP	0.58	Yes	No	41	$(?, ?, 2k_F)$
TTF—TCNQ	0.55—0.59	Yes	Yes	54	$(\frac{1}{2}, 2k_F, 0)$
				49	$(q_a(T), 2k_F, 0)$
				38	$(\frac{1}{4}, 2k_F, 0)$
TSF—TCNQ	0.63	Yes	No	29	$(\frac{1}{2}, 2k_F, 0)$
				50	
HMTTF—TCNQ	0.72	Yes	Yes	'43'	$(0.42, 2k_F, 0)$
HMTSF—TCNQ	0.74	Yes	No	24	$(0, 0, 2k_F)$

(a) *TTF—TCNQ and its family* (see Table I) *with two electrons to share.* In that case the number of electrons transferred to the acceptor chain is equal to the number of holes created on the donor chain and $\pm k_F$ is common to the two bands as shown in Figure 14(d). The charge transfer per molecule ρ is thus equal to $4k_F$. Experimental values of this quantity, obtained from the measurement of the wave vector of the $2k_F$ (and $4k_F$ — see Section 3.1.3) instability at ambient pressure are given in Table I.

(b) *NMP—TCNQ with one electron to share.* In that case the Fermi wave vector differs on the acceptor and donor chains (see Figure 14(e)), but they are related by the relation $\rho_{NMP} + \rho_{TCNQ} = 1$, with $2k_F^{NMP} = \rho_{NMP/2}$ and $2k_F^{TCNQ} = \rho_{TCNQ/2}$. Experimentally it is found that $\rho_{NMP} = \frac{1}{3}$ and $\rho_{TCNQ} = \frac{2}{3}$ [19].

Table I gives also the temperature of the Peierls transition, T_p, of some charge transfer salts and the wave vector of the modulation established below T_p at ambient pressure. Above T_p, structural fluctuations are observed in temperature ranges greater than 100 K. In some compounds like HMTSF—TCNQ (Figure 3), they remain still well defined at room temperature. In TTF—TCNQ the $2k_F$ fluctuations are mostly 1D in nature, the crossover temperature T^* occurring at 58—60 K, a few degrees only above T_p. In others compounds like TMTSF—DMTCNQ the regime of purely 1D $2k_F$ fluctuations is not observed.

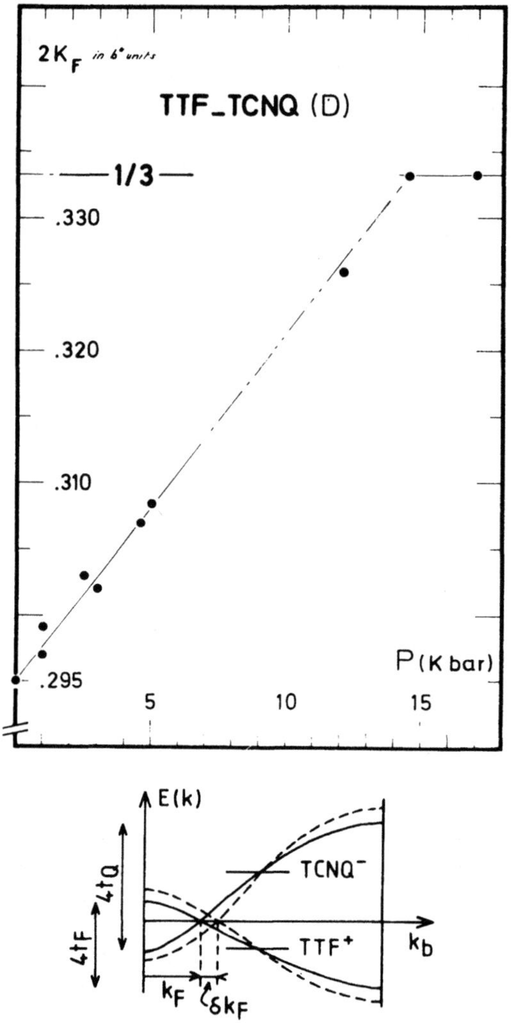

Fig. 18. Effect of pressure on the $2k_F$ value. Neutron scattering results (a) showing the continuous variation of $2k_F$ with pressure below 14 K bars and its locking at $\frac{1}{3}$ (b^* units) (after Megtert *et al.* [22]). This variation can be qualitatively explained in (b) by an increase of the bandwidth.

In these organic salts the charge transfer ρ, related to the electronic structure, changes in temperature, under pressure or upon alloying. Let us first examine the case of TTF—TCNQ. Table I shows that at ambient pressure ρ increases from 0.55 at room temperature to 0.59 at low temperature [20, 21]. Figure 18(a) shows that the low-temperature value of $2k_F$ ($\equiv\rho/2$), increases continuously with pressure [22]. These effects can be related to the decrease of the chain parameter b, either for T decreasing or for P increasing. In the simplest picture, one can easily imagine that with decreasing b the intrastack transfer integrals will increase, changing the crossing points $\pm k_F$ of the acceptor and donor bands, as schematically shown in Figure 18(b). Figure 18(a) shows also that at about 14 kbar, $2k_F$ reaches the commensurate value $\frac{1}{3}b^*$ where it locks for higher pressure values. The locking of $2k_F$ at a commensurate value suppresses a CDW collective contribution to the electrical conductivity, above the Peierls transition [23]. The change of ρ upon alloying is probably best illustrated by the study of the solid solution $NMP_xPhen_{1-x}TCNQ$. Phen is a neutral molecule which has no electron to give. Thus for a concentration x in NMP, there is only x electron to share between

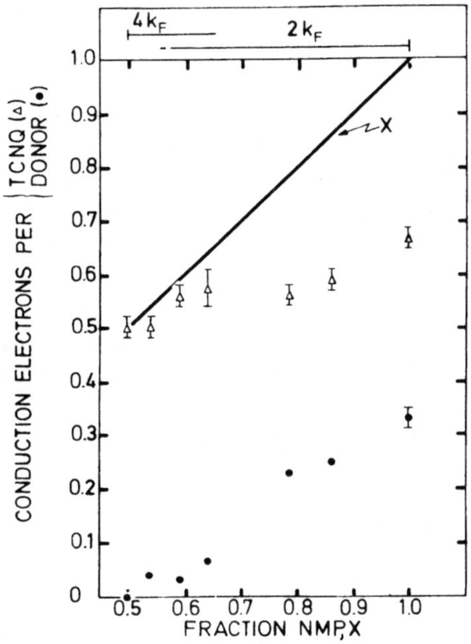

Fig. 19. Number of conduction electrons per TCNQ (Δ) (determined from the $2k_F$ and $4k_F$ wave vector values) and per donor molecules (●) (obtained by subtraction from the total number of electrons x) in $NMP_xPhen_{1-x}TCNQ$. (After Epstein *et al.* [24].)

donor (NMP, Phen) and acceptor (TCNQ) stacks. Figure 19 gives the variation of the charge per molecule for each stack as a function of x [24]. In particular, one sees that as x decreases there are fewer and fewer electrons on the donor stacks. Almost no conduction electron remains on the donor stacks for $x \lesssim \frac{2}{3}$.

3.1.3. *The* $4k_F$ *Instability*

Perhaps one of the most surprising results in the study of organic salts was (in addition to the $2k_F$ instability present in all the two-chain conductors) the dis-

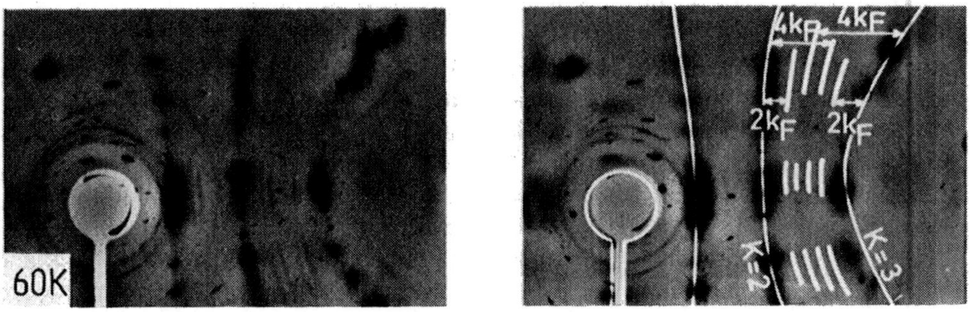

Fig. 20. X-ray monochromatic Laue photographs of TTF—TCNQ. At 60 K (a) one observes diffuse lines separated from the lines of Bragg reflections by $2k_F$ and $4k_F$ reduced wave vectors as schematized in (b).

covery of a $4k_F$ instability in some of these salts [25, 20, 21] (see Table I). In the X-ray pattern taken from TTF—TCNQ, and shown in Figure 20, the $4k_F$ instability gives rise to X-ray diffuse sheets comparable in intensity to that shown by the $2k_F$ instability. They are also related to an instability of the 1D electronic gas. Until now the $4k_F$ scattering has not been observed in inorganic 1D metals.

In order to give a simplified explanation of the $4k_F$ instability, let us consider once more the case of independent electrons. Figure 14(a) shows that with ρ electrons per repeat distance of the chain, the band is occupied from $-k_F$ to $+k_F$, where k_F, expressed in unit of the reciprocal repeat distance, is given by $\rho = 2 \times 2k_F$, with a factor 2 coming from the fact that each k state is occupied by two electrons of opposite spin. On the other hand, if the electrons experience very strong intrasite Coulomb repulsions, each site can be at most occupied by a single electron. We are thus faced with a problem of spinless fermions. Loosing the spin degree of freedom, the same number of electrons will now fill the 'conduction band' from $-2k_F$ to $+2k_F$, keeping unchanged the definition of k_F [26]. The Peierls instability of the strongly correlated electrons gas will now occur for the $4k_F$ wave vector. More sophisticated explanations of the $4k_F$ scattering can be found in reference [27].

Thus, the observation of a $4k_F$ instability shows that electron—electron interactions are very important in organic charge transfer salts and must be explicitly taken into account for a correct description of their electronic properties.

3.1.4. *Fluctuations above T_p*

(a) *Dynamics—Kohn anomaly.* The dynamical origin of the $2k_F$ X-ray diffuse scattering is the formation of a Kohn anomaly in the phonon spectrum at the same wave vectors (see Section 2). The best illustration is provided by the Krogmann salt $K_2Pt(CN)_4Br_{0.3}$, xH_2O which shows at room temperature a giant Kohn anomaly at $2\mathbf{k}_F = 0.3\mathbf{c}^*$ in the acoustic branch polarized along the chain direction [8, 28—30] (Figure 21). In TTF—TCNQ a very weak $2k_F$ Kohn anomaly is observed

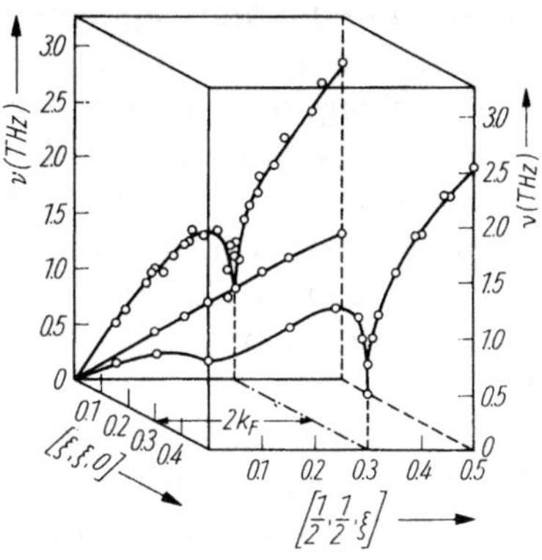

Fig. 21. Room temperature dispersion surface of KCP showing the giant Kohn anomaly in the acoustic branch polarized along the chain direction. (After Renker et al. [28]).

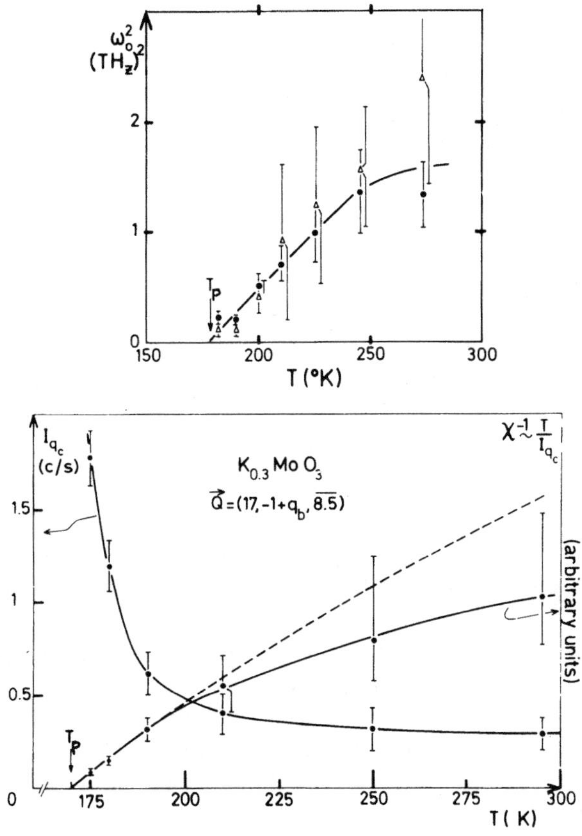

Fig. 22. Softening of the Kohn anomaly frequency with (a) temperature in $K_{0.30}MoO_3$ (after Pouget *et al.* [32]); (b) temperature dependence of I_{q_c} and of the inverse susceptibility χ^{-1}. χ^{-1} follows a Curie–Weiss law near T_c (after Pouget *et al.* [18]).

in the acoustic branch polarized in the transverse direction c^* [31]. The dynamical origin of the $4k_F$ scattering has not been found until now. In the blue bronze $K_{0.3}MoO_3$ the Kohn anomaly belongs to an optical and nearly nondispersive mode, which might be an internal mode of vibration of the cluster of MoO_6 octahedra [32]. It has been verified that the frequency of the bottom of the Kohn anomaly of $K_{0.3}MoO_3$, ω_{q_c}, softens as T_p is approached from above. More explicitly, Figure 22(a) shows that $\omega_{q_c}^2$ decreases roughly like $T - T_p$, as expected in the simplest theory of a phonon instability (see Section 1.4.1) [32].

(b) *Order parameter fluctuations.* This quantity can be obtained directly by X-ray diffuse scattering measurements. At the critical wave vector the diffuse intensity is proportional to the mean square of the order parameter. As shown in Section I, this quantity, divided by the thermal population factor, kT, in the high-temperature limit ($kT \gg \hbar\omega_c$ where ω_c is the characteristic frequency of the dynamics of the structural phase transition, here that of the Kohn anomaly), gives the susceptibility χ associated with the order parameter.

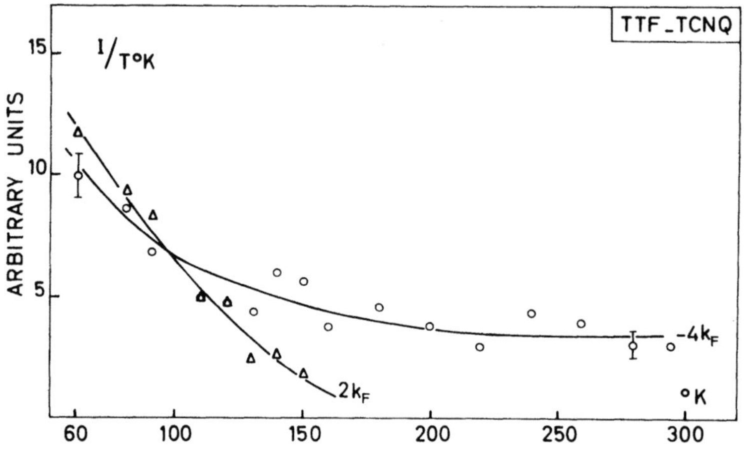

Fig. 23. Temperature dependence of the susceptibility χ_{1D} in TTF—TCNQ derived from the intensity of the $2k_F$ and $4k_F$ diffuse scatterings. These variations correspond roughly to power-law behaviors. Note the stronger divergence of $\chi_{1D}^{2k_F}$. (After Khanna *et al.* [21].)

Let us first give with TTF—TCNQ an example of fluctuations in the 1D regime ($T > T^*$). Figure 23 compares the temperature dependence of the susceptibility $\chi_{1D} \sim I_{1D}/T$ associated with the $2k_F$ and $4k_F$ 1D fluctuations [21]. The $4k_F$ instability is already present at room temperature while the $2k_F$ instability appears below 150 K. For decreasing temperatures, both susceptibilities diverge, following roughly a power law of the temperature:

$$\chi_{1D} \sim T^{-\alpha},$$

with $\alpha \sim 0.7$ for the $4k_F$ instability and $\alpha \sim 1.7$ for the $2k_F$ instability [33]. This power law behavior is predicted by theories including explicitly electron—electron interactions in the CDW fluctuations [34]. Figure 23 shows that the $2k_F$ susceptibility presents at low temperature a stronger divergence than the $4k_F$ susceptibility. This, and the establishment of lateral coupling between $2k_F$ fluctuations below $T^* \sim 60$ K, means that the 54 K Peierls transition of TTF—TCNQ is driven by the $2k_F$ instability. An experimental study of the dynamics of the 1D fluctuations accompanying the divergence of the $2k_F$ and $4k_F$ response functions has not yet been performed.

Let us now present with $K_{0.3}MoO_3$ an example of fluctuations in the temperature range where the chains are coupled ($T_p < T < T^*$). Figure 22(b) shows the strong increase of the maximum of the intensity of the $2k_F$ diffuse scattering I_{q_c}, as T_p is approached from above. Corrected by the thermal population factor, the $2k_F$ 3D susceptibility $I_{q_c}/T \sim \chi_{3D}$ diverges as the temperature decreases. More clearly, Figure 22(b) shows that in the vicinity of T_p, χ_{3D}^{-1} behaves like $T - T_p$ Curie Weiss law — see section 1.3) with, however, some deviation from this simple law at high temperature [18]. This temperature dependence recalls that of the square of the soft phonon frequency of the Kohn anomaly (Figure 22(a)). The fact that χ_{3D}^{-1} behaves like $\omega_{q_c}^2$ means that, except for the vicinity of T_p where a central peak in energy develops rapdily, the dynamical origin of the order parameter fluctuations lies in the softening of a phonon mode.

(c) *Correlation lengths.* This is mostly by the measurement of correlation lengths related to the increase of the HWHM of the X-ray diffuse scattering (corrected by the resolution) that the spatial anisotropy, and thus the regime of structural fluctuations, can be established.

Let us begin with the 1D regime of fluctuations ($T > T^*$) with TTF—TCNQ. Figure 24 gives the temperature dependence of the HWHM of the $2k_F$ and $4k_F$ scatterings in the chain direction for TTF—TCNQ and its alloy $TTF_{0.97}TSF_{0.03}TCNQ$ [21, 35]. In pure TTF—TCNQ the rate of increase of the HWHM of the $2k_F$ scattering above the experimental resolution ($\sim \xi_{1D}^{-1}$) increases with increasing temperature. Such a behavior for ξ_{1D}^{-1} is expected (see Figure 25) below the mean field temperature T_{MF}, of the 1D electronic gas, when 1D fluctuations are taken into account [36]. Typically, $\xi_{1D}^{2k_F} \sim 100$ Å near 60 K and $\xi_{1D}^{2k_F} \sim 5$ Å near 150 K for TTF—TCNQ. A weaker rate of increase of the HWHM of the $4k_F$ scattering, corresponding to weaker temperature-dependent $4k_F$ fluctuations, is observed in TTF—TCNQ; typically $\xi_{1D}^{4k_F} \sim 100$ Å near 60 K and $\xi_{1D}^{4k_F} \sim 10$ Å at room temperature. In the alloy $TTF_{0.97}TSF_{0.03}TCNQ$, the rate of increase of $(\xi_{1D}^{2k_F})^{-1}$ is not appreciably changed from its value in TTF—TCNQ, but now $(\xi_{1D}^{4k_F})^{-1}$ saturates at low temperature. Upon alloying, the in-chain extent of the $4k_F$ fluctuations is now limited to about $\xi_{1D} \sim 75 \pm 25$ Å, a distance comparable to the average distance 125 Å between substituted TSF molecules on the donor stack. The simplest interpretation of such a result is that the TCNQ and TTF stacks drive the $2k_F$ and $4k_F$ fluctuations, respectively.

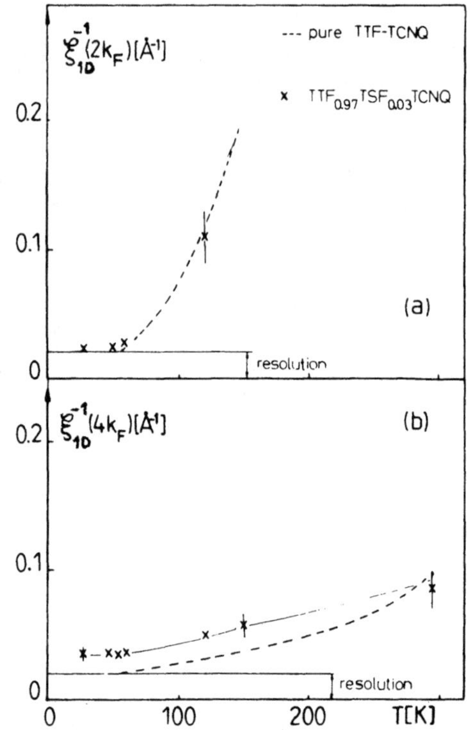

Fig. 24. Temperature dependence of the $2k_F$ and $4k_F$ inverse correlation lengths corrected from experimental resolution ξ_{1D}^{-1} (≡ HWHM) in TTF—TCNQ and $TTF_{0.97}TSF_{0.03}TCNQ$. The $2k_F$ fluctuations are more dependent on temperature (ξ_{1D}^{-1} increases more rapidly for $2k_F$). Although the $2k_F$ fluctuations are not affected by alloying, the correlation length of the $4k_F$ scattering is reduced. Thus, one can associate the $2k_F$ and $4k_F$ fluctuations to the TCNQ and TTF stacks respectively. (After Forro *et al.* [35].)

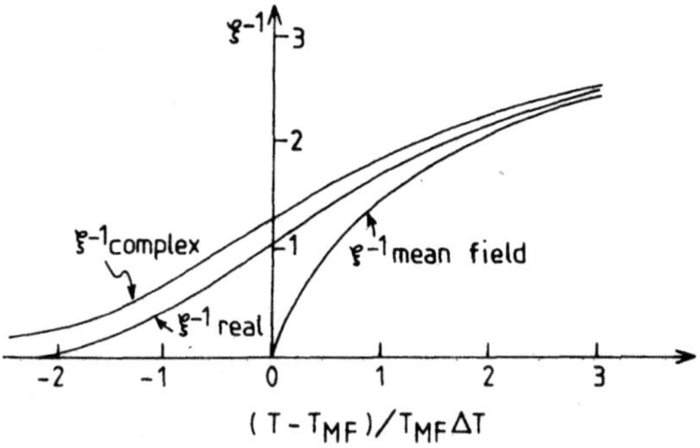

Fig. 25. Temperature dependence of the inverse correlation length ξ^{-1} calculated without fluctuations (mean-field) and for commensurate Peierls distortions ('real') and incommensurate Peierls distortions ('complex'), showing the effect of the fluctuations. (After Berlinsky, *Contemp. Phys.* **17**, 331 (1976).)

A somewhat different temperature dependence of the correlation lengths is observed in the regime where the $2k_F$ fluctuations are coupled between chains ($T_p < T < T^*$). This is illustrated by Figure 26 which gives, as a function of temperature, the HWHM of the diffuse scattering of $K_{0.3}MoO_3$ in the chain direction, the

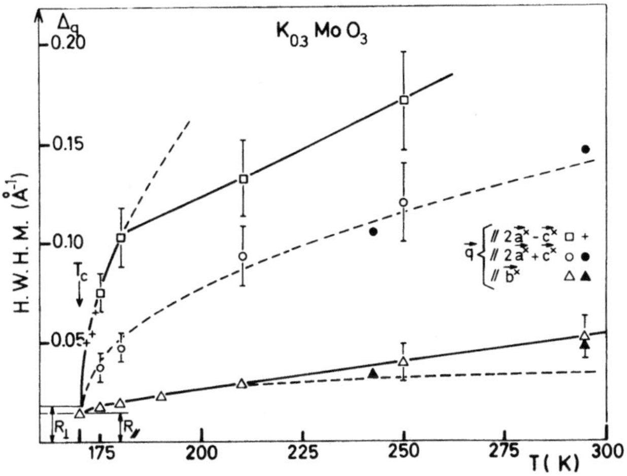

Fig. 26. Temperature dependence of the HWHM of the diffuse scattering in $K_{0.3}MoO_3$ for the chain direction b and two transverse directions. Mean-field behaviors are well followed ($\xi_{3D}^{-1} \propto \sqrt{T - T_p}$) except when the correlation length becomes too short. (After Pouget *et al.* [18].)

transverse direction of the layer of MoO_6 octahedra and the direction perpendicular to the layers [18]. The dashed line gives the $\xi_{3D}^{-1} \sim \sqrt{T - T_c}$ law characteristic of a mean field behaviour close to T_p (Equation (8) Sec. 1.4). This dependence is relatively well followed along the chain direction (with, however, some deviation at high temperature which can be explained in a more refined theory [18]) and in the transverse direction of the layer. In the direction perpendicular to the layers a strong deviation from this law occurs when the correlation length become comparable to the distance ~8 Å, between neighboring layers of MoO_6 octahedra. This indicates that structural fluctuations between layers become decoupled at high temperature.

(d) *Lattice modulation.* In a tight-binding description of the 1D electron gas, the $2k_F$ structural instability due to the electron—phonon coupling occurs for atomic displacements which modulate, to first order, the intrastack transfer integrals. Such is the case for the atomic displacements perpendicular to the $Pt(CN)_4$ complexes in $K_2Pt(CN)_4Br_{0.3}$, xH_2O (Figure 27(a)) [9]. In the case of organic charge transfer salts, where molecules are tilted along the stacking direction, a modulation of the transfer integrals can occur for translations of the molecules in the symmetry plane of the stack. In the case of TSF—TCNQ [37] and HMTSF—TCNQ [38] the analysis of the structure factor of the $2k_F$ diffuse scattering shows that translations of

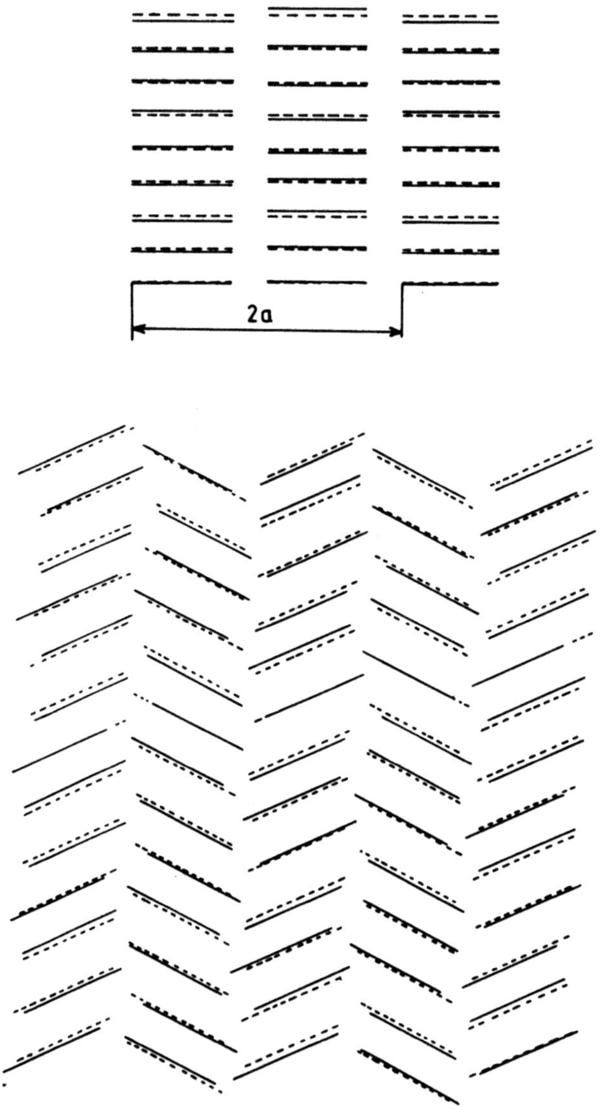

Fig. 27. Schematic pattern of the lattice modulations associated with the CDWs. (a) In $K_2Pt(CN)_4Br_{0.3}$, xH_2O the displacements are perpendicular to the $Pt(CN)_4$ complexes. (b) In TSF—TCNQ the displacements are mainly along the long axes of the molecules. (After Yamaji *et al.* [37].)

the donor molecules occur mainly along their long direction, as shown for the TSF stack in Figure 27(b).

3.1.5. *CDW Modulated Phases*

Below the Peierls transition, T_p, the $2k_F$ CDWs are long-range ordered. From the intensity of the satellite reflections one can get the amplitude of the structural modulation, a quantity which is proportional to the amplitude of the CDW or of the Peierls gap; each of these quantities can be chosen as the amplitude of the order parameter η_0. More precisely the temperature dependence of the satellite intensity generally behaves like the square of the order parameter. Figure 28(a) and (b), respectively, show such a dependence for the charge transfer salts

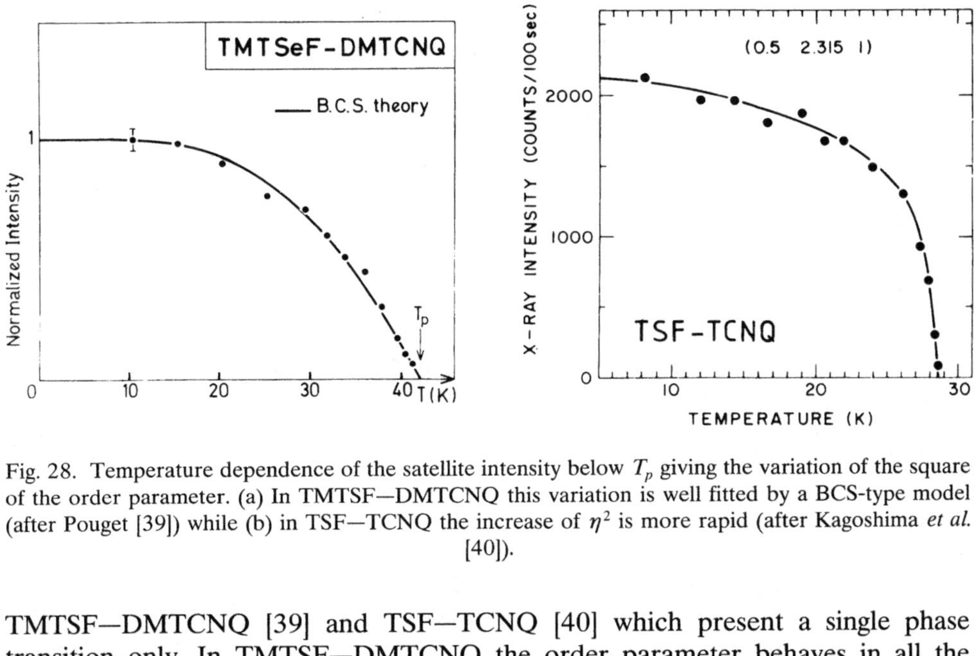

Fig. 28. Temperature dependence of the satellite intensity below T_p giving the variation of the square of the order parameter. (a) In TMTSF—DMTCNQ this variation is well fitted by a BCS-type model (after Pouget [39]) while (b) in TSF—TCNQ the increase of η^2 is more rapid (after Kagoshima *et al.* [40]).

TMTSF—DMTCNQ [39] and TSF—TCNQ [40] which present a single phase transition only. In TMTSF—DMTCNQ the order parameter behaves in all the temperature range as predicted by a mean-field theory of the Peierls transition (BCS-like theory) [41]. In TSF—TCNQ, $|\eta_0|^2$ increases with temperature more rapidly than predicted by the mean-field theory, which means that structural fluctuations have a dominant influence on the establishment of the 3D order. A mean-field behavior of the order parameter is also observed below 180 K in the blue bronzes [18, 42] and below the 144 K phase transition of $NbSe_3$ [12]. In particular, in the vicinity of T_p, the mean-field theory predicts that the satellite intensity grows like $(T_p - T)^{2\beta}$, with $\beta = \frac{1}{2}$ (see Section 1.3). In all these compounds the order parameter goes continuously to zero at T_p, showing the second-order nature of the Peierls transition.

The position of the satellite reflections, with respect to the main Bragg reflections of an average lattice, gives the wave vector of the structural distortion. The in-chain component is of course $2k_F$, a quantity related to the band filling (see Section 3.1.2). The transverse components give the phase shift from chain to chain for the $2k_F$ modulations if one assumes that the amplitude of the order parameter does not vary spatially (see Section 3.1.1).

Several kinds of coupling have been proposed to explain the transverse phase shift between CDWs and the two most important mechanisms were introduced above (Section 2.4).

Both mechanisms have been involved to explain the transverse wave vector components $\frac{1}{2}a^*$ observed along the tetragonal directions of $K_2Pt(CN)_4Br_{0.3}$, xH_2O at low temperature.

In the case where either several bands cross the Fermi surface or several CDW instabilities develop on different chains, the lateral ordering is more difficult to explain. we shall discuss in the following the case of organic charge transfer salts in

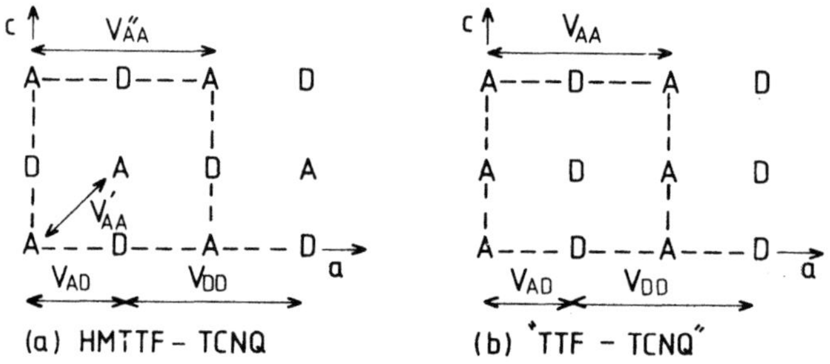

Fig. 29. Schematic pattern of the acceptor and donor stacks in (a) HMTTF—TCNQ and (b) TTF—TCNQ in the orthorhombic approximation, showing the different transverse couplings in these typical cases.

which each sublattice of donor (D) or acceptor (A) molecules tends to develop its own $2k_F$ instability. In the case of an orthorhombic lattice of chains (Figure 29), it has been shown [43, 44] that several types of ordering can occur, depending of the strength of the intersublattice (V_{AD}) and intrasublattice (V_{AA} and V_{DD}) couplings and of the temperature at which each sublattice becomes unstable. Let us examine three types of ordering, observed experimentally.

(a) For dominant intersublattice coupling, the two sublattices order at the same temperature, with the same temperature dependence of the order parameter on both stacks (Figure 30(a)) and a lateral order in opposition of phase between the D and A neighboring CDWs (giving $q_a = q_c = 0$ for the structures shown in Figure 29 with two stacks along the a and c directions per unit cell). This situation is observed in HMTSF—TCNQ, in TMTTF—TCNQ, and in TTF—TCNQ under high enough pressure.

(b) For intermediate values of the intersublattice and intrasublattice coupling, the two sublattices still order at the same temperature, but now an incommen-

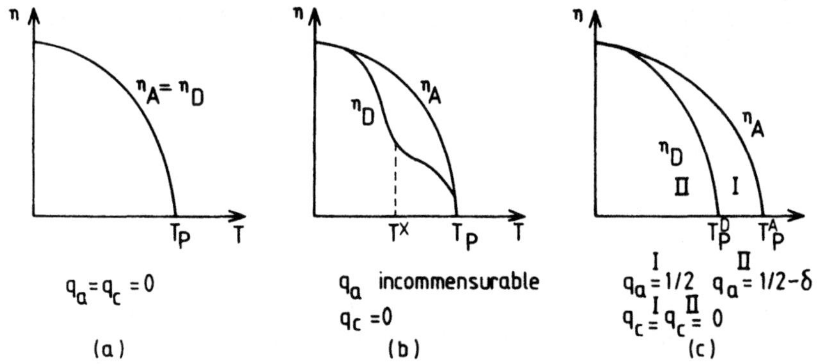

Fig. 30. Behavior of the order parameters of the donor and acceptor CDWs for different types of couplings: (a) strong intersublattice coupling; (b) intermediate inter- and intrasublattice coupling strengths; (c) weak intersublattice coupling.

surate transverse ordering between the *D* and *A* sublattices can be established. The situation shown in Figure 30(b) corresponds to the case of HMTTF—TCNQ, where a temperature independent and incommensurate modulation of wave vector component $q_a = 0.42$ is established below 50 K. This value is predominantly determined by the coupling between TCNQ chains in the diagonal and the *a* direc-

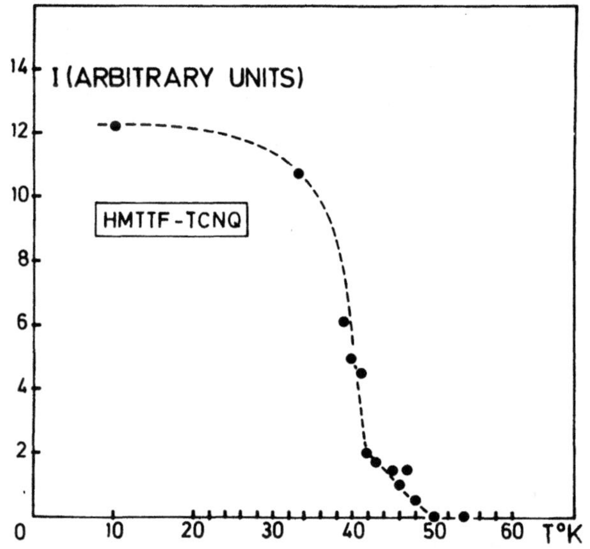

Fig. 31. Temperature dependence of the satellite intensity in HMTTF—TCNQ showing a sharp increase at 43 K. The HMTTF stacks dominate the intensity and this behavior illustrates the case of intermediate couplings. (See Figure 30(b)). (After Megtert *et al.* in *Molecular Metals* (ed. W. E. Hatfield), Plenum, New York, p. 87 (1979).

tions $\cos(\pi q_a) \simeq \frac{1}{2}(V'_{AA}/V''_{AA})$, (see Figure 29(a)) [44]. The different dependence in temperature of the order parameter η_A and η_D (Figure 30(b)) entering into the satellite intensity under the combination

$$I_s = |F_A\eta_A + F_D\eta_D|^2$$

(where F_A and F_D are, respectively, the structure factors of the acceptor and donor sublattice), gives the unusual temperature dependence shown in Figure 31. The sharp increase of the satellite intensity at about 43 K traduces the anomaly at T^* in the temperature dependence of the order parameter η_D on the HMTTF donor stacks, whose structure factor dominates the satellite intensity [45].

(c) For weak intersublattice coupling and sublattices with relatively well separated critical temperatures, the two sublattices can order separately (Figure 30(c)). This situation is basically that observed is TTF—TCNQ at ambient pressure. Below $T_p^A = 54$ K, the CDWs develop on the TCNQ sublattice in opposition of phase between neighboring stacks. This order keeps the lattice periodicity c ($q_c = 0$ because there are two TCNQ stacks per c unit) and doubles the lattice periodicity a ($q_a = \frac{1}{2}$). In the orthorhombic approximation, the CDW TTF sublattice experiences no interaction from the CDWs ordered in opposition of phase on the TCNQ sublattice. At $T_p^D = 49$ K the TTF sublattice begins to order itself, and tends to be in opposition of phase with the order of the TCNQ sublattice, which is itself in opposition of phase. This results in a linear coupling between the

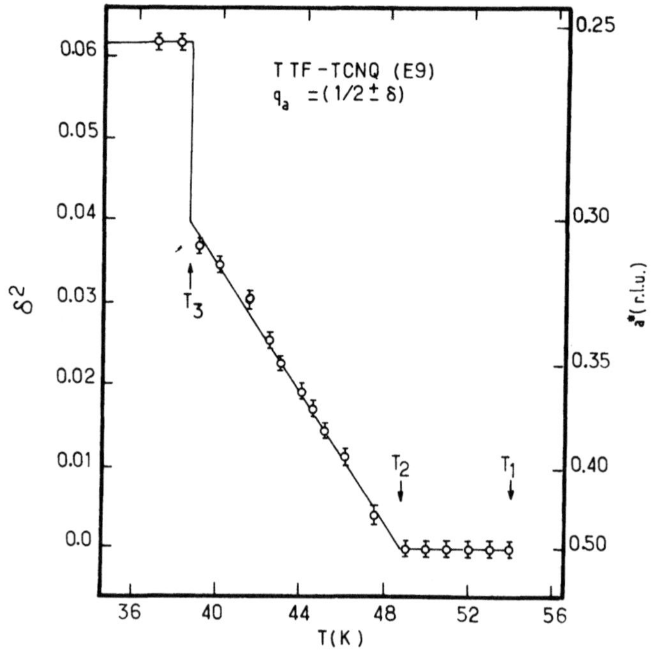

Fig. 32. Temperature dependence of q_a in TTF—TCNQ. δ^2 ($q_a = \frac{1}{2} - \delta$) is plotted as a function of T to illustrate the linear dependence of η_D^2 with T below 49 K. (After Ellenson *et al.* [48].)

order parameters η_A and η_D which induces a sliding of the wave vector q_a from the high-temperature value $q_a = \frac{1}{2}$. It is found [46, 47] that the shift is proportional to the amplitude of the TTF order parameter η_D; i.e.

$$q_a = \frac{1}{2} - \delta, \qquad \text{with } \delta \sim \frac{V_{AD}}{V_{AA}} \frac{\eta_D}{\eta_A},$$

where V_{AD} and V_{AA} are defined in Figure 29(b).

Figure 32 gives the observed q_a component as a function of temperature [48]. In particular, with η_A which has grown since 54 K and thus which does not change drastically in temperature below 49 K, and η_D which begins to grow at this temperature with the mean-field behavior $\eta_D \sim (T_p^D - T)^{1/2}$, one gets the observed temperature variation $\delta^2 \sim (T_p^D - T)$.

The description of the phase diagram of TTF—TCNQ that we have given is in fact oversimplified. To have a flavor of its richness, additional comments can be made:

(a) The real lattice is monoclinic, with an (**a**, **c**) angle which deviates substantially by about 30° from 90°. This means that between 49 and 54 K the resulting field due to the ordering of the TCNQ sublattice does not vanish by symmetry on the TTF sublattice [49]. But provided that the induced Peierls gap on the TTF stacks is less than kT in this temperature range, which is confirmed by the electronic properties, the above explanation of the 54 and 49 K phase transitions remains valid. The most important coupling, taking explicitly into account the monoclinic symmetry, are discussed in Reference [50].

(b) Figure 32 shows that TTF—TCNQ undergoes a first-order phase transition at 38 K, where q_a locks at the commensurate value $q_a = \frac{1}{4}$. Interpretations of this

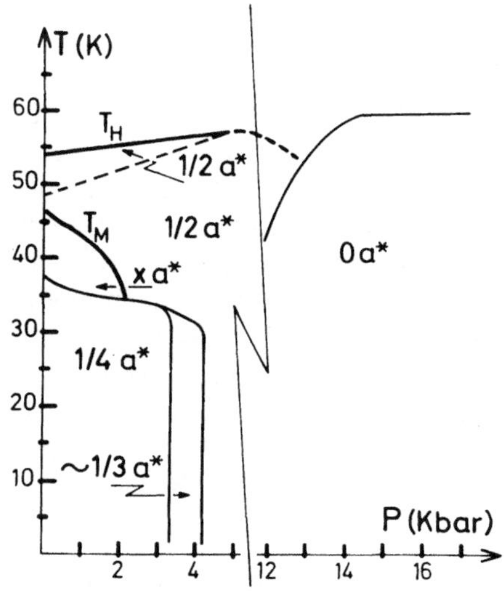

Fig. 33. Pressure—temperature phase diagram of TTF—TCNQ showing several domains which differ by the q_a values ($q_c = 0$). (After Megtert *et al.* [52].)

incommensurate—commensurate transition on the q_a component are beyond the scope of this review. They can be found in References [46, 47, 51].

(c) Figure 33 shows that the phase diagram of TTF—TCNQ changes drastically under pressure [52]. In particular, the two modulated phases observed below 49 and 38 K disappear for modest pressures. This figure shows also that at 10 K, as the pressure increases, q_a takes the following sequence of commensurate or nearly commensurate values $q_a = \frac{1}{4}$, $q_a \sim \frac{1}{3}$, $q_a = \frac{1}{2}$ and at high pressure $q_a = 0$. Pressure effects on the phase diagram of TTF—TCNQ are discussed in Reference [50].

(d) The above description of the phase diagram of TTF—TCNQ considers only the ordering between the $2k_F$ CDWs and ignores the $4k_F$ CDWs which seem to order below about 49 K in TTF—TCNQ. However, as this sequence of three phase transitions is observed at 50, 45, and 30 K in the alloy $TTF_{0.97}TSF_{0.03}TCNQ$, where there is no 3D ordering of the $4k_F$ CDWs [35, 53], it seems reasonable to ignore the coupling between $4k_F$ CDWs in a first interpretation of the phase diagram of TTF—TCNQ.

3.2. Anion Ordering Transitions in the $TMTS(T)F_2X$ Series

The previous section has been devoted to quasi-1D organic and inorganic conductors where the Peierls instability leads to charge density waves coupled to periodic lattice distortions. However, as was indicated in Section 2, in a 1D electron gas with interactions, there is a competition between charge density wave, spin density wave, and superconducting instabilities. In the compounds we are considering now, the charge density wave instability does not prevail over the two others and, furthermore, a new type of fluctuation is encountered, as will be shown below [54].

3.2.1. *Structure and Band Filling*

TMTSF (tetramethyltetraselenafulvalene) and TMTTF (the sulfur analogue) are quasi-planar donor molecules deriving from the TSF and TTF ones by addition of four methyl groups. They combine with monovalent anions X(PF_6^-, AsF_6^-, ClO_4^-, ReO_4^-, NO_3^-, SCN^-) to form triclinic arrangements of molecular zig-zag stacks extending along *a*, separated by (*a*, *b*) planes of anions [55, 56] (Figure 34). There is a complete transfer of one electron to each anion and, corresponding to the 2 : 1 stoichiometry, this creates 0.5 hole per donor molecules. With two molecules per period along the chain axis, the overlap of the π molecular orbitals gives rise to a 1D half-filled upper band, as shown in Figure 35. Thus the $2k_F$ wave vector is equal to $a^*/2$. There is also a significant overlap along the *b* direction, especially in $(TMTSF)_2X$ salts, due to Se—Se contacts [57].

3.2.2. *Peierls versus Magnetic and Superconducting Instabilities*

The competition between these different instabilities can be illustrated by the case of $(TMTSF)_2PF_6$. It was shown in Section 2 that Peierls (charge density waves)

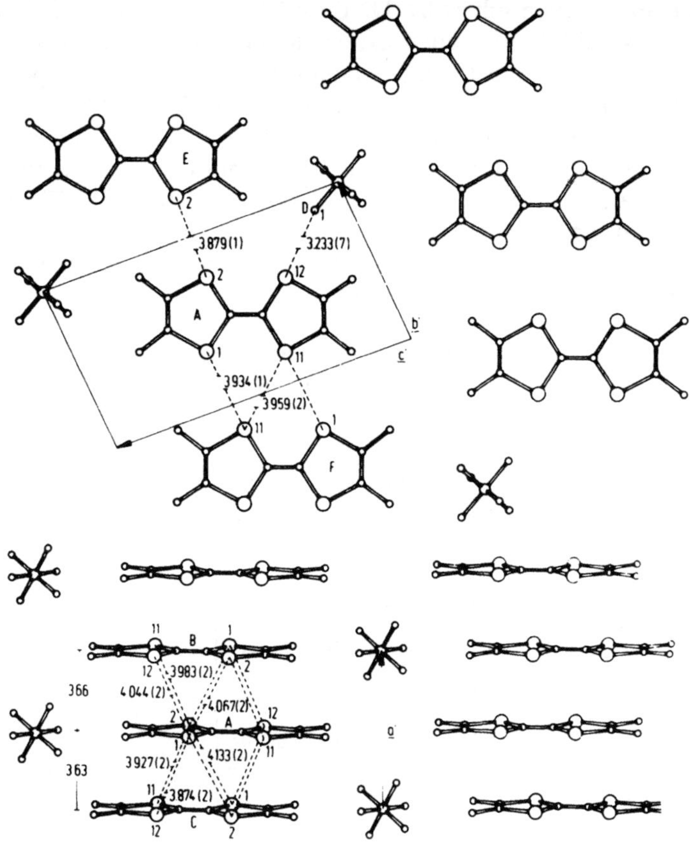

Fig. 34. Crystal structure of $(TMTSF)_2PF_6$ (a) view along the stacking axis a, (b) side view of the diagonal plane of the unit cell showing the molecular stacks. (a = 7.29 Å, b = 7.71 Å, c = 13.52 Å, α = 83.39°, β = 86.27°, γ = 71.01°, space group $P\overline{1}$). (After Thorup *et al.* [56].)

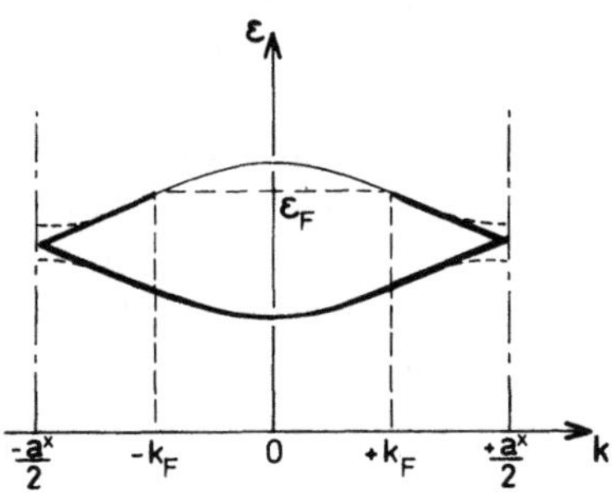

Fig. 35. One-dimensional band structure of the $(TMTSF)_2X$ and $(TMTTF)_2X$ compounds for two molecules per period along **a**. With 0.5 hole per molecule the upper band is half-filled and $2k_F$ = $a^*/2$. Note that a slight dimerization removes the degeneracy at the zone boundary, as shown by the dashed lines.

fluctuations manifest themselves by 1D diffuse scattering at the $2k_F$ wave vector. Therefore, one would for instance expect to observe, at reduced wave vectors $q = a^*/2$, X-ray diffuse sheets whose intensity grows as the temperature is lowered. Actually such effects are not observed in the $(TMTSF)_2X$ or $(TMTTF)_2X$ compounds. However, in the case of $(TMTSF)_2PF_6$ some very weak quasi-1D scattering has been detected below about 125 K, but instead of developing as T is reduced this scattering disappears below about 35 K [58] (Figure 36). This demonstrates that although Peierls fluctuations are potential in these one-chain compounds they do not diverge to induce charge density waves in contrast with two-chain organic or inorganic conductors.

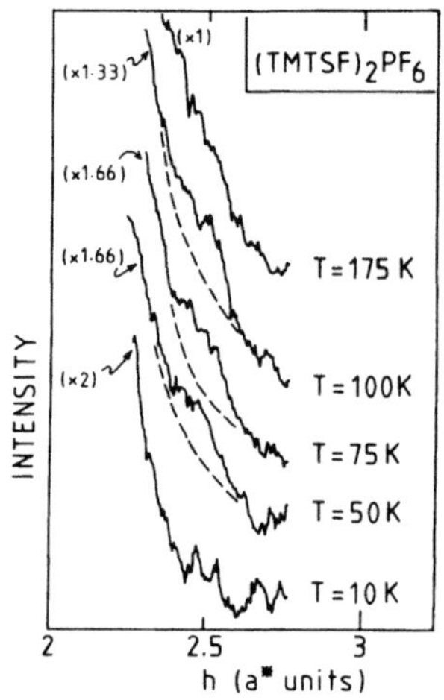

Fig. 36. Microdensitomer traces of stationary photographs in $(TMTSF)_2PF_6$ showing the very weak diffuse scattering at $2k_F = a^*/2$. This effect disappears on cooling below about 35 K. (After Pouget *et al.* [58].)

Instead of presenting a Peierls insulating state at low temperature, many of these compounds present magnetic or superconducting ground states. Thus, at ambient pressure, $(TMTSF)_2PF_6$ exhibits an antiferromagnetic (spin density wave) state below 11.5 K [59] (see Figure 37). The spin modulation period being twice that of the TMTSF stack, i.e. $2a$, opens gaps at $q = a^*/2$ on the Fermi surface and the compound is a magnetic insulator at low temperature. It should be noted that this spin density wave still remains to be characterized directly by neutron scattering studies, in spite of several attempts which failed to detect the corresponding magnetic satellites. Owing to the expected weakness of these satellites and to the unknown transverse wave vector components which may be incommensurate, this determination has become a real challenge.

Several of the $(TMTSF)_2X$ salts also present superconductivity ($X = PF_6^-$ (Figure 37) [60], AsF_6^-, SbF_6^-, TaF_6^-, ReO_4^-, . . .) under pressure (even at ambient pressure for $(TMTSF)_2ClO_4$ below $T = 1.2$ K) [61] and this is the first class of

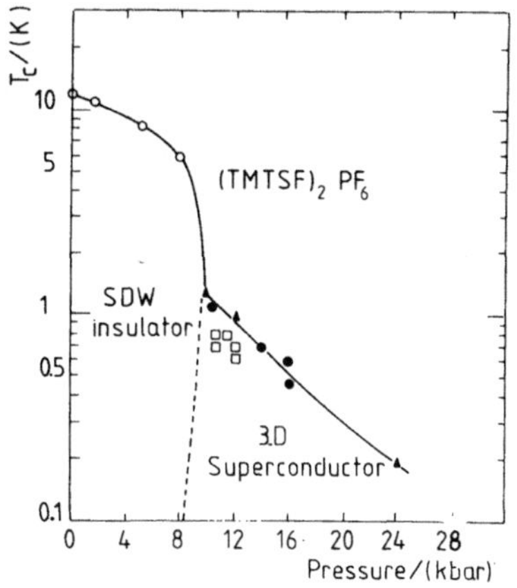

Fig. 37. Phase diagram of $(TMTSF)_2PF_6$ showing the conducting, SDW insulating and superconducting stability domains. (After Jerome and Schulz [54(c)].)

molecular compounds where superconductivity has been observed. Although there is, of course, no structural effect directly associated with the superconducting state, its occurrence is conditioned by an order—disorder transition involving the anions X, in the case of noncentrosymmetric ones.

3.2.3. *Anion Ordering Transitions*

(a) *General features.* At room temperature the anions are located at inversion centers in the $P\overline{1}$ space group of these structures. Centrosymmetric anions like octahedra (PF_6^-, AsF_6^-, . . .) exhibit in principle a unique orientation which minimizes their interactions with the neighboring molecules and anions. This is not possible for noncentrosymmetric anions; that is, when their symmetry is tetrahedral (ClO_4^-, ReO_4^-, BF_4^-, . . .), triangular (NO_3^-), or linear (SCN^-). In such cases the inversion symmetry is an average effect resulting from an orientational disorder between two (at least) different orientations related by inversion. This opens up the possibility for an order—disorder transition at low temperature. Actually such structural phase transitions have been identified in all the salts where the anions have no inversion center, leading to superstructures. The superstructures are formed by the arrangement of like or alternate orientations of the anions along the *a*, *b*, and *c* axes and four types have been found so far. The corresponding reduced wave vector **q** are listed in Table II, together with the transition temperatures and the order of the transitions (first- and second-order (or continuous) transitions are observed) [62—65].

Some general features of these transitions are worth mentioning:

(1) There is a correspondence between the anion symmetry and the type of superstructure (with the exception of $(TMTSF)_2ClO_4$, as discussed later).

TABLE II

		$(TMTSF)_2X$				$(TMTTF)_2X$			
Anion X	Geometry	T_c (K)	Order	q	Ref.	T_c (K)	Order	q	Ref.
ClO_4^-		24	2nd	0, ½, 0	71	70	1st	½, ½, ½	58
ReO_4^-		177	1st	½, ½, ½	67	160		½, ½, ½	64
BF_4^-	Tetrahedral	39		½, ½, ½	65—68	40		½, ½, ½ + uniden. satellites	62
BrO_4^-		250		½, ?, ?	a				
FSO_3^-	Tet. dipolar	87.5	1st	½, ½, ½	62				
NO_3^-	Triangular	41	2nd	½, 0, 0	62	50		½, 0, 0	62
SCN^-	Linear	90	2nd	0.48, 0.65, 0.1	b	160	2nd	0, ½, ½	c
$H_2F_3^-$		63		½, ½, ½	63				

[a] S. Tomic, J. P. Pouget, D. Jerome, K. Bechgaard, and J. M. Williams, *J. Physique* **44**, 375 (1983).
[b] R. Moret, J. P. Pouget, S. S. P. Parkin *et al.* (to be published).
[c] C. Coulon, A. Maaroufi, J. Amiell, E. Dupart, S. Flandrois, P. Delhaes, R. Moret, J. P. Pouget, and J. P. Morand, *Phys. Rev.* **B26**, 6322 (1982).

(2) The intensity of the superstructure reflections are, on average, only one order of magnitude weaker than that of the fundamental reflections and therefore two or three orders of magnitude stronger than that of CDW satellites in two-chain organic conductors (Figure 38) [66].
(3) Above the transition temperature, precursor effects in the form of broad and weak reflections are observed. They are nearly isotropic and show no 1D character [67].

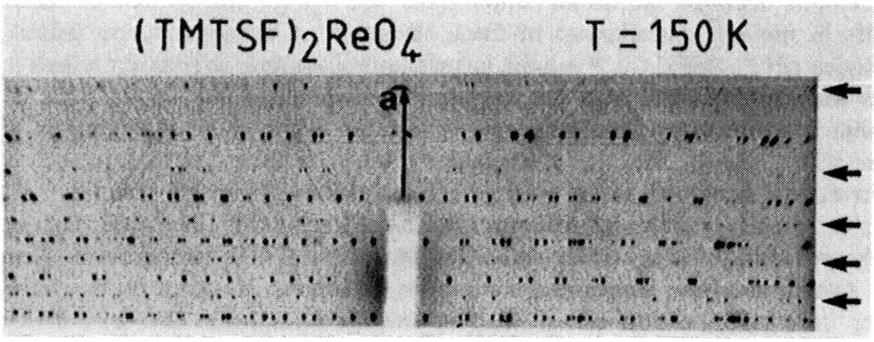

Fig. 38. X-ray oscillating crystal photograph of $(TMTSF)_2ReO_4$ at $T = 150$ K in the low-temperature ordered phase. The lines of superstructure reflections are easily visible (indicated by arrows) and their intensity is only about 1/10 of the fundamental reflections. (After Moret *et al.* [66].)

These last two properties show unambiguously that these transitions are not driven by instabilities of the 1D electron system alone, though the molecular stacks may be involved, as shown later.

The anion ordering transitions can be characterized by an Ising order parameter $\eta = P_+ - P_-$, where P_+ and P_- are the probabilities to have one of the two inverse orientations of, say a ReO_4^- tetrahedron. At high temperature $\eta \equiv 0$ since $P_+ = P_- = \frac{1}{2}$, while in the ordered phase the value of η varies from site to site between $+\eta$ and $-\eta$. This is expressed by:

$$\eta_j = \eta \, e^{i\mathbf{q}_s \cdot \mathbf{r}_j},$$

$\mathbf{q}_s$ being the wave vector of the superstructure and $\mathbf{r}_j$ the vector to site j.

The temperature dependence of η can be obtained from the intensity of the superstructure reflections since $I_{\text{sup. ref.}} \propto \eta^2$. In Figure 39 the intensity of such a reflection is plotted versus temperature for $(TMTTF)_2ReO_4$ which exhibits an anion ordering transition at 155 K [64]. Below this temperature, the curve

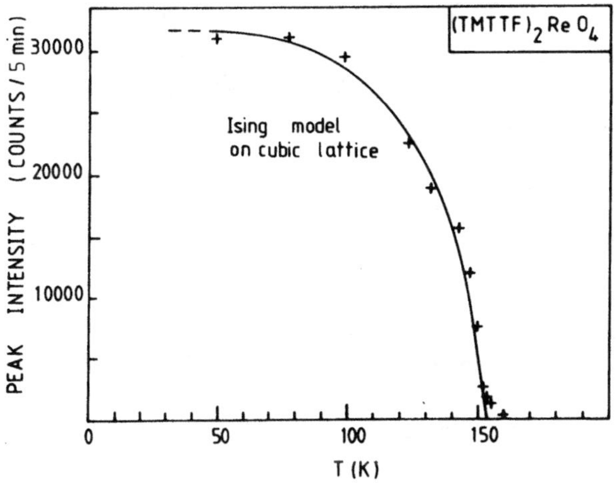

Fig. 39. Superstructure reflection intensity versus temperature in $(TMTTF)_2ReO_4$. The line is a fit of the data to the theoretical $\eta^2 = f(T)$ dependence calculated for the Ising model on a cubic lattice. (After Ravy [68].)

can be well fitted using the temperature dependence of the order parameter calculated for the Ising model on a simple cubic lattice [68]. Given the complexity and anisotropy of these compounds, the use of this model may be considered as a crude approximation although the quasi-isotropic character of the short-range order diffuse scattering observed above these order—disorder transitions show that the spatial fluctuations are not far from being isotropic.

The anion ordering transitions have also important consequences on the low-temperature properties and ground states of these materials. This can be understood simply if one notices that when the a^* component of the wave vector is equal to $\frac{1}{2}$ (doubling of the period along a) the anion potential can open a gap at the Fermi level thus inducing a metal-insulator transition. Actually this is the case for many of the transitions given in Table II. On the other hand, when q_{a^*} is equal to 0 there is no change of periodicity along a and the metallic character can be kept below the transition. Good examples of these two contrasted behaviors are provided by $(TMTSF)_2ReO_4$ and $(TMTSF)_2ClO_4$.

(b) *$(TMTSF)_2ReO_4$*. In $(TMTSF)_2ReO_4$ the anions order at 177 K with a wave vector $(\frac{1}{2}, \frac{1}{2}, \frac{1}{2})$ [67] via a first-order metal insulator transition which opens a large electrical gap (0.2 eV) [69]. The alternate orientation of the ReO_4^- tetrahedra along the three axes does not minimize direct electrostatic interaction and the

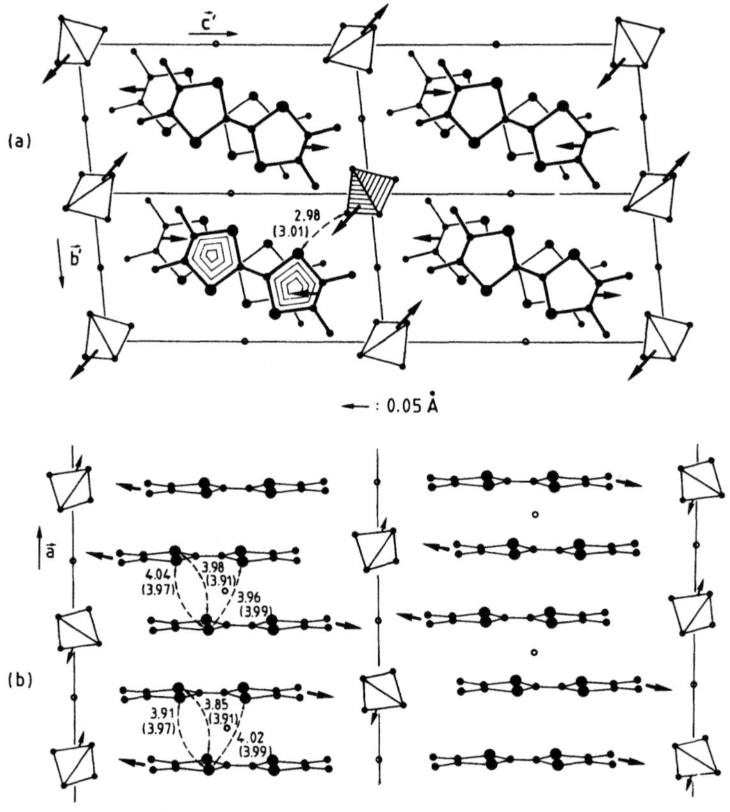

Displacements (Å)

	ReO_4	TMTSF
ΔX	0.03 (0.015)	-0.01 (0.015)
ΔY	0.06 (0.01)	0 (0.01)
ΔZ	-0.05 (0.01)	-0.055 (0.01)

Fig. 40. Displacement pattern in the low-temperature ordered phase of $(TMTSF)_2ReO_4$ (a) view along a; (b) side view of the stacks. The arrows indicate the direction of the displacements whose values are given in the table. Note the zig-zag shifts of the TMTSF molecules. All the TMTSF stacks are equivalent. (After Moret *et al.* [66].)

coupling between the anion sublattice and the quasi-1D electron gas is crucial. This coupling was clearly demonstrated by structural studies of the low-temperature phase. Figure 40 shows two projections of the structure revealing that, apart from ordering of the anions, the transition also implies significant displacements of the Re atoms and, most important, quite large distortions of the TMTSF stacks in the form of displacements oriented mainly along **c**. These distortions modulate the overlap between the molecular orbitals along the stacks with a period $2a$.

(c) *(TMTSF)*$_2$*ClO*$_4$. The anion-ordering transition of $(TMSTF)_2ClO_4$ occurs at a much lower temperature (24 K) and is second order [71]. The wave vector is $(0, \frac{1}{2}, 0)$ so that the ClO_4^- tetrahedra have the same orientation in a given (a, c) plane while the orientation alternates from plane in the b direction. The period along **a** being preserved, one does not expect a gap opening at T_c, which is verified since the conductivity increases even slightly faster below T_c [72], leading to a superconducting transition at 1.2 K and ambient pressure [61].

Structure determinations have shown that the ClO_4^- anion, being much smaller than the ReO_4^- one, the Se—O distance between anions and TMTSF molecules is larger in the perchlorate salt [73]. This and other considerations indicate that interactions between anions and organic stacks are weak in $(TMTSF)_2ClO_4$ and this is confirmed by negligible distortions of the TMTSF stacks in the low-temperature ordered phase (Figure 41) [74, 75]. Since the 1D electron gas does not seem

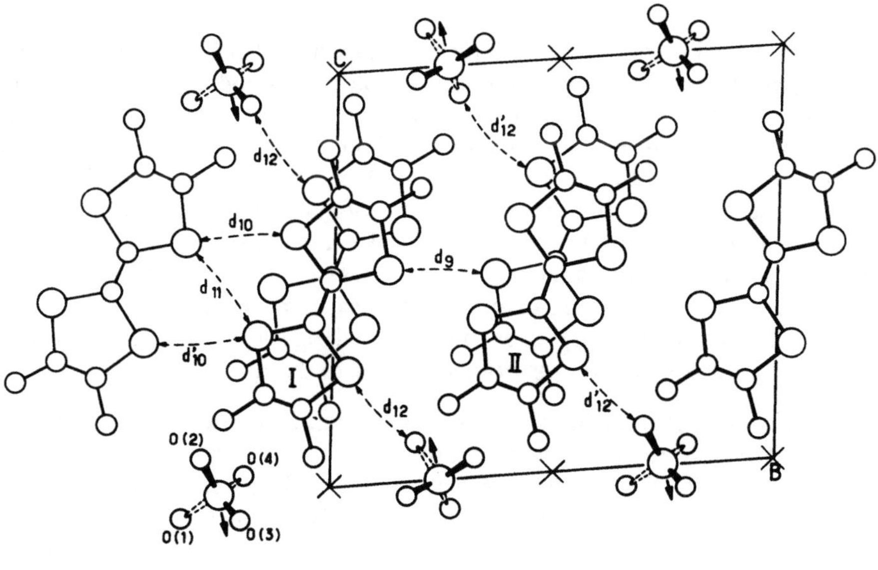

Fig. 41. Schematic projection of two unit cells of $(TMTSF)_2ClO_4$ in the low-temperature ($T = 11$ K) ordered phase. Only the Cl atoms undergo a significant shift from the inversion centers but no distortion of the organic sublattice is observed. Note that the two TMTSF stacks have different interactions with the surrounding anions. (After Gallois *et al.* [75].)

to play a key role in the stabilization of the ordered state, a possible minimization of the electrostatic interaction between the ClO_4^- tetrahedra can be considered. However, although the identical orientation taken in the *a* direction is undoubtedly favorable it is not clear whether the alternation of the orientation along the *b* direction is also favorable. The situation is even more complex in the *c* direction where the organic stacks necessarily intervene given the long distance between (*a*, *b*) planes of anions.

We also mention that the anion-ordering transition in $(TMTSF)_2ClO_4$ is highly dependent on the rate of cooling and that the anion disorder can be quenched [76, 4], resulting in the superconductivity being suppressed [77] at the expense of an antiferromagnetic ground state [78]. This is an interesting new phenomenon which opens up this field for kinetic studies. Such studies are not feasible in second-order Peierls transitions.

Finally, we point out that in the recently discovered $(BEDT\text{—}TTF)_nX_m$ materials, the first sulfur based compounds which display superconducting properties [79], new types of structural problems are posed.

For instance in $(BEDTTTF)_2ReO_4$ where the anions are already ordered at room temperature [79, 80], a metal-insulator transition occurs at 90 K with the appearance of weak superstructure reflections whose origin is still unclear [81].

4. Concluding Remarks

We have tried to illustrate how structural studies have been essential in the progress accomplished in the understanding of quasi-1D conductors. This is mainly due to the fact that they allow a direct characterization of the peculiar phase transitions which occur in these materials. Concerning Peierls transitions, they have been invaluable in the determination of charge transfers or in the analysis of the dimensionality of the fluctuations. For the organic charge transfer salts of the $(TMTSF)_2X$ and $(TMTTF)_2X$ series, structural studies have revealed a variety of order—disorder transitions which can be closely associated with the rich spectrum of properties and phase diagrams displayed by these compounds.

However, there are still large gaps in our knowledge of the structural properties of quasi-1D conductors. Let us just point out some of them:

(1) The low-temperature 3D structures of the charge density wave modulations are completely unknown in most cases.
(2) The dynamics of the Peierls transitions has only been studied in very few compounds due to the lack of large crystals (as needed for inelastic neutron experiments). With the recent interest for new inorganic materials like the blue bronzes ($K_{0.3}MoO_3$) and the chalcogenides (ITa_2Se_8, $I(NbSe_4)_{3.33}$ such studies become feasible.
(3) The influence of external parameters like pressure or electric field on the structural properties is still poorly documented. With the development of

experimental techniques (for instance, diamond-anvil pressure cells and the availability of intense X-ray sources) the situation should improve in the coming years.

This offers hopeful prospects for the efforts of people involved in the structural studies of these materials.

Acknowledgements

We are grateful to our colleagues and to visiting scientists in Orsay for their fruitful collaboration and especially to the members of the X-ray group P. A. Albouy, R. Comès, S. Megtert, A. H. Moudden, and S. Ravy. We are also much indebted to the many people and groups who have collaborated with us at various stages of our work on the 1D conductors.

References

1. More detailed information about phase transitions can be found in M. E. Stanley, *Introduction to Phase Transitions and Critical Phenomena*, Clarendon Press (1971); F. Gauthier, 'Introduction to the Study of Phase Transitions' in *Solids State Transformation in Metals and Alloys*, Aussois 78, Editions de Physique (1980); R. A. Cowley and A. D. Bruce, 'Structural Phase Transitions', in *Adv. in Physics* **29**, 1 (1980); and N. Boccara, *Symétries brisées*, Hermann, Paris (1976).
2. Some general review articles and books on the Peierls transition are: G. A. Toombs, 'Quasi-One-Dimensional Conductors', *Phys. Rep.* **C40**, 181 (1978); A. J. Berlinsky, 'One-Dimensional Metals and Charge Density Wave Effects in These Materials', *Rep. Prog. Phys.* **42**, 1243 (1979); J. T. Devreese and V. E. Van Doren (eds), *Highly Conducting One-Dimensional Solids*, Plenum, New York (1979); and J. Solyom, 'The Fermi Gas Model of One-Dimensional Conductors', *Adv. in Physics* **28**, 201 (1979).
3. R. Moret and J. P. Pouget (unpublished results).
4. J. P. Pouget, S. Kagoshima, C. Schlenker, and J. Marcus, *J. Physique Lett.* **44**, L113 (1983).
5. J. L. Hodeau, M. Marezio, C. Roucau, R. Ayroles, A. Meerschaut, J. Rouxel, and P. Monceau, *J. Phys. C: Solid State Phys.* **11**, 4117 (1978).
6. R. Comès, M. Lambert, H. Launois, and H. R. Zeller, *Phys. Rev.* **B8**, 571 (1973).
7. R. Comès, M. Lambert, and H. R. Zeller, *Phys. Stat. Sol.* **58**, 587 (1973).
8. B. Renker, L. Pintschovius, W. Gläser, H. Rietschel, R. Comès, L. Liebert, and W. Drexel, *Phys. Rev. Lett.* **32**, 836 (1974).
9. J. W. Lynn, M. Izumi, G. Shirane, S. A. Werner, and R. B. Saillant, *Phys. Rev.* **B12**, 1154 (1975).
10. D. W. Bullet, *J. Phys. C*, **12**, 277 (1979).
11. R. Hoffmann, S. Shaik, J. C. Scott, M. H. Whangbo, and M. J. Foshee, *J. Solid State Chem.* **34**, 263 (1980).
12. R. M. Fleming, D. E. Moncton, and D. B. McWhan, *Phys. Rev.* **B18**, 5560 (1978).
13. F. Devreux, *J. Physique* **43**, 1489 (1982).
14. S. J. Hillenius, R. V. Coleman, R. M. Fleming, and R. J. Cava, *Phys. Rev.* **B23**, 1567 (1981).

15. R. Moret, J. P. Pouget, A. Meerschaut, and L. Guemas, *J. Physique Lett.* **44**, L93 (1983).
16. J. P. Pouget, R. Moret, A. Meerschaut, L. Guemas and J. Rouxel, *J. Physique* **44**, C3-1729 (1983).
17. M. H. Whangbo and L. F. Schneemeyer (to be published).
18. J. P. Pouget, C. Noguera, A. H. Moudden, and R. Moret (J. Physique, October 1985).
19. J. P. Pouget, S. Megtert, R. Comès, and A. J. Epstein, *Phys. Rev.* **B21**, 486 (1980).
20. S. Kagoshima, T. Ishiguro, and H. Anzai, *J. Phys. Soc. Japan* **41**, 2061 (1976).
21. S. K. Khanna, J. P. Pouget, R. Comes, A. F. Garito, and A. J. Heeger, *Phys. Rev.* **B16**, 1468 (1977).
22. S. Megtert, R. Comès, C. Vettier, R. Pynn, and A. F. Garito, *Solid State Commun.* **37**, 875 (1981)
23. R. H. Friend, M. Miljak, and D. Jerome, *Phys. Rev. Lett.* **40**, 1048 (1978); A. Andrieux, H. J. Schulz, D. Jerome and K. Bechgaard, *Phys. Rev. Lett.* **43**, 227 (1979).
24. A. J. Epstein, J. S. Miller, J. P. Pouget, and R. Comès, *Phys. Rev. Lett.* **47**, 741 (1981).
25. J. P. Pouget, S. K. Khanna, F. Denoyer, R. Comès, A. F. Garito, and A. J. Heeger, *Phys. Rev. Lett.* **37**, 437 (1976).
26. J. Bernasconi, M. J. Rice, W. R. Schneider, and S. Strassler, *Phys. Rev.* **B12**, 1090 (1975); A. A. Ovchinnikov, *Sov. Phys. — JETP* **37**, 176 (1973).
27. V. J. Emery, *Phys. Rev. Lett.* **37**, 107 (1976); J. B. Torrance, *Phys. Rev.* **B17**, 3099 (1978); J. Kondo and K. Yamaji, *J. Phys. Soc. Japan* **43**, 424 (1977); J. Hubbard, *Phys. Rev.* **B17**, 494 (1978).
28. B. Renker, H. Rietschel, L. Pintschovius, W. Gläser, P. Bruesch, D. Kuse, and M. J. Rice, *Phys. Rev. Lett.* **30**, 1144 (1973).
29. R. Comès, B. Renker, L. Pintschovius, R. Currat, W. Gläser, and G. Scheiber, *Phys. Stat. Sol.* **71**, 171 (1975).
30. K. Carneiro, G. Shirane, S. A. Werner, and S. Kaiser, *Phys. Rev.* **B13**, 4258 (1978).
31. G. Shirane, S. M. Shapiro, R. Comès,, A. F. Garito, and A. J. Heeger, *Phys. Rev.* **B14**, 2325 (1976).
32. J. P. Pouget, C. Escribe-Filippini, B. Hennion, R. Currat, A. H. Moudden, R. Moret, J. Marcus, and C. Schlenker, *Proceeding ICSM 84 Mol. Cryst. Liq. Cryst.* **121**, 111 (1985).
33. S. Megtert and J. P. Pouget (in preparation).
34. J. Solyom, *Adv. in Physics* **28**, 201 (1979); S. Barišić and A. Bjeliš (to be published in *Electronic properties of Materials with Quasi-1D Structures* (ed. H. Kamimura), D. Reidel, Dordrecht).
35. L. Forro, S. Bouffard, and J. P. Pouget, *J. Physique Lett.* **45**, L543 (1984).
36. D. J. Scalapino, M. Sears, and R. A. Ferrell, *Phys. Rev.* **B6**, 3409 (1972).
37. K. Yamaji, S. Megtert, and R. Comès, *J. Physique* **42**, 1327 (1981).
38. K. Yamaji, J. P. Pouget, R. Comès, and K. Bechgaard, *J. Physique* **44**, C3-1321 (1983).
39. J. P. Pouget (unpublished results).
40. S. Kagoshima, T. Ishiguro, T. D. Schultz, and Y. Tomkiewicz, *Solid State Commun.* **28**, 485 (1978).
41. M. J. Rice and S. Strassler, *Solid State Commun.* **13**, 125 (1973).
42. R. M. Fleming, L. F. Schneemeyer, and D. E. Moncton, *Phys. Rev.* **B31**, 899 (1985).
43. A. Bjeliš and S. Barišić, *Lecture Notes in Physics* **65**, 291 (1977).
44. A. Bjeliš and S. Barišić, *J. Physique Lett.* **39**, L347 (1978).
45. S. Megtert, J. P. Pouget, R. Comès, A. F. Garito, K. Bechgaard, J. M. Fabre, and L. Giral, *J. Physique Lett.* **39**, L118 (1978); J. P. Pouget, S. Megtert, and R. Comès, *Lecture Notes in Physics* **95**, 14 (1978).
46. P. Bak and V. J. Emery, *Phys. Rev. Lett.* **36**, 978 (1976).
47. P. Bak, *Phys. Rev. Lett.* **37**, 1071 (1976); S. Barišić, and A. Bjeliš, *J. Physique* **38**, C3-254 (1977).
48. W. D. Ellenson, R. Comès, S. M. Shapiro, G. Shirane, A. F. Garito, and A. J. Heeger, *Solid State Commun.* **20**, 53 (1976); W. D. Ellenson, S. M. Shapiro, G. Shirane, and A. F. Garito, *Phys. Rev.* **B16**, 3244 (1977).

49. T. D. Schultz and R. A. Craven, *Highly Conducting One-Dimensional Solids*, Plenum, New York, p. 147 (1979); E. Abrahams, J. Solyom, and F. Woynarovich, *Phys. Rev.* **B16**, 5238 (1977).
50. S. Megtert, A. Bjeliš, J. Przystawa, and S. Barišić, *Phys. Rev.* (submitted).
51. A. Bjeliš and S. Barišić, *Phys. Rev. Lett.* **37**, 1517 (1976).
52. S. Megtert, R. Comès, C. Vettier, R. Pynn, and A. F. Garito, *Mol. Cryst. Liq. Cryst.* **85**, 159 (1982).
53. S. Kagoshima, H. Anzai, T. Ishiguro, E. M. Engler, T. D. Schultz, and Y. Tomkiewicz, *Lattice Dynamics*, Flamarion Sciences, Paris, p. 591 (1978).
54. General references concerning the series of $(TMTSF)_2X$ and $(TMTTF)_2X$ conductors can be found in reviews and conference proceedings: (a) K. Bechgaard and D. Jerome, 'Organic Superconductors', *Scient. American*, p. 52 (July 1982); (b) J. Friedel and D. Jerome, 'Organic Superconductors: The $(TMTSF)_2X$ Family', *Contemp. Phys.* **23**, 583 (1982); (c) D. Jerome and H. J. Schulz, 'Organic Conductors and Superconductors', *Adv. in Physics* **31**, 299 (1982); (d) *Colloque International sur la physique et la chimie des métaux synthétiques et organiques*, Les Arcs, France, Déc. 1982: *J. Physique*, Colloque C3, **44** (June 1983); (e) *International Conference on Physics and Chemistry of Low-Dimensional Synthetic Metals*, Abano Terme, Italy, June 1984 (to be published in *Mol. Cryst. Liq. Cryst.*).
55. J. L. Galigne, B. Liautard, S. Peytavin, G. Brun, J. M. Fabre, E. Toreilles, and L. Giral, *Acta Cryst.* **B34**, 620 (1978).
56. N. Thorup, G. Rindorf, H. Soling, and K. Bechgaard, *Acta Cryst.* **B37**, 1236 (1981).
57. M. A. Beno, J. M. Williams, M. M. Lee, and D. O. Cowan, *Solid State Commun.* **44**, 1195 (1982).
58. J. P. Pouget, R. Moret, R. Comès, K. Bechgaard, J. M. Fabre, and L. Giral, *Mol. Cryst. Liq. Cryst.* **79**, 129 (1982).
59. H. J. Pedersen, J. C. Scott, and K. Bechgaard, *Solid State Commun.* **35**, 207 (1980); A. Andrieux, D. Jerome, and K. Bechgaard, *J. Physique Lett.* **42**, L87 (1981).
60. D. Jerome, A. Mazaud, M. Ribault, and K. Bechgaard, *J. Physique Lett.* **41**, L95 (1980).
61. K. Bechgaard, K. Carneiro, M. Olsen, and F. B. Rasmussen, *Phys. Rev. Lett.* **46**, 852 (1981).
62. R. Moret, J. P. Pouget, R. Comès, and K. Bechgaard, *J. Physique* **44**, C3-957 (1983).
63. K. Mortensen, C. S. Jacobsen, A. Lindegaard-Andersen, and K. Bechgaard. *J. Physique* **44**, C3-963 (1983).
64. S. S. P. Parkin, J. J. Mayerle, and E. M. Engler, *J. Physique* **44**, C3-1105 (1983).
65. P. C. W. Leung, A. J. Schultz, H. H. Wang, T. J. Emge, G. A. Ball, D. D. Cox, and J. M. Williams, *Phys. Rev.* **B30**, 1615 (1984).
66. R. Moret, J. P. Pouget, R. Comès, and K. Bechgaard, *Phys. Rev. Lett.* **49**, 1008 (1982).
67. J. P. Pouget, R. Moret, R. Comès, and K. Bechgaard, *J. Physique Lett.* **42**, 543 (1981).
68. S. Ravy, DEA Report. Orsay (1984) (unpublished).
69. C. S. Jacobsen, H. J. Pedersen, K. Mortensen, G. Rindorf, N. Thorup, J. B. Torrance, and K. Bechgaard, *J. Phys. C* **15**, 2651 (1982).
70. G. Rindorf, H. Soling, and N. Thorup, *Acta Cryst.* **C40**, 1137 (1984).
71. J. P. Pouget, G. Shirane, K. Bechgaard and J. M. Fabre, *Phys. Rev.* **B27**, 5203 (1983).
72. D. U. Gubser, W. W. Fuller, T. O. Poehler, J. Stokes, D. O. Cowan, M. Lee, and A. N. Bloch, *Mol. Cryst. Liq. Cryst.* **79**, 225 (1982).
73. G. Rindorf, H. Soling, and N. Thorup, *Acta Cryst.* **B38**, 2805 (1982).
74. R. Moret, J. P. Pouget, R. Comès, and K. Bechgaard, *J. Physique* **16**, 1521 (1985).
75. B. Gallois, A. Meresse, J. Gaultier, and R. Moret, *Mol. Cryst. Liq. Cryst.* (submitted).
76. S. Kagoshima, T. Yasunaga, T. Ishiguro, M. Anzai, and G. Saito, *Solid State Commun.* **46**, 867 (1983).
77. P. Garoche, R. Brusetti, and K. Bechgaard, *Phys. Rev. Lett.* **49**, 1346 (1982).
78. T. Takahashi, D. Jerome, and K. Bechgaard, *J. Physique Lett.* **43**, L565 (1982).
79. S. S. P. Parkin, E. M. Engler, R. R. Schumaker, R. Lagier, V. Y. Lee, J. C. Scott, and R. L. Greene, *Phys. Rev. Lett.* **50**, 270 (1983).
80. J. M. Williams, M. A. Beno, H. Wang, P. E. Reed, L. J. Azevedo, and J. E. Schirber, *Inorg.*

Chem. **23**, 1790 (1984).
81. S. Ravy, S. Megtert, R. Moret, J. P. Pouget, and R. Comès, *Proc. American Chemical Society Symp. Hawaii Mol. Cryst. Liq. Cryst.* **125**, 91 (1985).
82. C. Noguera, *J. Phys. C* **18**, 1647 (1985).

CHARGE DENSITY WAVE INSTABILITY AND NONLINEAR TRANSPORT IN THE QUASI-ONE-DIMENSIONAL BLUE BRONZES $K_{0.30}MoO_3$ AND $Rb_{0.30}MoO_3$

CLAIRE SCHLENKER* and JEAN DUMAS

*Laboratoire d'Etudes des Propriétés Electroniques des Solides,** CNRS, BP 166, 38042, Grenoble Cédex, France*

1. Introduction

The so-called blue bronzes are ternary molybdenum oxides of formula $A_{0.30}MoO_3$ where the alkali metal A can be K or Rb. The potassium blue bronze was synthesized more than twenty years ago by Wold *et al.* [1]. While the crystal structure was found by Graham and Wadsley to be a layered type structure [2], physical studies by Bouchard *et al.* [3] established that the blue bronze shows a semiconductor-to-metal transition in the vicinity of 180 K. Later, Perloff *et al.* noticed a large anisotropy of the electrical conductivity in the plane of the layers [4]. Detailed studies of the transport properties were then performed by Fogle and Perlstein [5] (Figure 1). They especially measured the low-temperature behavior of the conductivity σ and reported a non-ohmic behavior at $T < 20$ K, when the sample is insulating ($\sigma \sim 10^{-14}\ \Omega^{-1}\ \mathrm{cm}^{-1}$). They proposed a model of an excitonic insulator for the semiconducting state.

It is only recently that the anisotropy of the conductivity was rediscovered and studied in greater details by Brusetti *et al.* [6]. Optical reflectivity measurements showed indeed that $K_{0.30}MoO_3$ is a quasi-one-dimensional (quasi-1D) metal in the high-temperature phase [7]. Later, X-ray diffuse scattering studies by Pouget *et al.* [8] led to the conclusion that the metal-to-semiconductor transition is a Peierls transition towards an incommensurate charge density wave (CDW) state. At the same time, in a search for non-ohmic conductivity in the temperature range where the conductivity is not vanishingly small ($T > 50$ K), Dumas *et al.* [9] establish that the blue bronze shows nonlinear transport, very likely due to the sliding of the CDW.

The blue bronzes belong to the large family of the oxide bronzes of general formula A_xTO_m where A is an alkali metal and T a transition metal. The designation of bronze has been introduced to describe the metallic luster of some of these compounds, such as the sodium tungsten bronze Na_xWO_3. The properties of the tungsten and vanadium bronzes had been reviewed by Sienko [10]. A more exten-

* This review was partly written during a sabbatical stay of four months of one of us (C.S.) at the IBM San José Research Laboratory, and thanks are due to I.B.M. for its financial support.

** Laboratoire associé à l'Université Scientifique et Médicale de Grenoble.

J. Rouxel (ed.), Crystal Chemistry and Properties of Materials with Quasi-One-Dimensional Structures, 135—177.

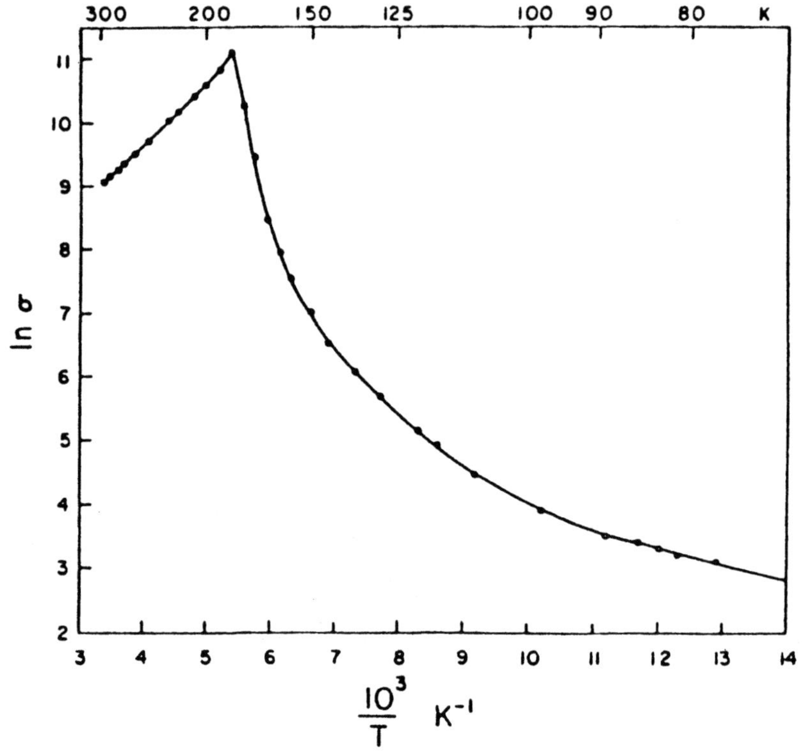

Fig. 1. Log of the conductivity versus inverse temperature for $K_{0.30}MoO_3$ (current along the **b** axis). (After Reference [5].)

sive review, including the molybdenum bronzes, was made by Hagenmuller [11]. In these compounds, the alkaline metal donates its outer electron to the conduction band, which is often empty in the oxide TO_m. While the tungsten bronzes, at all temperatures, are metals, the vanadium bronzes, such as $Na_xV_2O_5$-β, rather show a semiconducting behaivor. The molybdenum bronzes including $K_{0.30}MoO_3$ exhibiting a metal-to-semiconductor transition, seem therefore to be intermediate, as far as the $4d$ electron localization and the width of the conduction band are concerned, between the W and V bronzes. But, while the vanadium, and especially the tungsten, bronzes have been extensively studied in the last two decades, it is only very recently that a considerable effort has been concentrated on the molybdenum bronzes in relation with the low-dimensional and nonlinear properties. This article will review only the blue bronzes, although other molybdenum bronzes such as the so-called purple bronze $A_{0.9}Mo_6O_{17}$ [12], closely related to the molybdenum oxides η-Mo_4O_{11} [13] and γ-Mo_4O_{11} [14] is also a quasi-two-dimensional (quasi-2D) metal.

The properties of all these compounds are mostly determined by anisotropic Fermi surfaces with nesting properties leading to CDW instabilities. These instabilities are now comparatively well understood in the layered transition metal dichalcogenides [15]. The properties of the quasi-1D systems had been reviewed, for example by, Toombs [16].

The possible sliding of incommensurate CDW had been predicted long ago by Fröhlich and is known as the Fröhlich mechanism [17]. Nonlinear conductivity attributed to this mechanism was first reported for niobium triselenide in 1976 by Monceau *et al.* [18]. Since then, a considerable amount of work has been devoted to this property and to the related phenomena in $NbSe_3$ and TaS_3 [19]. In the CDW state of these compounds, the conductivity was found to be non-ohmic above a sharp threshold electric field E_t. In the nonlinear regime, an anomalously large noise voltage including well-defined frequencies was also found [19]. The origin of these frequencies, in all cases proportional to the excess CDW current, is not yet clear [20]. In addition to the trichalcogenides, some transition metal tetrachalcogenides show similar properties [23], but the blue bronze belongs to a different class of materials. This, together with the possibility of growing comparatively large single crystals, has now induced numerous studies (see, for example, References [20]—[22]).

2. Structural Aspects

Single crystals of $K_{0.30}MoO_3$ are grown by the electrolytic reduction of a K_2MoO_4—MoO_3 melt at 550 °C. The details of the process are reported in Reference [1]. The same technique has been used for the preparation of the rubidium blue bronze, first as a powder [24] and, later, as single crystals [25]. The as-grown crystals are usually platelets, elongated along the monoclinic *b* axis, of typical size 5 × 2 × 1 mm³. They can easily be cleaved parallel to the platelet plane. The chemical formula obtained by Graham and Wadsley from chemical analysis was

TABLE I
Lattice parameters obtained by Ghedira *et al.* [26] in the I body centered unit cell.

	a	*b*	*c*	*β*
$K_{0.30}MoO_3$	16.2311(7)	7.5502(4)	9.8614(4)	94.895(4)
$Rb_{0.30}MoO_3$	16.361(3)	7.555(1)	10.094(2)	93.87(5)

$K_{0.28}MoO_3$. However, the crystal refinement recently performed by Ghedira *et al.* is consistent with the formula $K_{0.30}MoO_3$ [26].

The crystal strucrture of $K_{0.30}MoO_3$ has been found by X-ray refinement to be monoclinic, space group $C2/m$ with 20 formulae per unit cell [2]. $Rb_{0.30}MoO_3$ has the same structure with very similar lattice parameters, as shown in Table I [26]. The structure is built with clusters of ten distorted MoO_6 octahedra. These clusters are linked together via corner sharing along the *b* axis and the [102] direction (C-centered unit cell) and therefore form infinite slabs (Figure 2). The alkali atoms lie between the slabs (Figure 3). The crystal structure is primarily a layered type

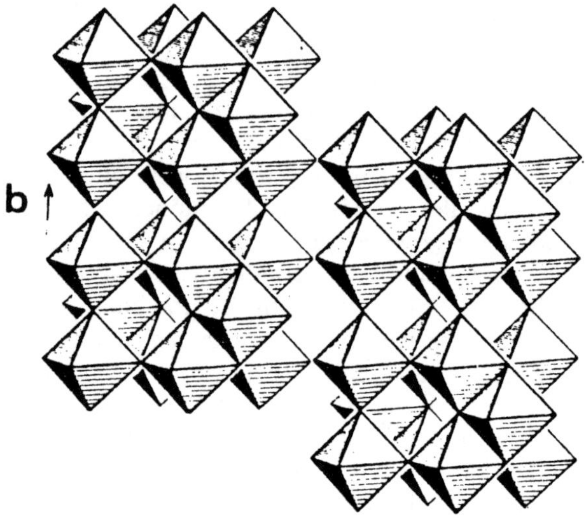

Fig. 2. Portion of an infinite slab built upon MoO_6 octahedra parallel to *b* and [102]. The portion contains four clusters of ten MoO_6 octahedra. (After Perloff *et al.* [4].)

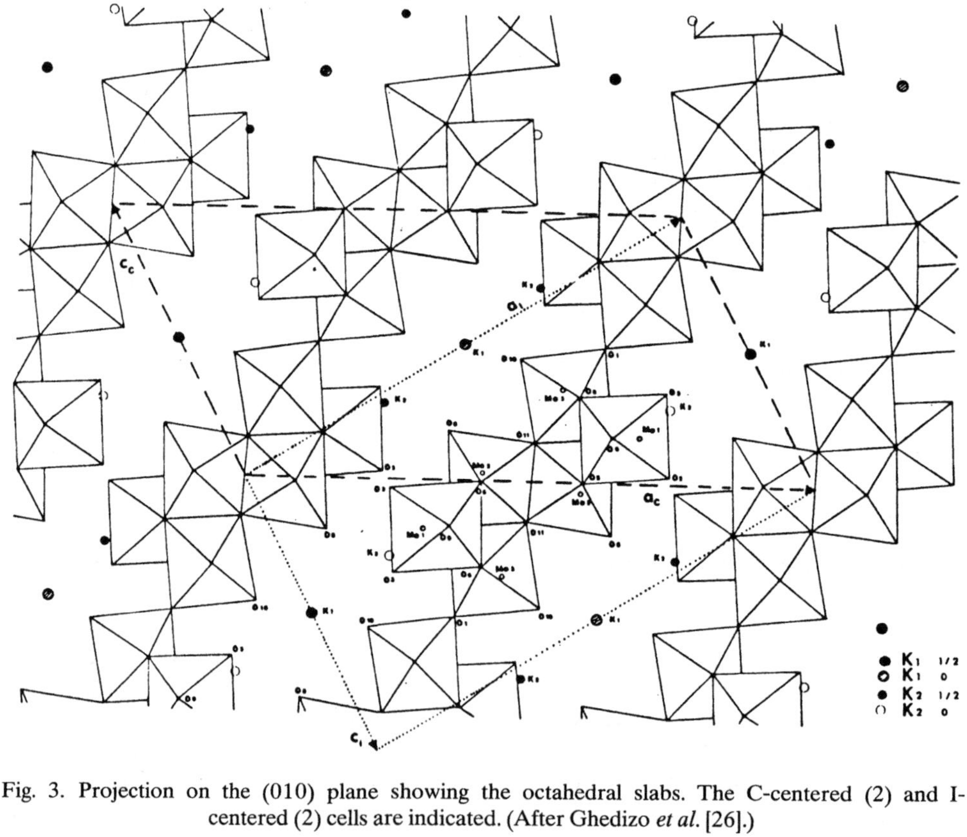

Fig. 3. Projection on the (010) plane showing the octahedral slabs. The C-centered (2) and I-centered (2) cells are indicated. (After Ghedizo *et al.* [26].)

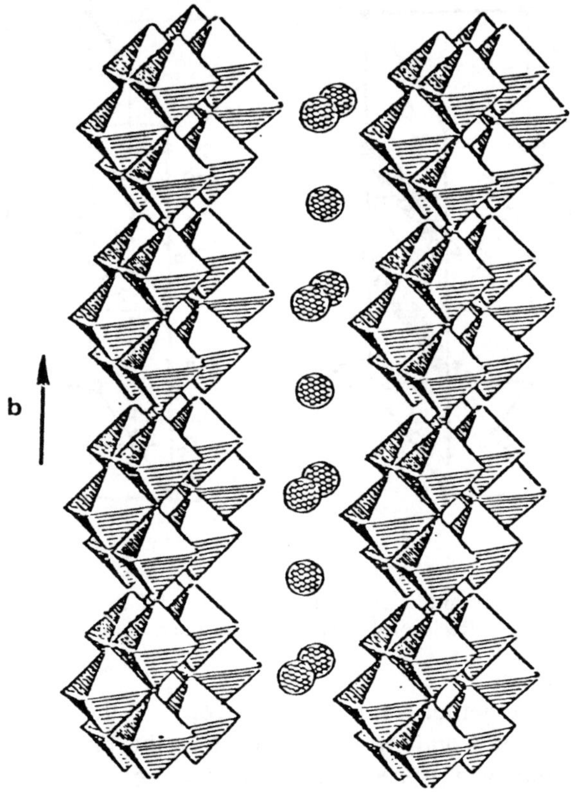

Fig. 4. Crystal structure showing the infinite slabs separated by the alkaline ions (●) and the infinite chains of MoO_6 (Mo(2) and Mo(3)) octahedra parallel to the *b* axis.

structure. It contains three independent Mo sites (Figure 3) and the 4*d* electron distribution over the 20 Mo sites [4 Mo(1), 8 Mo(2), 8 Mo(3)] of the unit cell are 10, 45, and 45% for $K_{0.30}MoO_3$ and 14, 43, and 43% for $Rb_{0.30}MoO_3$ [25]. The 4*d* electrons are therefore mainly located on the Mo(2) and Mo(3) sites. One should note that these sites are involved in infinite chains of MoO_6 octahedra, parallel to the *b* axis, as shown in Figure 4.

X-ray diffuse scattering studies have been performed by Pouget *et al.* through monochromatic Laue and Weissenberg photographs [8]. They found at room temperature diffuse streaks in shape of flattened cigars, normal to **b** and along the $2\mathbf{a}^*$—$\mathbf{c}^*$ reciprocal direction (Figure 5(a)). These streaks are situated at $\pm q_b \mathbf{b}^*$ from planes with odd *k* indices. The value obtained for $q_b = 0.28$ at 292 K is incommensurate with the crystal lattice. Below 180 K, these streaks condense in well-defined superstructure reflections, with wave vector $\mathbf{q} = 0\mathbf{a}^* + q_b\mathbf{b}^* + 0.5\mathbf{c}^*$ with $q_b = 0.26 \pm 0.01$ at 110 K (Figure 5(b)). Detailed analysis of these data indicate that one should rather consider that the satellites are located at $\pm(1 - q_b)\mathbf{b}^*$ from the main Bragg reflections. Similar results were found for $K_{0.30}MoO_3$ and $Rb_{0.30}MoO_3$ thus showing that the transition is not driven by the alkaline

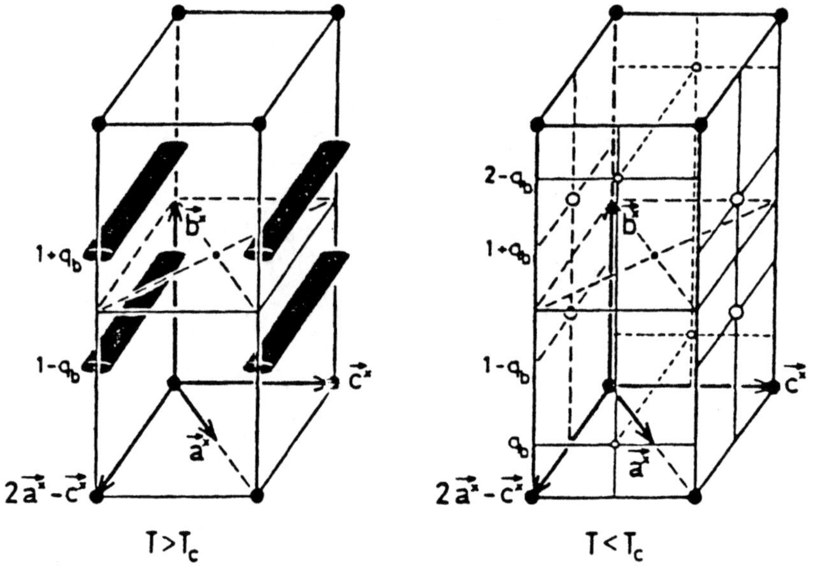

Fig. 5. Schematic representation in the reciprocal space of the diffuse streaks for $T > T_c$ and of the satellite reflections for $T < T_c$. Reciprocal directions correspond to the c-centered cell of Reference [2].

metal. These results, together with the anisotropy of the electrical and optical properties (see Sections 3 and 4), established that the periodic lattice distortion which accompanies the metal-to-semiconductor transition is of Peierls type.

Pouget *et al.* concluded from their results that the MoO_6 octahedra were mainly affected by the transition. A model for the structure factor of the CDW has also been given by Sato *et al.* on the basis of the intensities of the superlattice points collected with an X-ray four-circle diffractometer [27]. They deduced that the displacements of the Mo ions are mainly transverse (perpendicular to **b**), with a wave vector component $q_b\mathbf{b}^*$, and that the modulations of neighboring octahedra layers are antiphase.

Mössbauer studies performed on ^{57}Fe doped $K_{0.30}MoO_3$, indicating that Fe is substituted on the Mo sites, also show that the electric field gradient axis at the transition metal sites Mo(2) and Mo(3) are rotating at the Peierls transition [28]. This may be consistent with the analysis by Sato *et al.* [27]. Chen *et al.* showed that these satellites appear for extremely small times of electron beam irradiation only [29]. Short exposures (a few seconds with aceleration voltages of 60 to 200 kV) of the crystals to the electron beam completely destroy the superlattice, as it has also been found by Roucau *et al.* [30]. More detailed results have been obtained by Mutka *et al.* who studied the effect of electron irradiation on the electrical properties and especially on the transition temperature [31]. This was found to decrease from 180 to 110 K for an irradiation dose of the order of 200 mC cm^{-2}.

The temperature dependence of the **q** wave vector has been studied by neutron diffraction by Sato *et al.* [32] and by X-ray diffraction by Fleming and Schneemeyer [33] and by Pouget *et al.* [34]. All authors have found that $1 - q_b$ increases with

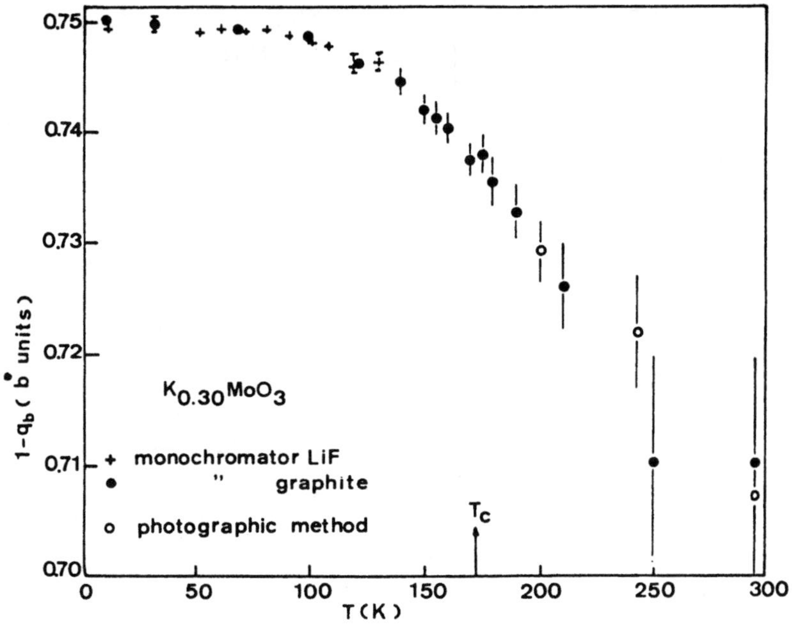

Fig. 6. Temperature dependence of $1 - q_b$, the **b*** component of the superstructure wave vector. (After Pouget *et al.* [34].)

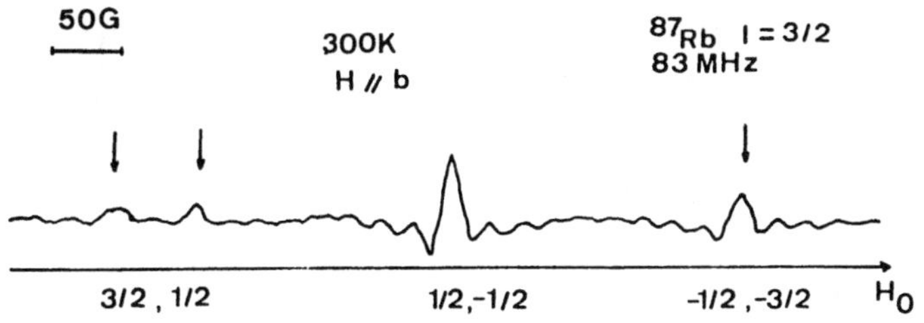

Fig. 7. ^{87}Rb NMR spectrum of a $Rb_{0.30}MoO_3$ single crystal at room temperature. (After Butaud *et al.* [35].)

decreasing temperature and approaches the commensurate value of 0.75 in the vicinity of 100 K (Figure 6). But while Sato *et al*. report a constant value of 0.748 at low temperatures, Fleming and Schneemeyer consider that $K_{0.30}MoO_3$ undergo a lock-in incommensurate to commensurate transition at ~110 K [33]. This is consistent with the data of Pouget *et al.* [34] and with preliminary ^{87}Rb NMR data in a $Rb_{0.30}MoO_3$ single crystal by Butaud *et al.* [35]. In Figure 7 is shown the full spectrum at 300 K of ^{87}Rb. The two Rb sites give rise to two separated first-order satellites only for the low field ones. In Figure 8 is shown the temperature dependence of the line shape at 102, 95, and 90 K. A sudden narrowing of the line at 90 K corresponding to a finite number of sites, not resolved in the present data, is clearly seen.

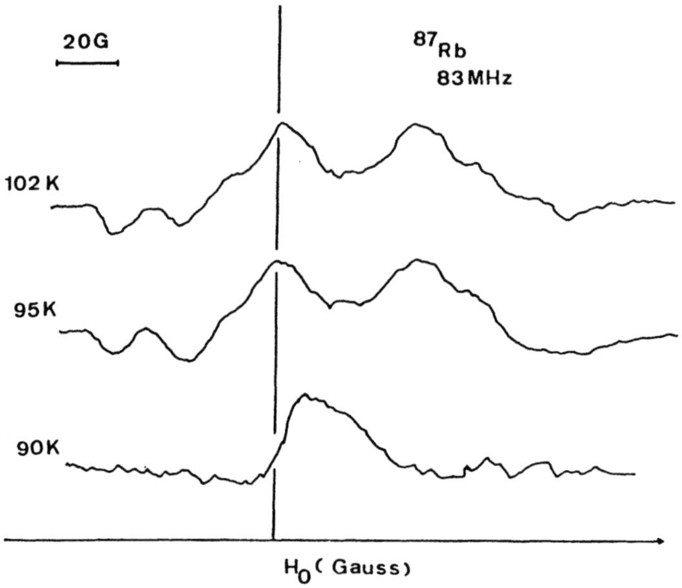

Fig. 8. Temperature dependance of the NMR line shape at 102, 95, and 90 K. (After Butaud *et al.* [35].)

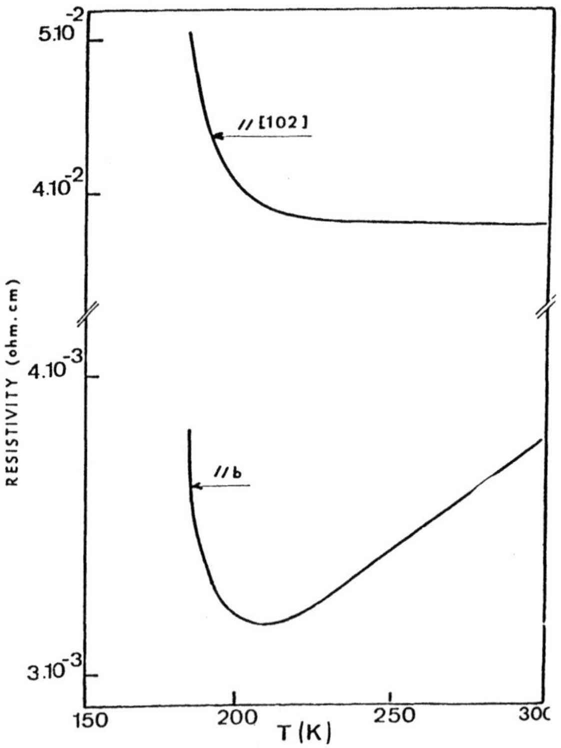

Fig. 9. Electrical resistivity (linear scale) versus temperature in the vicinity of the transition for the crystallographic directions **b** and [102] of the octahedra layers. (After Brusetti *et al.* [6].)

3. Ohmic Transport

The electrical resistivity of $K_{0.30}MoO_3$ above 180 K is metallic-like along the *b* axis (Figures 1 to 9) and is one order of magnitude at least larger along [102] than along [010]. Similar results are found for $Rb_{0.30}MoO_3$, as shown in Figure 10. As expected, the resistivity is still higher along the [201] axis perpendicular to the layers of MoO_6. The temperature of the Peierls transition, as obtained from the maximum of the curve of d log $\rho/\mathrm{d}T$ versus $1/T$, is found to be more exactly T_p = 183 ± 1 K for both $Rb_{0.30}MoO_3$ and $K_{0.30}MoO_3$ [36].

At low temperatures, the resistivity of the blue bronzes reaches extremely large values (typically 10^{14} Ω cm near 20 K). Below 20 K, anomalous results have been found: ρ is temperature independent or decreases with increasing temperature [5, 37] (Figure 11). In the same temperature range a nonlinear behavior was found. However, these results would require further studies: they could be related to the large value of ρ and involve either contacts effects or the presence of a surface layer with a conductivity higher than that of the bronze below 20 K.

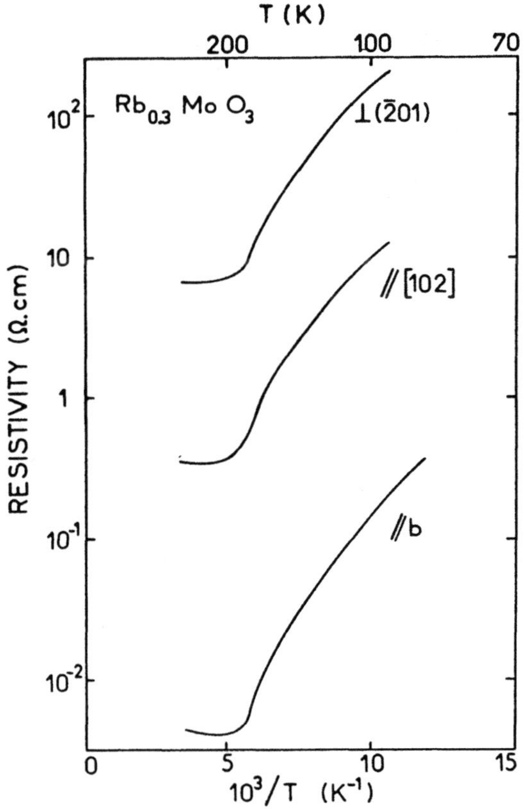

Fig. 10. Electrical resistivity (logarithmic scale) of $Rb_{0.30}MoO_3$ versus inverse temperature along **b**, [102] and perpendicular to the octahedra layers $(\overline{2}01)$ plane. (After References [8] and [37].)

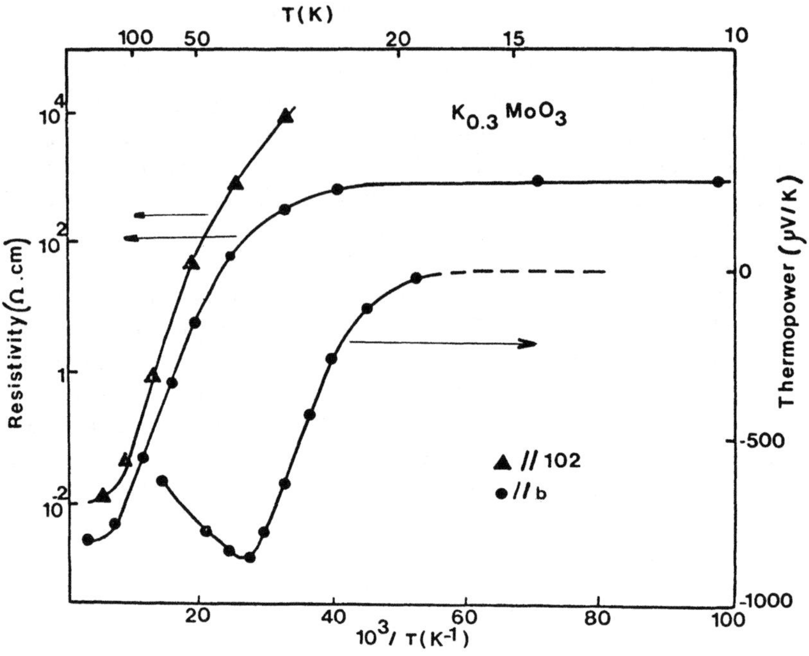

Fig. 11. Resistivity and thermopower versus inverse temperature at $T < 20$ K for $K_{0.30}MoO_3$. (After Reference [37].)

The thermopower of $K_{0.30}MoO_3$ is shown as a function of temperature in Figure 12 [6]. The anisotropy found above 180 K corroborates that the transport properties are metallic-like along **b** and semiconductor-like along [102]. In the low-temperature phase, the peak found at $T \simeq 30$ K may indicate a change in the transport process from a band type at $T > 30$ K to a hopping type, possibly extrinsic with carriers being polarons below 30 K. However, Figure 11 shows that below 20 K, the small (nonmeasurable) thermopower is related to the temperature-independent resistivity [36].

The Hall effect, as measured by Fogle and Perlstein [5] showed a change of behavior from *p*-type in the metallic phase to *n*-type in the semiconducting phase. Figure 13 shows more complete data obtained by Bervas *et al.* [37, 38], for the Hall constant R_H as a function of temperature. One should note that while the thermopower changes sign in the vicinity of 260 K, R_H becomes negative at the Peierls transition.

The effect of impurities on the electrical properties has been studied by different authors. It seems surprising that the Peierls transition is not significantly displaced by nonisoelectronic transition metal impurities of the first series such as vanadium [6] and iron [28]. Since the potassium and rubidium bronzes have very similar properties it is expected that the substitution of K by Rb does not strongly

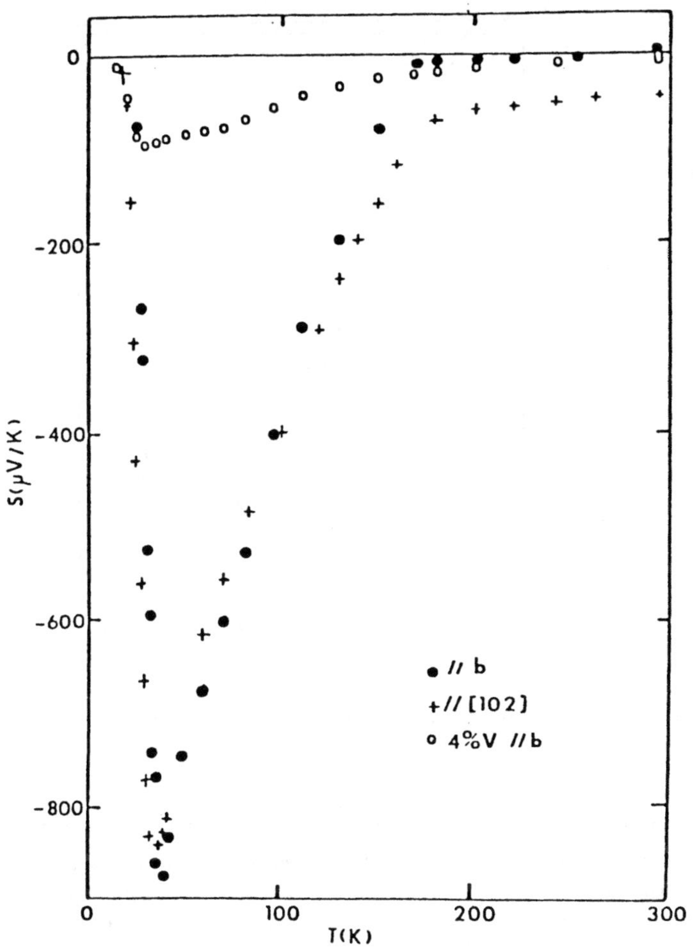

Fig. 12. Thermopower as a function of temperature with the temperature gradient either along **b** or along [102] for $K_{0.30}MoO_3$ and V-doped $K_{0.30}MoO_3$. (After Reference [6].)

modify the physical properties [39]. However, the substitution of Mo by isoelectronic W impurities has a dramatic effect and decreases sharply the transition temperature [39]. The role played by the different transition metal impurities is at the moment not clear. One may suggest that doping by nonisoelectronic impurities is compensated by a stoichiometry defect, in such a way that the 4*d* electron concentration is kept approximately constant.

4. Optical Properties

The reflectivity data obtained by Travaglini *et al.* with polarized light first established that, above 180 K, $K_{0.30}MoO_3$ is a truly quasi-1D metal [7]. At 300 K, the

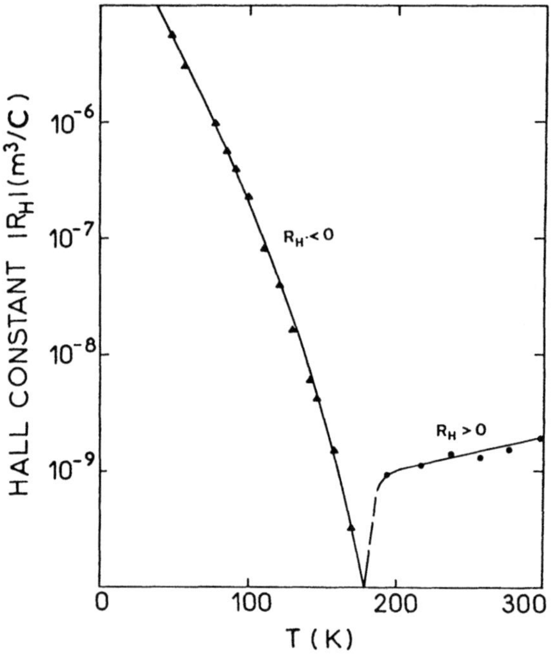

Fig. 13. Hall constant as a function of temperature for $K_{0.30}MoO_3$. (After Bervas *et al.* [37, 38].)

blue bronze shows a metallic behavior with a plasma edge for light polarized along **b**, while it shows a semiconducting type behavior with an absorption edge for light polarized along [102] (Figure 14). From Kramers—Kronig analysis, the plasma frequency was estimated to be $\omega_P = 2.7$ eV along **b** and much smaller than 0.03 eV along [102]. The value of 2.7 eV corresponds to a carrier concentration resulting from a complete charge transfer from the outer electrons of the K into the conduction band, if the effective mass m^* is assumed to be the free electron mass. In the Peierls state, the band gap was estimated from the same data to be in the range of 0.15 eV. Partly similar results have been obtained for the reflectivity data on the rubidium blue bronze by Jandl *et al.* [40].

Raman scattering studies by Travaglini *et al.* [41] then by Dierker *et al.* [42] show that a Raman line of Ag symmetry appears at about 50 cm^{-1} in the semiconducting phase. The anomalous temperature dependence of this line below 180 K indicates that it is related to the CDW state (Figure 15). The authors have attributed it to the amplitude mode A^+ of the CDW (see, for example, Reference [43]), although the softening of this line is only 13% between 10 K and the Peierls transition.

Reflectivity measurements in the far infrared region have also been performed at low temperature [44]. A giant peak has been found below 8 meV at 5 K for light polarized along **b**. This structure has been attributed to the phase A^- mode of the

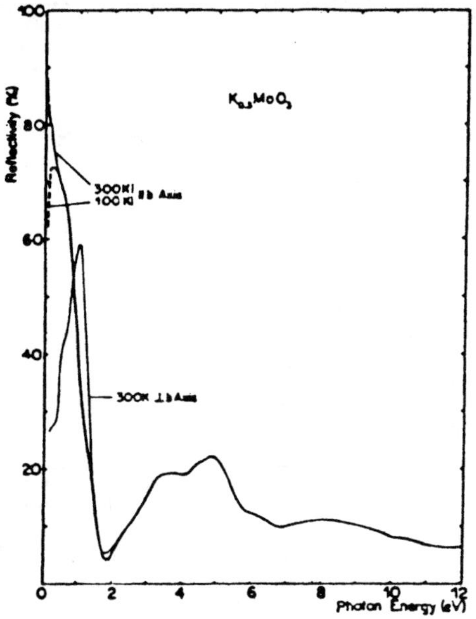

Fig. 14. Reflectivity of $K_{0.30}MoO_3$ with light polarized either along **b** or along [102] at 300 and 100 K. (After Reference [7].)

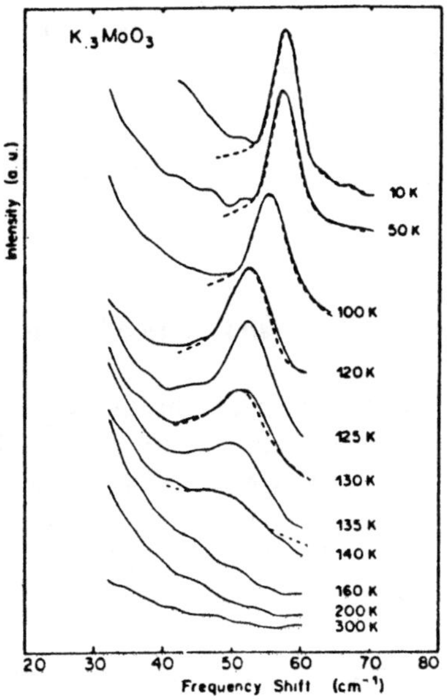

Fig. 15. Raman spectrum of $K_{0.30}MoO_3$ showing the CDW A^{+} mode at different temperatures. (After Reference [41].)

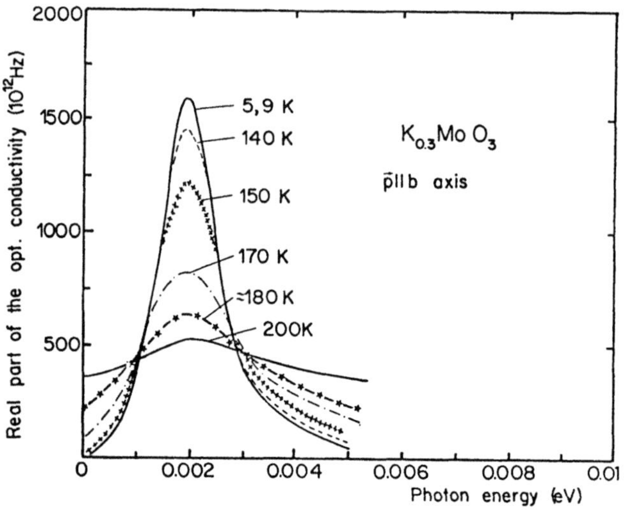

Fig. 16. Temperature dependence of the optical conductivity in the far-infrared region showing the CDW A^- mode at different temperatures. (After Reference [44].)

CDW. The intensity of this peak rapidly decreases as a function of temperature about 140 K and becomes very broad near 200 K (Figure 16). However, no noticeable softening was observed in all this temperature range. Information can be obtained from these data on the pinning potential which acts on the CDW. This will be discussed again later in Section 8. Analysis of these data leads to a value of 2000 to 3000 for the static dielectric constant and to an effective mass for the CDW of $910m_e$ (m_e is the free electron mass).

It should also be noted that in the course of the same studies, a strong peak was found at 0.07 eV and was attributed to a mid-gap absorption and possibly to solitons viewed as dislocations in the CDW lattice.

5. Specific Heat and Lattice Dynamics

From low-temperature specific heat studies, the Debye temperature for the potassium and the rubidium bronzes was found to be close to 320 K (320 ± 10 K for $K_{0.30}MoO_3$ and 310 ± 10 K for $Rb_{0.30}MoO_3$) [45]. This value is similar to what has been obtained either for MoO_3 (340 ± 10 K), for the purple bronze $K_{0.9}Mo_6O_{17}$ and for the oxides η-Mo_4O_{11} and γ-Mo_4O_{11} [13, 45]. This shows that the Debye temperature is mainly determined by the presence of the MoO_6 octahedra in all these materials.

A specific heat anomaly, found in both bronzes between approximately 150 and 190 K, corresponds to an enthalpy of ~25 J mole^{-1} and an entropy change of the order of 150 mJ mole^{-1} K^{-1} [45] (Figure 17). This leads to a value for the density of states in the metallic phase of the order of 0.3 eV^{-1} mole^{-1}, if all the entropy is

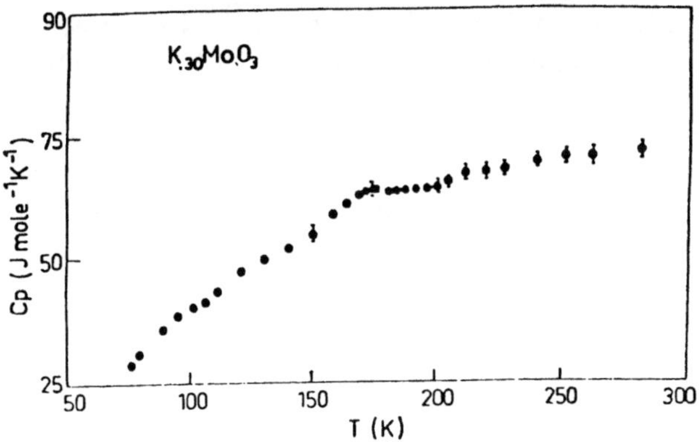

Fig. 17. Specific heat of $K_{0.30}MoO_3$ as a function of temperature in the vicinity of the Peierls transition. (After Reference [45].)

attributed to the conduction electrons. One also obtains from these data a value of roughly 0.24 R for the discontinuity in the electronic specific heat at the transition. This, compared to the entropy change, suggests that 1D fluctuations are important in this system [45, 49]. The high-temperature values of the specific heat are, at

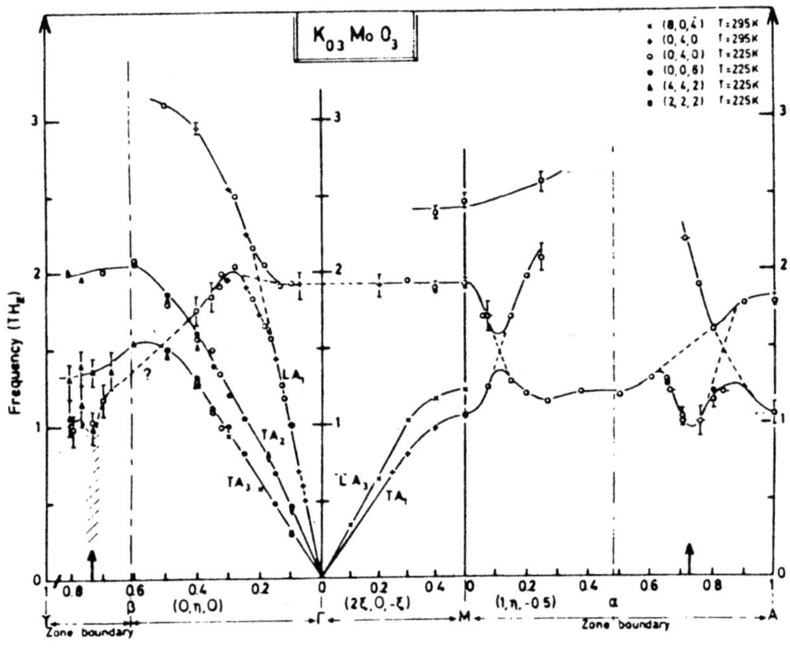

Fig. 18. Phonon spectrum of $K_{0.30}MoO_3$ as obtained by inelastic neutron scattering at T = 295 K and T = 225 K. The Kohn anomaly is visible at **q** (1, 0.72, —0.5). (After Reference [47].)

400 K, 11.0 and 11.3 R for $Rb_{0.30}MoO_3$ and $K_{0.30}MoO_3$ respectively, therefore slightly below the classical Dulong—Petit value of 12.9 R expected with a chemical formula $A_{0.30}MoO_3$.

The lattice dynamics has been studied in the blue bronze by Raman scattering as reported in Section 4 [42] and also by inelastic neutron scattering. Sato *et al.* first reported the existence of a soft phonon in $K_{0.30}MoO_3$ above the Peierls transi-

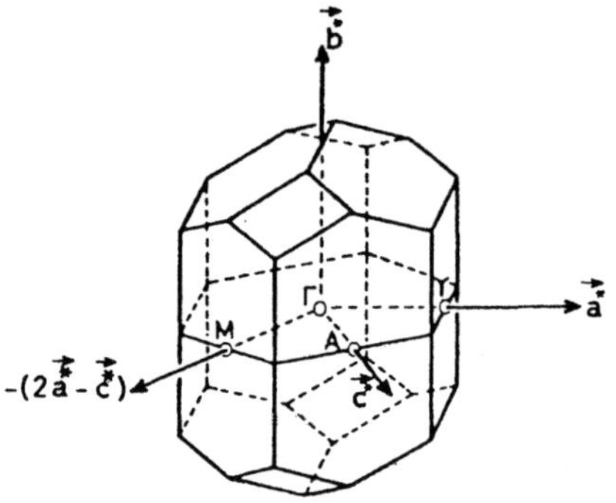

Fig. 19. First Brillouin zone of the blue bronze. (After Reference [47].)

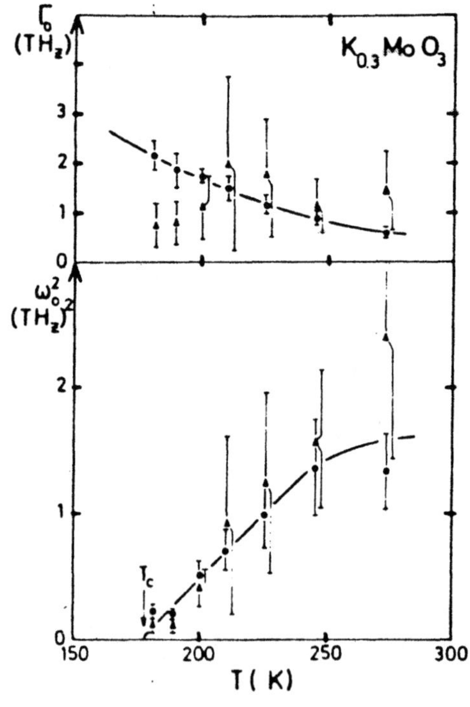

Fig. 20. Damping coefficient Γ_0 and square of the frequency ω_0^2 of the soft mode as a function of temperature. (After Reference [34].)

tion [46]. More accurate data were obtained by Pouget *et al.* [34, 47]. Figure 18 shows the phonon spectrum above the Peierls transition, including the Kohn anomaly found for the **q** vector (1, 0.72, −0.5). The corresponding Brillouin zone is shown in Figure 19. The analysis of the data obtained as a function of temperature leads to the values of soft phonon frequency and damping coefficient shown in Figure 20. There is an almost complete phonon softening at the Peierls transition. Sato *et al.* [46] also report the observation of the amplitude mode below the transition, at an energy of approximately 6 to 7 meV, in good agreement with the Raman data [42].

6. Magnetic Properties

The magnetic susceptibility of $K_{0.30}MoO_3$ has been measured by Bouchard *et al.* [3], by Bang and Sperlich [48], Brusetti *et al.* [6], by Johnston [49] and by Schneemeyer *et al.* [39, 50]. All authors find that the dsusceptibility χ is approximately temperature independent below 100 K, increases rather steeply between 100 and 180 K, and smoothly above 180 K (Figure 21). The increase in χ be-

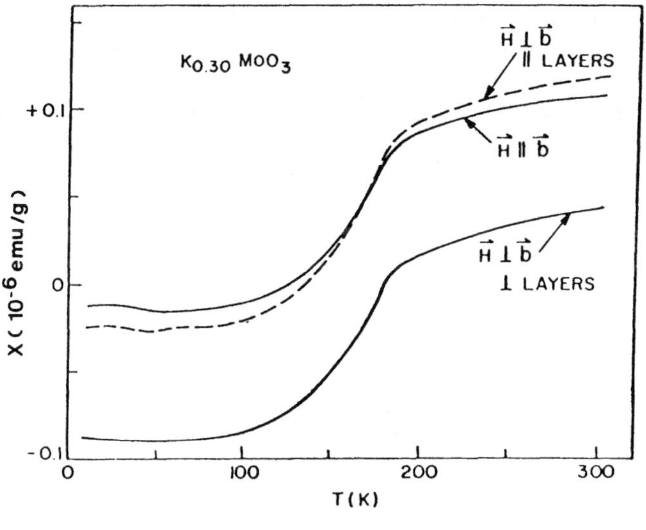

Fig. 21. Magnetic susceptibility of $K_{0.30}MoO_3$ as a function of temperature for different orientations of the magnetic field. (After Reference [50].)

tween 100 and 180 K is attributed to the decrease of the Peierls gap and to the contribution of the Pauli paramagnetism. It should, however, be noted that the orbital Van Vleck paramagnetism may not be negligible and is expected to be also related to the value of the Peierls gap. Johnston has calculated the Peierls gap $2\Delta_0$ and its temperature variation that one deduces from the susceptibility data, if one assumes that the Van Vleck paramagnetism is negligible. He obtains $\Delta_0 \simeq 1100$ K at $T = 0$, in good agreement with the optical data [40]. The effective bandgap of 400 K found above the Peierls transition should include the effect of fluctuations. A larger anisotropy of the susceptibility between the plane of the layers and the perpendicular orientation has also been reported [50]. This may well be related to the anisotropy of the Van Vleck paramagnetism.

Electron paramagnetic resonance spectra have been obtained in the semiconducting phase of $K_{0.30}MoO_3$ below ≈ 70 K [48, 51]. Three lines with g values close to 2.00, corresponding to Mo^{5+} ions very likely on the three molybdenum sites, are found. The intensity of these lines is weak and they are due to $4d$ electrons localized on Mo sites in the vicinity of a crystal defect, vacancy or impurity. The concentration of these centers, also found in $Rb_{0.30}MoO_3$, is in the range of 100 to 1000 ppm and could give an indication on the stoichiometry defects of the crystals [51].

7. Band Structure and the Peierls Transition

To get an insight into the nature of the Peierls transition, one would need a detailed band calculation which had not been performed when this article was written, due to the complexity resulting from 43 atoms per primitive cell ($K_3Mo_{10}O_{30}$). However, Figure 22 shows the type of band structure one obtains as a first approximation, on the basis of a model corresponding to the oxide ReO_3, due to Goodenough [52] and first applied to the bronzes by Dickens and Neild [53]. The relevant part of the band structure is built on hybridized Mo and O states that one can first obtain in a single MoO_6 octahedron. The σ and π bonding bands resulting from hybridization be-

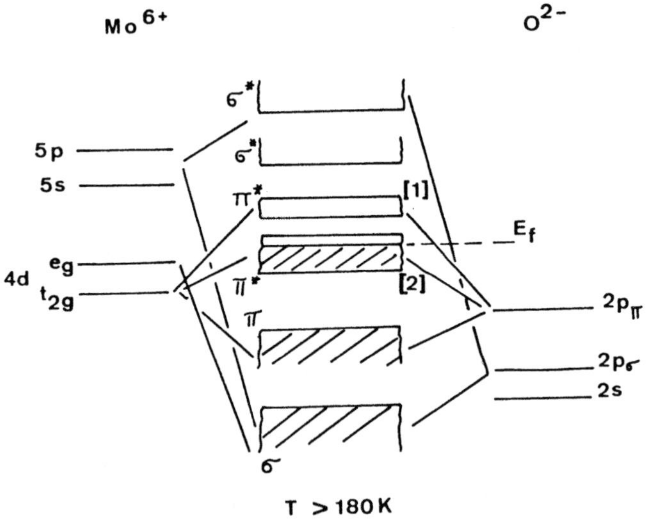

Fig. 22. Possible band structure of the blue bronze above the Peierls transition, showing the three-quarter-filled orbital doubly degenerate Π* conduction band. (After Reference [37].)

tween the oxygen $2s$, 2_{p_σ}, 2_{p_π} and the molybdenum $4d$, t_{2g}, and e_g orbitals are filled bonding bands, while the corresponding σ^* and π^* antibonding bands would be empty if there was no charge transfer from the alkali metal. The s band of either the K or the Rb lies far above and is not shown on Figure 20. The π^* band would have a trifold degeneracy in a cubic crystal such as ReO_3. In the blue bronze, due to the low monoclinic symmetry, this degeneracy must be at least partially lifted and the lowest π^* subband partially filled as a consequence of the charge transfer from the alkali metal.

The value of the Fermi wave vector, with $2k_F \simeq \frac{3}{4}\mathbf{b}^*$, is indeed consistent with three electrons per primitive cell in an orbital doubly degenerate 1D conduction band [8]. This corresponds to a complete charge transfer from the K (or Rb) into the conduction band since the primitive cell contains ten formula units ($K_3Mo_{10}O_{30}$) and to a conduction band either doubly degenerate or resulting from the overlap of two nondegenerate subbands. This is in agreement with the band structure calculated for a single chain of MoO_6 octahedra, with a LCAO method [44] and using band parameters calculated for a perovskite of formula $KMoO_3$ [54]. This band calculation also yields a density of states at the Fermi level of 0.6 eV^{-1} $mole^{-1}$, intermediate between the values obtained from specific heat [45] and susceptibility data [49].

The corresponding Fermi surface is now expected to be quasi-1D and to show large portions perpendicular to the **b** axis and distant of $2k_F$. The origin of such an anisotropic Fermi surface is not really clear in the absence of any detailed band calculation. It is probably related to a small transfer integral along the [102] direction between different clusters of octahedra (see Section 2). The correlation length found by X-ray studies in the metallic phase corresponds roughly along [102] to the size of a cluster, thus indicating that the blue bronzes should rather be considered as consisting of chains of clusters of octahedra and not of single octahedra. Nevertheless, the dimensionality is low enough to provide nesting at the Fermi surface and therefore the CDW instability leading to the opening of the Peierls gap below 180 K. The giant Kohn anomaly found for $q_b = 2k_F$ above the Peierls transition and the almost complete softening of this phonon at 180 K shows that the blue bronze is a model 1D system. Several of its properties are indeed very similar to those found in KCP [$K_2Pt(CN)_4Br_{0.30}$, $3H_2O$] some time ago (see, for example, Reference [55] and references therein).

As far as the nature of the transition is concerned, most properties, especially X-ray diffuse scattering and specific heat, seem to indicate that fluctuations, possibly quasi-1D, are predominant above the transition. A simple molecular field model therefore should not apply.

It seems now well established that in the best quality samples, an incommensurate to commensurate lock-in transition takes place at $T \simeq 100$ K by decreasing temperature. As mentioned above, the commensurate wave vector is consistent with a model of complete charge transfer in a stoichiometric compound of chemical formula $A_{0.30}MoO_3$. The temperature-independent incommensurate value found below 100 K in some samples may be due to stoichiometry deviation or to some impurities content. The origin of the temperature-independent incommensurate wave vector at $T > 100$ K, in all crystals, is not clear at the moment. It could be related to a nonperfectly 1D Fermi surface and to a change of the nesting condition as a function of temperature. One should also point out that several results show that the Peierls gap is only weakly temperature dependent below 100 K and decreases steeply above (see Reference [46] and [49]).

8. Nonlinear Transport and Related Properties

As mentioned earlier, nonlinear conductivity found in quasi-1D materials below the temperature of the onset of a CDW instability, above comparatively small electric threshold fields, has been attributed to the sliding of the CDW condensate. Below the threshold field, the CDWs are assumed to be pinned by impurities or defects. A variety of phenomena associated to this nonlinear transport have been observed. These include both broad band and so-called narrow band noise, anomalous frequency-dependent conductivity and memory and hysteresis properties [19].

Several theoretical models have been proposed to account for the CDW motion and the related properties. Phenomenological models consider the CDW as a rigid [56] or deformably [57] medium with an overdamped motion in a periodic pinning potential. Bardeen proposed a Zener type tunneling of the CDW across potential barriers [58]. Metastable states have been incorporated in more recent theories [59]. Ong *et al.* have proposed that the periodic voltage response to a driving dc current (seen as well-defined frequencies in the noise and also called narrow band noise) is induced below the contacts and due to the nucleation and motion of phase vortices [60]. This noise is attributed to a bulk effect, possibly related to the finite size of the samples by others [61].

All the above properties have now been found in the blue bronzes [9, 20, 21, 22, 62]. In addition they show very low frequency (in the Hertz range) phenomena, which seem to be more typical of this class of compounds.

8.1. THRESHOLD ELECTRIC FIELD

The threshold electric field above which the nonlinearity of the voltage versus current curve occurs depends on the samples and, for a given sample, on temperature. Three types of behavior are found: smooth threshold; switching from the ohmic to the non-ohmic regime with precursor voltage pulses when one sweeps the dc current; and mixed regime with a smooth threshold followed by a switching. Figure 23 shows switching in the V—I curve with precursor voltage pulses, the system hopping back and forth between the two states, ohmic and non-ohmic. The amplitude of the pulses is found, in this sample, to increase linearly with the voltage. Switching is observed in approximately 10% of the samples investigated only. Figure 24 shows a mixed behavior with two thresholds E_{t_1}, E_{t_2}, together with the onset of the broad band noise voltage at E_{t_1} and E_{t_2}, respectively. We must point out that substantial variations in the values of the threshold fields are found: E_t lies in the range 50—500 mV at 77 K depending on the samples. The threshold

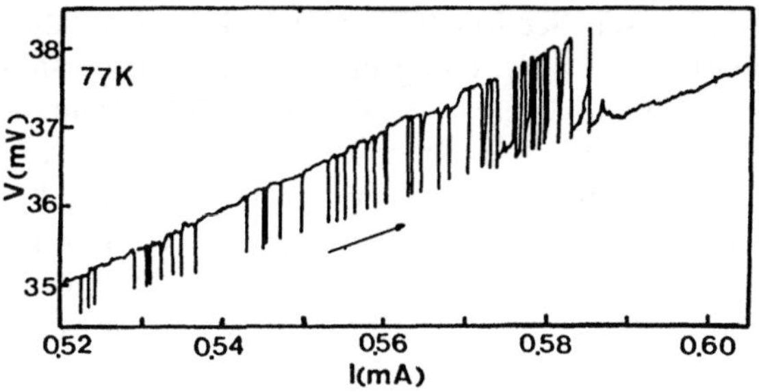

Fig. 23. DC V–I characteristic of $K_{0.30}MoO_3$ showing precursor voltage pulses ($\Delta V < 0$) and switching.

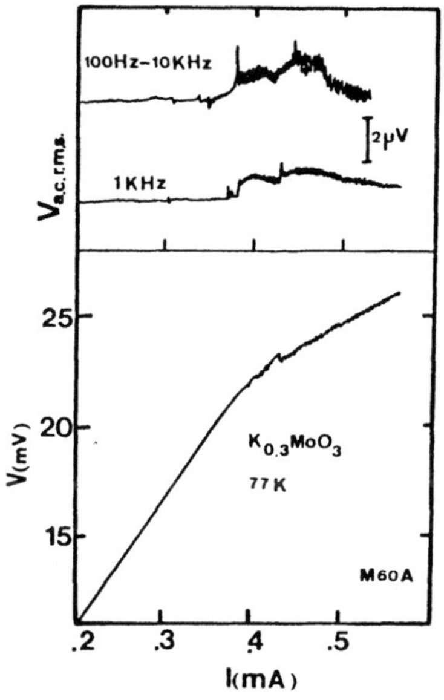

Fig. 24. Bottom curve: V–I curve showing two thresholds E_{t_1}, E_{t_2}; upper curves: Onset of broad band noise at E_{t_1} then E_{t_2}, in the frequency range 100 Hz to 10 kHz and at 1 kHz.

fields for switching or for smooth onset of the nonlinearity are comparable. No clear correlation between E_t and the resistivity values at 300 or 77 K as well as the resistivity ratio $\rho(300\text{ K})/\rho(77\text{ K})$ has been found. These results seem to indicate that the nature of the defects, impurities, dislocations or grain boundaries play an important role in the value of the threshold fields.

The various behaviors near E_t suggest that the onset of the non-ohmic regime is related to two processes: first, a nucleation of filaments parallel to the high conductivity **b** axis inside which the CDW is depinned, then a transverse propagation of these filaments. The threshold behavior of electron irradiated samples may corroborate this picture (see Section 8.7 below). The irradiation turns a smooth onset of the nonlinearity into a switching process and, at the same time, raises the threshold field value. The existence of sliding CDW filaments has also been proposed recently by Joos and Murray [63]. In this context, two threshold fields can be defined: a nucleation field E_t^p and a propagation field E_t^p. Depending on the relative magnitude of E_t^n and E_t^p, different behaviors are possible as has been experimentally found [21]. The recent data obtained by Hall and Zettl [64] indicate that, for $NbSe_3$, both thresholds have different temperature dependences. One should also note that multiple switchings with complicated hysteresis have been observed in Fe-doped $NbSe_3$ [65].

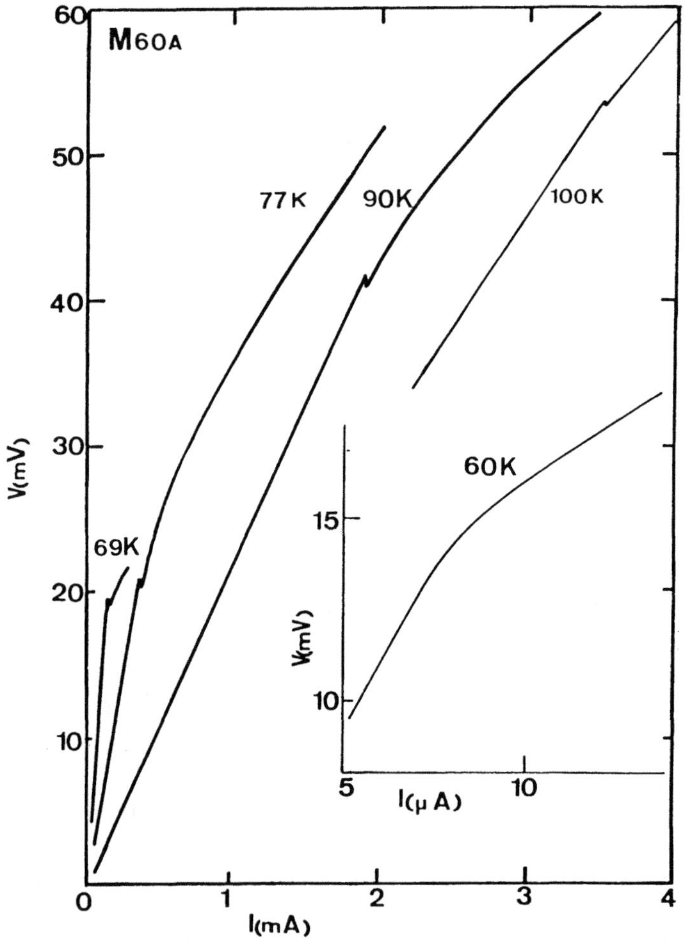

Fig. 25. V—I curves at different temperatures for a $K_{0.30}MoO_3$ crystal exhibiting switchng below 100 K.

The temperature dependence of E_t is not fully elucidated. In our samples, E_t increases from 60 to 110 K and decreases above that temperature. Below 60 K, the high impedance of the samples gives rise probably to contact effects and E_t cannot be measured easily. Some authors report a continuous rise of E_t up to 150 K [66] while others report a plateau above 100 K [67]. These discrepancies might be due to the fact that the threshold becomes very smooth above ~100 K and cannot be defined accurately. The switching, when observed, is found in a limited temperature range only, as shown in Figure 25. One should note that these different behaviors might be related to the value of the q_b component below 100 K either commensurate or incommensurate.

Preliminary Hall effect studies have also shown a threshold field with a decrease of the Hall constant above E_t [38]. These results seem similar to what has been found in $NbSe_3$ [68].

8.2. High-Frequency Voltage Oscillations

Figure 26 shows the Fourier analysis of the noise voltage of $K_{0.30}MoO_3$ for dc currents above threshold. This coherent noise voltage is found in a few samples only. The noise spectrum consists of one fundamental frequency F_1 defined as the smallest frequency in the noise spectrum, and its harmonics. As in $NbSe_3$ and TaS_3, the frequency increases linearly with the CDW current [62], as shown in Figure 27, with slopes $F_1/J_{CDW} = 0.1$ MHz/(A cm^{-2}) for $K_{0.30}MoO_3$ and 1 MHz/(A cm^{-2}) for $Rb_{0.30}MoO_3$. The slopes F/J_{CDW} are found independent of the slope of the V–I curve near the threshold (switching of smooth onset of nonlinearity). AC conductivity measurements by Cava *et al.* [69] in $K_{0.30}MoO_3$ are in agreement with our

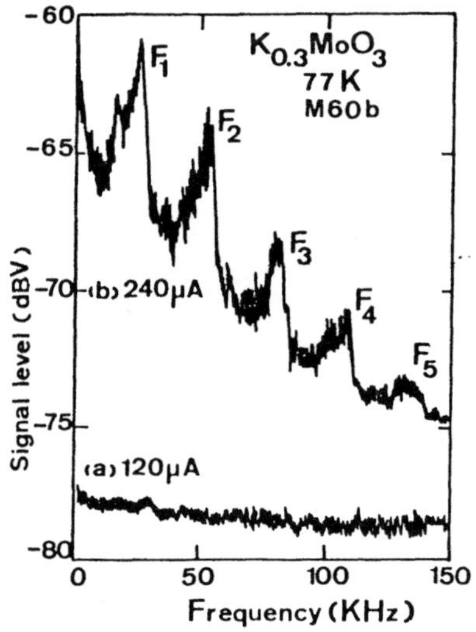

Fig. 26. Fourier analysis of the narrow band noise voltage of $K_{0.30}MoO_3$ at 77 K. Threshold current: $I_t = 130$ μA.

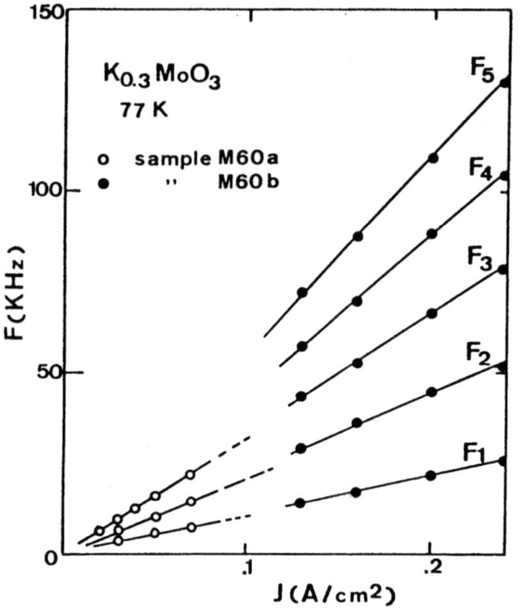

Fig. 27. Noise frequencies as a function of the excess CDW current at 77 K.

F/J_{CDW} values, as described in more detail below. Fleming *et al.* have also observed transient oscillations in the edge of the response to a current step. The frequency of these oscillations corresponds to the 'narrow band noise' [70].

Monceau *et al.* [71] have shown that in $NbSe_3$ the slope F/J_{CDW} was given by $F/J_{CDW} = 1/ne\lambda$, where n is the concentration of electrons condensed in the CDW and λ the superlattice period. However, our results do not seem consistent with this prediction [62]. The order of magnitude of this slope is found to be 20 to 200 times

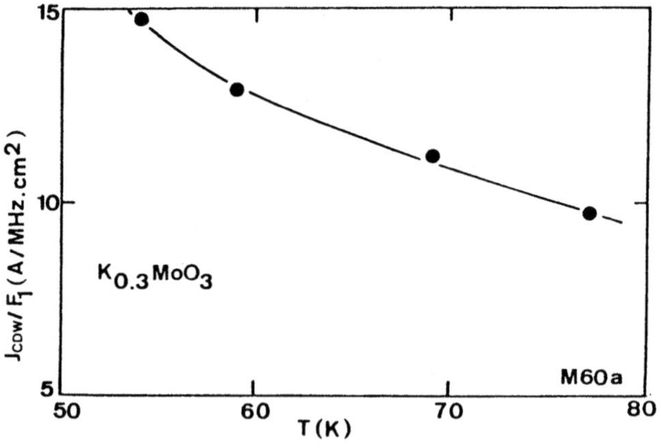

Fig. 28. Temperature dependence of the J_{CDW}/F_1 ratio.

larger than the value obtained by assuming that all the conduction electrons are condensed in the CDW and with $\lambda = 4b$. These discrepancies might be related to the complicated band structure, unknown at the present time, of the blue bronze.

Figure 28 also shows that this slope is temperature dependent, even at temperatures lower than 80 K where the wave vector of the superstructure is constant. This variation could be due to a decrease of the density of electrons condensed in the CDW with increasing temperature.

Just as in other CDW materials, the origin of the narrow band noise, bulk or contact effect, is not clear at the present time.

8.3. FREQUENCY-DEPENDENT CONDUCTIVITY

Cava *et al.* have measured the ac conductivity of $K_{0.30}MoO_3$ between 40 and 101 K in the frequency range 5 Hz to 13 MHz [69] (Figure 29). They have found that the dielectric response of the blue bronze should be described by a relaxation mechanism rather than as an overdamped oscillator. A distribution of relaxation

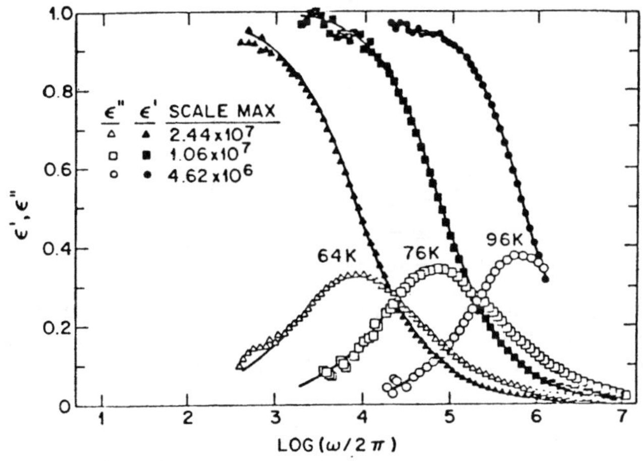

Fig. 29. Frequency dependence (log scale) of the real and imaginary parts of the dielectric constant for the pinned CDW in $K_{0.30}MoO_3$ for three representative temperatures. (From Reference [69].)

times not symmetric about the mean relaxation value τ_0 has to be involved. These relaxation times follow an activated temperature dependent behavior. The static dielectric constant ε_0 deduced from these data is of the order of 10^7 and also follows an Arrhenius-type law. There are some discrepancies between these results and those obtained from far infrared reflectivity studies, which lead to a value of ε_0 of the order of 2000 to 3000 [44]. AC conductivity measurements have also been performed by Hall *et al.* [72] and Reagor *et al.* [73]. Their analyses are not entirely consistent with those of Reference [69].

Cava *et al.* have also measured the dielectric constant as a function of a bias dc current. For dc currents larger than the threshold value, they have observed dips at given frequencies as a function of the dc current. These dips were attributed to interference effects between the external frequency and the internal noise frequencies.

8.4. Very Low Frequency Phenomena

At constant current, pulses of transient voltage corresponding to a sudden decrease of the dc voltage value V_{dc} are produced just below and in some samples above E_t [9]. The amplitude of the pulses is ~2% of V_{dc} and their shape is reproducible. Below E_t, these pulses are vanishing as a function of time without decay of their amplitude. In Figure 30 is plotted the number of events per time

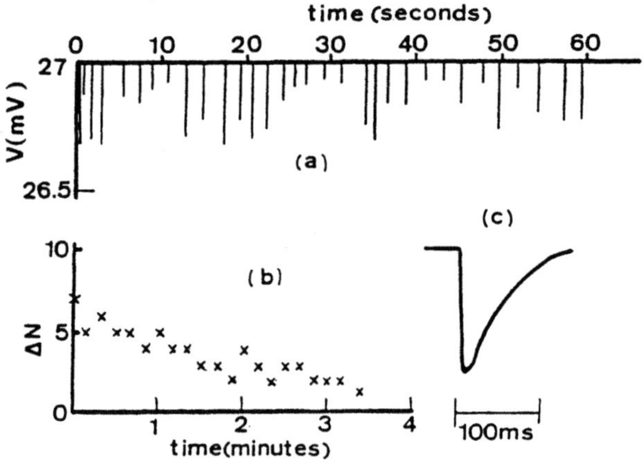

Fig. 30. Intermittent regime found for current slightly smaller than I_t. (a) Voltage pulses as a function of time; (b) number of events ΔN during a given time interval (10 s) versus time, showing the vanishing of this regime after a few minutes; (c) typical pulse shape (oscillogram trace).

unit (10 s) as a function of time, at 77 K. The average time between two pulses increases versus time. In this sample, complete extinction of the pulses occurs after ~10 min. After a slight increase of $I(I < I_t)$ the regime of pulses is found again. In Figure 31 are shown examples of voltage pulses recorded as a function of time for different values of the current just below the threshold value.

One should note that these results have some analogy with those concerning the displacement of a single Bloch wall in a ferromagnetic crystal [74]. The Barkhausen noise recorded when the wall is constrained to move with a constant velocity is very similar to the regime of pulses in the case of CDW transport in $K_{0.30}MoO_3$.

We had therefore suggested that these pulses might be related to the existence of CDW domain boundaries or possible dislocations in the CDW lattice. The existence of such charged CDW dislocations had been proposed several years ago by Lee and Rice [75]. Their displacement in the presence of impurities or crystal defects could induce such pulses.

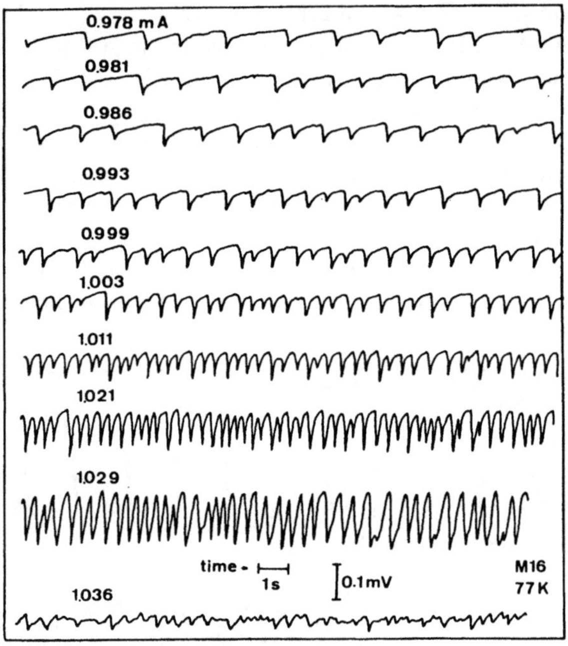

Fig. 31. Voltage pulses as a function of time at indicated currents, for a $K_{0.30}MoO_3$ sample at 77 K. Switching occurs for $I_t = 1.030$ mA in this sample.

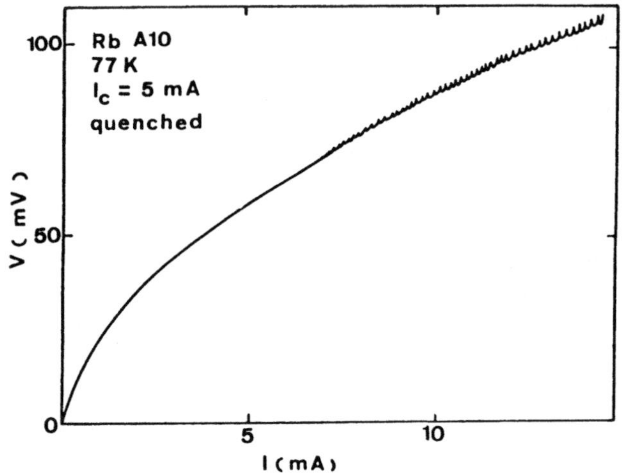

Fig. 32. V–I curve showing voltage fluctuations well above the threshold for a quenched $Rb_{0.30}MoO_3$ sample ($I_t = 0.4$ mA). Current applied during cooling $I_c = 5$ mA.

We have also found [76] that coherent low-frequency (~1 Hz) voltage fluctuations could be generated by quenching a sample from 300 to 77 K with an applied dc current I_c. These effects occur especially in $Rb_{0.30}MoO_3$ and in some pure, V-doped or Fe-doped $K_{0.30}MoO_3$ samples. Figure 32 shows a V–I curve for a quenched $Rb_{0.30}MoO_3$ sample. The voltage fluctuations generated well above the threshold are superimposed on the V_{dc} voltage and can be directly visualized. Depending on the cooling rate, two types of voltage oscillations appear across the voltage leads, as shown in Figure 37. For a fast cooling (~5 s) with a dc current, periodic voltage pulses corresponding to a sudden increase in voltage at fixed current are found. The amplitude of the pulses is nearly constant and is

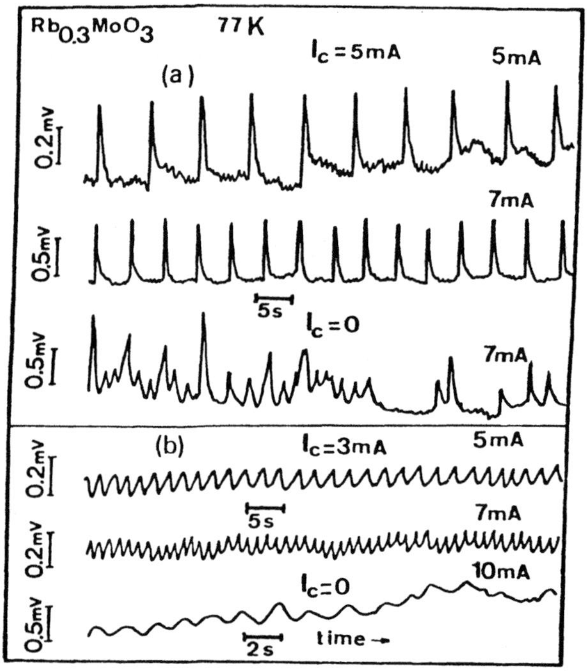

Fig. 33. Voltage fluctuations in quenched $Rb_{0.30}MoO_3$. (a) Regime of pulses found after a fast cooling (~5 s). For the two upper curves, the current applied during quenching was I_c = 5 mA. The measuring currents are 5 and 7 mA, respectively. Lower curve: chaotic oscillations for I_c = 0. (b) Regime of oscillations found after a slower cooling (~30 s).

~1% of V_{dc}. From the pulse area, an electric charge Q ~ 30 μC is deduced indicating that the charged objects responsible for these pulses are of macroscopic size. For a slower cooling rate (~60 s), quasi-periodic voltage oscillations are found. The frequencies of these oscillations are different from those of the pulses. The voltage oscillations amplitude is ~0.5% of V_{dc}. These voltage fluctuations are very stable with time, for at least several hours. When the sample is cooled in zero current only erratic voltage spikes (fast cooling) or erratic voltage oscillations

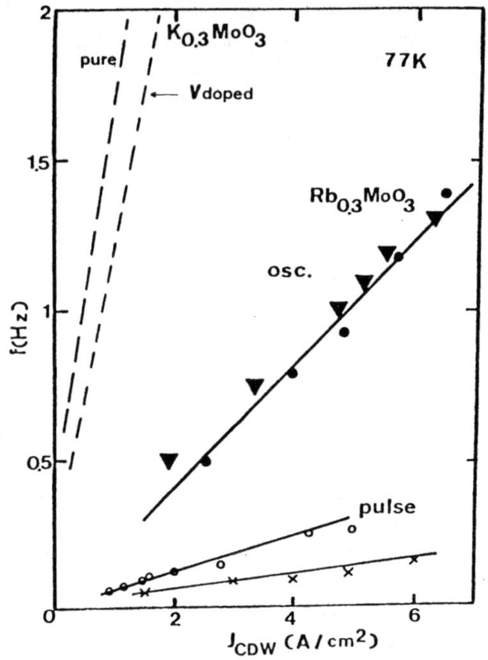

Fig. 34. Frequency of the voltage oscillations as a function of the excess CDW current for $Rb_{0.30}MoO_3$ and $K_{0.30}MoO_3$. ●, ▼: $Rb_{0.30}MoO_3$; oscillations for $I_c = 3$ and 8 mA, respectively. ○, ×: pulses for $I_c = 5$ and 3 mA, respectively. Dashed lines: voltage oscillations frequencies as a function of the excess CDW current for pure and V-doped $K_{0.30}MoO_3$ (V concentration: 2 at.%).

(slow cooling) are found. The quasi-periodic voltage fluctuations frequencies are found to increase almost linearly with the excess CDW current as shown in Figure 34. The slope f/J_{CDW} is 0.06 Hz/(A cm^{-2}) for voltage pulses and 0.2 Hz/(A cm^{-2}) for oscillations. In the case of $K_{0.30}MoO_3$, the slope f/J_{CDW} is ~ 1 Hz/(A cm^{-2}). One should note that these results are similar to the onset of serrations in stress—strain curves of Al alloys for example (also called Portevin—Le Chatelier effect or jerky flow) [77].

8.5. Other Metastability Phenomena

In the nonlinear state, the V—I curve is not always stable as a function of time. For $K_{0.30}MoO_3$, the excess CDW current may decrease logarithmically with time while for $Rb_{0.30}MoO_3$ and V- or Fe-doped $K_{0.30}MoO_3$ no significant drift occurs over several hours, as shown in Figure 35. The drift of the CDW current is not a monotonous function of time, it exhibits positive and negative steps [20] as shown in Figure 36. These time-dependent effects have been attributed to a progressive pinning of CDW domains boundaries as a function of time. Metastability phenomena with a time scale of ~ 1 ms have also been found by Fleming and Schneemeyer [78]

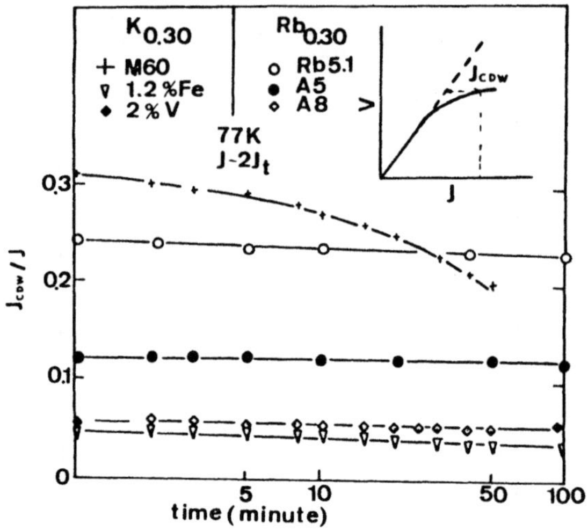

Fig. 35. Ratio of the CDW current density J_{CDW} to the total current density I measured versus time (logarithmic scale). J is kept constant during the experiment; J_t is the threshold current at 77 K. The upper right inset shows how J_{CDW} is defined.

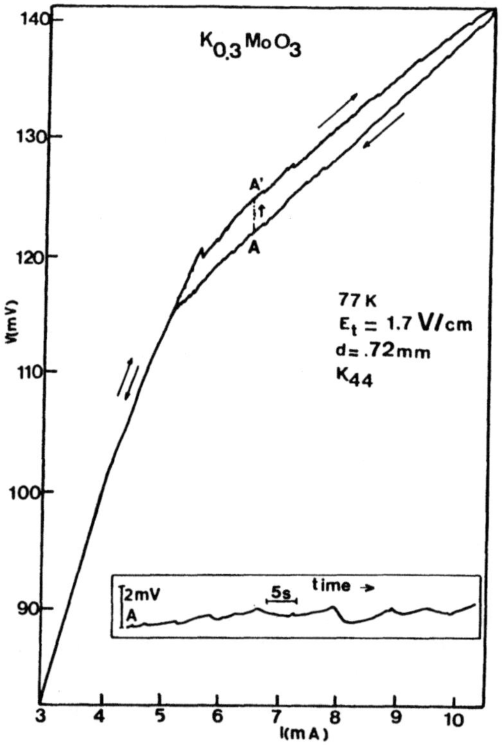

Fig. 36. Voltage versus current curve showing large hysteresis. The inset shows the drift of the voltage from A to A' at constant current.

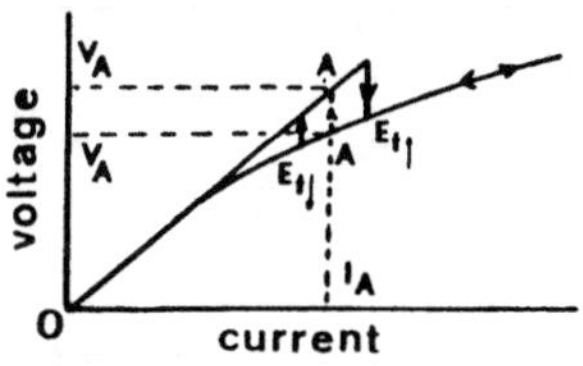

Fig. 37. V—I curve showing hysteresis near E_t. $E_{t\uparrow}$ (resp. $E_{t\downarrow}$) are the threshold values found on increasing (resp. decreasing) the dc current.

in their studies of the transient response to current pulses. Hysteresis is found near the threshold. Figure 37 shows the thresholds $E_{t\uparrow}$ obtained when the current I is swept up to I_{max} and $E_{t\downarrow}$ obtained when I is swept back to zero at the same sweeping rate. If the sweeping rate is much slower when the current is decreased from I_{max}, then $E_{t\downarrow}$ is closer to $E_{t\uparrow}$. After decreasing the current from I_{max}, if one fixes the current at a value I_A, the voltage will drift slowly from V_A to $V_{A'}$. Metastability phenomena are more pronounced in electron irradiated samples (see Section 8.7) and W-doped samples [50]. Hysteresis in V—I curves has also been found by Fleming and Schneemeyer [78].

The existence of metastable states is also proved by the properties of the low field resistance [22]. Figure 38 shows that the low field resistance exhibits sudden steps when one sweeps the dc current.

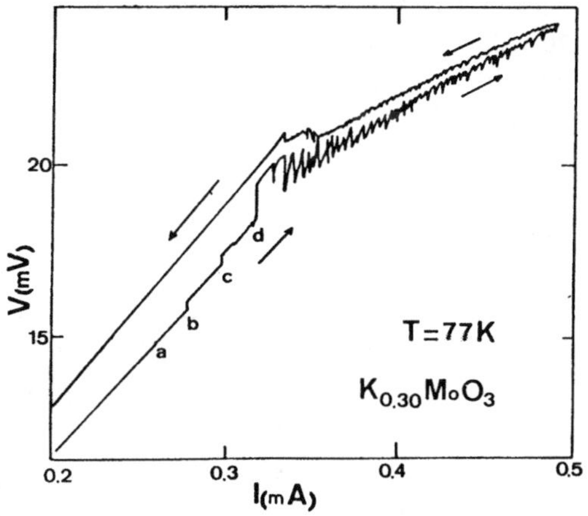

Fig. 38. DC V—I curve showing steps (a, b, c, d) in the low field resistance.

Significant changes in the low field resistance values are found when a thermal cycling has been performed. Let R_0 be the resistance found after cooling the sample from 300 to 77 K. After heating up to a given temperature T, then cooling again to 77 K, a new resistance value $R(T)$ is found at 77 K as shown in Figure 39. For each point of this figure, the sample has been heated again up to 300 K before starting a

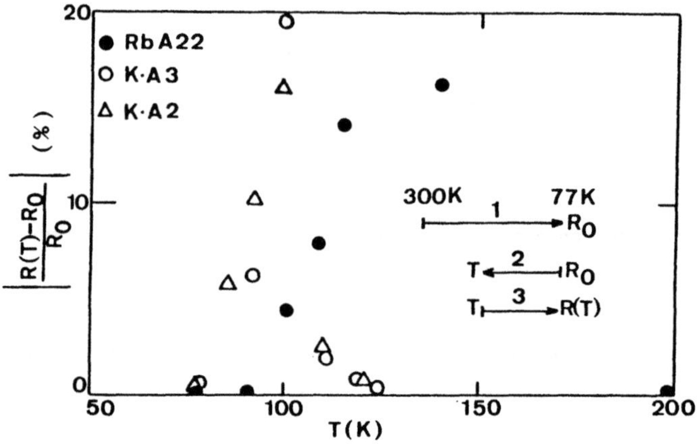

Fig. 39. Absolute relative change of $R(T) - R_0/R_0$ for successive temperature cycles for $K_{0.30}MoO_3$ and $Rb_{0.30}MoO_3$ (see text). $R(T)$ is measured by ac lock-in technique with currents less than I_t (77 K)/100.

new cycle in order to achieve a virgin state. $R(T)$ is found to increase noticeably when $T \sim 100$ K; $R(T)$ is found always smaller than R_0. The results for $K_{0.30}MoO_3$ and $Rb_{0.30}MoO_3$ are similar.

Thermal hysteresis has also been observed by Higgs and Gill [79] in orthorhombic TaS_3 between 55 and 215 K. The wave vector in this compound seems to lock to a commensurate value at ~130 K [80]. This hysteresis has been attributed to a nonequilibrium distribution of discommensurations interacting with crystal defects. The hysteresis observed near 100 K in the blue bronzes may have a similar origin.

For both $K_{0.30}MoO_3$ and $Rb_{0.30}MoO_3$, the low field resistance depends not only on the thermal but also on the electrical history of the sample at a given

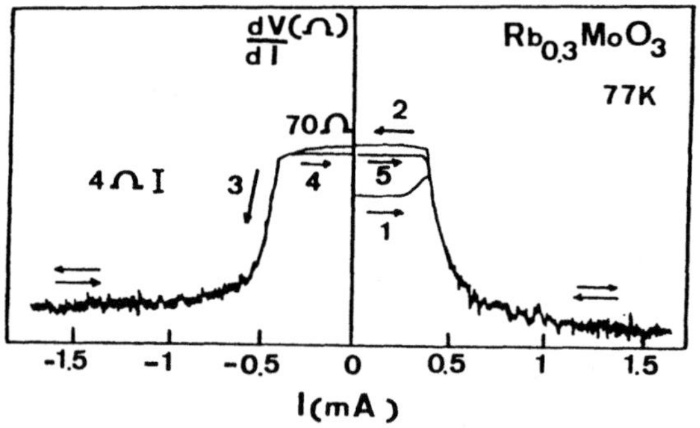

Fig. 40. Differential resistance dV/dI (at 33 Hz) as a function of the dc current for $Rb_{0.30}MoO_3$ at 77 K. 1 refers to the virgin state.

temperature. Figure 40 shows the differential resistance dV/dI measured by an ac lock-in technique at a frequency of 33 Hz as a function of the driving current for $Rb_{0.30}MoO_3$. A hysteresis in the low field resistance is found when the current has been swept above the threshold value I_t up to a given value I_{max}. If one labels R_1 the low field resistance in the virgin state and R_2 the resistance after the current has been swept above I_t, one can define an isothermal resistance (IRR) as $\Delta R/R = (R_2 - R_1)/R_1$. After a full cycle (2—5), the following cycles are nearly reproducible if one keeps the same value for I_{max}. R_2 does not seem to drift with time over several hours. One should also note that if the current is not reversed, curve 2 is fully reproducible.

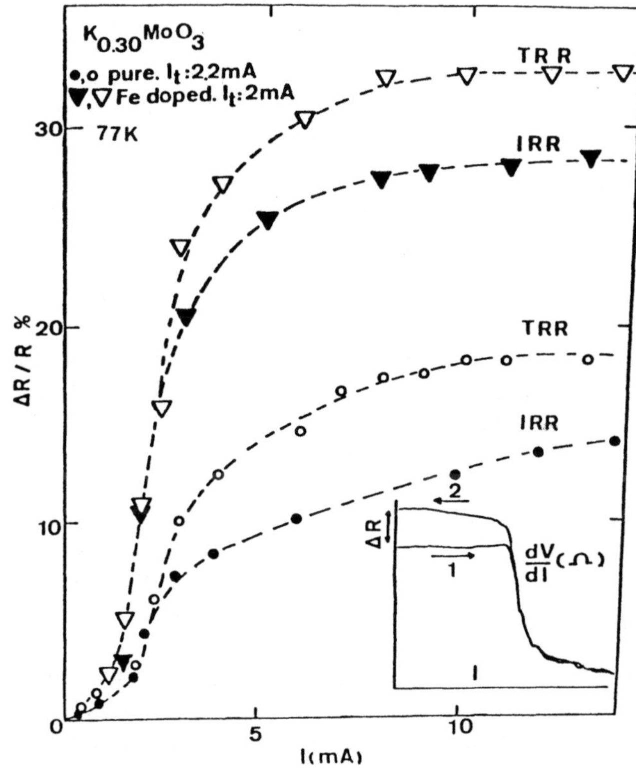

Fig. 41. Thermoremanent (TRR) and isothermal remanent resistance (IRR) of $K_{0.30}MoO_3$ at 77 K. The threshold current I_t at 77 K is 2.2 mA. The inset shows the hysteresis in the differential resistance measurement (1 corresponds to the first sweeping from a virgin state). The horizontal axis corresponds to the current applied during cooling for the TRR and to I_{max} values for the IRR.

Figure 41 summarizes the effect of the thermal and electrical history of the sample on the low field resistance. When the sample is cooled from 300 to 77 K with an applied dc current, the ohmic resistance R_{th} is found larger than the resistance R_1 found under zero current cooling. One can note this increase as $\Delta R/R$

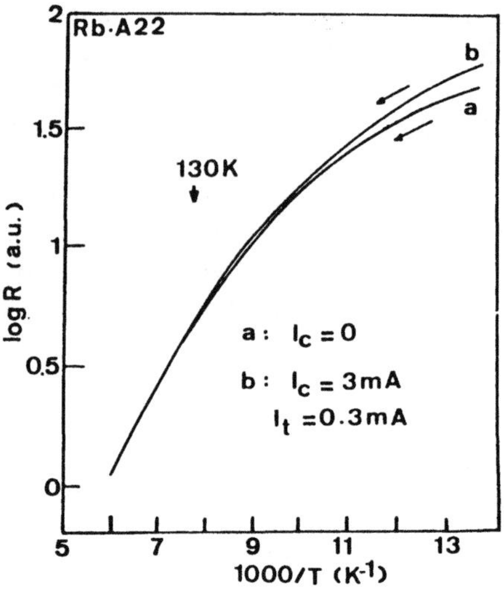

Fig. 42. Resistivity curves as a function of inverse temperature measured with an ac lock-in detection for a $Rb_{0.30}MoO_3$ crystal cooled in zero current (a) and with a current (b) well above the threshold value I_t at 77 K ($I_t = 0.3$ mA).

$= (R_{th} - R_1)/R_1$, the thermoremanent resistance (TRR). The TRR increases noticeably when the current applied during cooling is larger than the threshold current I_t at 77 K. Both the TRR and IRR are found larger in Fe-doped $K_{0.30}MoO_3$ than in pure samples [28]. The TRR becomes vanishingly small near 130 K as shown in Figure 42. In this figure we have plotted the low field resistance as a function of the inverse temperature obtained after a zero current cooling and after cooling with a current large enough to saturate the TRR at 77 K. The TRR and IRR vary slightly from sample to sample but are found systematically larger in $Rb_{0.30}MoO_3$ than in $K_{0.30}MoO_3$. These TRR and IRR have some analogy with the remanent magnetizations of spin glasses. Similar results have been obtained by Tsutsumi *et al.* [21] and Mihaly *et al.* [81] on $K_{0.30}MoO_3$, by Hutiray *et al.* on TaS_3 [82], and by Gill on $NbSe_3$ [83].

Cava *et al.* [84] have recently shown that $K_{0.30}MoO_3$ can act as a current source on heating after the CDW has been frozen into a nonunique metastable state by cooling down to 42 K with an applied dc electric field. The weak thermally stimulated depolarization current generated by heating might be due to the relaxation of the CDW from its metastable states acquired under cooling. One should also note that the depolarization current depends on the heating rate and shows a maximum in the vicinity of 30 K. This maximum could be related with the peak found in the thermopower at the same temperature [6].

8.6. X-RAY STUDIES UNDER ELECTRIC FIELD

Tamegai *et al.* [85] have studied the wave vector of the superstructure as a function of the electric field applied along *b*. They find for some samples a change of one of the transverse components of the wave vector perpendicular to the 1D axis (along **2a*—c***) (Figure 43). This change corresponds to phase shifts between CDW in neighboring octahedra layers. The authors propose that this may be due to inhomogeneity in the electric field. In any case, these experiments suggest some deformation of the CDW under an electric field, in the blue bronzes.

8.7. Effects of Irradiation on the Nonlinear Properties

The effects of electron irradiation on the transport properties of the blue bronzes $K_{0.30}MoO_3$ and $Rb_{0.30}MoO_3$ have been investigated by irradiation with 2.5 MeV electrons by Mutka *et al.* [31]. As reported in Section 3, this technique allows the concentration of defects to be increased in a controlled manner and is a useful tool for learning about the role of defects in the collective CDW transport. The threshold field E_t and the metastability phenomena are found very sensitive to irradia-

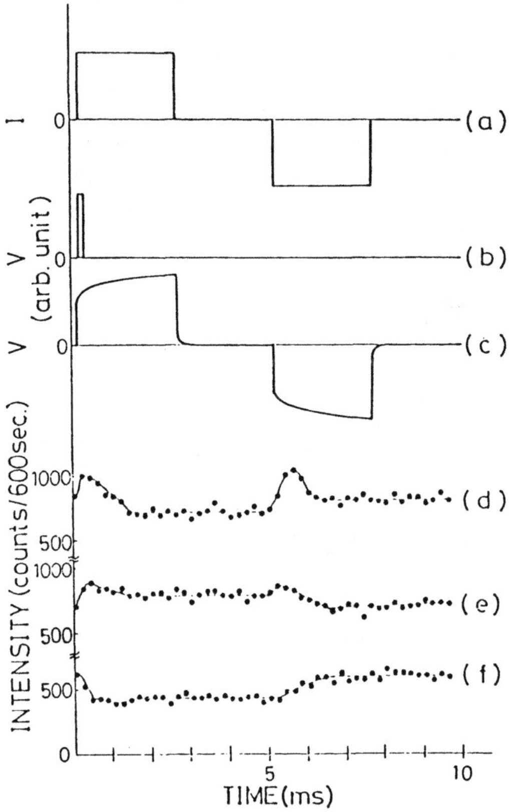

Fig. 43. Current pulses, voltage responses, and scattering intensities as functions of time: (a) alternating current pulses; (b) the multichannel scaling start pulse is synchronized to the current pulse; (c) voltage response of the sample which is time dependent with transient time of order of 1 ms. Time resolved scattering intensities: (d) at peak position; (e) +0.0014 (**2a*—c***) apart from peak position; (f) −0.0014 (**2a*—c***) apart from peak position. (After Reference [85].)

tion [86]. The defect concentration has been estimated from the intensity of a 4*d* electron Mo^{5+} electron paramagnetic resonance signal. It is found that a dose of 1 mC cm^{-2} of 2.5 MeV incident electrons increases the concentration of electron paramagnetic resonance active defects by 10^{-5} atomic fraction. It is believed that the production rate of nonmagnetic defects is of the same order of magnitude. Under these irradiations conditions, the most effective mechanisms for the creation of defects is a displacement process.

At very low doses (~ 100 μC cm^{-2}), the threshold field increases linearly with defect concentration as expected for strong pinning of the CDW. Similar results have been found on orthorhombic TaS_3 [87]. In this respect, $K_{0.30}MoO_3$ and $Rb_{0.30}MoO_3$ behave exactly in the same manner. This indicates that the alkaline metal plays a minor role in this process and corroborates the fact that the active defects seem to be Mo vacancies in the [MoO_6] octahedra running along the monoclinic high-conductivity **b** axis. Doping $K_{0.30}MoO_3$ with Fe or V impurities does not substantially increase the threshold. These impurities act as weak pinning centers.

In pure samples exhibiting a smooth onset of nonlinearity in the *V*–*I* curve, irradiation turns this behavior into a switching process with an increase of the threshold field as shown in Figure 44. Before observing switching, voltage pulses just below the threshold are found. These voltage pulses grow with the irradiation dose and at higher doses (~ 100 μC cm^{-2}) they finally turn into a switching process between the linear and nonlinear state. For these doses, neither the low

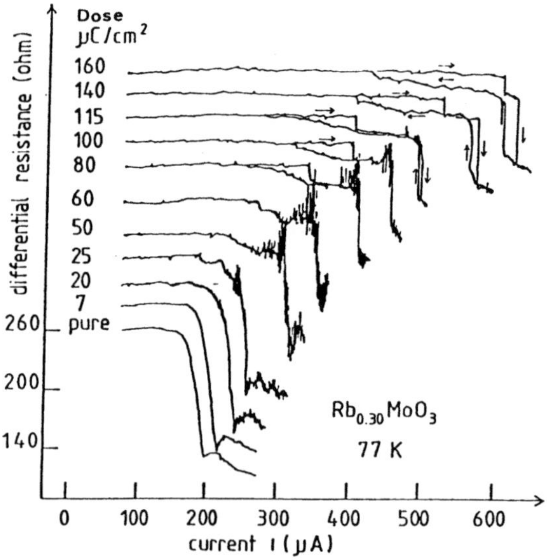

Fig. 44. Differential resistance d*V*/d*I* versus applied current. The curves have been shifted for clarity; the ohmic resistance is unaffected by irradiation. Irradiations (2.5 MeV electrons) and in-situ measurements have been made in liquid nitrogen.

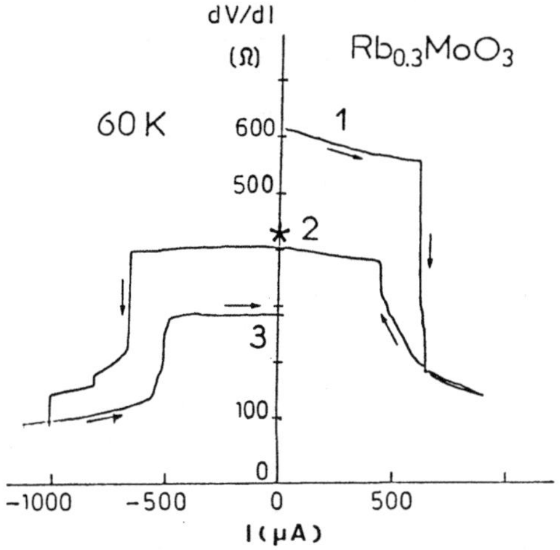

Fig. 45. Differential resistance as a function of current after a dose of 480 μC cm^{-2} for $Rb_{0.30}MoO_3$. (*) Ohmic resistance before irradiation, the initial state after irradiation is modified by the first current cycle (1). Curves (2) and (3) are reproducible.

field resistance nor the Peierls transition temperature T_P are affected. At very large doses, T_P decreases with substantial smearing of the transition: $T_P \sim 100$ K for a dose of 200 mC cm^{-2}.

The enhancement of metastability phenomena by irradiation is illustrated in Figure 45. This figure shows the differential resistance as a function of the dc current I after an irradiation dose of 480 μC cm^{-2}. The irradiation increases the ohmic resistance by a factor of 2. The first current cycle 1—2 restores approximately the initial value of the resistance. The cycles 2—3 are reproducible.

8.8. Comparison between $K_{0.30}MoO_3$ and $Rb_{0.30}MoO_3$

We have summarized in Table II the properties of $K_{0.30}MoO_3$ and $Rb_{0.30}MoO_3$. It can be seen that none of the classical transport measurements allows us to distinguish between these two bronzes. The only differences found up to now are the slopes F/J_{CDW} of the high and very low frequency phenomena. The origin of these different time scales in the potassium and in the rubidium compounds is not known at the moment. Apart from this point, Table II corroborates that the main role of the alkali metal is to provide the electrons which fill the conduction band.

8.9. Discussion

Several questions related to CDW transport now remain unanswered. A puzzling question, especially for the blue bronzes, concerns the nature of the CDW state. It seems that, at least in some crystals, this state is commensurate below 100 K. If the experiments corroborate that the commensurate crystals indeed show CDW

TABLE II

	$K_{0.30}MoO_3$	$Rb_{0.30}MoO_3$
Lattice parameters (Å)	$a = 16,23$; $b = 7,55$ $c = 9,861$	$a = 13,36$; $b = 7,55$ $c = 10,094$
Transition temperature	183 ± 1 K	183 ± 1 K
Room temperature resistivity	~ 10^{-3} Ω cm	~ 10^{-3} Ω cm
Ratio $\rho_{300\,K}/\rho_{77\,K}$	~ 200	~ 200
Anisotropy $\rho_{\|b}/\rho_{\perp\,\text{layers}}$	600	1500
Debye temperature	320 ± 10 K	310 ± 10 K
Enthalpy of the transition	6 cal mole^{-1}	6 cal mole^{-1}
Wave vector of the distortion	$q_b = 0.75$ at $T < 100$ K (commensurate in some samples)	id.
Threshold fields	50—500 mV cm^{-1}	50—500 mV cm^{-1}
High-frequency noise voltage F/J_{CDW}	0.1 MHz/(A cm^{-2})	1 MHz/(A cm^{-2})
Very low frequency voltage oscillations	1 Hz/(A cm^{-2})	0.2 Hz/(A cm^{-2})
Thermal and isothermal remanent resistance (saturated values)	$\Delta R/R$ ~ 15%	$\Delta R/R$ ~ 20%
Electron paramagnetic resonance signal at 4 K	Due to defects (Mo^{5+})	Weaker signal

transport, then the simple Frölich mechanism cannot be involved. The intrinsic threshold field estimated from the optical studies of the phase mode, may be in the range of 170 kV cm^{-1} [44]. This value, several orders of magnitude larger than the experimental threshold field (typically 0.1 V cm^{-1}), could well be related to lattice pinning. In this case, CDW transport would be due to the nucleation and motion of defects in the CDW lattice, either planar such as discommensurations or (and) linear such as CDW dislocations. One should note that these defects must be charged and therefore coupled to an electric field. At this point, one may wonder whether the type of the onset of nonlinearity, with or without switching, is related to the commensurability or incommensurability of the CDW state. It seems to be well established that above 100 K, when the CDW is clearly incommensurate, the threshold field is not well defined, thus indicating that the coupling between different CDW domains is weaker than below 100 K.

We now want to discuss mainly the origin of metastability related to CDW transport. As a general comment, one should note that, in the case of an incommensurate CDW state, an intrinsic source of metastability lies in the competition between the lattice periodicity and the Fermi wave vector periodicity, as has been worked in greater details by Aubry [88]. One may then expect frustration phenomena, analogous to the properties of spin glasses, for example [89].

However, when the CDW is commensurate, another mechanism should be involved. We believe that it is related to the presence of impurities and/or crystal defects coupled to the CDW. To account, for example, for various types of memory effects, one has to assume that these defects are more or less mobile during the cooling process through the Peierls transition and may reach different possible configurations in the low-temperature CDW state, depending on the thermal and electrical parameters of the cooling process.

In this context, let us first discuss the origin of the low-frequency ($f \sim 1$ Hz) voltage fluctuations. Thcy can be attributed either to the excess CDW current or, as has been suggested by Gill [90], to the so-called ohmic current. We rather propose that in our case they are due to the motion of the CDW in a medium containing some distribution of defects. One can then estimate a characteristic length L from the formula $f = v_d/L$, where v_d is the drift velocity of the CDW. If one believes the quasi-classical picture for the high-frequency (F) noise with $F = v_d/\lambda$, where λ is the superlattice period, the order of magnitude of L is found to be in the range of 30 μ. As the detailed mechanism of these low-frequency fluctuations is not understood, it is difficult to decide whether L gives some indication of the CDW domain size.

Another puzzling property, also found in TaS_3 [91], concerns the hysteresis found in the low field resistivity [21, 22]. A somewhat detailed mechanism may be proposed to account for this in the case of the blue bronzes. In the semiconducting state of these compounds, the ohmic resistance is very likely extrinsic and due to nonstoichiometry and/or impurities levels in the Peierls gap. In the best ('pure') samples, these levels correspond to Mo^{5+} localized donor centers. These centers are paramagnetic and have been observed by EPR spectroscopy. They may be located on the Mo sites labeled Mo(2) or Mo(3). The ohmic resistivity is therefore controlled by the population of these two levels and/or by their position in the Peierls gap. An obvious way of changing the ohmic resistivity would be to change the relative populations of the two levels. We suggest that, for example after a fast cooling process, these populations are metastable. If now these defects are coupled to the CDW, the motion of the CDW may induce a redistribution of the d electrons between these two levels. One should point out that such a redistribution only involves jumps of $4d$ electrons between neighboring Mo(2) and Mo(3) sites, and is therefore comparatively easy. Another possibility would be a variable distortion due to the CDW displacement which would lead to some displacements of the levels in the gap; this would be less likely.

Such a picture may account for the effect of the thermal and electrical history of the sample on the low field resistivity, although the nature of the coupling between the defects and the CDW is not known at the present time. Such a coupling has also been invoked by Abe *et al.* [92] to explain the enhancement of the backscattering yield of 1 MeV He^+ ions by the CDW motion.

9. Conclusion

At the time when this article is being written, a great number of studies are being performed on the blue bronzes in different laboratories. It is therefore too early to draw definite conclusions. However, it is now well established that these compounds are model quasi-one-dimensional systems, showing a Peierls transition towards an incommensurate state at 180 K and very likely an incommensurate—commensurate transition near 100 K. The nonlinear transport properties found in the Peierls semiconducting state are reasonably attributed to CDW transport. But below 100 K, nucleation and motion of CDW discommensurations and/or dislocations must be a predominant mechanism. This is probably responsible for the slow time scale shown by low-frequency voltage fluctuations, hysteresis, time-dependent and memory effects. The CDW pinning must therefore rather be viewed as pinning of discommensurations or dislocations through impurities and crystal defects. These slow phenomena, however, do not hinder the fast phenomena or narrow band noise, common to all CDW transport materials and not clearly understood at the moment.

It is now obvious that further important progress in this class of materials would require direct observation of CDW domains and of their rearrangement as a function of electric field. Unfortunately, in the case of the blue bronzes, due to electron irradiation damage, electron microscopy does not seem to provide the suitable tool. Other means of viewing these domains should now be looked for.

Acknowledgements

The studies performed at CNRS Grenoble on the blue bronzes have benefited from many contributions and collaborations. The crystals are grown by J. Marcus. The X-ray studies are due to J. P. Pouget and A. H. Moudden at Laboratoire de Physique des Solides, Orsay. The neutron inelastic scattering studies have been performed by C. Escribe-Filippini, J. P. Pouget, R. Currat, and B. Hennion at Laboratoire Léon Brillouin and Institut Laue Langevin, the Mössbauer studies by R. Chevalier, J. Y. Veuillen, and D. Salomon, the Hall effect measurements by E. Bervas and R. B. Cochrane at Université de Montréal and the NMR studies by C. Berthier, P. Segransan, and P. Butaud at Laboratoire de Spectrométrie Physique, Grenoble. The authors also acknowledge numerous discussions, especially with P. Monceau and with so many other colleagues that cannot be listed here. They acknowledge W. Fogle, M. Marezio, G. Travaglini, R. M. Fleming, and S. Kagoshima for their kind permission to reproduce some of their results (W. Fogle, Figure 1; M. Marezio, Figure 3; G. Travaglini, Figures 14, 15, 16; R. M. Fleming, Figures 21 and 29; S. Kagoshima, Figure 43).

References

1. A. Wold, W. Kunnmann, R. J. Arnott, and A. Ferretti, *Inorg. Chem.* **3**, 545 (1964).
2. J. Graham and A. D. Wadsley, *Acta Cryst.* **20**, 93 (1966).
3. G. H. Bouchard Jr J. H. Perlstein, and M. J. Sienko, *Inorg. Chem.* **6**, 1682 (1967).
4. D. S. Perloff, M. Vlasse, and A. Wold, *J. Phys. Chem. Solds* **30**, 1071 (1969).
5. W. Fogle and J. H. Perlstein, *Phys. Rev.* **B6**, 1402 (1972).
6. R. Brusetti, B. K. Chakraverty, J. Devenyi, J. Dumas, J. Marcus, and C. Schlenker, in *Recent Development in Condensed Matter Physics* (eds J. T. Deevreese, L. F. Lemmens, V. E. Van

Doren, and J. Van Royen), Plenum, New York, Vol. 2, p. 181 (1981).
7. G. Travaglini, P. Wachter, J. Marcus, and C. Schlenker, *Solid State Commun.* **37**, 599 (1971).
8. J. P. Pouget, S. Kagoshima, C. Schlenker, and J. Marcus, *J. Physique Lett.* **44**, L113 (1973).
9. J. Dumas, C. Schlenker, J. Marcus, and R. Buder, *Phys. Rev. Lett.* **50**, 757 (1973).
10. M. J. Sienko, in *Non-Stoichiometric Compounds, Advances in Chemistry Series* (ed. R. F. Gould), American Chem. Soc., Washington, D.C., p. 224 (1963).
11. P. Hagenmuller, in *Progress in Solid State Chemistry* (ed. H. Reiss), Pergamon, Vol. 5, p. 71 (1971).
12. R. Buder, J. Devenyi, J. Dumas, J. Marcus, J. Mercier, C. Schlenker, and H. Vincent, *J. Physique Lett.* **43**, L59 (1982); C. Escribe-Filippini, R. Almairac, R. Ayroles, C. Roucau, K. Konaté, J. Marcus, and C. Schlenker, *Phil. Mag.* **B50**, 311 (1984).
13. H. Guyot, C. Escribe-Filippini, G. Fourcaudot, K. Konaté, and C. Schlenker, *J. Phys. C: Solid State Phys.* **16**, L1227 (1983).
14. H. Guyot, G. Fourcaudot, and C. Schlenker, *Int. Conf. on Charge Density Wave in solids, Budapest (Sept. 84), Lecture Notes in Physics*, Springer-Verlag (p. 133).
15. See, for example, the series *Physics and Chemistry of Materials with Layered Structures* (ed. F. Lévy), Vol. 2 (1976); (eds T. J. Wieting and M. Schluter), Vol. 3 (1979); (ed. P. A. Lee), Vol. 4 (1976), Reidel, Dordrecht.
16. G. A. Toombs, *Phys. Rep.* **C40**, 181 (1978).
17. H. Fröhlich, *Proc. Roy. Soc. London* **A223**, 296 (1954).
18. P. Monceau, N. P. Ong, A. M. Portis, A. Meerschaut, and J. Rouxel, *Phys. Rev. Lett.* **37**, 602 (1976).
19. For reviews, see: R. M. Fleming in *Physics in One Dimension*, Vol. 23 of Springer Series in Solid State Sciences (eds J. Bernasconi and T. Schneider) Springer, Berlin (1981); N. P. Ong, *Canad. J. Phys.* **60**, 757 (1982); P. Monceau, J. Richard, and M. Renard, *Phys. Rev.* **B25**, 1931 (1982); J. Richard, P. Monceau, and M. Renard, *Phys. Rev.* **B25**, 948 (1982); G. Grüner, *Comments in Solid State Phys.* **10** (1983); G. Grüner, *Physica* **8D**, 1 (1983); *Electronic Properties of Inorganic Quasi-One-Dimensional Compounds* (ed. P. Monceau) in *Physcis and Chemistry of Materials with Low-Dimensional Structures, Series B: Quasi-One-Dimensional Structures*, Reidel, Dordrecht (1985); G. Grüner and A. Zettl, *Phys. Rep.* **119**, 117, (1985).
20. *Proc. Int. Symp. on Non-Linear Transport and Related Phenomena in Inorganic Quasi-One-Dimensional Conductors, Sapporo, Japan* (Oct. 1983), Hokkaido University.
21. *Proc. Int. Conf. on the Physics and Chemistry of Low Dimensional Synthetic Metals* (ICSM 84), *Abano Terme (italy), June 1984* (Vol. 121, 1985, *Molecular Crystals and Liquid Crystals*, Gordon and Breach).
22. *Proc. Int. Conf. on Charge Density Waves in Solids, Budapest, Hungary 3—7 Sept. 1984; Lecture Notes in Physics*, Springer-Verlag (ed. Gy. Hukiray and J. Solyom).
23. Z. Z. Wang, M. C. Saint-Lager, P. Monceau, M. Renard, P. Gressier, A. Meerschaut, L. Guemas, and J. Rouxel, *Solid State Commun.* **46**, 325 (1983); M. Maki, M. Kaiser, A. Zettl, and G. Grüner, *Solid State Commun.* **46**, 497 (1983); see also A. Meerschaut in Reference 20, p. 231, and J. Rouxel in Reference 21.
24. J. M. Réau, C. Fouassier, and P. Hagenmuller, *J. Solid State Chem.* **1**, 326 (1970).
25. P. Strobel and M. Greenblatt, *J. Solid State Chem.* **36**, 331 (1981).
26. M. Ghedira, J. Chenavas, M. Marezio, and J. Marcus, *J. Solid State Chem.* **57**, 300 (1985).
27. See M. Sato in Reference 22.
28. See J. Y. Veuillen, R. Chevalier, D. Salomon, J. Dumas, J. Marcus, and C. Schlenker in Reference 22.
29. C. H. Chen, L. F. Schneemeyer, and R. M. Fleming, *Phys. Rev.* **B29**, 3765 (1984).
30. C. Roucau, R. Ayroles, and C. Escribe-Filippini (unpublished).
31. See H. Mutka, S. Bouffard, M. Sanquer, J. Dumas, and C. Schlenker in Reference 21.
32. M. Sato, H. Fujishita, and S. Hoshino, *J. Phys. C: Solid State Phys.* **16**, L877 (1983).
33. R. M. Fleming and L. F. Schneemeyer, *Bull. Am. Phys. Soc.* **29**, 470 (1984).
34. See J. P. Pouget, A. H. Moudden, R. Moret, C. Escribe-Filippini, B. Hennion, J. Marcus, and C. Schlenker in Reference 21.

35. P. Butaud, P. Segransan, C. Berthier, J. Dumas, and C. Schlenker, *Phys. Rev. Lett.* **55**, 253, (1985).
36. J. Dumas and A. Arbaoui (unpublished).
37. E. Bervas, Thèse d'Ingénieur-Docteur, Université de Grenoble (February 1984).
38. E. Bervas, R. W. Cochrane, J. Dumas, J. Marcus, and C. Schlenker (to be published).
39. L. F. Schneemeyer, F. J. Di Salvo, S. E. Spengler, and J. V. Waszczak, *Phys. Rev.* **B30**, 4297 (1984).
40. S. Jandl, J. Marcus, and J. Y. Veuillen (to be published).
41. G. Travaglini, I. Mörke, and P. Wachter, *Solid State Commun.* **45**, 289 (1983); G. Travaglini and P. Wachter, in *Proc. 15th Conf. Physics of Semiconductors, Kyoto 1980; J. Phys. Soc. Japan Suppl. A,* **49**, 869 (1980).
42. S. B. Dierker, K. B. Lyons, and L. F. Schneemeyer, *Bull. Am. Phys. Soc.* **29**, 469 (1984).
43. P. A. Lee, T. M. Rice, and P. W. Anderson, *Solid State Commun.* **14**, 703 (1976).
44. G. Travaglini and P. Wachter, *Phys. Rev.* **B30**, 1971 (1984); see also Reference 21.
45. K. Konaté, Thèse de 3ème cycle, Université de Grenoble (July (1984).
46. M. Sato, H. Fujishita, and S. Hoshino, *J. Phys. C: Solid State Phys.* **16**, L877 (1983); also in Reference 20; M. Sato, H. Fujishita, S. Sato, and S. Hoshino (submitted to *J. Phys. C*).
47. See C. Escribe-Filippini, J. P. Pouget, R. Currat, B. Hennion, and J. Marcus in Reference 22.
48. G. Bang and G. Sperlich, *Z. Phys.* **B22**, 1 (1975).
49. D. C. Johnston, *Phys. Rev. Lett.* **52**, 2049 (1984).
50. L. F. Schneemeyer, F. J. Di Salvo, R. M. Fleming, and J. V. Waszczak, *J. Solid State Chem.* **54**, 358 (1984).
51. J. Dumas, C. Escribe-Filippini, and C. Schlenker (to be published).
52. J. B. Goodenough, *Czech, J. Phys.* **B17**, 304 (1967).
53. P. G. Dickens and D. J. Neild, *Trans. Faraday Soc.* **64**, 13 (1968).
54. L. F. Mattheis, *Phys. Rev.* **B6**, 4718 (1972).
55. P. Bruesch, S. Strassler, and H. R. Zeller, *Phys. Rev.* **B12**, 219 (1975).
56. G. Gruner, A. Zawadowski, and P. M. Chaikin, *Phys. Rev. Lett.* **46**, 511 (1981); see also P. Monceau *et al.* in Reference 19.
57. L. Sneddon, M. C. Cross, and D. S. Fisher, *Phys. Rev. Lett.* **49**, 292 (1982).
58. J. Bardeen, *Phys. Rev. Lett.* **24**, 1978 (1980).
59. D. S. Fisher, *Phys. Rev. Lett.* **50**, 1486 (1983); H. Matsukawa and H. Takayama, *Solid State Commun.* **50**, 283 (1984); P. M. Littlewood and T. M. Rice, *Phys. Rev. Lett.* **48**, 44 (1982).
60. N. P. Ong, G. Verma, and K. Maki, *Phys. Rev. Lett.* **52**, 663 (1984).
61. G. Mozurkewich and G. Grüner, *Phys. Rev. Lett.* **51**, 2206 (1983).
62. See References 9, 20, 21 and 22; see also J. Dumas and C. Schlenker, *Solid State Commun.* **45**, 885 (1983).
63. B. Joos and D. Murray, *Phys. Rev.* **B29**, 1004 (1984).
64. R. P. Hall and A. Zettl, *Solid State Commun.* **50**, 813 (1984).
65. M. P. Everson and R. V. Coleman, *Phys. Rev.* **B28**, 6659 (1983).
66. See K. Tsutsumi, T. Tamegai, and S. Kagoshima in Reference 21.
67. R. M. Fleming (private communication).
68. N. P. Ong and P. Monceau, *Solid State Commun.* **26**, (1978); M. P. Everson, G. Eiserman, A. Johnson, and R. V. Coleman, *Phys. Rev.* **B30**, 3582 (1984).
69. R. J. Cava, R. M. Fleming, P. Littlewood, E. A. Rietman, L. F. Schneemeyer, and R. F. Dunn, *Phys. Rev.* **B30**, 3228 (1984); R. J. Cava, R. M. Fleming, R. G. Dunn, E. A. Rietman, and L. F. Schneemeyer (to be published).
70. R. M. Fleming, L. F. Schneemeyer, and R. J. Cava, *Phys. Rev. B* **31**, 1181 (1985).
71. P. Monceau, J. Richard, and M. Renard, *Phys. Rev. Lett.* **45**, 43 (1980).
72. R. P. Hall, M. Sherwin, and A. Zettl in Reference 22.
73. D. Reagor, S. Sridhar, and G. Grüner in Reference 21.
74. J. C. Cotillard, D. Guillet and J. L. Porteseil, *Phys. Lett.* **88A**, 219 (1982).
75. P. A. Lee and T. M. Rice, *Phys. Rev.* **B19**, 3970 (1979).
76. J. Dumas, A. Arbaoui, H. Guyot, J. Marcus, and C. Schlenker, *Phys. Rev.* **B30**, 2249 (1984).

77. G. G. Saha, P. G. McCormick, and R. Rama Rao, *Mat. Science Eng* **62**, 187 (1984).
78. R. M. Fleming and L. F. Schneemeyer, *Phys. Rev.* **B28**, 6996 (1983).
79. A. W. Higgs and J. C. Gill, *Solid State Commun.* **47**, 737 (1983).
80. C. Roucau, *J. Physique* **11**, C3-1725 (1983); Z. Z. Wang, H. Salva, P. Monceau, M. Renard, C. Roucau, and R. Ayroles, *J. Physique Lett.* **44**, L311 (1983).
81. L. Mihaly, Ting Chen, B. Alavi, and G. Grüner in Reference 22.
82. Gy. Hutiray, G. Mihaly, and L. Mihaly, *Solid State Commun.* **47**, 121 (1983).
83. J. C. Gill, *Mol. Cryst. Liq. Cryst.* **81**, 73 (1982).
84. R. J. Cava, R. M. Fleming, E. A. Rietman, R. G. Dunn, and L. F. Schneemeyer, *Phys. Rev. Lett.* **53**, 1677 (1984).
85. T. Tamegai, K. Tsutsumi, S. Kagoshima *et al. Solid Commun.* **51**, 585 (1984); see also in Reference 22.
86. H. Mutka, S. Bouffard, J. Dumas, and C. Schlenker, *J. Physique Lett.* **45**, L729 (1984); S. Bouffard, M. Sanquer, H. Mutka, J. Dumas, and C. Schlenker in Reference 22.
87. H. Mutka, S. Bouffard, G. Mihaly, and L. Mihaly, *J. Physique Lett.* **45**, L113 (1984).
88. P. Y. Le d'Haeron and S. Aubry, *J. Phys. C: Solid State Phys.* **16**, 4827 (1983).
89. P. Bak, *Rep. Prog. Phys.* **45**, 587 (1982).
90. See J. C. Gill in Reference 20, p. 139.
91. See A. Janossy, G. Mihaly, and L. Mihaly in Reference 22.
92. See Y. Abe, T. Haga, T. Kimura, Y. Tajima, and K. Imai in Reference 20.

NEW TRANSITION METAL HALIDES AND OXIDES WITH EXTENDED METAL—METAL BONDING

JOHN D. CORBETT and ROBERT E. McCARLEY

Department of Chemistry and Ames Laboratory-DOE,
Iowa State University, Ames, IA 50011, U.S.A.

1. Introduction

In the last fifteen years or so, synthetic studies in metal-rich systems involving the chalcogenides or halides of approximately the first half of the transition metals have led to some remarkable discoveries of new types of strong, extended metal—metal bonding [1]. The lower dimensional members of this family will be the subject of this article, thus excluding the equally remarkable but more isotropically bonded Chevrel phases $M_xMo_6Ch_8$ (Ch = S, Se, Te) which otherwise logically relate to the phases to be described. In addition, although compounds containing the one-dimensional finite and infinite chain members $(Mo_3Ch_3)_nCh_2^{m-}$ and $(Mo_3Ch_3^-)$, respectively, are appropriate members of the family of compounds under discussion, these are described separately in this volume [2] as well as elsewhere [3]. Subsets of the present article have appeared elsewhere [4—8].

The metals known to form the new extended phases considered here involve ternary oxides of molybdenum together with binary halides of zirconium and hafnium in group IV, the preceding Y and La in group III, several of the lanthanides, and the first member of the 3*d* series, scandium. Examples involving the intermediate group V elements Nb and Ta are not known although their eventual discovery seems likely.

The transition elements exhibit bonding patterns in their compounds which reflect those in the metals. For the elements in groups III—VI all electrons in the valence shell (Cr excepted) readily enter into bond formation, whether this be in metal—metal bonds in the free metal, or in compounds like YCl_3, $ZrCl_4$, $NbCl_5$ or MoF_6 where all valence electrons are involved in metal—halogen bonds. It is natural then that upon reduction of the oxidation state of the metal and consequent lowering of the nonmetal/metal (N/M) ratio, electrons formerly involved in M—N bonding should then be given over to M—M bonding. (The presence of localised paramagnetic states in the elements under consideration here is quite rare.) Within the range of N/M ratios where M—M bonding is observed, the M—M bonding is relatively weak or sparse at the higher ratios and becomes dominant at the lower ratios. Principally this trend is determined by both steric and electronic effects. As the N/M ratio decreases, the number of nearest-neighbor nonmetal atoms decreases, making close approach to neighboring metal atoms geometrically probable, and at the same time an increasing number of electrons become available for M—M bonding. Thus, in the series $NbCl_4$, Nb_3Cl_8, Nb_6Cl_{14}, and Nb_6I_{11}

J. Rouxel (ed.), Crystal Chemistry and Properties of Materials with Quasi-One-Dimensional Structures, 179—204.

metal—metal bonding becomes increasingly important, with 1, 2.33, 2.67, and 3.17 electrons per metal atom (*e/M*), respectively, available for this purpose.

The radius of the nonmetal atoms also is an important factor in determining structure. In cases where the N/M and *e*/M ratios are comparable, different structures may be dictated when the N atoms have greatly different radii. For example, the structures observed for oxides and sulfides are nearly always different even when N/M and *e*/M ratios are comparable. This is perhaps most evident for the MO_2 and MS_2 compounds, where the former often have rutile-derived structures, and the latter have composite SMS close-packed layered structures [9]. In relation to the oxide structures to be discussed in subsequent sections, we note that none of the reduced ternary oxides of molybdenum has yet been found in the same structures as those of the ternary sulfides $M_xMo_6S_8$, $M_2Mo_6S_6$, etc. [3, 10]. Because of the smaller oxide anion radius, condensation of metal cluster units into infinite chains becomes possible, whereas discrete cluster units are retained in compounds of the larger sulfide anion having comparable N/M ratios. This difference is exemplified in a striking way by the disparate structures of $NaMo_4O_6$ and $PbMo_6S_8$. Likewise, iodides seldom if ever form the new extended metal structures of the sort to be described here for chlorides and a few of the less-well-investigated bromides. However, fluorides, in contrast to oxides, generally are not known to form any phases containing extended metal—metal bonding.

The radius of the metal atom relative to that of the anion also is important [11]. If the bonding radius of M becomes comparable to the van der Waals radius of the anion, extended M—M bonded structures are favored. Thus, infinite chains related to those in $NaMo_4O_6$ are found in Sc_5Cl_8, Gd_2Cl_3, and other lanthanide subhalide compounds, in spite of the much larger van der Waals radius of chlorine. The low M—M bond order dictated by low *e*/M ratios and the larger radius of the group III metal atoms permits commensurate M—M bonding and X···X nonbonding distances along the chain direction. Stability of these infinite chain structures under widely different M—M bonding electron counts is strikingly demonstrated by the comparison between the Gd_4Cl_6 chains of Gd_2Cl_3 with 6 electrons per repeat unit, and the $Mo_4O_6^-$ anion chains of $NaMo_4O_6$ with 13 electrons per unit.

2. Background

Prior information regarding related examples of extended, lower dimensional metal—metal bonding is not very extensive. Halides of the 3*d* elements other than scandium seem to be devoid of structurally significant metal—metal bonding. At first sight the β-$ZrCl_3$ (or β-$TiCl_3$) structure type known for the chloride, bromides, and iodides (X) of Ti, Zr and, partially, Hf would appear to provide an optimal arrangement for good metal—metal bonding and a half-filled band. The structure consists of hcp halogen layers with the metal occupying every third chain of octahedral interstices that run perpendicular to the layers, thus producing infinite chains of MX_6 trigonal antiprisms sharing opposite faces [9]. However, the strength of the metal—metal interactions through the common faces appears to be minimal even though the metal—metal separations are quite short, 3.07 Å in $ZrCl_3$ versus 3.23 and 3.18 Å in the hcp metal. Instead, this particular compound is a complex paramagnet with $\sigma \sim 10^{-5}\,\Omega^{-1}\,cm^{-1}$ [12, 13].

Layered compounds with strong metal—metal bonds within fairly isotropically bonded, single metal layers are of only peripheral interest in the present article. The repeat distances within the metal layers always appears to be dictated or limited by van der Waals contacts within the parallel and commensurate nonmetal layers. Metallic or semiconducting compounds with fairly narrow metal-based valence bands are well known in these cases, e.g. for NbS_2 and MoS_2, respectively, in numerous polytypes, in $ZrCl_2$, and in many intercalated, layered compounds, such as Li_xTiS_2 [14]. The novel Ag_2F with close-packed layers sequenced AgFAg... (anti-CdI_2 type) exhibits relatively long Ag—Ag distances and does not appear to be particularly well understood. An unusual set of layered metallic diiodides MI_2, M = La, Ce, Pr, with a somewhat intermetallic-like $MoSi_2$ layered structure together with the related GdI_2 (2H—MoS_2) and ThI_2 (4H—NbS_2) phases all appear to have considerable involvement of iodine in the make-up of the conduction band as the large anion requires quite large metal—metal separations; analogous chlorides and bromides are not known [4].

A similar nonmetal participation appears to be important in the metallic character of those dioxides that possess the undistorted rutile structure, VO_2 (at high temperature), TaO_2 and RuO_2 for example, where the M—M distances are also relatively long. The case with $LiNbO_2$ (intercalated 2H—MoS_2, d(Nb—Nb) = 2.90 Å) [15] is probably comparable to the layered halides discussed above but is not well studied.

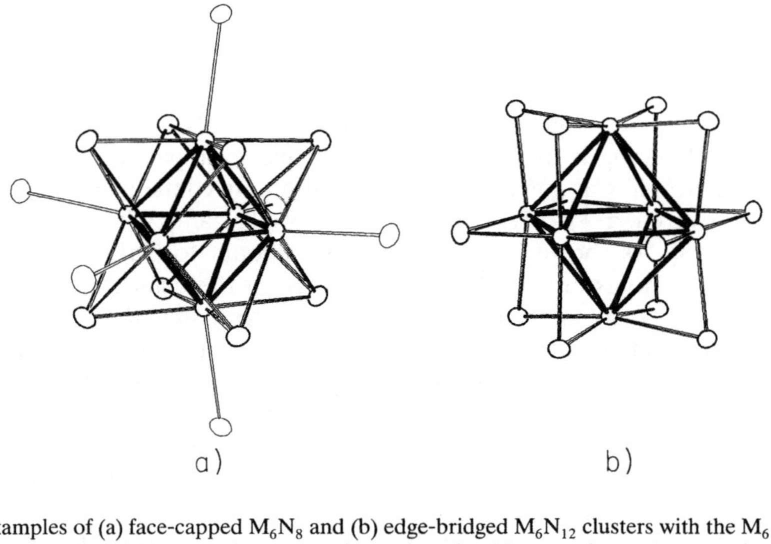

Fig. 1. Examples of (a) face-capped M_6N_8 and (b) edge-bridged M_6N_{12} clusters with the M_6 octahedra in heavy outline. The M_6N_8 example (a) also shows the other N atoms that occupy terminal positions on the metal octahedron.

This article concerns itself then with structurally novel halides and oxides in which low-dimensional metal—metal bonding is clearly much more robust than that in the foregoing examples; in fact, the presence of a structurally dominant metal network is a common impression. Moreover, it is remarkable that the conceptual construction of infinite metal chains, double chains and double-metal-layered phases can nearly always be based on octahedral metal clusters as fundamental building blocks. Examples of the isolated clusters have been known for a long time in diverse solids, e.g. as $(Nb, Ta)_6X_{12}^{n+}$, X = F, Cl, Br, I (but not in all cases), $n = 2, 3, 4$, and $Mo_6X_8^{4+}$, $X \neq F$ [9, 16]. Fluorides generally participate in this sort of chemistry only rarely and only one example is known for an oxide, $Mg_3Nb_6O_{11}$, where oxide is shared between clusters [17].

Because of the importance of these cluster units in visualizing the extended arrays, an example of each type is shown in Figure 1, an M_6N_8 (or 6—8) type cluster (a metal octahedron face-capped by N) in (a) and an M_6N_{12} (or 6—12) unit (edge-bridged) in (b). The left view illustrates a further maxim regarding the clusters and the structures of all compounds considered here, namely, that metal atoms at all accessible vertices of the metal framework are also always bonded to a terminal or exo halide, oxide, etc. The latter may come from several sources, an inner (face or edge-positioned) nonmetal in another cluster, one that functions only in bridging between units, or one that is only terminal to that vertex. Only the first two modes of bonding nonmetal at metal vertices are important in the present metal-rich compounds where these N atoms also serve to link the extended arrays together to form the three-dimensional structures.

3. Infinite Metal Chains

3.1. ZIG-ZAG CHAINS AND FUSED RHOMBOIDS

Single atom chains are relatively rare to date, presumably because more extensive metal—metal bonding is usually possible. A linear example is found (along with other chain types) in $Ca_{5.45}Mo_{18}O_{32}$ (below) but others contain zig-zag chains. The recently discovered α-, β-ZrI_2 occur in the β-$MoTe_2$ and WTe_2 structure types, respectively [18]. As shown in Figure 2, these amount to CdI_2-like structures in which the metal atoms are regularly displaced from the octahedral sites between the iodine layers so as to bond to two others through shared edges of the MX_6 octahedra. Puckering of the iodine layers accompanies the metal displacement so that the three-layer slabs can be packed in alternate ways to give the two structure types. Differences between these within the ZrI_2 sheets are very small and the two forms readily intergrow. These structures probably represent the simplest means by which a d^2 ion can form two regular metal—metal bonds; it may well be that further metal—metal aggregation is precluded by the anions, which are as large as any usually involved. The magnitude of the matrix effects provided by the nonmetal array that limits the approach of the zirconium atoms is suggested by the length

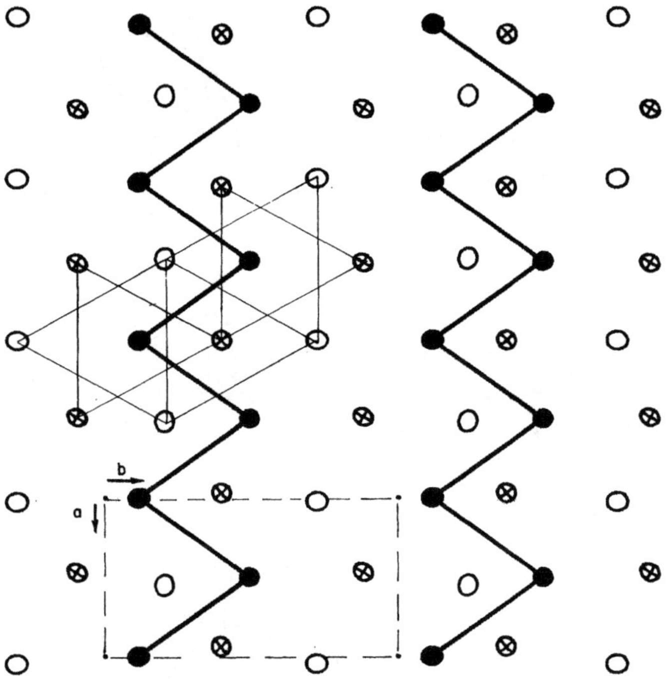

Fig. 2. A projection of the puckered I—Zr—I layers in α- and β-ZrI_2. Open and crossed circles are iodine at $\pm \sim z$; solid circles are zirconium atoms that have been displaced from the nominal octahedral centers through Zr—Zr bond formation.

of the metal—metal bonds, which are only of order 0.31. The iodides are semiconductors with an electrical band gap of ~ 0.1 eV while β-$MoTe_2$ is a semimetal.

In compounds in which more than two *d* electrons become available for M—M bonding, additional interactions between atoms along the chains result in further clustering. A zig-zag chain is constructed from two parallel rows of metal atoms with the M—M distance between members of the same row large compared to that between an atom and its two neighbors in the adjacent row. With higher *d*-electron count, further clustering occurs by contraction of the repeat distance between atoms within each row. In $Ba_{1.14}Mo_8O_{16}$ [19] and $K_2Mo_8O_{16}$ [20] discrete rhomboidal clusters result, while in $NaMo_2O_4$ infinite chains of edge-fused rhomboids are formed, Figure 3 [21]. The former pair occur in a hollandite-like structure while the latter has a CdI_2-like layer structure with at least three polymorphic modifications. The clustering in the related chalcogenides ReS_2, $ReSe_2$, and MMo_2S_4 (M = V, Cr, Fe, Co), each with three electrons per metal, gives chains of rhomboidal cluster units with only one M—M bond from each unit to each of its neighbor units in the chain [22]. An exception to this is found in β-ReO_2 where simple zig-zag chains persist in spite of the higher electron count on rhenium (d^3) [23].

Only two examples are now known in which the clustering results in fused rhomboidal units with equal spacing between metal atoms along the same row of

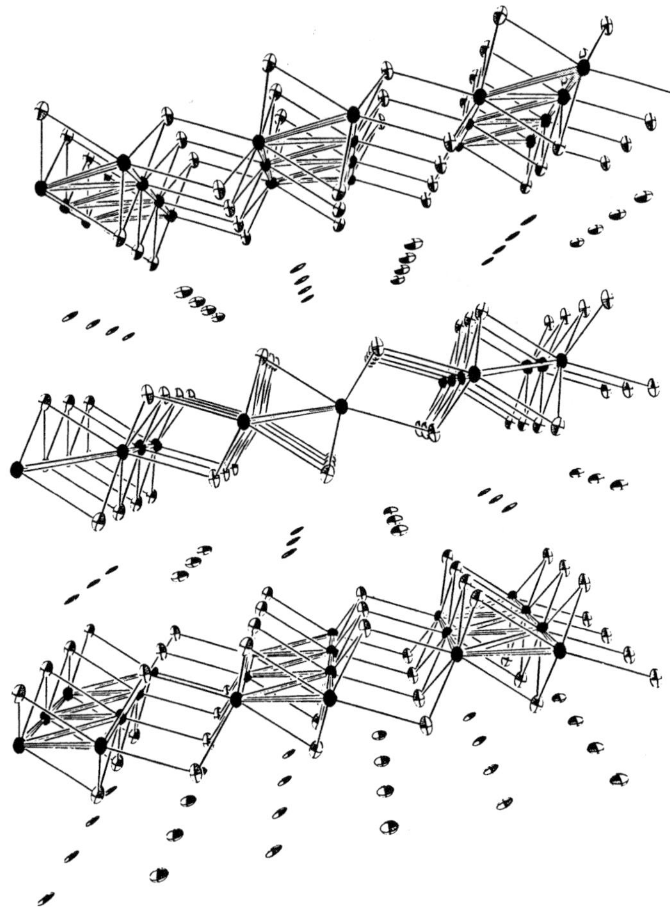

Fig. 3. A view down the crystallographic *b* axis of the structure of $NaMo_2O_4$. Molybdenum atoms are filled ellipsoids, sodium atoms unconnected ellipsoids.

the infinite chains. The structure of $NaMo_2O_4$ shown in Figure 3 may be regarded as the sodium intercalate of CdI_2-like, MO_2 close-packed slabs. If each sodium position were fully occupied the composition would be $Na_2Mo_2O_4$. However, both the structural refinement and microprobe analyses of individual crystals indicate the composition $Na_xMo_2O_4$ with $x \sim 1$. That each sodium position is thus only about half occupied suggests that the stoichiometry may be variable over some range $0 \leqslant x \leqslant 2$. Interestingly, resistivity measurements of a crystal of $NaMo_2O_4$ show it to be a semiconductor which, in turn, suggests that the particular composition observed with $x \sim 1$ results from the stability of a filled metal valence band [21, 24]. The second example of a compound containing fused rhomboidal clusters is $Ca_{5.45}Mo_{18}O_{32}$, a phase that is particularly interesting because it contains not one, but three different kinds of infinite chains: linear (as in the rutile structure), fused rhomboids, and edge-shared octahedra [21]. Its formulation as $Ca_{5.45}(MoO_3)_2(Mo_2O_{3.5})_4(Mo_4O_6)_2$ reflects the presence of these respective structural units. This will be described and discussed further in a later section.

A theoretical examination of the clustering modes available to metal atoms as a function of *d*-electron count in the oxides M'_xMO_2 has been made. Although very useful insights to bonding interactions were gained, reliable prediction of the clustering mode versus *d*-electron count could not be made because of the structural complexities [25].

3.2. Chains of Octahedra

Infinite metal—metal bonded chains derived by condensation of metal octahedra through sharing of opposite (*trans*) edges are the most common structural element known to date, and these occur in a diverse collection of structure types and compounds. The condensation concept is illustrated in Figure 4 in terms of the

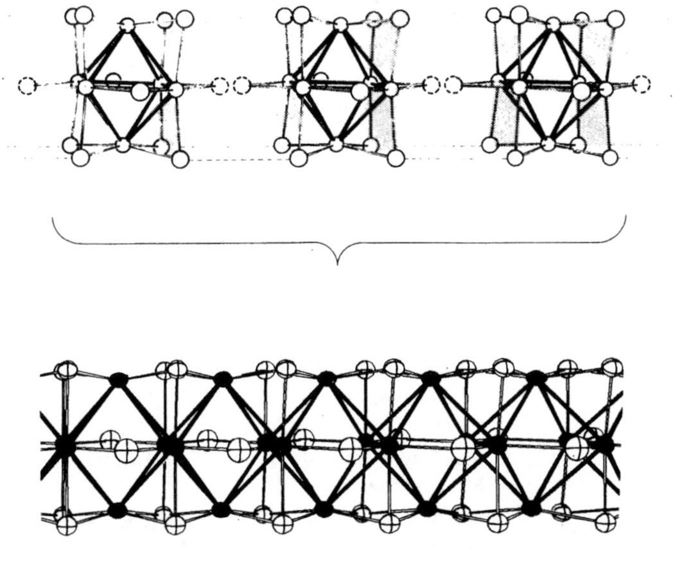

Fig. 4. The conceptual condensation of M_6N_{12} clusters to form an infinite chain of composition M_4N_6. The shaded faces of the M_6N_{12} units are shared and the dashed N atoms are eliminated in the process.

formation of an M_4N_6 chain from M_6N_{12}-type clusters through sharing of the opposite, approximately rectangular, M_2N_4 faces of individual clusters. The structure types known to contain infinite chains and known members of each group are listed in Table I [26—42]. In all cases the nonmetal sheathing of these chains may again be categorized in terms of either a 6—8 (face-capping) or a 6—12 (edge-bridging) parentage (see Figure 1). The diverse modes by which these chains are interconnected, always through bridging nonmetal, accounts for some of the variety of structure types and compositions. The question of interstitials in some of the halides listed at the bottom will be considered later.

TABLE I
Infinite octahedral chain structures

Parent	Type	Structural features	Other examples	Ref.
Gd_2Cl_3	6—8	$[Gd_4Cl_4Cl_{4/2}]$ chains	Cl, Br of Y, Tb, Er, Tm, Lu	26—28
Sc_5Cl_8	6—12	$[(Sc^{III}Cl_2^+)(Sc_4Cl_6^-)]$ double chains	Gd_5Br_8, Tb_5Br_8	29, 30
Sc_7Cl_{10}	6—8	$[(Sc^{III}Cl_2^+)(Sc_6Cl_8^-)]$, pairs of Gd_4Cl_6-type chains sharing edges		31
MMo_4O_6 (M = Na, Li, K, Rb, In)	6—12	$[Mo_4O_6^-]$ chains	$Ba_{0.62}Mo_4O_6$ $Pb_{0.77}Mo_4O_6$ $Sn_{0.9}Mo_4O_6$	19, 21, 32
$Mn_{1.5}Mo_8O_{11}$	6—12	$[Mo_4O_6^-]$ with a different connectivity		33
$M_xM'_{2-x}Mo_4O_7$ (M = Sc, Fe, Al, Ti; M′ = Zn, Fe)	6—12	$[Mo_4O_6^-]$ in sheets cross-linked by MO_6 and $M'O_4$ polyhedra		7, 34
$Ca_{5.45}Mo_{18}O_{32}$	6—12	Separate $[Mo_4O_6^-]$, fused rhomboid, and linear units		21
$LiMo_8O_{10}$	6—12	Perpendicular, non-intersecting $[Mo_4O_6^-]$	$ZnMo_8O_{10}$	34
Interstitial Nonmetals				
Sc_4Cl_6B	6—12	B-centered $[Sc_4Cl_6]$	Sc_4Cl_6N	35
$Sc_7Cl_{10}C_2$ [a]	6—12	Sc_7Cl_{10} type but edge-bridged, C-centered	Er_7I_{10}	36, 37
Gd_4I_5C [a]	6—12	Sc_5Cl_8 type without M^{III} side chain	Ce_4Br_5, Er_4I_5	37—40
$Gd_6Br_7C_2$ [a]	6—12	$Sc_7Cl_{10}C_2$ type without the M^{III} side chain	$Gd_6I_7C_2$, Tb_6Br_7, Er_6I_7	39, 41, 42
$Gd_{12}I_{17}C_6$	6—12	Zig-zag chains of *cis*- and *trans*-edge-shared octahedra, C_2 in each		38

[a] See text regarding these classes.

3.2.1. *Chains Based on M_6N_8-Type Clusters*

Only a few examples of condensation of 6—8 type clusters to a single chain are known. The first structure discovered with this mode of condensation was that of Gd_4Cl_6, as shown in Figure 5. Chlorine atoms cap all exposed faces of the metal

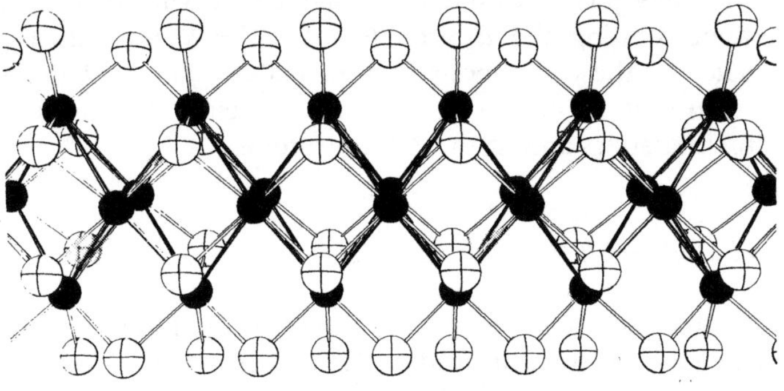

Fig. 5. A portion of the infinite chain in Gd_2Cl_3 (= $Gd_4Cl_4Cl_{4/2}$), with the metal atoms as solid circles. A duplicate set of chlorine atoms at the top and bottom bridge to other chains.

octahedra to give Gd_4Cl_4 chain units while two additional chlorines per repeat unit bridge between metal vertices in separate chains to form sheets in the plane of the page. (A duplicate set of these chlorines is shown at the top and bottom of the figure.) The vacant positions terminal to the shared Gd_2 edges are further occupied by face-bridging chlorines in other chains to link the aforementioned sheets together. The metal chain appears to be quite well sheathed by chloride, all Cl—Cl nearest-neighbor separations thereabout ranging between 3.2 and 3.6 Å, the latter being the customary sum of van der Waals radii. The nominal metal octahedra in this chain are in fact distorted, a feature common to all of those with such relatively low electron counts. Here the repeat distance along the chain is 3.90 Å in contrast to 3.35 Å for the shared Gd—Gd edge and intermediate separations of 3.71 to 3.78 Å between the waist and apex atoms. (The 3.90 Å repeat along the chain has been somewhat arbitrarily omitted from the figure construction.) The Gd—Gd single bond distance computed from the distances in gadolinium metal is 3.24 Å. The phase Sc_7Cl_{10} contains pairs of chains of the same 6—8 construction that have been further condensed through sharing of Sc—Sc side edges. This chain will be illustrated later in connection with that in the related carbide.

This M_4X_6 phase is known for the chloride and/or the bromide of Y, La, Gd, Tb, Er, Tm, Lu and can possibly be obtained for others as well. It is significant that this condensation of M_6X_8 type clusters to such chains has only been observed for metals of larger radii and low *e*/M ratio. This mode of condensation requires that half of the nonmetal atoms be displaced from faces of the metal octahedron towards bridging terminal positions for steric reasons, namely those directly above the octahedral faces hinged on the shared edges along the chain. There appears to be both a loss of M—N bonding and a nonbonding radius restriction that may impede closer approach of the M atoms along the chain. Although these factors are not at all well understood, high M—M bond orders resulting from increased *e*/M ratios would appear to favor chain formation from the M_6X_{12} condensation, as discussed below. Resistivity and photoelectron spectral measurements on Gd_4Cl_6 and Tb_4Cl_6 show the phases to be semiconductors with a gap of approximately

1 eV [43, 44]. A theoretical treatment is in agreement with this, the metal—metal bonding in effect splitting the *d* orbitals [45]. The semiconducting and highly reduced character of these compounds is probably important in the inability to prepare intercalation derivatives of the compound even though there are suitably sized octahedral interstices among the chlorine atoms.

3.2.2. *Chains Based on M_6N_{12}-Type Clusters*

A particularly remarkable and significant aspect of the infinite chain structures is the predominance of the edge-bridged M_4N_6 repeat unit derived by condensation of M_6N_{12} clusters. An example of this was shown in Figure 4. The construction applies to all but two of the 6—12 type chain structures listed in Table I, one of these ($Sc_7Cl_{10}C_2$) involving double chains formed by further condensation of these same M_4N_6 single chains. Furthermore, the M_4N_6 examples are seen to span a considerable range of stoichiometries and elements, the latter ranging from scandium chlorides through several lanthanide halides to the molybdenum oxides. The rela-

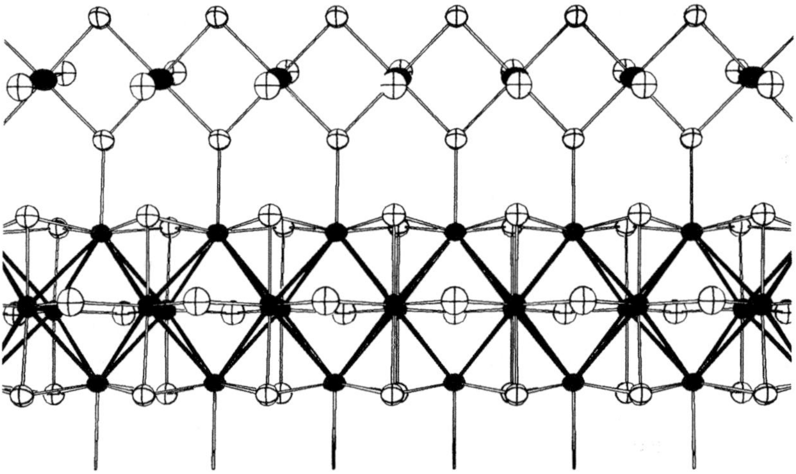

Fig. 6. A portion of the infinite chain in $Sc_5Cl_8 \cong {}^1_\infty[Sc^{III}Cl_2^+Sc_4Cl_6^-]$, with scandium atoms as solid ellipsoids. Dangling bonds at the bottom go to chlorine atoms at the top in other chains.

tive dimensions of these chains do vary through the series, the smaller and more regular metal octahedra appearing in some of the electron-richer molybdenum compounds.

One variety of infinite structures based on M_4N_6 chains also contain a second chain of isolated, normal-valent metal atoms. Such an example is illustrated in Figure 6 for Sc_5Cl_8, where a second commensurate chain of centered $Sc^{III}Cl_6$ octahedra sharing chlorine edges bridge to vertices of the $Sc_4Cl_6^-$ chain [29]. Dangling bonds at the bottom and the chlorine atoms at the top of the figure likewise bridge to other units to generate sheets. Further interconnectivity of these involves the chlorine atoms that bridge the side edges of the metal chains, these also being the apices of chlorine octahedra in other sheets. All chlorine are thus three-coordinate, a characteristic of all of these structures with low N/M ratios in which all nonmetal atoms are well bound, serving if necessary to fill outward-pointing metal orbitals as it were. It is surmised that the donation of additional electrons to what would otherwise be a rather electron-poor Sc_4Cl_6 chain is one reason for the presence of the isolated chain containing scandium(III) atoms. This assessment is indicated by the formulation $^1_\infty[ScCl_2^+Sc_4Cl_6^-]$. It is noteworthy that yttrium does not seem to form compounds with either the Sc_5Cl_8 or Sc_7Cl_{10} structures while the Y_2Cl_3 structure (Gd_2Cl_3, Figure 5) has not been found for scandium.

Figure 7 illustrates the remarkable $NaMo_4O_6$ chain structure, now in a view along the chains [32]. Although it is not obvious in this presentation, the Mo_4O_6 units have the same structure and interconnectivity as those in the chemically distant $Sc_4Cl_6^-$ units in Sc_5Cl_8, the isolated chains enclosing Sc^{III} atoms in the latter now being replaced by two more $[Mo_4O_6]$ strings. The bridging oxygen atoms are now planar and the sodium cations are well caged in relatively large cubic oxide polyhedra. The repeat along the molybdenum chains is appropriately less than in the scandium phases, decreasing from 3.52 to 2.86 Å in response to both the

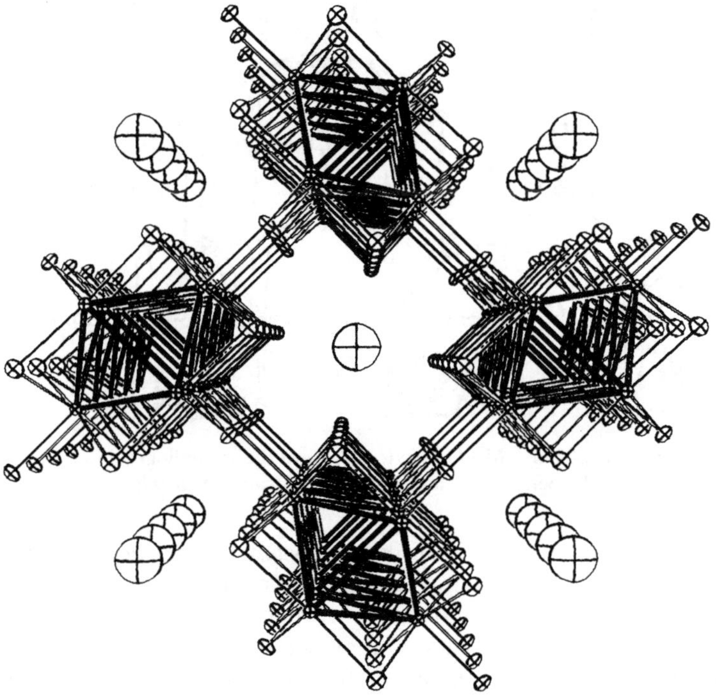

Fig. 7. The structure of tetragonal $NaMo_4O_6$ as viewed down the crystallographic c axis. The Mo—Mo separations within the octahedral cluster units are shown by heavy filled lines, Mo—O bonds by light unfilled lines. Unconnected ellipsoids represent sodium atom positions.

smaller metallic radius of molybdenum and an increase in the number of electrons bonding the metal chain from 7 to 13 per formula unit. These factors also require that the nonmetal atoms involved be smaller than chlorine since these have the same repeat distance along the chains as does the metal framework. This match between electron count, metallic radius, and counter atom size is doubtlessly of paramount importance in determining or limiting structures of this type to certain elements (Section 5.1). The $NaMo_4O_6$ compound is stable for hours in air at 300 °C while the indium and tin derivatives are clearly metallic [32—34]. There was no precedent for either this structure among oxides or for this oxidation state for molybdenum (+2.75), all previously known and structurally characterized molybdenum oxides corresponding to oxidation states of +4 or higher.

The same Mo_4O_6 chains are persistent building blocks in a considerable variety of compositions and structures (Table I). The different interconnectivity found in $Sc_{0.75}Zn_{1.25}Mo_4O_7$ and related compounds is illustrated in Figure 8 [7]. Here parallel $[Mo_4O_6]$ chains running in and out of the plane of the page are connected into

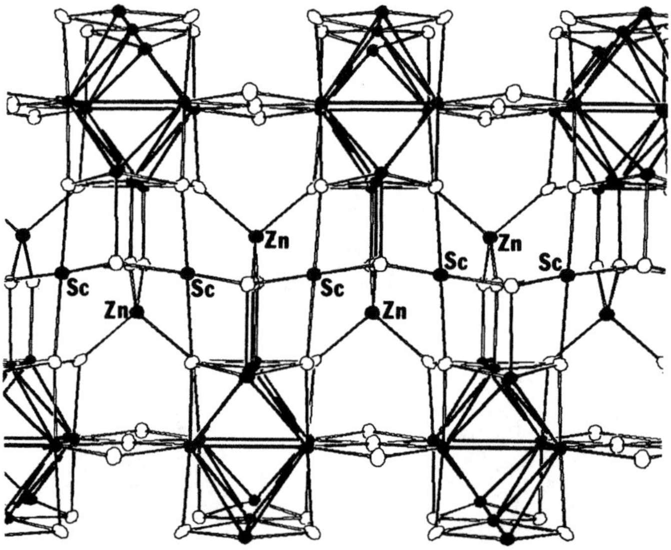

Fig. 8. The structure of $Sc_{0.75}Zn_{1.25}Mo_4O_7$ as viewed down the crystallographic *c* axis. The Mo—Mo edges of the octahedral cluster units are shown by heavy filled lines.

horizontal sheets by square-planar bridging oxygen. These sheets are in turn interbonded by cations within tetrahedral (Zn) and octahedral (Sc) polyhedra composed of oxygen atoms that are also bound to vertices of the molybdenum chain. A number of relatively high field cations may serve in the latter role, these usually showing somewhat mixed occupancy of the two sites or mixed valency as with Fe(II, III) in $(Fe_{1.89}Mo_{0.11})Mo_4O_7$. Figure 9 shows another version of this theme in $LiMo_8O_{16}$ where the same Mo_4O_6 chains can be seen running normal to and within the plane of the page [34].

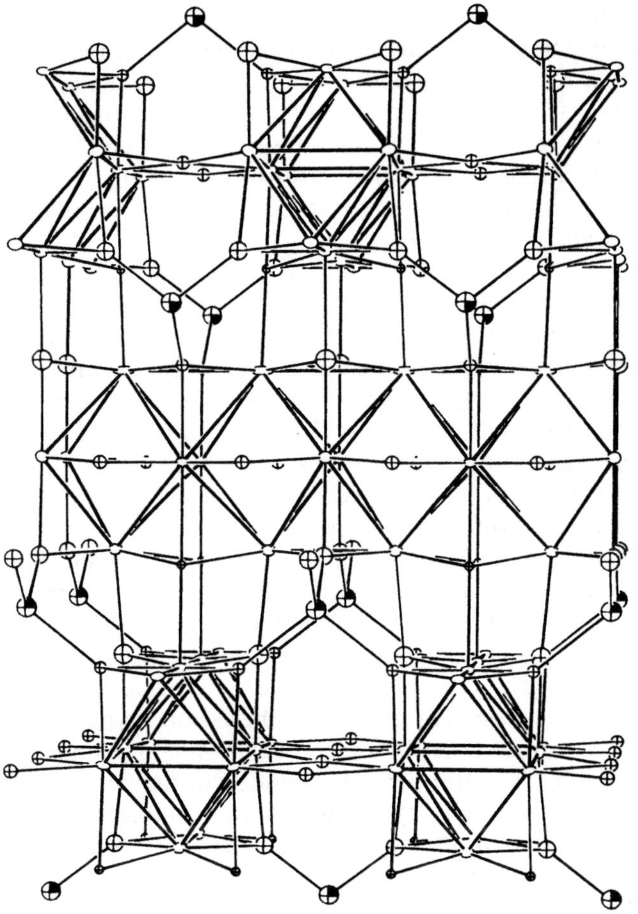

Fig. 9. A section of the tetragonal structure of $LiMo_8O_{10}$ as viewed down the *a* axis with *c* vertical. Notice that the octahedral cluster chains in the upper and lower third of the figure run into and out of the plane of the paper while the cluster chain in the middle third runs horizontally in the plane of the page. Open ellipsoids are Mo, crossed ellipsoids O, and shaded spheres Li.

3.3. Distortions in Chains

Significant distortions of the octahedral units within these Mo_4O_6 chains occur as a function of structure type or electron count. In $NaMo_4O_6$ (Figure 7), the most symmetric of these structures, all metal atoms occupy special positions on crystallographic mirror planes along and normal to the chain with the result that only one distance is found for each of the apex—apex, apex—waist, and waist—waist edges. Although $Sn_{0.9}Mo_4O_6$ belongs to the same space group and is isomorphous with $NaMo_4O_6$, apex—apex pairing occurs in the tin compound (in disordered arrangement from one chain to another) such that alternately short (2.61 Å) and long (3.06 Å) repeat distances result along the chain. The tin atoms are displaced from the center of the surrounding cube of oxygen atoms (where Na^+ is found in $NaMo_4O_6$) to near the cube face and are thus four-coordinate, square planar [46]. All other structural features remain the same, and the distortion apparently results from the higher electron count (13.8) in the tin compound. The pairing of apex atoms along the chains, of course, also causes shortening of one set and lengthening of another set of the apex—waist bonds to 2.74 and 2.85 Å, respectively, from the value of 2.78 Å found in $NaMo_4O_6$.

Pairwise distortions of the chains in the $Sc_{0.75}Zn_{1.25}Mo_4O_7$ structure type are also quite evident. Here the pairing between apex atoms occurs in ordered fashion with doubling of the repeat distance in the chain direction. The difference between alternating short and long apex—apex distances is greatest in $Sc_{0.75}Zn_{1.25}Mo_4O_7$, 0.508 Å, and covers the range down to 0.200 Å for the various members of the $M_xM'_{2-x}Mo_4O_7$ series listed in Table I. Electron counts range from approximately 14.5 to 14.75 per repeat unit in this series, again with the Sc, Zn member having the greatest value. There is thus a direct but not clearly understood correlation of increasing electron count with enhanced pairing of the apex atoms along the chains.

Distortions in the octahedral cluster units among the remaining oxide structures again are most pronounced in the compounds having the higher electron counts, viz. $LiMo_8O_{10}$ and $ZnMo_8O_{10}$, with 14.5 and 15.0 electrons per Mo_4 unit, respectively. In these structures not only apex—apex pairing, but also strong apex—waist and waist—waist alternation is observed, resulting in maximum bond differences of 0.60, 0.16, and 0.24 Å, respectively, within these bond types [34]. Consideration of the observed distortions over the range of known compounds leads to the conclusion that $NaMo_4O_6$ is nearly perfectly bonded in the symmetric structure with 13 electrons per unit, and that beyond this, additional electrons cause distortions which result in the filling of additional bonding states and the avoidance of anti-bonding states in the more symmetric structure [47].

Finally, we return to the complex compound $Ca_{5.45}Mo_{18}O_{32}$ with the structure represented in Figure 10 [21]. The Mo_4O_6 chains in this compound are found at the center and midpoints of two edges of the cell outlined in the figure. In contrast to the very evident distortions in the octahedral units of other structures with more than 13 electrons per unit, distortions in the units of $Ca_{5.45}Mo_{18}O_{32}$ are not pronounced. Here apex—apex and waist—waist distances are equal at 2.847(1) Å, and an insignificant difference in apex—waist distances is observed, 2.758(2) and 2.762(2) Å, each four times per unit. From Mo—O bond order sums (*vide infra*) in this compound, an estimate of 14.6 electrons per Mo_4O_6 unit has been deduced. Unfortunately, the true distortions of the octahedral units in this case are obscured by the effects of disorder. The structure was solved in a subcell with a repeat distance of only 2.85 Å along the chain direction. Consequently, positional disorder of the Ca and Mo positions results in statistical occupany of the Ca positions (5.45/8 occupation factor), while the Mo atoms were constrained to special positions which would be relaxed in an ordered structure with larger cell dimension in the chain direction. The 'average structure' thus appears to be essentially un-

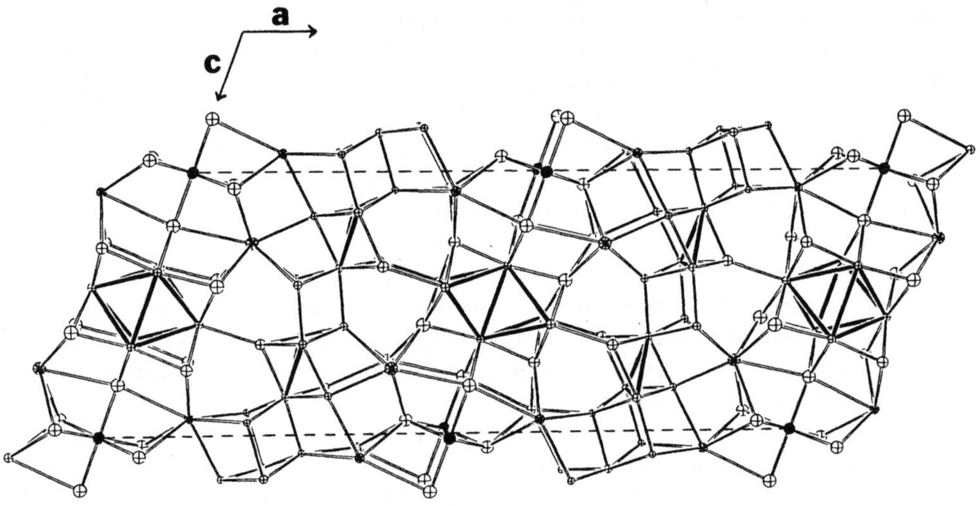

Fig. 10. A section of the structure of $Ca_{5.45}Mo_{18}O_{32}$ as viewed down the crystallographic *b* axis with close Mo—Mo separations in heavy outline. The repeat units in the Mo_4O_6 cluster chains are centered at the positions (0, ½, ½) and (½, 0, ½). The fused rhomboid chains are centered at positions equivalent to (¼, 0, ¼), and the single Mo atom chains (solid atoms) are at the equivalent positions (0, ½, 0). Calcium is marked ⊕.

distorted, but the Mo—Mo bond order sums nevertheless reflect the enhanced bonding resulting from the higher electron count of 14.6 electrons in each Mo_4O_6 unit.

An assessment of the electron count available in each type of chain, as derived from Mo—O and Mo—Mo bond order summations (below), provides a charge estimate for the fused rhomboidal chains about the same as that found in the $NaMo_2O_4$. Apparently not only the *d* bands arising from the fused rhomboidal chains, but also those of the other two chains are filled, because the material is a semiconductor as determined from single-crystal, four-probe resistivity measurements over the range 30—300 K. Thus, again, the particular composition $Ca_xMo_{18}O_{32}$ with $x = 5.45$ may result from the limiting electron count for collective band filling of the three chains.

3.4. Interstitial Atoms in Chains

In the past few years it has become increasingly clear that certain (but by no means all) metal—metal bonded clusters, chains, or sheets may take up or even be uniquely stabilized by small nonmetal atoms bound within interstices in the metal arrays, mainly with compounds of the earlier transition metals. Molybdenum structures appear to provide insufficient space within the cluster units, there being no anomalous structural results to suggest otherwise, while for niobium and tantalum only hydrogen is known to occur within two Nb_6I_8-type clusters. The situation with the larger metals in transition groups III and IV is more variable. There are some compounds which clearly contain such interstitial atoms, others for which the correct formulation is not known at present, and some which appear to be true binary phases.

The presence of unrecognized interstitial atoms arising from adventitious impurities is often signaled by low and erratic yields of a phase, and, in favorable cases, by an undefined residual peak within the cluster in a Fourier synthesis map following a structure refinement. Attainment of a high yield when the correct impurity is added to the reactants as well as specific identification of the interstitial — for example, via core and valence levels in PES or microprobe data — are often used to clarify the situation [36].

On the other hand, there is no good evidence that impurity stabilization or contamination pertains to the Gd_2Cl_3 or Sc_7Cl_{10} type structures in that these can be regularly prepared in moderate (> 20—30%) to high yields, by more than one route, and even when extreme care is used to ensure high-purity systems. The phase Tb_5Br_8, isostructural with Sc_5Cl_8, has also been structurally characterized and found to exhibit no electron density in the centers of the clusters comprising the metal chains [30]. Even so, some caution is still appropriate since the isostructural Sc_5Cl_8N has recently also been synthesized; in this case the metal chains have somewhat larger dimensions than found earlier in Sc_5Cl_8 [35]. Also, the magnetic susceptibility and spin resonance data for Sc_7Cl_{10} are somewhat unusual, suggesting either unusual phenomenology or low levels of interstitial impurities [48]. However, a different structure is obtained for $Sc_7Cl_{10}C_2$ (below) on purposeful doping with carbon while addition of nitrogen, boron, oxygen, or hydrogen to the reactions favor other types of products. Hydrogen is generally the most difficult impurity to detect and to really eliminate as it is readily derived from metals, the pervasive water, etc., in these highly reduced systems (see the monohalide problems below).

In the cases of $Sc_7Cl_{10}C_2$, Gd_4I_5C, and $Gd_6Br_7C_2$, the indicated compositions with a single carbon more or less centered in each metal octahedron appear well established [36, 39]. On the other hand, other binary halides were earlier reported with virtually the same atom positions and/or closely comparable lattice constants — viz. Er_7I_{10}, Er_4I_5, and Er_6I_7 — and these usually also showed residual electron densities in the centers of the condensed clusters. For the purposes of this review, it is assumed that the latter were probably also carbides and they are so classified in Table I. However, other nonmetal impurities in these can certainly not be ruled out.

The chain structures and connectivities achieved in the presence of known interstitials are very similar to those already presented for binary phases. The Sc_4Cl_6B unit is an orthorhombic version of the tetragonal $NaMo_4O_6$ structure (Figure 7) but lacking the intercalated sodium and with the chains slightly tilted so as to give nonplanar bridging chlorine atoms between chains. The rather complex structure of $Sc_7Cl_{10}C_2$ is simplified in Figure 11 in two views of only the infinite metal framework, pairs of Sc_4Cl_6-type chains further condensed through shared edges, plus a carbon atom in each nominal octahedron. The structure of $Sc_7Cl_{10}C_2$

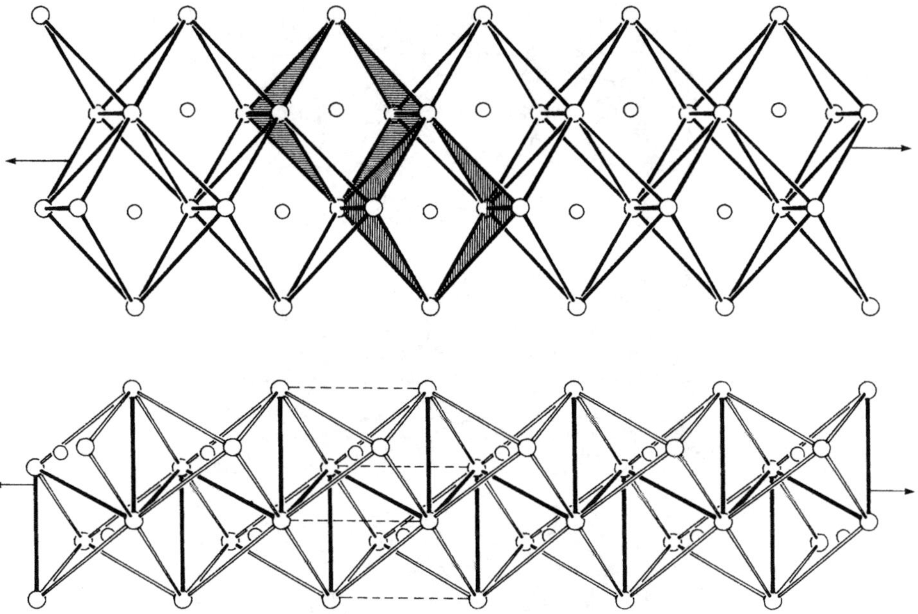

Fig. 11. Two approximately normal views of the double chain of metal octahedra in $Sc_7Cl_{10}C_2$ together with the centered carbon atoms. Shared edges between octahedra are in heavy outline in the lower view.

is related to that of Sc_7Cl_{10} in a simple and direct way. Occupation of the nominal metal octahedra in Sc_7Cl_{10} would place carbon unreasonably close (~ 2.9 Å) to the chlorine atoms that cap exposed faces of the clusters (i.e. 6—8 type). Carbide formation is therefore accompanied by displacement of the metal chain relative to the chloride by half the repeat length so as to convert the halide to the more distant edge-bridging functionality [36]. The metal halide framework in the Gd_4I_5C and $Gd_6Br_7C_2$ structure types may be derived from Sc_5Cl_8 and Sc_7Cl_{10} types, respectively, by removal of the side chain of isolated atoms from the latter two [8, 37]. An example of this side chain is shown at the top of Figure 6.

4. Related Sheet Structures

The foregoing one-dimensional examples of infinite chains are remarkably specific in that they have been found to date only among group III halides or group VI oxides, specifically with molybdenum. The N/M range that is possible in such structures is severely restricted, being just sufficient to provide nonmetal to sheath the metal chain and to fill terminal positions either with the same or additional nonmetal atoms. Furthermore, the nonmetal size clearly must be commensurate with the chain repeat for the sorts of compounds shown in Figures 3 to 11 where

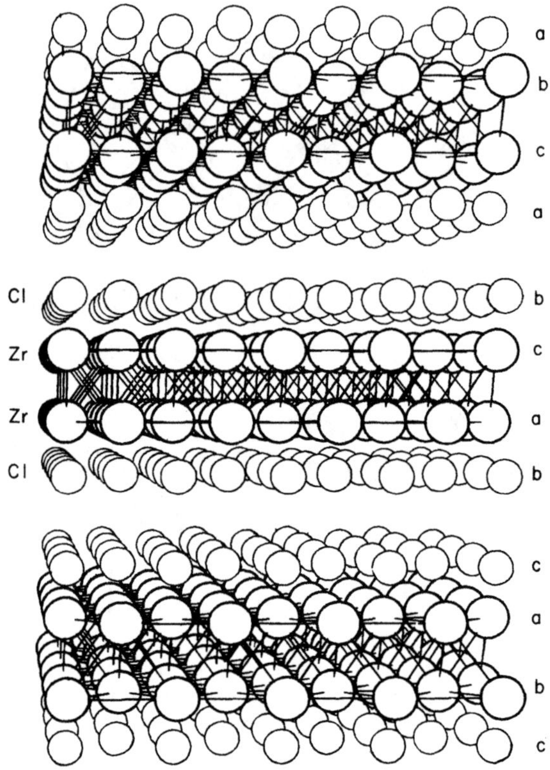

Fig. 12. A section of the rhombohedral structure of ZrCl. The metal—metal separations in each four-layer slab are 3.42 Å within and 3.09 Å between the close-packed zirconium layers.

the chain repeat length also correlates with the electron count (below). As will be considered more generally later, these appear to be some of the more important reasons for the absence of analogous chains with group IV halides. Instead, the only low-dimensional binary structures known for these with N/M < 2 are those of the remarkable monohalides ZrCl and ZrBr (and the hafnium analogs) which contain the infinite double-metal-layered slabs shown in Figure 12 [49—51]. (The chloride and bromide structures represent different sequences for rhombohedral stacking of the three slabs.) The close-packed double-metal layers within these are in fact the end result of a continued condensation of infinite chains, the beginnings of which were seen in Figure 11 (metal atoms only), while metal condensation in the third dimension from ZrX would lead directly to the (ccp) metal. In fact, ZrBr has been found to grow epitaxially on the metal. The microscopic properties of monohalides appear to be appropriately those of a two-dimensional metal while the compounds macroscopically look and act like graphite.

The period within the layers is apparently dictated largely by the size of the nonmetal, 3.42 and 3.50 Å in ZrCl and ZrBr, respectively, (3.37 Å in HfCl) while the separation between the two metal layers reflects the electron count and stronger bonding, namely 3.09 and 3.13 Å for the respective zirconium phases. (The average distance in the hcp metal is 3.20 Å.) The interstitial polyhedra between the metal layers are thus trigonally compressed. The formation of a layered structure suggests considerable Zr—Cl covalency in order to leave only a relatively low charge on the chlorine layers, while photoelectron spectra and theory indicate the metal conduction band involves very little halogen contribution. The phases appear to be metallic by the same techniques [55, 57, 58].

The interstitial compounds also formed by these monohalides provide an interesting facet and a prolific chemistry that are pertinent to such derivatives of the infinite chain compounds already discussed. Known phases are listed in Table II in terms of the interstitial positions the nonmetals occupy between the double metal layers [52—56]. The tetrahedral derivatives appear to be appreciably non-

TABLE II
Interstitial derivatives of the double-metal-layered monohalides for group III and IV metals

Tetrahedral	Ref.	Octahedral	Ref.
3R—ScClH, ScBrH, YClH[a]	35, 52	1T—ZrXH	52
3R—$GdXH_z$[b]	39, 53		
		1T—Sc_2Cl_2N, Zr_2Cl_2N	35, 55
3R—$ZrClO_{0.43}$, $ZrBrO_{0.35}$[a]	54		
		1T—M_2Cl_2C[c]	35, 55, 56
		3R—Sc_2Cl_2C	35
3R—$ZrClN_{1-x}$	55	1T—Gd_2Br_2C	39, 42
		$Gd_2Br_2C_2$[d]	39, 42
		1T—Zr_2Cl_2B	55

[a] Upper limits of a wide one-phase region; $0.25 \lesssim y \leqslant 1.0$ for $MClH_y$.
[b] X = Cl, Br, I, $z = 0.8$—0.9.
[c] M = Sc, Y, Zr.
[d] Monoclinic; C_2 interstitials.

stoichiometric while those with octahedral occupancy are substantially line compounds. Introduction of a nonmetal into the intermetal layer region gives rise to one (with H) or two ($2s$, $2p$) anion valence bands and correspondingly reduces the population of the metal band; thus Sc_2Cl_2C, presumably a valence compound, is reddish colored and a semiconductor by UPS, while Zr_2Cl_2C (d^1) is metallic according to both PES and theory [35, 55]. The phases $ZrXH_y$ are also metallic by PES for $y = 0.5$ and probably with $y = 1$ [58].

It now appears that the previously reported monochlorides and monobromides of Sc, Y, La and of many lanthanides are actually hydrogen stabilized [35, 39, 52]. The lower limit of hydrogen content in equilibrium with the metal is about 0.4H/MCl for M = Sc and Y [35, 52].

5. General Correlations

The early transition metals in group III to VI appear to possess several characteristics favorable to the formation of interesting and novel clusters and extended arrays, namely atomic orbitals of sufficient radial extension to form metal—metal bonds of substantial strength at separations appropriate to the size of the nonmetals that sheath the metal structures. A structural dominance of the metal arrays in the resulting structures seems evident, a feature which is probably related to the minimal involvement of the halide or oxide in the metal—metal valence or conduction bands. This characteristic will be contrasted with that of the better known chalcogenides, phosphides, etc., later (Section 5.4). A remarkable facet of most of the new structures described here is that they utilize octahedral metal clusters as building blocks, units for which a considerable amount of cluster chemistry also exist. Condensation of these clusters appears as not just a response to the decrease in N/M ratios and the loss of M—N bonds but as a thermodynamic necessity, namely, the replacement of M—N bonding by M—M bonds of comparable strength. Even at this stage we recognize another characteristic of metals, namely, the gain of a larger number of relatively long M—M interactions to replace the lost M—N bonds. This can be seen in coordination numbers achieved by the metal in these arrays. Thus, all the normal-valent compounds $ScCl_3$, $ZrCl_4$, or MoO_3 with these nonmetals have six nonmetal neighbors octahedrally disposed about the metal atom. On reduction, these nonmetals are progressively replaced by a greater number of metal neighbors, e.g. substituting nine metal for three chlorine neighbors about each metal in the formation of ZrCl or 4.5 to 6.5 metals for 1.5 nonmetals (average) in the chains in Sc_5Cl_8 and $NaMo_4O_6$, respectively, the range depending on the relative magnitude of the chain repeat length. This condensation process will ultimately give the metals themselves with coordination numbers of 12 or 14.

5.1. THE METALS AND NONMETALS INVOLVED

The selectivity among structure types based on the metals and the number of electrons involved in metal—metal bonding is particularly striking. The infinite chains of octahedra in chlorides or bromides of group III metals are achieved with ratios of 1.5 to 1.8 e/M and 1.33 to 1.5 N/M in the chains, Table I, while with the smaller oxide, similar chains (some with distortions) are obtained for 3.25 to 3.75 e/M but with a comparable ratio of 1.25 to 1.75 N/M. On the other hand, the structural chemistry of the intermediate group IV and group V elements to date appears to be devoid of these chains. For zirconium and niobium chains at a similar N/M ratio of 1.5, the bonding in halides would involve 2.5 and 3.5 e/M, respectively. One may surmise that these larger numbers together with smaller M—M bond lengths would make these chains too strongly bound to be commensurate with the size of even a sheathing chloride, i.e. there would be a serious matrix effect [11] originating from the nonmetal that would preclude achieving anywhere near the full degree of M—M bonding in an extended structure. Appropriately lower electron counts to allow for the steric effects with halide would naturally mean higher X/M ratios and these would probably preclude condensation. Alternatively, the use of oxide or sulfide anions at these compositions with zirconium or niobium wou'd leave only one or two electrons per metal, respectively, for chain bonding. The first is lower than that known in any one-dimensional compound (but the same as that in NbO_2 and NbS_2) while the group V model with only two electrons per metal would be somewhat weakly bound compared with the molybdenum oxide chains with about 3.5. Nevertheless, the former electron count in niobium sulfides would be comparable to that achieved in the group III metal chain compounds (Table I) with a nonmetal of comparable size, but it should also be noted that the metallic radius of niobium is signficantly less, about 0.5 Å versus Y and 0.2 Å versus Sc, but 0.1 Å larger than that for Mo. In the end, niobium and tantalum analogs as either sulfides or oxides cannot be convincingly ruled out. No relevant examples have been found in synthetic explorations of the ternary niobium sulfide systems [59]; rather several other reduced niobium sulfides provide alternative stabilities with which to contend, e.g. intercalated Nb_3S_4 [60] or $Nb_{21}S_8$, $Nb_{14}S_5$, and NbS [61], some of these exhibiting interesting but interlinked chains of metal atoms. More pertinent is Nb_5Se_4 which contains isolated chains derived from Nb_6Se_8-type clusters condensed by sharing metal vertices and face-capping selenide, i.e. $Nb_4Nb_{2/2}Se_{8/2}$ [8].

In contrast to the foregoing problems, real or imagined, associated with the formation of one-dimensional structures for group IV and V elements, known reduced zirconium and hafnium halides illustrate that an alternative structure better matches the electron count with nonmetal size, namely, the two-dimensional monohalide structure containing double metal layers that was shown in Figure 12. In this case there is no matrix effect from the nonmetal to restrict the bonding between the metal layers, and the Zr—Zr separations in this direction are appreciably less than within the sheets (by 0.33 Å in ZrCl). Presumably similar structures for NbX (d^4) and MoX (d^5) are not found because the matrix effect within the layers become more severe as the metal sizes decrease, although rigid-band EHMO calculations do not clearly rule out the former for electronic reasons [55]. Synthesis is of the character, however, that one can never be certain that a given compound *does not* exist. Interestingly, d^3 Hf_2S in which the metal is isoelectronic with that in HfCl exhibits analogous double metal layers, but these are now separated by single sulfur layers and the result is much less anisotropic [60]. There are a few other ways around matrix effects though: ZrI_2 and $M^IMo_3Ch_3$ with 2 and 4.3 electrons per metal, respectively, achieve one-dimensional metal structures with M—M repeats appreciably less than that of the anions by respectively forming zig-zag chains (Figure 2) or chains of the so-called confacial octahedra — that is, chains of triangular Mo_3Ch_3 units stacked with alternate rotations of 60° [2].

5.2. Bonding and Bond Orders

It seems significant that many of the compounds discussed here achieved a total amount of metal—metal bonding that is clearly comparable to that in the metals themselves when allowance is made for the reduced number of electrons available for that purpose. The easiest way to compare these is with the aid of Pauling Bond Orders (PBO) defined as $d_n = d_1 - 0.6 \log n$, the calibration coming via the single bond distance d_1 derived for the metal itself with the experimental distances d_n and bond order n calculated from the number of valence electrons n_e divided by the coordination number. In this sense a reduced compound is also metal-like if

$$\left(\sum \text{PBO}\right)/n_e\,(\text{compound}) \simeq \left(\sum \text{PBO}\right)/n_e\,(\text{metal}) = 1,$$

the second equality being the calibration just described. In fact, the first ratio is close to unity, and the compounds therefore apparently metal-like, for many examples described here, the indicated bond order sums per electron being 1.04 in Sc_5Cl_8, 0.96 in Y_2Cl_3, 0.81 in ZrCl, and 1.00 in $NaMo_4O_6$. Recognizable matrix effects give lower results, e.g. 0.66 for ZrBr, but there do not appear to be many examples of this character in stable extended compounds [11].

Bond order relations also may be used to estimate with some accuracy metal valence states in the oxide systems [21]. An empirical relation of the form $s = [d(\text{M—O})/d_1]^{-n}$, with $d_1 = 1.882$ Å and $n = 6$ (where s is the M—O bond valence (in valence units)), has been particularly well established for M = Mo [62]. A summation over all Mo—O bonds about Mo then gives the net valence (oxidation state), $\sum s(\text{M—O})$ = valence. The utility of these relations for the compounds discussed here comes from their use to estimate electron counts in the metal chains, even in those complex cases where other evidence for degree of electron transfer from cations to anion chains cannot be readily obtained. For example, in $(Fe_{1.89}Mo_{0.11})Mo_4O_7$ what electron count can be assigned to the metal atoms in the Mo_4O_7 chains? In this complex material the iron atoms exist in a mixture of oxidation states and thus the degree of electron transfer from these atoms to the Mo_4O_7 chains cannot be straightforwardly inferred in any other way. Application of these relations using the observed Mo—O bond distances gives an average valence of about 2.4 for Mo, an anion charge $Mo_4O_7^{4.3-}$, and approximately 14.3 electrons per Mo_4O_7 unit for Mo—Mo bonding. As stated in the discussion of structures, the compound $Ca_{5.45}Mo_{18}O_{32}$ presents a particularly interesting problem because it contains not one, but three different types of infinite chains. Application of these relations leads to the formulation $Ca_{5.45}(MoO_3^{2.2-})_2(Mo_2O_{3.5}^{0.24})_4(Mo_4O_6^{2.6-})_2$ with the internal check that the sum of cation and anion charges, $\sum q^+(\text{Ca})$ and $\sum q^-(\text{anion chains})$, 10.9 and 10.6, respectively, are indeed in good agreement. In this way we obtain a reliable estimate of 14.6 electrons per Mo_4O_6 in the octahedral cluster chains, and 2.6 electrons per metal in the fused rhomboid ($Mo_2O_{3.5}$) chains. The latter estimate compares well with the value of 2.5 electrons each in the fused rhomboid chains in $NaMo_2O_4$. The third type of chain, indicated as $MoO_3^{2.2-}$ in the above formula, is like the single atom chains along the c direction of the rutile structure; even the pairing of metal atoms here is essentially the same as in MoO_2. With an estimated 2.2 electrons per Mo in these chains, the M—M bonding is thereby quite analogous to that in MoO_2 with 2.0 electrons.

5.3. Interstitial Atoms

Interstitial nonmetal atoms are found to provide additional strong bonding in structures where the cavities are sufficiently large. Alternatively, these same effects may provide the means for stabilization of electron-poor species that otherwise would not exist. An example of the former appears to be $Sc_7Cl_{10}C_2$ (Figure 10) while Sc_4Cl_6B and perhaps Gd_4I_5C and $Gd_6Br_7C_2$ are examples of the latter. The interstitials may in different cases be B, C, N, O, or H in the two-dimensional examples (Table II), and a greater variety of interstitials may ultimately be found within the chain compounds as well (Table I). (An even greater diversity of interstitial atoms can be bonded within isolated zirconium clusters where the electronic factors are better understood [63, 64].) Photoelectron spectroscopic measurements and band calculations on $Sc_7Cl_{10}C_2$ and some of the layered examples in Table II make clear the character of the bonding involved. With carbon, for example, there are strong interactions of the valence 2*s* and 2*p* orbitals with zirconium 4*d*, analogous to that in say ZrC (NaCl-type), to produce two new valence bands at *ca.* 20 and 4 eV below E_F, respectively. The metal-based conduction band is correspondingly depleted of electrons, for example, from three electrons per metal in ZrCl to one in Zr_2Cl_2C. A substantial M—C covalence is clearly indicated by a 3—4 eV reduction in the metal core (2*p*) binding energy in $Sc_7Cl_{10}C_2$ and M_2Cl_2C, M = Sc, Zr, relative to that in $ScCl_3$ or $ZrCl_3$. The interstitial radii observed in these compounds are well defined and quite reproducible, those in $Sc_7Cl_{10}C_2$ and Zr_2Cl_2C, for example, being closely comparable to those in the respective binary metal carbides [36, 55].

The configuration of the nonmetal atoms about the metal chains or double metal layers containing bonding interstitials is always such as to avoid close interstitial-halide approaches. This results in the conversion from face-capping chlorine in Sc_7Cl_{10} to the edge-bridging in $Sc_7Cl_{10}C_2$ while the octahedral interstitials in the layered monohalides (Table II) all occur in a 1T-structure in which the individual slabs have rearranged from cubic close-packed (6—8 type) as shown in Figure 11 to one in which the halide does not lie above the octahedral holes, as in 6—12 type clusters. Thus, in 1T—M_2Cl_2C the layers stack AB(*c*)AB [56], while in 1T—ZrBrH *two* hydrogens are bound within each (enlarged) octahedral cavity in the same arrangement, viz. AB(*cc*)AB [52].

5.4. Other Electronic Factors

Band calculations provide an illuminating treatment of the $NaMo_4O_6$ structure, its bonding, and the distortions encountered at higher electron counts [47]. It was shown that the doubled repeat distance along the chains of $Sc_{0.75}Zn_{1.25}Mo_4O_7$ is consonant with an electronically driven distortion for >13 electrons per unit which results in a tilting of octahedral units such that apex—apex distances are alternately lengthened and shortened. In the $NaMo_4O_6$ structure this distortion would result in a symmetry-forbidden band crossing involving primarily orbitals on the apex atoms. With the doubled repeat and reduced symmetry along the chains of $Sc_{0.75}Zn_{1.25}Mo_4O_7$, this crossing would occur near the Fermi level, and a band gap is opened up by lowering of a band formerly crossing E_F. Additional bonding states are thereby made available and these are reflected in the short apex—apex distance, 2.62 Å. Because of the presence of a third band near E_F and the lack of full three-dimensional character in the calculations, accurate prediction of limiting electron counts (for band filling) and direct correlation of electron count with apex—apex bond distance could not be made.

Note has already been made of theoretical treatments at various levels of sophistication for Gd_2Cl_3, ZrCl, and M_2Cl_2C, M = Sc, Zr. These all confirm the presence of a limited mixing between nonmetal and metal valence levels in the compounds discussed herein, a characteristic that is also evident from the narrow and well separated halogen *p* and metal *d* valence or conduction bands observed in PES studies. On the other hand, similar chalcogenides (oxides excluded) show much broader, overlapping bands. This greater mixing of metal and chalcogenide valence orbitals appears to be a significant factor in the different structural types and more isotropic bonding that are observed with chalcogenides. This structural contrast is found on almost any basis of comparison, for example, with comparable formula types, ZrCl versus $Zr_{1-x}S$ (NaCl), for isoelectronic examples, ZrCl versus NbS (MnP type) or for the same oxidation state but with fewer nonmetal atoms in ZrX versus Zr_2Ch (Ta_2P type). In all cases the halides appear to give notably lower dimensional structures. Comparable observations can be made regarding the molybdenum oxides, the only similarity to the structures described herein being the $Mo_3Ch_3^-$ chain structures. It would appear that these differences reflect the lower charges (on Cl, Br), lower polarizabilities (Cl, O) and the generally higher valence state ionization energies for these nonmetals relative to sulfide, selenide, phosphide, etc. These factors produce less mixing with higher lying metal orbitals and, probably, less screening and more contracted metal orbitals. It will be recalled that only the layered metallic diiodides among the halides really appear very similar to analogous sulfides structurally and electronically [4].

6. Outlook

Two additional points should be made about the types of compounds described here. First, it is quite evident from the foregoing discussion that these constitute a very new and incompletely developed collection of new compounds. The relatively poor understanding of factors affecting bonding interactions, especially as they relate to stability of one structure versus another, and our inherently limited vision of what structures are possible make predictions of additional structural types or of new combinations having one of the known types a matter of low success. Experimental approaches are another complicating factor, in that the failure to obtain a compound of given structural type may be dictated by lack of proper experimental conditions and not an inherent lack of stability. Thus, new compounds in this realm will continue to be found, but infrequently by prediction followed by designed synthesis. Secondly, because of the recent evolution of this chemistry, not much is yet known of the physical and chemical properties of compounds discussed here. Studies to elucidate properties should become a major thrust as well-defined materials become available in sufficient purity and quantity.

Acknowledgement

This research was supported in part by the Office of Basic Energy Sciences, Materials Sciences Division. The Ames Laboratory is operated for the U.S. Department of Energy by Iowa State University under contract No. W-7405-Eng-82.

References

1. J. D. Corbett, *Pure Appl. Chem.* **56**, 1527 (1984).
2. R. Chevrel and M. Sergent (this volume, pp. 315–373).
3. R. Chevrel, P. Gougeon, M. Potel, and M. Sergent, *J. Solid State Chem.* **57**, 25 (1985).
4. J. D. Corbett, *Adv. Chem. Ser.* **186**, 329 (1980).
5. J. D. Corbett, *Acc. Chem. Res.* **14**, 239 (1981).
6. R. E. McCarley, *Phil. Trans. Roy. Soc. Lond.* **A308**, 141 (1982).
7. R. E. McCarley, *Am. Chem. Soc. Symp. Ser.* **211**, 273 (1983).
8. A. Simon, *Angew Chem., Int. Ed. Engl.* **20**, 1 (1981).
9. A. F. Wells, in *Structural Inorganic Chemistry* (4th edn), Clarendon Press, Oxford (1975).
10. K. Yvon, in *Current Topics in Material Science* (ed. E. Kaldis), North-Holland, Amsterdam, Vol. 3, p. 83 (1979).
11. J. D. Corbett, *J. Solid State Chem.* **39**, 56 (1981).
12. R. G. Clemmer, Ph.D. Thesis, Univ. of Wisconsin (1977); *Dissert. Abst.* **38B**, 674 (1977).
13. J. D. Corbett and J. W. Anderegg, *Inorg. Chem.* **19**, 3822 (1980).
14. M. S. Whittingham, *Prog. Solid State Chem.* **12**, 41 (1978).
15. G. Meyer and R. Hoppe, *J. Less Common Met.* **46**, 55 (1976).
16. H. Schäfer and H.-G. von Schnering, *Angew. Chem.* **76**, 833 (1964).
17. B.-O. Marinder, *Chem. Scripta* **11**, 97 (1977).
18. J. D. Corbett and D. H. Guthrie, *Inorg. Chem* **21**, 1747 (1982).
19. C. C. Torardi and R. E. McCarley, *J. Solid State Chem.* **37**, 393 (1981).
20. C. C. Torardi and J. C. Calabrese, *Inorg. Chem.* **23**, 328 (1984).
21. R. E. McCarley, K.-W. Lii, P. A. Edwards, and L. F. Brough, *J. Solid State Chem.* **57**, 17 (1985).
22. P. J. Guillevic, J.-Y. LeMorouille, and D. Grandjean, *Acta Crystallogr.* **B30**, 111 (1974).
23. A. Magnéli, *Acta Chem. Scand.* **11**, 28 (1957).
24. R. E. McCarley and L. Aleandri (unpublished research).
25. J. Burdett and T. Hughbanks, *Inorg. Chem.* **24**, 1741 (1985).
26. D. A. Lokken and J. D. Corbett, *Inorg. Chem.* **12**, 556 (1973).
27. A. Simon, N. Holzer, and Hj. Mattausch, *Z. Anorg. allg. Chem.* **456**, 207 (1979).
28. Hj. Mattausch, J. B. Hendricks, R. Eger, J. D. Corbett,and A. Simon, *Inorg. Chem.* **19**, 2128 (1980).
29. K. R. Poeppelmeier and J. D. Corbett, *J. Am. Chem. Soc.* **100**, 5039 (1978).
30. Hj. Mattausch, R. Eger, and A. Simon, *Rev. Chim. Miner.* **17**, 516 (1980).
31. K. R. Poeppelmeier and J. D. Corbett, *Inorg. Chem.* **16**, 1107 (1977).
32. C. C. Torardi and R. E. McCarley, *J. Am. Chem. Soc.* **101**, 3963 (1979).
33. L. F. Brough, P. A. Edwards, C. D. Carlson, and R. E. McCarley (unpublished research).
34. K.-H. Lii and R. E. McCarley (unpublished research).
35. S.-J. Hwu and J. D. Corbett (unpublished research).

36. S.-J. Hwu, J. D. Corbett, and K. R. Poeppelmeier, *J. Solid State Chem.* **57**, 43 (1985).
37. K. Berroth and A. Simon, *J. Less Common Met.* **76**, 41 (1981).
38. A. Simon and E. Warkentin, *Z. Anorg. allg. Chem.* **497**, 79 (1983).
39. A. Simon; *J. Solid State Chem.* **57**, 2 (1985).
40. A. Simon and E. Warkentin (unpublished research).
41. K. Berroth, Hj. Mattausch, and A. Simon, *Z. Naturforsch.* **B35**, 626 (1980).
42. U. Schwanitz and A. Simon (unpublished research).
43. W. Bauhofer and A. Simon, *Z. Naturforsch.* **A37**, 568 (1982).
44. G. Ebbinghaus, A. Simon, and A. Griffith, *Z. Naturforsch.* **A37**, 564 (1982).
45. D. W. Bullett, *Inorg. Chem.* **19**, 1780 (1980).
46. B. A. Aufdembrink and R. E. McCarley (unpublished research).
47. T. Hughbanks and R. Hoffmann, *J. Am. Chem. Soc.* **105**, 3528 (1983).
48. F. J. DiSalvo, J. V. Waszczak, W. M. Walsh, L. W. Rupp, and J. D. Corbett, *Inorg. Chem* (submitted).
49. S. I. Troyanov, *Vestn. Mosk. Univ., Khim.* **28**, 369 (1973).
50. D. G. Adolphson and J. D. Corbett, *Inorg. Chem.* **15**, 1820 (1976).
51. R. L. Daake and J. D. Corbett, *Inorg. Chem.* **16**, 2029 (1977).
52. S. Wijeyesekera and J. D. Corbett, *Solid State Commun.* **54**, 657 (1985).
53. Hj. Mattausch, W. Schramm, A. Simon, and R. Eger (unpublished research).
54. L. M. Seaverson and J. D. Corbett, *Inorg. Chem.* **22**, 3202 (1983).
55. R. Ziebarth and J. D. Corbett (unpublished research).
56. J. E. Ford, J. D. Corbett, and S.-J. Hwu, *Inorg. Chem.* **22**, 2789 (1983).
57. J. Marchiando, B. N. Harmon, and S. A. Liu, *Physica* **99B+C**, 259 (1980).
58. J. D. Corbett and H. S. Marek, *Inorg. Chem.* **22**, 3194 (1983).
59. E. Garcia, J. D. Corbett, and R. E. McCarley (unpublished research).
60. R. Schöllhorn and W. Schramm, *Z. Naturforsch.* **B34**, 697 (1979).
61. H. F. Franzen, *Progr. Solid State Chem.* **12**, 1 (1978).
62. J. C. J. Bart and V. Ragaini, *Inorg. Chim. Acta* **36**, 261 (1979).
63. R. P. Ziebarth and J. D. Corbett, *J. Am. Chem. Soc.* **107**, 4571 (1985).
64. J. D. Smith and J. D. Corbett, *J. Am. Chem. Soc.* (in press).

PSEUDO-ONE-DIMENSIONAL MX_3 AND MX_4 TRANSITION METAL CHALCOGENIDES

ALAIN MEERSCHAUT and JEAN ROUXEL

Laboratoire de Chimie des Solides, Associé au CNRS,
2, rue de la Houssinière, F. 44072 Nantes Cédex, France.

Introduction

Transition metal dichalcogenides have provided us with some of the best examples for the illustration of both the physics of low-dimensional compounds and intercalation chemistry. Their structure is made of MX_2 layers built up from $[MX_6]$ octahedra as in TiS_2, or $[MX_6]$ trigonal prisms as in NbS_2 or $NbSe_2$. The latter compounds allowed for the first time an experimental observation of a charge density wave (CDW) [1].

The preparation and characterization of $NbSe_3$ in 1975 [2] led to a new class of low-dimensional materials based upon transition metal trichalcogenides, especially the niobium and tantalum trichalcogenides. Very recently this field was extended to new chalcogen-rich chalcogenides such as Nb_2Se_9, the tetrachalco-halogenides $(MX_4)_nY$ (M = Nb, Ta; X = S, Se; Y = Halogen) and pentatellurides MTe_5 (M = Zr, Hf).

In this article we shall focus on the tri- and tetrachalcogenides which are known today, with a particular interest to their structural types and their transport properties.

1. Trichalcogenides of IVA, VA, and VIA Elements

Transition metal trichalcogenides of IVA and VA elements are obtained by direct combination of the metal and chalcogen. The reaction is carried out in silica tubes sealed under vacuum, the temperature ranging usually from 500 to 700 °C. At higher temperature partial decomposition may occur, leading to the dichalcogenides. The unit cells are generally monoclinic as shown in Table I.

All of the transition metal trichalcogenides have a common structural framework, i.e. $[MX_3]$ trigonal prismatic chains made up of $[MX_6]$ units. Irregular $[MX_6]$ trigonal prisms are stacked on top of each other to form the $[MX_3]$ chains parallel to the *b* axis. This building scheme, represented in Figure 1(a) where *b* is the M—M distance along the chains, is responsible for the one-dimensional character of the trichalcogenides and plays an important role concerning their physical properties. The shortest M···X distances between metal atoms of one chain and chalcogen atoms of its nearest neighbor chains (translated from $b/2$), are too short to exclude any bonding. These interchain M···X interactions lead to two-dimensional slabs made up of $[MX_3]$ chains, as shown in Figure 1(b).

J. Rouxel (ed.), Crystal Chemistry and Properties of Materials with Quasi-One-Dimensional Structures, 205—279.

TABLE I
Unit cell dimensions for MX_3 compounds

Compound	a (Å)	b (Å)	c (Å)	$\beta°$	Symmetry	Ref.
TiS_3	4.958	3.4006	8.778	97.32	Monoclinic	5
ZrS_3	5.1243	3.6244	8.980	97.28	Monoclinic	5
$ZrSe_3$	5.4109	3.7488	9.444	97.48	Monoclinic	5
$ZrTe_3$	5.8939	3.9259	10.100	97.82	Monoclinic	5
HfS_3	5.0923	3.5952	8.967	97.38	Monoclinic	5
$HfSe_3$	5.388	3.7216	9.428	97.78	Monoclinic	5
$HfTe_3$	5.879	3.9022	10.056	97.98	Monoclinic	
NbS_3	4.963	2 × 3.365	9.144	97.17 ($\alpha = \gamma = 90°$)	Triclinic	7
$NbSe_3$	10.006	3.478	15.626	109.30	Monoclinic	2
TaS_3	36.804	3.340	15.173		Orthorhombic	55
TaS_3	9.515	3.3412	14.912	109.99	Monoclinic	54
$TaSe_3$	10.402	3.495	9.829	106.26	Monoclinic	55

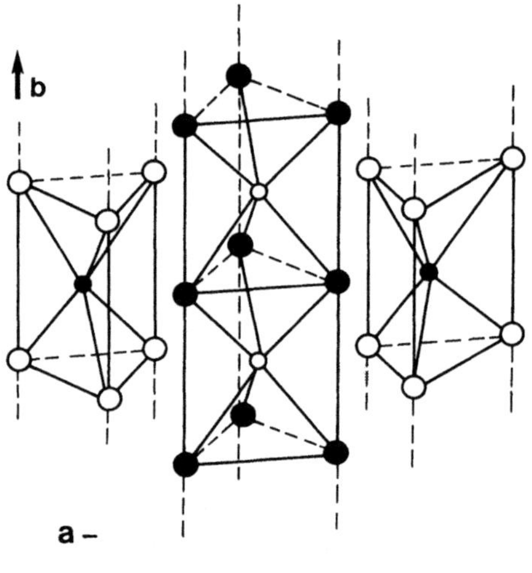

Fig. 1(a). MX_3 chains running along the growing axis.

This structural feature is an essential characteristic of transition metal trichalcogenides. The $[MX_3]$ chains of these compounds may not be equivalent due to the differences in the chalcogen—chalcogen bonds of the triangular base of the trigonal $[MX_3]$ prisms. Based upon this difference, transition metal trichalcogenides may be classified into three distinct structural types. Up to now it has been found that an MX_3 compound may consist of one, two, or three kinds of different chains. This classification was first developed by Wilson [3]. It is a very convenient way to review these materials because the physical properties are directly related to such structural features.

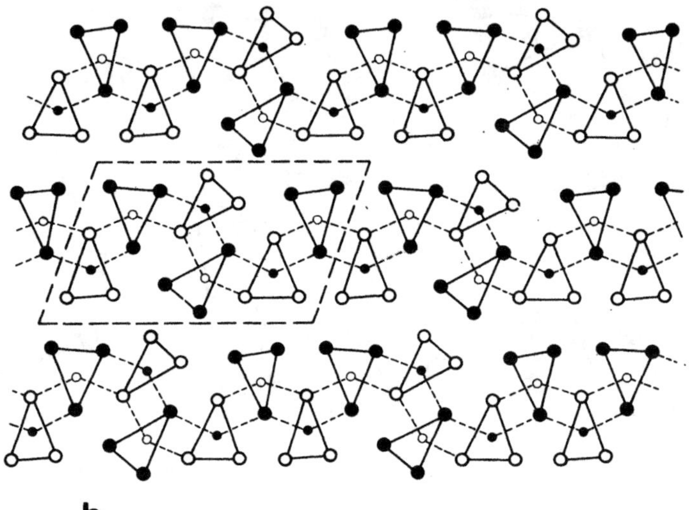
b -

Fig. 1(b). Slabs of MX_3 chains after the $NbSe_3$ example.

1.1. $ZrSe_3$ STRUCTURAL TYPE

Only one type of $[MX_3]$ chain is present in the $ZrSe_3$ structure determined by Kröniert and Plieth [4]. $ZrSe_3$ crystallizes in fibrous strands or filamentary ribbon-shaped platelets. Figure 2 shows a projection of the structure along the [010] direction. Fractional coordinates are listed in Table II, and interatomic distances in Table III.

Each zirconium atom is bonded to six selenium atoms at the corners of a trigonal prism and also to selenium atoms of the adjacent chains. Thus, each $[MX_3]$ chain is shifted with respect to the adjacent one by $b/2$. This arrangement forms layers two prisms thick, and these layers are linked together by weak van der Waals (VDW) bonds (e.g. the Se—Se distances across the VDW gap are equal to 3.87, 3.99, and 4.03 Å). Consequently, it is easy to cleave the crystal along the a—b plane, i.e. parallel to the layers.

As previously reported [4], a possibility of twinning is to be considered. Indeed, recent structural determinations of group IVA MX_3 compounds (M = Ti, Zr, Hf; X = S, Se, Te) have revealed that the $ZrSe_3$ structural type, in addition to the type described above (variant A), could occur in another form (variant B as typified by TiS_3) [5]. The typical type A and B variants are schematically shown in Figure 3 by full and dashed lines, respectively. The shape of the triangular base is closer to an isosceles triangle in the type A variant than in the type B variant. In addition, the interchain M···X distances are equivalent in the type A variant, while two inequivalent M···X distances exist in the type B variant. It is evident from the data of Kjekshus *et al.* [5] (see Table III) that the chalcogen polyhedron around the metal of the variant B is not as regular as that of the variant A. This appears clearly

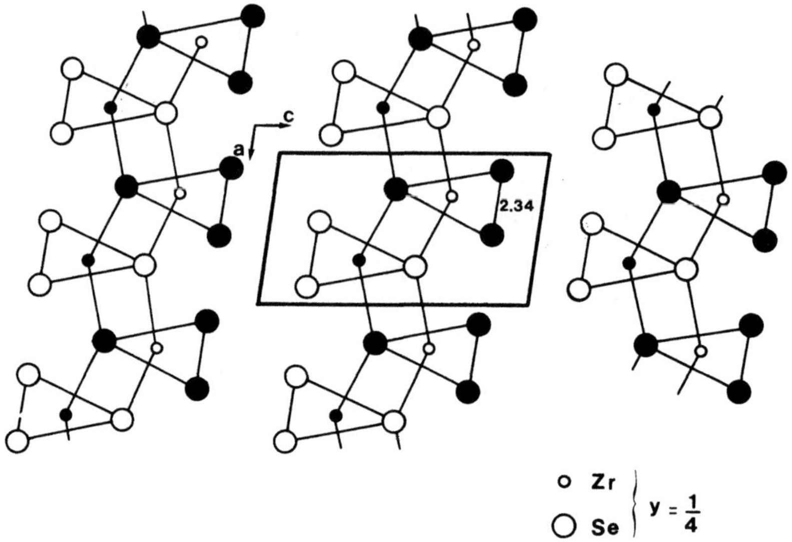

Fig. 2. The structure of $ZrSe_3$ projected along [010]. (After [4].)

TABLE II
$ZrSe_3$ structure, monoclinic $P2_1/m$, $Z = 2$
All atoms in 2(*e*): $\pm(x, \frac{1}{4}, z)$
$a = 5.4109$ Å, $b = 3.7488$ Å, $c = 9.444$ Å, $\beta = 97.48°$. (After [5])

	Zr	Se(1)	Se(2)	Se(3)
x	0.715	0.762	0.456	0.888
y	$\frac{3}{4}$	$\frac{1}{4}$	$\frac{1}{4}$	$\frac{1}{4}$
z	0.344	0.554	0.174	0.169

TABLE III
Interatomic distances in $ZrSe_3$ (after [4]) and comparison of interatomic distances in the structures of TiS_3 [5] (variant B) and ZrS_3 [5] (variant A) (Å)

	$ZrSe_3$	Variant B TiS_3	Variant A ZrS_3
M—2X(1)	2.72(1)	2.496(4)	2.602(3)
M—2X(2)	2.73(1)	2.667(4)	2.602(3)
M—2X(3)	2.74(1)	2.358(4)	2.605(3)
M—1X(1)	2.87(1)	2.416(6)	2.724(4)
M—1X(1)	2.87(1)	2.855(6)	2.707(4)
X(2)—X(3)	2.34(1)	2.038(7)	2.090(5)

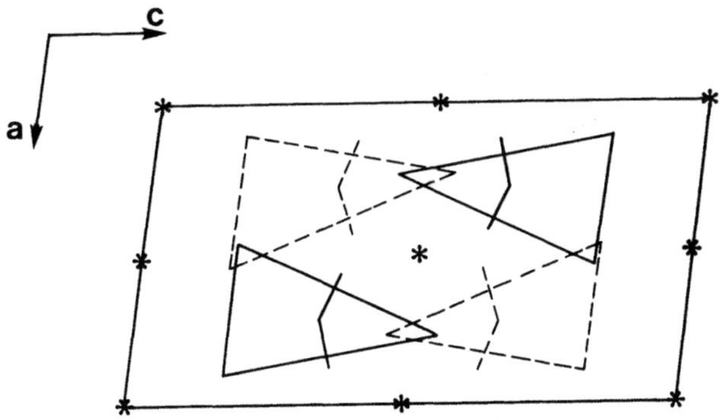

Fig. 3. Schematic representation of the two variants of the $ZrSe_3$ structural type (variant A, solid lines; variant B, dashed lines).

when comparing the M—X interatomic distances for TiS_3 (variant B) and ZrS_3 (variant A). The structural features of both variants and the XPS measurements [6] are consistent with the presence of a $(X_2)^{2-}$ pair in the triangular base of the trigonal prism (S—S ≈ 2.05 Å; Se—Se ≈ 2.35 Å), while conductivity and susceptibility measurements indicate a diamagnetic semiconducting behavior. This would lead to the oxidation formalism as $M^{4+}(X_2)^{2-}X^{2-}$ for this compound, although the compound is not as ionic as implied by this formulation.

All the trichalcogenides of group IVA transition metals belong to variants A or B of the $ZrSe_3$ structural type (Table I). NbS_3 can also be considered as a member of this group. The needle-shaped crystals of NbS_3 have a triclinic unit cell (space group $P\bar{1}$). twinned samples are generally formed when they grow from the starting elements, and this raises difficulties in the structural determinations. However untwinned crystals have been obtained by a thermal decomposition (580 °C) of NbS_2Br_2 and NbS_2Cl_2, and the crystal structure has been determined by Rijnsdorp and Jellinek [7]. The fractional coordinates are listed in Table IV. The structure is closely related to that of $ZrSe_3$. The only difference is that the niobium atoms are shifted by 0.16 Å from the pseudo mirror planes of the coordination polyhedra to form Nb—Nb pairs along the *b* axis (see Figure 4). Thus, short (3.037 Å) and long (3.693 Å) Nb—Nb distances alternate. Interatomic distances are summarized in

TABLE IV
NbS_3 structure, triclinic $P\bar{1}$, $Z = 4$
All atoms in $2(i)$: $\pm(x, y, z)$
$a = 4.963$ Å, $b = 6.730$ Å, $c = 9.144$ Å
$\alpha = 90°$, $\beta = 97.17°$, $\gamma = 90°$. (After [7])

	Nb I	Nb II	S(1)	S(2)	S(3)	S(4)	S(5)	S(6)
x	0.7156(3)	0.7154(3)	0.7632(7)	0.7636(7)	0.8747(8)	0.8830(7)	0.4631(8)	0.4708(8)
y	0.8526(3)	0.4013(3)	0.6163(4)	0.1327(4)	0.6288(4)	0.1248(4)	0.6274(5)	0.1224(4)
z	0.3509(1)	0.3464(1)	0.5618(4)	0.5453(3)	0.1552(4)	0.1837(4)	0.1610(4)	0.1873(4)

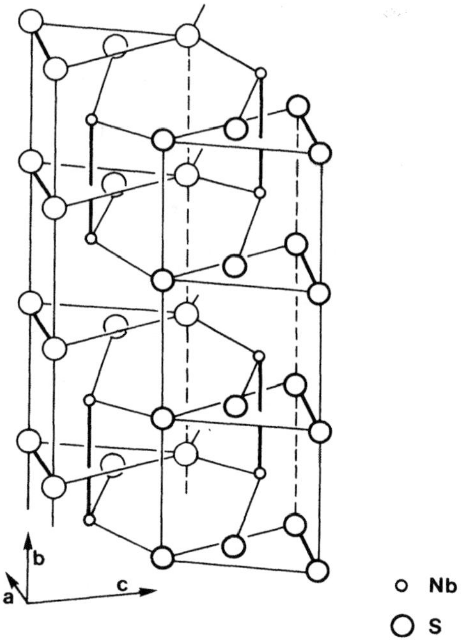

Fig. 4. The structure of NbS_3. (After [7].)

Table V. Short S—S distances of 2.05 Å indicate the formation of $(S_2)^{2-}$ pairs. XPS studies of NbS_3 [8, 9] reveal the presence of both $(S_2)^{2-}$ pairs and S^{2-} isolated ions. Like in the case of $ZrSe_3$, NbS_3 could be formulated as $Nb^{4+}(S_2)^{2-}S^{2-}$ so that the Nb—Nb pairing is a consequence of a 'Peierls distortion'. This Nb—Nb pairing is fully developed at room temperature. Nevertheless, upon heating crystals in the electron microscope, Cornelissens *et al.* [10] found that satellite spots associated with the $2b$ superstructure vanish into streaks. This observation suggests a decor-

TABLE V
Interatomic distances in NbS_3 (Å) (after [7])

Triangular base	Nb I—S(1)	2.488(4)	Nb II—S(1)	2.432(4)
	—S(3)	2.539(5)	—S(3)	2.523(5)
	—S(5)	2.518(4)	—S(5)	2.500(4)
	Nb I—S(2)	2.582(3)	Nb II—S(2)	2.554(3)
	—S(4)	2.589(4)	—S(4)	2.584(4)
	—S(6)	2.561(3)	—S(6)	2.584(3)
'Caps' of the trigonal prism	Nb I—S(2)	2.642(4)	Nb II—S(1)	2.619(6)
	—S(2)	2.670(5)	—S(1)	2.621(5)
	Short Nb I—Nb II	3.037(3)	Short S(3)—S(5)	2.050(8)
	Long Nb I—Nb II	3.693(3)	Short S(4)—S(6)	2.050(8)

relation between chains; a complete disappearance of the diffuse lines which would have meant a change towards a unique Nb—Nb average distance was not observed. In the course of this study, a new type of NbS_3, called type II, was found. It seems to differ from 'normal' NbS_3 only in the nature of the superstructure along the chain axis. In type II crystals, two narrowly spaced diffuse lines are observed at room temperature at around $\pm b^*/3$ on each sides of the main spot rows. On cooling these diffuse lines turn into rows of spots. Boswell *et al.* [11] reported very similar results. The position of the extra spots were measured at:

$$\text{first doublet} \begin{cases} 0.297b^* \\ 0.352b^* \end{cases} \qquad \text{second doublet} \begin{cases} 0.648b^* \\ 0.703b^* \end{cases}$$

After Boswell *et al.* [11], a double periodic distortion along the metal chains, could give rise to such extra spots. Taking into account a long period spacing of $17b$, the doublet positions agree well with the fifth, sixth, eleventh, and twelfth orders of diffraction. But the fact that these extra rows are observed only in the vicinity of $\frac{1}{3}$ and $\frac{2}{3}$ of the main row could be due to a second distortion of period $3b$. Thus, the Nb atoms would be inclined to form clusters of three atoms. At this point, we have to notice that these investigations are only based on the observation of electron diffraction patterns. No X-ray determinations on this hypothetical phase were attempted since such crystals have never been isolated.

However, Roucau *et al.* [12] reinvestigated the electron diffraction patterns of the type II crystal of NbS_3. The indexation of these patterns was not possible by means of the NbS_3 type I lattice parameters, but with those of a bigger cell as given by $a' = 2a$, $b' = b/2$, $c' = 2c$, and $\beta' = \beta$ (a, b, c, β being referred to the type I). In addition, the double rows of the superlattice spots appearing at around $\pm b^*/3$ have not only a b^* component, but also an a^* component for the two wave vectors $\mathbf{q}_1$ $(\frac{1}{2}, 0.297, 0)$ and $\mathbf{q}_2$ $(\frac{1}{2}, 0.353, 0)$.

Resistivity measurements on NbS_3 ($2b$ superstructure) were performed up to 300 °C. The conductivity is thermally activated with an activation energy of 0.44 eV [13]. Unfortunately it was not possible to reach the metal-insulator transition

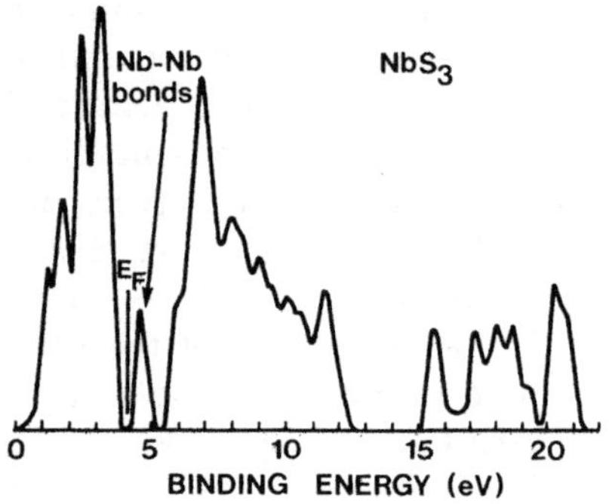

Fig. 5. Valence band density of states of NbS_3. (After [14].)

before the sample decomposition. The semiconducting behavior results from a splitting of the metal d_{z^2} band, which is half filled for the structure with equidistant Nb—Nb separation. The Nb—Nb pairing along the b axis stabilizes the structure by introducing a band gap at the Fermi level, which splits the d_{z^2} band into bonding and antibonding subbands. According to Bullett [14], this results in an energy gap of about 0.5 eV. This electronic investigation (Figure 5) agrees well with the experimental results. On the contrary, a high-pressure NbS_3 variety which does not exhibit such an Nb—Nb pairing should be metallic. However Kikkawa [15] reported that this high-pressure form (a = 9.68 Å, b = 3.37 Å, c = 14.83 Å, β = 103.9°) behaves as a semiconductor ($\rho_{300\,K} \approx 10^2$ Ω cm) with an activation energy of 0.18 eV.

A more detailed study on the conductivity behaviors of various NbS_3 polytypes was carried out by Izumi *et al.* [16]. Metallic crystals were obtained by annealing the sulfur-rich or nearly stoichiometric crystals between 600 and 650 °C in vacuum. The chemical analysis indicates a wide range of homogeneity centered around the NbS_3 composition. This NbS_3 polytype undergoes a superconducting transition at around 2 K. Another sulfur-rich phase of NbS_3 [16], which is believed to be $NbS_{3.5-4.0}$ in chemical composition and consists of two different types, seems to be different from the above reported types. However, it appears rather difficult to consider such a drastic nonstoichiometry for the MX_3 structural type, and a great care should be taken in discussing these new NbS_3 forms. According to Izumi *et al.* [16], the resistivity curve as a function of temperature for the first group (type A) shows an anomaly around 140 K in the semiconducting state. For the second group (type B) there are two successive anomalies at around 140 and 235 K. The resistivity anomaly observed for type A could be similar to that reported for the type III phase of a new NbS_3 by Zettl *et al.* [17]. Once again, neither the X-ray structural data nor the chemical analysis is convincing enough to ensure that type III NbS_3 is a new polytype.

1.2. $TaSe_3$ STRUCTURAL TYPE

The crystal structure of $TaSe_3$ has been determined by Bjerkelund *et al.* [18]. The needle-shaped crystals of $TaSe_3$ present a monoclinic unit cell (space group $P2_1/m$). Figure 6 shows a projection of the structure along [010]. The fractional coordinates are listed in Table VI and interatomic distances in Table VII. Unlike the case of $ZrSe_3$, a unit cell of $TaSe_3$ has two types of chains. As noticed above, the chains may be distinguished by their short chalcogen—chalcogen distance in the triangular base. Thus, among the two types of chains, the shortest Se—Se distance of one type is 2.576 Å while that of the other type is 2.896 Å. These values exceed the normal bond length of the $[Se—Se]^{2-}$ pair (2.35 Å). As can be seen from Figure 6, there exist linear Se_4 groups (A—B—C—D) with successive Se—Se separations of 2.576 Å (A—B), 2.653 Å (B—C), and 2.896 Å (C—D). Both kinds of chains are consequently more or less coupled by a Se—Se bridge of 2.653 Å, to form blocks of four $TaSe_3$ units. The situation was quite different in

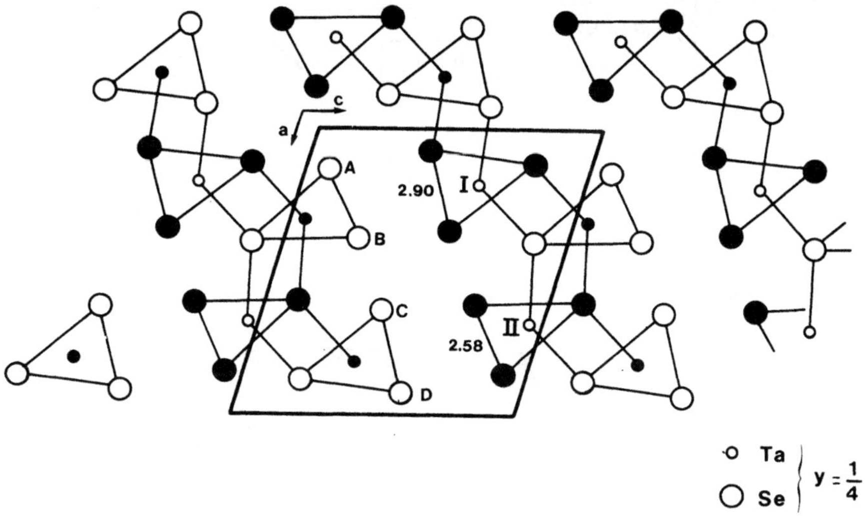

Fig. 6. The structure of $TaSe_3$ projected along [010]. (After [18].)

TABLE VI
$TaSe_3$ structure, monoclinic $P2_1/m$, $Z = 4$
All atoms in $2(e)$: $\pm(x, \frac{1}{4}, z)$
$a = 10.402$ Å, $b = 3.495$ Å, $c = 9.829$ Å, $\beta = 106.26°$. (After [18])

	Ta I	Ta II	Se(1)	Se(2)	Se(3)	Se(4)	Se(5)	Se(6)
x	0.1937(8)	0.6723(8)	0.915(2)	0.869(2)	0.151(2)	0.634(2)	0.383(2)	0.401(2)
y	$\frac{1}{4}$	$\frac{1}{4}$	$\frac{1}{4}$	$\frac{1}{4}$	$\frac{1}{4}$	$\frac{1}{4}$	$\frac{1}{4}$	$\frac{1}{4}$
z	0.6250(8)	0.9518(9)	0.576(2)	0.201(2)	0.075(2)	0.422(2)	0.258(2)	0.882(2)

TABLE VII
Interatomic distance in $TaSe_3$ (Å) (after [18])

Triangular base	Ta I—2Se(1)	2.64(2)	Ta II—2Se(3)	2.60(1)
	—2Se(2)	2.65(1)	—2Se(5)	2.64(2)
	—2Se(4)	2.64(1)	—2Se(6)	2.64(1)
'Caps'	Ta I—1Se(1)	2.80(2)	Ta II—1Se(2)	2.72(2)
	—1Se(6)	2.82(2)	—1Se(6)	2.72(2)
	Se(1)—Se(4)	2.90(2)	Se(3)—Se(5)	2.58(2)
	Se(1)—Se(2)	3.58(3)	Se(3)—Se(6)	3.62(2)
	Se(2)—Se(4)	3.70(2)	Se(5)—Se(6)	3.75(3)
	Ta I—Ta I = Ta II—Ta II = 3.495			

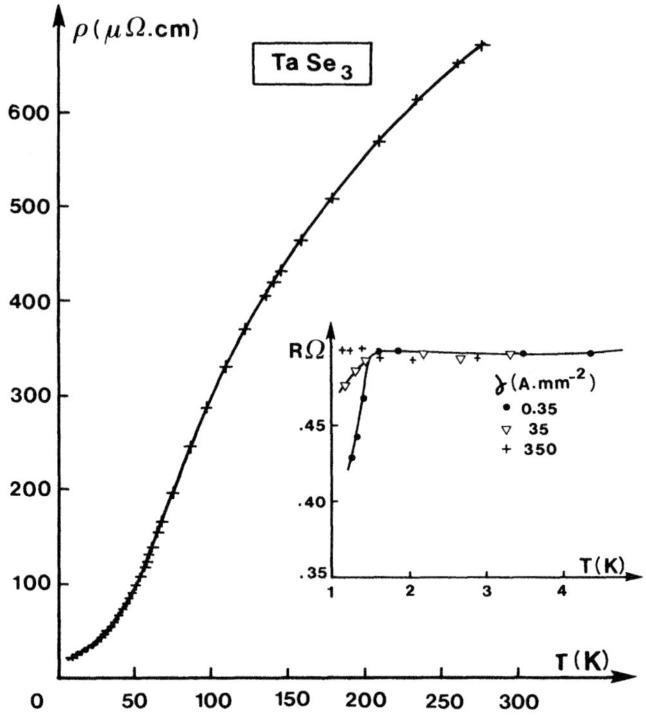

Fig. 7. Resistivity as a function of temperature for $TaSe_3$. (After [20].)

$ZrSe_3$, where the intercolumn Se—Se separation is equal to 3.07 Å. In both cases each MX_3 layer is two prisms thick, but $TaSe_3$ has a more distorted layer structure (compare Figures 6 and 2).

Resistivity measurements show a metallic behavior for $TaSe_3$. No resistivity anomaly was observed (Figure 7) down close to a superconducting transition, which appears at about 2 K [19—22]. At a very low temperature ($T < 4.2$ K), there exists a resistive anomaly dependent upon the current density [13, 23]. The resistance is non-ohmic below the superconducting transition $T_c = 2.3$ K but ohmic above, at all current densities. The temperature onset is unaffected by a change in the current density J. According to Tajima and Yamaya [23], this behavior could be associated with the superconducting state itself rather than with other phase transitions such as CDW and SDW types.

Very recently, Kikkawa *et al.* [24] have prepared a high-pressure $TaSe_3$ variety. The unit cell parameters ($a = 10.02$ Å, $b = 3.48$ Å, $c = 15.65$ Å, $\beta = 109.6°$) are very similar to those of $NbSe_3$ [2]. The transport properties are basically the same as those of the previous form ($TaSe_3$ with four [$TaSe_3$] chains), e.g. a metallic behavior down to about 2 K followed by a superconducting transition. Structural investigations on this high-pressure form are needed. A nonstoichiometric form of $TaSe_3$ was also announced by Kikkawa *et al.* [24] with the proposed formulation of

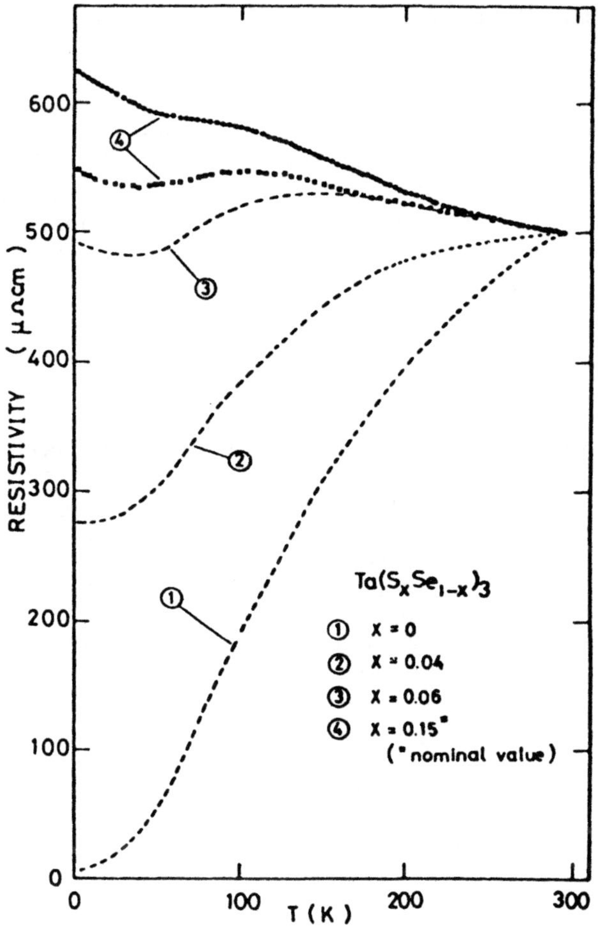

Fig. 8. The electrical resistivity as a function of temperature for samples of nominal composition $Ta(S_xSe_{1-x})_3$ (After [25].)

$TaSe_{2.80}$. Density measurements and physical studies are necessary in order to see whether the real formulation is $TaSe_{3-y}$ or $Ta_{1+x}Se_3$. However, this would require a careful preparation of the compound as a pure single phase, which has not yet been possible.

Yamaya *et al.*. [25] attempted to substitute some S for Se in the $TaSe_3$ phase. The effect of substitution is monitored by resistivity measurements (see Figure 8). From the samples with initial $Ta(Se_{1-x}S_x)_3$ of $0 < x < 0.15$, it was observed that the electrical resistivity changes gradually from a metallic to a semiconducting behavior, and the superconducting transition disappears as sulfur concentration is increased. Electron diffraction studies of $Ta(Se_{0.85}S_{0.15})_3$ reveal diffuse lines perpendicular to the b^* axis (at $\pm 0.283\ b^*$) above 150 K, which suggests the existence of a CDW.

1.3. $NbSe_3$ STRUCTURAL TYPE

1.3.1 *$NbSe_3$*

$NbSe_3$ appears generally in the preparation tubes as a cottony mass of filamentary crystals. The structure is monoclinic, space group $P2_1/m$. Up to now, $NbSe_3$ represents the most complicated structural type [2, 26]. The fractional coordinates are listed in Table VIII, and the interatomic distances are summarized in Table IX. Figure 9 shows a projection of the $NbSe_3$ structure along [010]. Three different kinds of chains are present in a unit cell. In the triangular bases of these three types

TABLE VIII
$NbSe_3$ structure, monoclinic $P2_1/m$, $Z = 6$
All atoms in $2(e)$: $\pm(x, \frac{1}{4}, z)$
$a = 10.006$ Å, $b = 3.478$ Å, $c = 15.626$ Å, $\beta = 109.30°$. (After [26])

	x	y	z		x	y	z
Nb I	0.6532(3)	$\frac{1}{4}$	0.1185(2)	Se(4)	0.1644(3)	$\frac{1}{4}$	0.0869(2)
Nb II	0.3162(3)	$\frac{3}{4}$	0.2059(2)	Se(5)	0.1788(3)	$\frac{1}{4}$	0.2759(2)
Nb III	0.2942(3)	$\frac{1}{4}$	0.4613(2)	Se(6)	0.5273(3)	$\frac{1}{4}$	0.2488(2)
Se(1)	0.4421(3)	$\frac{3}{4}$	0.0665(2)	Se(7)	0.0826(3)	$\frac{3}{4}$	0.4246(2)
Se(2)	0.8344(3)	$\frac{3}{4}$	0.0877(2)	Se(8)	0.4441(3)	$\frac{3}{4}$	0.4054(2)
Se(3)	0.8035(3)	$\frac{3}{4}$	0.2390(2)	Se(9)	0.2443(3)	$\frac{3}{4}$	0.5755(2)

TABLE IX
Interatomic distances in $NbSe_3$ (Å) (after [26])

Triangular base	Nb I—2Se(1)	2.647(3)	Se(1)—Se(2)	3.829(4)
	—2Se(2)	2.670(4)	Se(1)—Se(3)	3.722(4)
	—2Se(3)	2.636(3)	Se(2)—Se(3)	2.485(5)
'Caps'	Nb I—1Se(1)	2.726(4)		
	—1Se(6)	2.729(5)		
Triangular base	Nb II—2Se(4)	2.629(3)	Se(4)—Se(6)	3.670(4)
	—2Se(5)	2.668(4)	Se(5)—Se(6)	3.651(5)
	—2Se(6)	2.645(3)	Se(4)—Se(5)	2.909(5)
'Caps'	Nb II—1Se(1)	2.857(5)		
	—1Se(8)	2.949(4)		
Triangular base	Nb III—2Se(7)	2.652(3)	Se(7)—Se(8)	3.729(5)
	—2Se(8)	2.630(4)	Se(8)—Se(9)	3.822(5)
	—2Se(9)	2.657(4)	Se(7)—Se(9)	2.374(4)
'Caps'	Nb III—1Se(5)	2.738(4)		
	—1Se(8)	2.750(4)		
Short Se—Se bonds between chains:				
	Se(2)—Se(4)	2.733(5)		
	Se(6)—Se(9)	2.929(4)		

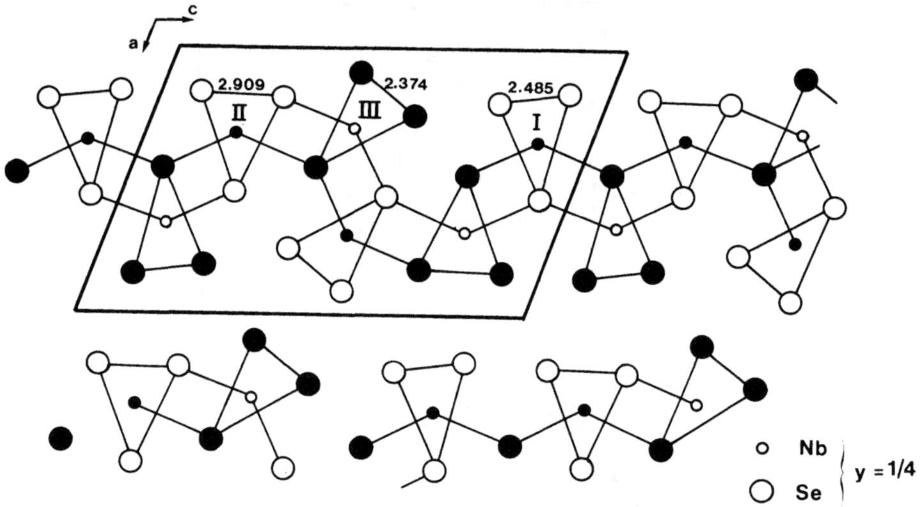

Fig. 9. The structure of $NbSe_3$ projected along [010]. (After [2].)

of chains, the shortest Se—Se distances are found to be 2.37, 2.48, and 2.91 Å. The $NbSe_3$ structure includes blocks of four chains similar to those of $TaSe_3$, which are separated by blocks of two chains of the $ZrSe_3$ type. In the four chain block, Se atoms are linked to form linear Se_4 unit (Se—Se = 2.48, 2.73, and 2.91 Å), as in $TaSe_3$. But, contrary to $TaSe_3$ in which the same blocks of four units are linked together by an inversion center at $(0, \frac{1}{2}, \frac{1}{2})$, the repetition of the four chain group in

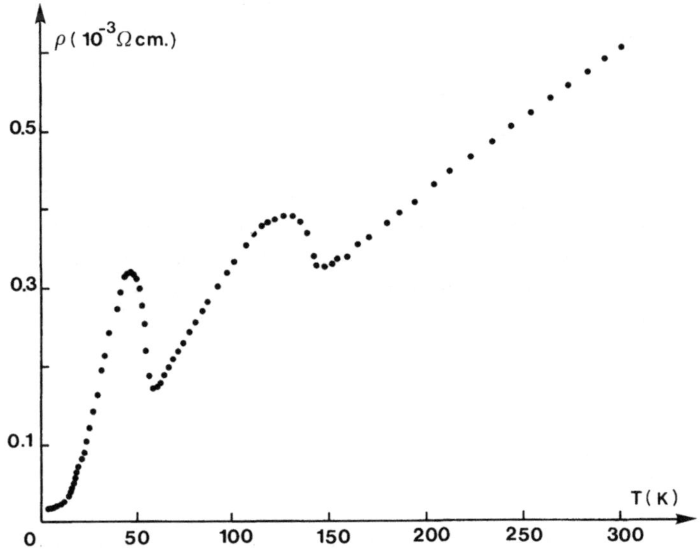

Fig. 10. Resistivity as a function of temperature for $NbSe_3$. (After [27].)

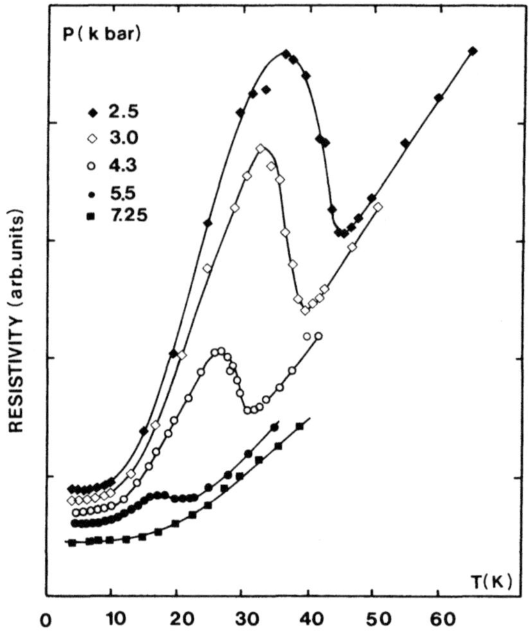

Fig. 11. Temperature variation of the resistivity for different pressures for the anomaly (at $T_2 = 59$ K) of $NbSe_3$. (After [30].)

$NbSe_3$ is interrupted by the two chain unit. The surface separating these blocks is parallel to the a—b plane. This explains why it is easy to cleave the crystal parallel to this plane as well as to the b—c plane (i.e. the plane of the slabs).

Resistivity measurements [27(a), (b)] versus temperature indicate a metallic behavior. But, the decrease of ρ with decreasing temperature is not monotonous. Two giant anomalies are observed at 145 and 59 K, where ρ increases abruptly and reaches maxima at 125 and 49 K (Figure 10). The resistivity anomalies are due to CDWs [28]. When a CDW forms, a band gap is opened at the Fermi level. The increase in resistivity is related to a decrease in the area of the Fermi surface. As usual in CDW systems, pressure removes the anomaly. Figure 11 shows the effect of pressure upon the anomaly at lower temperature. Both anomalies are displaced towards lower temperature and may be even suppressed. Such a behavior results from a competition between two terms in the expression of the free energy of the system: the strain energy, which increases with the formation of a superlattice distortion, and the gain in electronic energy resulting from the opening of band gaps. The latter term increases when the temperature decreases because the boundary of the Fermi surface is sharper at lower temperatures. Below a certain temperature, the electronic energy gain dominates over the strain energy so that a CDW may remain.

The effect of pressure on the temperature of the CDW formation (T_0) and on that of the superconducting transition (T_c) has been studied for the dichalcogenide

$NbSe_2$ by Jerome *et al.* [29]. It was found that increasing pressure results in a decrease of T_0 but an increase of T_c. Briggs *et al.* [30] observed the same phenomena with $NbSe_3$, as shown in Figure 12.

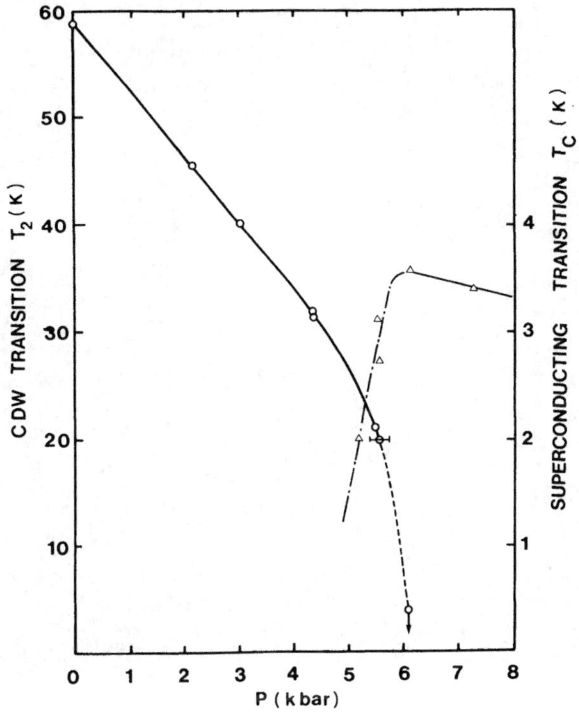

Fig. 12. Variation of the CDW temperature onset and the superconducting transition of $NbSe_3$. (After [30].)

According to Wilson [3], the CDW transitions occur in the chains exhibiting the two shortest Se—Se distances. Indeed, from electron diffraction measurements, Tsutsumi *et al.* [31] and Hodeau *et al.* [26] have determined the following components of the two wave vectors (see Photo 1):

at T_1 (145 K):

$$\mathbf{q}_1 \quad (0, 0.243(3), 0)$$

at T_2 (59 K):

$$\mathbf{q}_1 \quad (0, 0.241(3), 0)$$
$$\mathbf{q}_2 \quad (\tfrac{1}{2}, 0.259(3), \tfrac{1}{2})$$

Thus, the former distortion could be present in the $[NbSe_3]$ chains with the Se—Se length of 2.37 Å, i.e. the ones that form sheets parallel to the *a* axis. The second

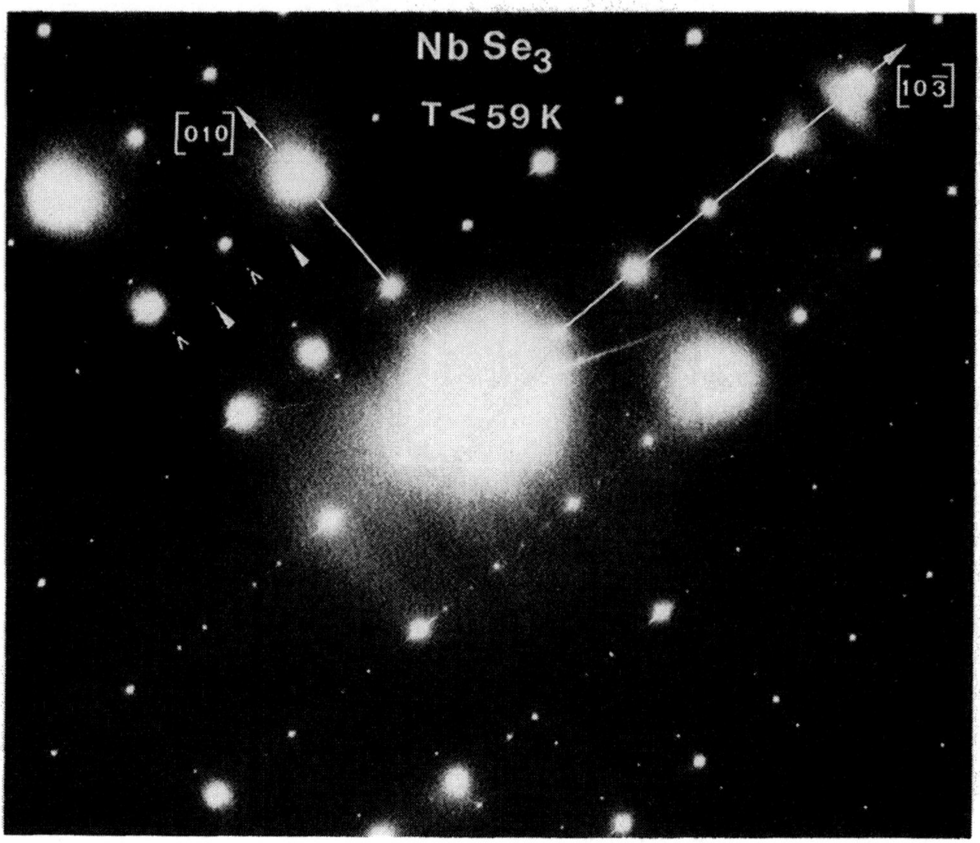

Photo 1. Electron diffraction patterns of $NbSe_3$ for $T < 59$ K. We observe the two sets of superlattice spots: the first set for $T < 145$ K at la^*, $(m \pm 0.241)b^*$, nc^* (open arrows); the second one for $T < 59$ K at $(l \pm 0.5)\,a^*$, $(m \pm 0.259)b^*$, $(n \pm 0.5)c^*$ (full arrows). (After [58].)

distortion could be present in the [$NbSe_3$] chains with 2.48 Å as Se—Se bond value, since they form sheets parallel to the [101] orientation. That this conclusion is correct is supported by the calculated $2k_F$ values along the b axis and the analysis of the band orbital character at the Fermi level [32]. X-ray diffraction studies [33] of the CDW states show that an electrical field has no measurable effect on either the CDW amplitude or the wave vector components. The experiments indicate that the two CDWs are independent as can be seen by the absence of variation in the intensities of the satellite spots of the upper transition (145 K) when the second one develops at lower temperature (59 K). X-ray diffuse scattering experiments [34] seem to indicate that pretransitional effect will take place well above the upper transition temperature. The presence of diffuse lines (pretransitional effect) can be interpreted as 1D fluctuations of these CDWs. Above the transition, their modulation pre-exists dynamically with phase incoherence between adjacent chains.

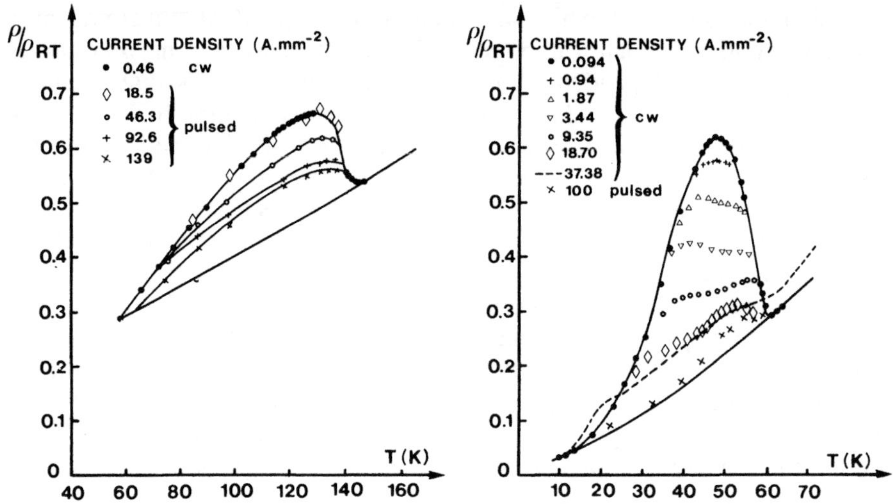

Fig. 13. Normalized resistivity as a function of temperature for different current densities with the $NbSe_3$ example. (After [35].)

The most striking transport properties are the non-ohmicity [35] and the behavior of the conductivity at microwave frequencies [36]. The resistivity anomalies are greatly suppressed by small electrical fields E, but the onset temperature is not affected (Figure 13). Below the transitions the conductivity was originally fitted by the expression:

$$\sigma(T) = \sigma_a(T) + \sigma_b(T)\, e^{-E_0/E}, \tag{1}$$

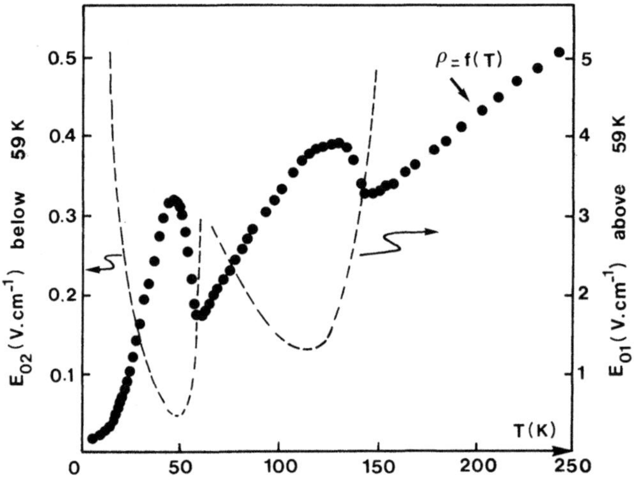

Fig. 14. Behavior of the characteristic field E_0 versus temperature for $NbSe_3$. (After [36].)

E_0 is a characteristic field which is temperature dependent, diverging at each transition and going through a minimum when the resistivity is maximum (Figure 14). The exp($-1/E$) dependence of the conductivity suggests that nonlinearity could be due to quantum tunneling processes. A first explanation by Monceau *et al.* [35] was based upon Zener breakdown through the gaps induced by the CDW. Others proposed a tunneling between impurity sites [37], raised the possibility of nonlinear excitations in the CDW condensate, and suggested the creation of soliton—antisoliton by quantum tunneling in a large electrical field [38]. In those mechanisms the energy could be supplied by kT as well as by the electrical field. Bardeen [39] indicated that the nonlinear conduction in $NbSe_3$ may be due to Fröhlich conductivity, a conductivity that occurs when the the CDW moves. Fleming and Grimes [40] found a threshold field for the depinning of the CDW which suggested an impurity pinning model based on sliding friction of the CDW as reported by Lee and Rice [41]. Above the threshold field the depinned CDW can slide as a whole. Thus Equation (1) was modified by Fleming [42] as follows:

$$\sigma(T) = \sigma_a(T) + \sigma_b(T)(1 - (E_T/E)) \exp[-E_0/(E - E_T)], \quad E > E_T, \tag{2}$$

where the second term on the right-hand side is operative only above the threshold field E_T.

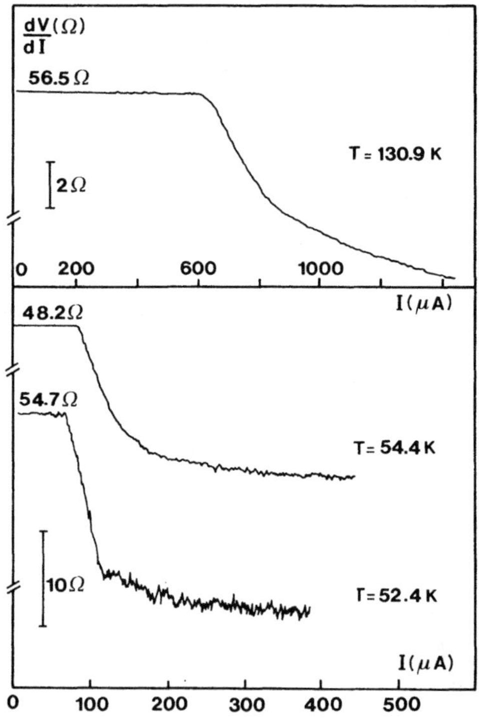

Fig. 15. Differential resistance dV/dI as a function of applied current for different temperatures. (After [43].)

Figure 15 represents typical variations of the differential resistance (dV/dI) at different temperatures, which clearly shows the nonlinear behavior. This plot allows us to measure the threshold field E_T. When a normalized differential resistance [42, 43] $(dV/dI)/R_0$ is plotted as a function of the electrical field, the break is well defined (Figure 16) near the resistivity maxima. However, this method is less reliable when E_0 is large. In that case one can use the appearance of a periodic noise in V. Indeed, above the critical field, a noise is generated in the crystal. Fleming and Grimes [40] and Richard *et al.* [43, 44] have shown that this noise is the superposition of a broad band noise and a periodic noise. The Fourier spectra of the periodic noise can be perfectly described with the help of three fundamental frequencies and their harmonics, as shown by Richard [44]. These frequencies were interpreted as the modulation of the velocity and consequently of the current

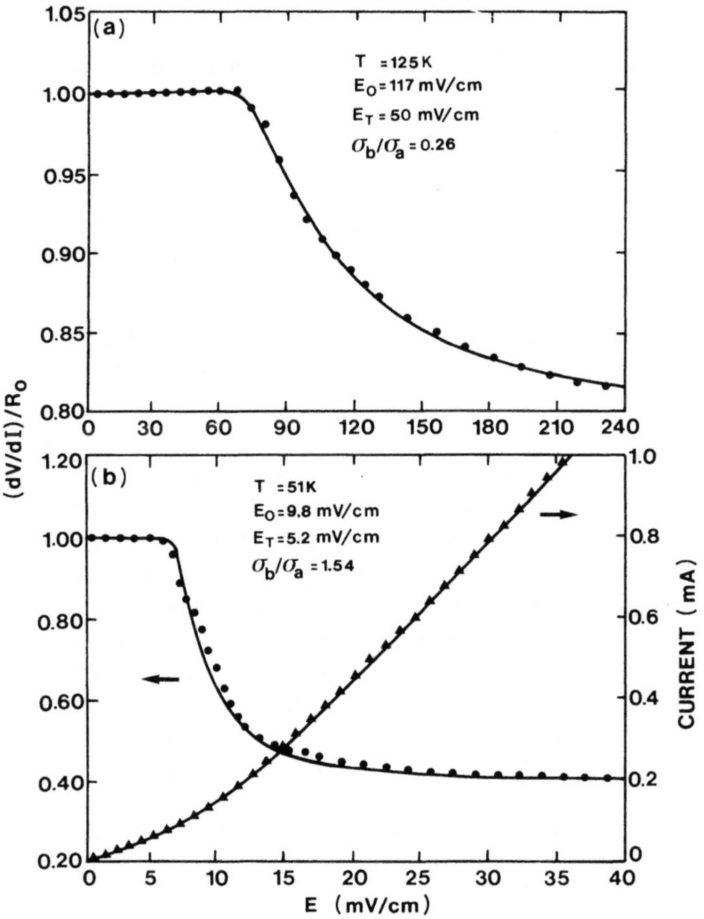

Fig. 16. Normalized differential resistance versus electric field for the $NbSe_3$ example as reported by Fleming [42].

carried by the CDW when it moves in the pinning potential. The fundamental frequency f increases with the electric field E. Monceau *et al.* [43] pointed out that the fundamental frequency f is linearly related to the CDW current, J_{CDW}:

$$J_{\mathrm{CDW}} = nev_d = nef\lambda_{\mathrm{CDW}},$$

where n is the number of the electrons condensed into the CDW, e is the electronic charge, and λ is the wavelength of the CDW; v_d is the drift velocity of the CDW.

Then, J_{CDW}/f is a direct measure of the number of electrons participating in the CDW state.

J_{CDW} is indirectly deduced from measurements in the nonlinear regime assuming that

$$J_{\mathrm{CDW}} = J(1 - R/R_N),$$

where J is the total current, R is the resistance for this current, and R_N is the ohmic resistance for $J \sim 0$.

In order to check the role of impurities and defects on the pinning mechanism, several authors [45—48] have performed detailed electrical resistivity measurements on doped and irradiated samples. The Nb atoms were substituted with either isoelectronic (Ta) or nonisoelectronic (Ti, Zr, . . .) impurities. The resistivity as a function of temperature was measured for several concentrations of the same impurity (from a few ppm to a few per cent impurities concentrations). From these experiments, it was noted that the threshold field depends upon the impurity concentration C as follows:

$E_T \propto C^2$ for a weak pinning (isoelectronic impurity)
$E_T \propto C$ for a strong pinning (nonisoelectronic impurity).

The effect of disorder created by an irradiation technique has also been studied in view of the pinning mechanism. A great advantage of the irradiation technique is that at low irradiation doses, the number of defects is always proportional to the irradiation dose. Thus, the physical properties as a function of defects are more reliable than those associated with the chemical impurities effects. Fuller *et al.* [46] have irradiated $NbSe_3$ with 2.5 MeV protons. On measuring the resistance versus temperature for several doses of radiation (leading up to 0.5% defects), they observed that the resistance increases almost linearly with the damage at low temperature. But the radiation damage produces a more different behavior than substitutional impurities in $NbSe_3$: the threshold field E_T increases linearly with defect concentration. However, even at large doses, the radiation does not destroy the CDW unlike in the case of chemical impurity doping. Monceau *et al.* [47] have also performed some resistivity measurements on samples irradiated with 3 Mev electrons, and found that the two CDW onset temperatures do not vary with the irradiation doses. On the other hand, the two maxima of the resistance (respectively 125 and 49 K for pure $NbSe_3$) are shifted towards lower temperatures (Figure 17). It can be noticed that at low temperature (4.2 K), the resistance increases as a function of the dose, in a similar manner found by Nunez—Regueiro [48] on sintered samples with various granulometry. From this analogy it seems that defects

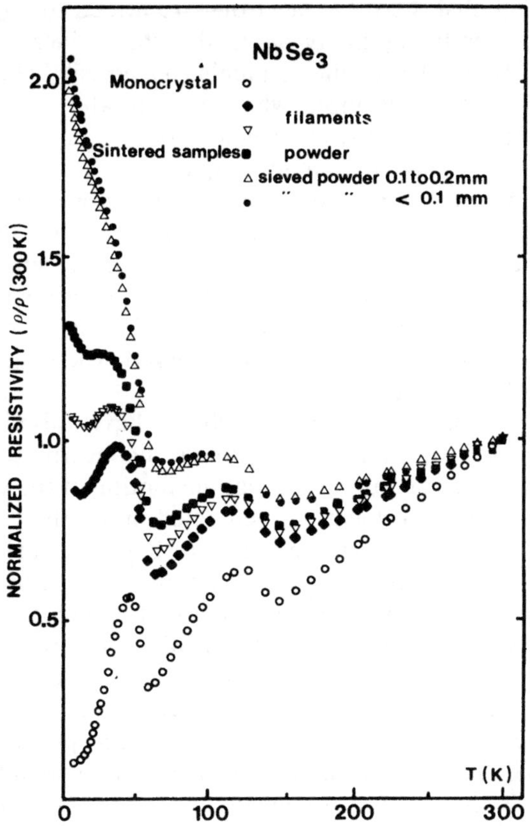

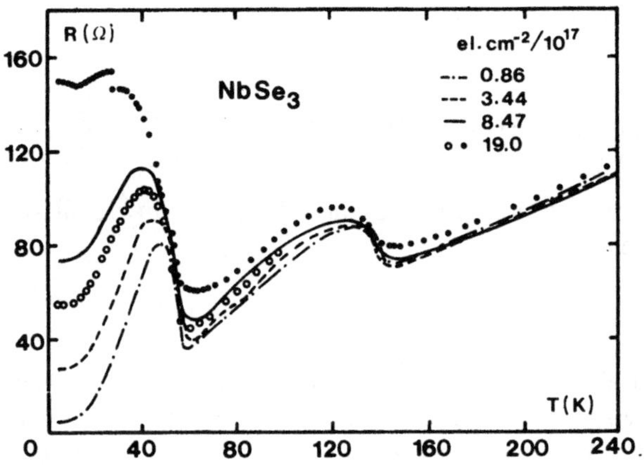

Fig. 17. Normalized resistivity versus temperature for sintered samples [48] and irradiated samples [47] of $NbSe_3$.

interrupt the conducting chains so that electrons are forced to jump on neighboring chains, and consequently resistivity along the chains increases. In addition, Monceau *et al.* [47] observed that the depinning electrical field is very weakly dependent on the defect concentration, which is somewhat different from what Fuller [49] observed. Thus, there arises a question concerning the nature of defects created by proton and electron irradiation techniques.

A common feature among the resistivity curves for samples containing chemical impurities, irradiation defects, or sintered polycrystals is that the resistivity increases sharply at lower temperatures (Figure 17). A similar behavior is found on $NbSe_3$ samples prepared under pressure [15]. Besides a residual impurity effect, these above features seem to be associated with a shortening of the lengths of conducting chains, since a similar feature is observed when the conductivity is measured on samples of small grains [48, 50].

The extraordinary physical properties of $NbSe_3$ have raised a considerable interest among physicists. Two models were proposed concerning the CDW transport in $NbSe_3$. Lee and Rice [51] suggested that the motion of the CDW could be a motion as a whole or through dislocations generated inside the CDW lattice. The necessity of a threshold field and also the effects of impurities are considered and predicted. Noise appears as a consequence of a nonuniform motion of the CDW passing the pinning centres. Another model, due to Bardeen [52], associates the non-ohmic conductivity with a tunneling of the CDW through the pinning potential. This model is based on a Zener-type tunneling through a small pinning gap. By introducing a correlation length L for the CDW across which the field E must be applied to be effective in accelerating the CDW, Bardeen [53] has explained the existence of a threshold field E_T. Thus, no tunneling is possible unless $e^*EL > \varepsilon_g$, where $\varepsilon_g = \hbar\omega_p$ (ω_p is the pinning frequency and e^* is an effective charge for CDW). The threshold field is derived from the above equation as: $E_T = \varepsilon_g/e^*L$.

The field dependent conductivity can be expressed as follows:

$$\sigma(E) = \sigma_a + \sigma_b P(E/E_T)$$

with

$$P(E/E_T) = (1 - E_T/E)\exp(-E_T/E) \quad \text{when } E > E_T$$
$$P(E/E_T) = 0 \quad \text{when } E \leqslant E_T.$$

Physical developments and theories are discussed in detail by Monceau elsewhere in these series.

1.3.2. *TaS_3 Monoclinic Form*

The $NbSe_3$ structural type has recently been found in a new variety of TaS_3, which has a monoclinic unit cell [54] and is different from the previously known orthorhombic TaS_3 [55]. As in the case of the other trichalcogenides, the monoclinic TaS_3 is prepared by direct combination of the elements in sealed silica tubes. At 650 °C, after a slow increase of temperature, single needle-shaped crystals of monoclinic TaS_3 are formed along with the fibrous orthorhombic form. This

TABLE X
TaS_3 (monoclinic) structure, $P2_1/m$, $Z = 6$
All atoms in $2(e)$: $\pm(x, \frac{1}{4}, z)$
$a = 9.515$ Å, $b = 3.3412$ Å, $c = 14.912(2)$, $\beta = 109.00°$. (After [57])

	x	y	z		x	y	z
Ta I	0.64153(4)	$\frac{1}{4}$	0.11909(3)	S(4)	0.1656(3)	$\frac{1}{4}$	0.0837(2)
Ta II	0.69498(4)	$\frac{1}{4}$	0.79811(3)	S(5)	0.1785(2)	$\frac{1}{4}$	0.2765(2)
Ta III	0.29736(4)	$\frac{1}{4}$	0.45911(3)	S(6)	0.5162(2)	$\frac{1}{4}$	0.2465(2)
S(1)	0.5649(2)	$\frac{1}{4}$	0.9366(2)	S(7)	0.9104(3)	$\frac{1}{4}$	0.5685(2)
S(2)	0.1693(2)	$\frac{1}{4}$	0.8971(2)	S(8)	0.5553(2)	$\frac{1}{4}$	0.5943(2)
S(3)	0.2007(3)	$\frac{1}{4}$	0.7640(2)	S(9)	0.7580(3)	$\frac{1}{4}$	0.4309(2)

method leads to a mixture of both varieties. Single crystals have to be separated carefully. A pure monoclinic TaS_3 has recently been obtained in the powder form by Kikkawa *et al.* [56] under pressure.

The structure of the monoclinic form of TaS_3 has been determined from single crystal X-ray diffraction data [57]. Fractional coordinates are listed in Table X, and Figure 18 shows a projection along the [010] direction. As in the case of $NbSe_3$, three different kinds of chains may be distinguished based upon the S—S pair bond lengths. In fact two chains are related to short S—S distances of 2.068 and 2.105 Å, corresponding to the usual distance in $(S_2)^{2-}$ anions (for example, S—S = 2.050 Å in NbS_3 [7]) which is somewhat smaller than the value of 3.315 Å found in 2H TaS_2. The interatomic distances in TaS_3 are summarized in Table XI.

Resistivity measurements on monoclinic TaS_3 as a function of temperature [58] show that it undergoes two transitions, the first around 240 K and the second at

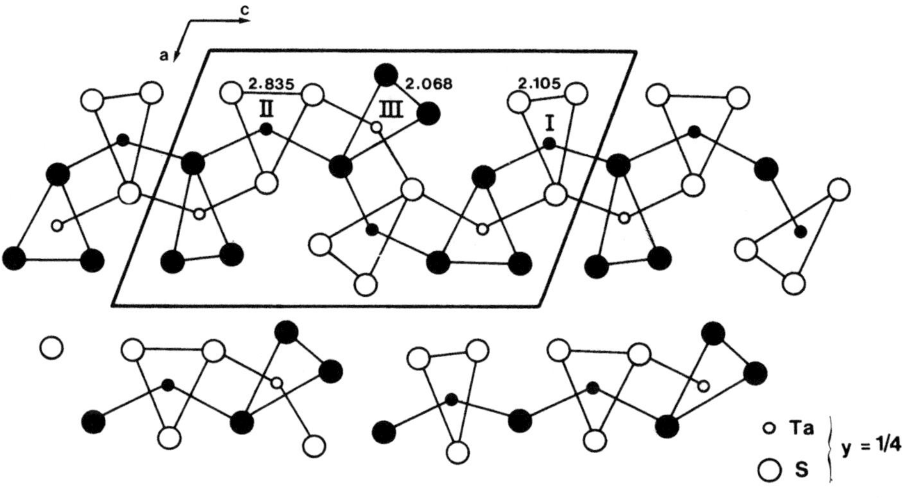

Fig. 18. The structure of TaS_3 projected along [010]. (After [57].)

TABLE XI
Interatomic distances in TaS_3 (monoclinic) (Å) (after [57])

Triangular base	Ta I—2S(1)	2.495(2)	S(1)—S(2)	3.605(3)
	—2S(2)	2.527(2)	S(1)—S(3)	3.541(3)
	—2S(3)	2.509(2)	S(2)—S(3)	2.105(4)
'Caps'	Ta I—1S(1)	2.564(2)		
	—1S(6)	2.566(3)		
Triangular base	Ta II—2S(4)	2.461(1)	S(4)—S(6)	3.389(3)
	—2S(5)	2.529(2)	S(5)—S(6)	3.392(4)
	—2S(6)	2.522(2)	S(4)—S(5)	2.835(3)
'Caps'	Ta II—1S(1)	2.747(3)		
	—1S(8)	2.870(2)		
Triangular base	Ta III—2S(7)	2.512(2)	S(7)—S(8)	3.530(4)
	—2S(8)	2.481(2)	S(8)—S(9)	3.588(4)
	—2S(9)	2.520(2)	S(7)—S(9)	2.068(3)
'Caps'	Ta III—1S(5)	2.566(2)		
	—1S(8)	2.589(2)		
Short S—S bonds between chains:				
	S(2)—S(4)	2.796(4)		
	S(6)—S(9)	2.920(3)		

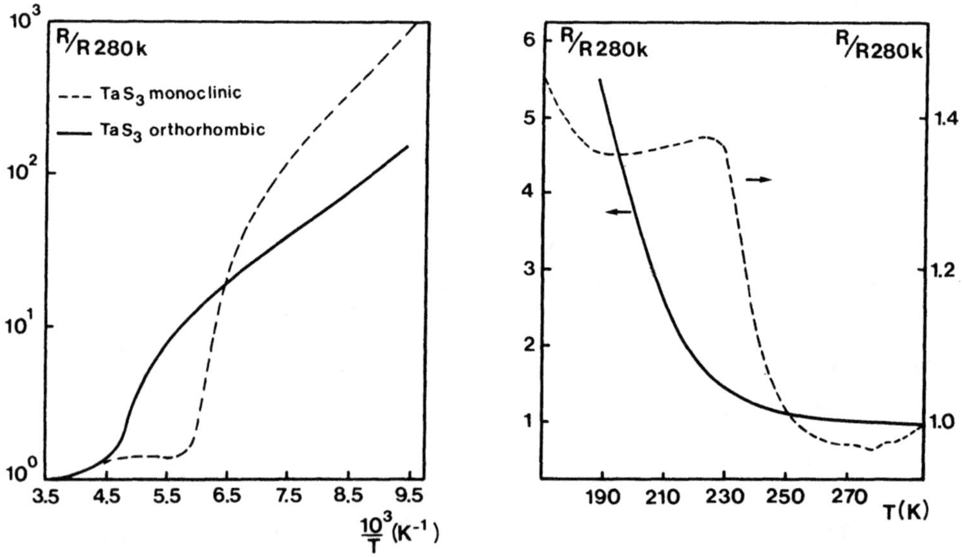

Fig. 19. Variation of the normalized resistances as a function of temperature for both varieties of TaS_3. (After [58].)

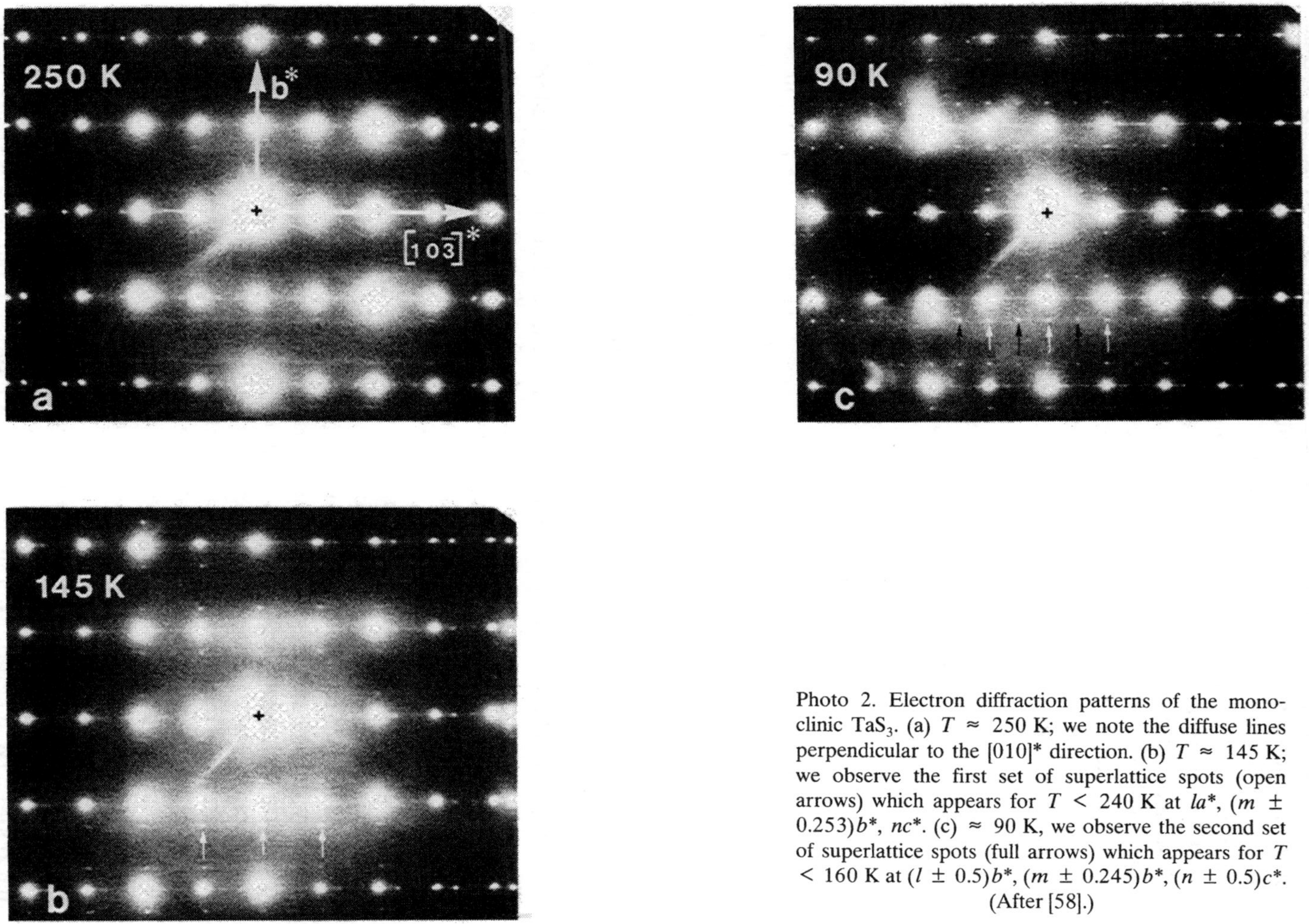

Photo 2. Electron diffraction patterns of the monoclinic TaS_3. (a) $T \approx 250$ K; we note the diffuse lines perpendicular to the [010]* direction. (b) $T \approx 145$ K; we observe the first set of superlattice spots (open arrows) which appears for $T < 240$ K at la^*, $(m \pm 0.253)b^*$, nc^*. (c) ≈ 90 K, we observe the second set of superlattice spots (full arrows) which appears for $T < 160$ K at $(l \pm 0.5)b^*$, $(m \pm 0.245)b^*$, $(n \pm 0.5)c^*$. (After [58].)

160 K (Figure 19). TaS_3 becomes semiconducting below the second CDW transition temperature (160 K), in contrast to $NbSe_3$ which remains metallic below its second CDW transition temperature (59 K).

Electron diffraction studies (Photo 2) at room temperature (i.e. above the first transition) show a pretransitional effect, which is indicated by the presence of diffuse lines perpendicular to the b^* axis at about $b^*/4$. These diffuse lines may imply incoherent fluctuations of CDWs between adjacent chains. On decreasing the temperature, a first set of superlattice spots appears at $T < 240$ K. The wave vector components of the associated distortion are (0, 0.254(3), 0). On further decreasing the temperature ($T < 160$ K), a second set of superlattice spots appears. The wave vector components of the distortion are ($\frac{1}{2}$, 0.245(3), $\frac{1}{2}$). The occurrence of these incommensurate superlattices is well related with the two resistivity anomalies. Table XII summarizes the wave vector components of the distortions in the transition metal trichalcogenides $NbSe_3$ and TaS_3.

TABLE XII
Distortion wave vectors of $NbSe_3$ and $TaSe_3$ (after [58, 67])

$NbSe_3$	$T_2 = 59$ K 0, 0.241(3), 0 $\frac{1}{2}$, 0.259(3), $\frac{1}{2}$	$T_1 = 145$ K 0, 0.243(3), 0	$T > 145$ K No visible diffuse scattering
TaS_3 (monoclinic)	$T_2 = 160$ K 0, 0.254(3), 0 $\frac{1}{2}$, 0.245(3), $\frac{1}{2}$	$T_1 = 240$ K 0, 0.254(3), 0	$T > 240$ K Diffuse scattering
TaS_3 (orthorhombic)	$T \leqslant 130$ K Commensurate	$T \leqslant 210$ K Incommensurate $\frac{1}{2}$, $\frac{1}{8}$, 0.250(3) (roughly)	$T > 240$ K Diffuse scattering

Both CDWs in monoclinic TaS_3 can be depinned under application of an electrical field [48, 59], but it appears that the threshold field (above which non-ohmicity is observed) is greater than in the case of $NbSe_3$ 0.5 V cm^{-1} at 150 K; the threshold field, of course, is sample and temperature dependent. The pinning of the CDW by defects and its effect on the transport properties have been studied by irradiation techniques [60]. Single crystals of TaS_3 (mono) have been irradiated with electrons (in a van de Graaf accelerator and in a 1 MeV electron microscope) and neutrons (in the triton nuclear reactor at Fontenay aux Roses, CEA (France)). Disorder due to random defects created by irradiation is expected to destroy the 3D ordering of the two CDWs. Indeed micro-diffraction patterns recorded in situ during irradiation at 7 K have revealed that the condensed spots give rise to diffuse lines for a defect content greater than 10^{-3} displacements per tantalum (d.p. Ta) [60]. Thus, there is no longer a 3D ordering at 7 K but only uncorrelated CDWs. The resistvity of irradiated samples, measured as a function of temperature, shows that both CDW phase transitions are almost suppressed, but the resistance values remain surprisingly unchanged even at low temperature. Due to this behavior, TaS_3 is shown to be the first low-dimensional conductor in which the pinning of CDW produces no change in the low-temperature resistivity [61].

1.3.3. *TaS_3 Orthorhombic Form*

The exact structure of the orthorhombic TaS_3 is still unknown. From the unit cell parameters [55], (a = 38.804 Å, b = 15.173 Å, c = 3.340 Å, V = 1865 Å^3, Z = 24, space group $C222_1$), it can be seen that $b_{mono} \simeq c_{ortho}$. This corresponds to the height of one $[TaS_6]$ trigonal prims or equivalently to a Ta—Ta distance along the chain axis. A hypothetical arrangement of such trigonal prisms has been proposed [58] and is represented in Figure 20. To date, attempts to make a complete structural determination have failed. Nevertheless, the trials were successful enough

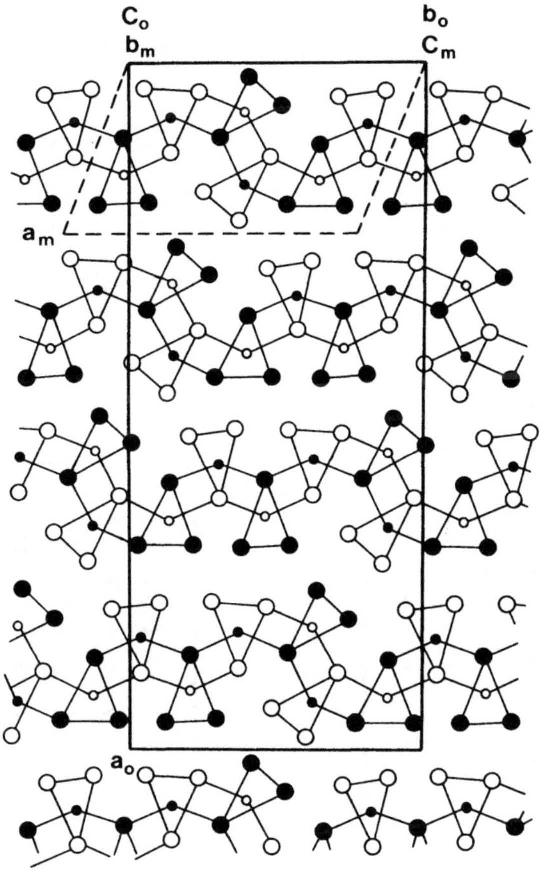

Fig. 20. Relative orientation of the monoclinic and orthorhombic unit cells in respect of each parameters set. (After [58].)

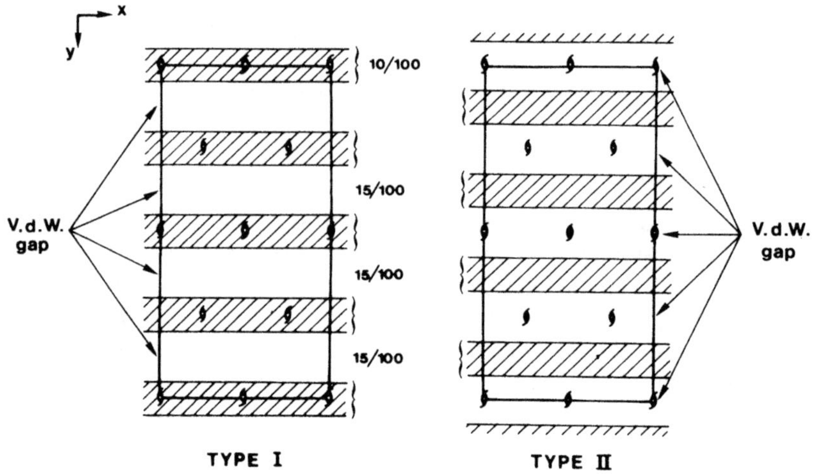

Fig. 21(a). Schematic representation of both types of TaS_3 (ortho) according to the position of the slabs against the 2_1 axes (see also Figure 21(b)).

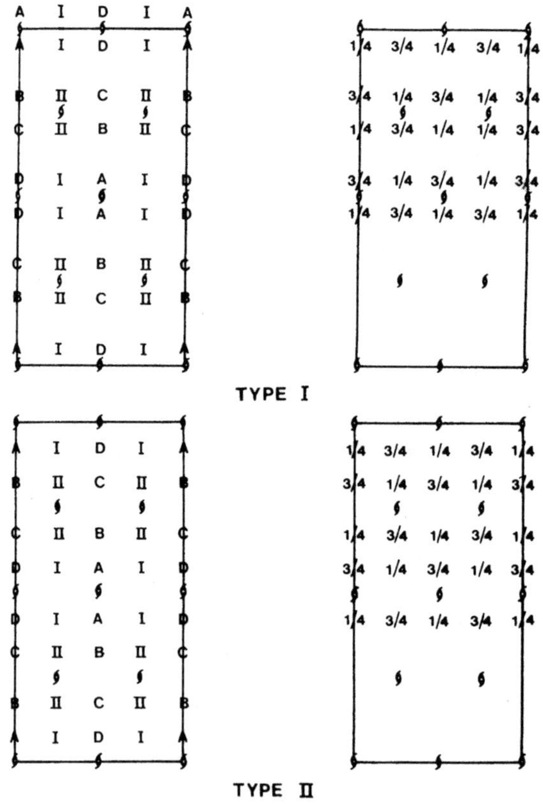

Fig. 21(b). Schematic representation of both types of TaS_3 (ortho) according to the position of the slabs against the 2_1 axes (see also Figure 21(a)).

to confirm the proposed arrangement [58]. Indeed, examination of the three-dimensional Patterson map leads us to assume that the TaS_3 structure is composed of four slabs [62]. Two possibilities may occur, which are referred to as type I and type II in Figures 21(a) and 21(b). In the former the 2_1 axis is located in the slab, whereas in the latter it is between the slabs. The tantalum positions are represented by the letters A, B, C, D, I, and II. Here A, B, C, D have four equivalent positions, while I and II have eight equivalent positions. This gives rise to 32 positions. Thus, eight of them should be removed. However, stacking faults along the *a* axis, which are revealed by diffuse lines on the Weissenberg photographs, indicate that a statistical distribution has to be considered. This study also allowed the *Cmcm* space group to be considered more suitable than the $C222_1$ frequently reported. To be in agreement with the extinction conditions, *a* and *b* parameters have to be inverted and additional condition $h0l$ ($l = 2n$) has to be considered, which gives the advantage that all *z* cordinate values are only $\pm\frac{1}{4}$ as is often the case with the parent MX_3 structures. Several arrangements of tantalum can be obtained by combining the A, B, C, D, I, and II positions (in fact, eight arrangements for each type as schematically shown in Figure 22(a)). Some arrangements present a tantalum distribution within a slab which is very similar to that observed for the TaS_3 monoclinic case. If tantalum atoms are surrounded by a trigonal prismatic frame of sulfur atoms, we can assume that the MX_3 chains are distributed, as shown in Figure 22(b).

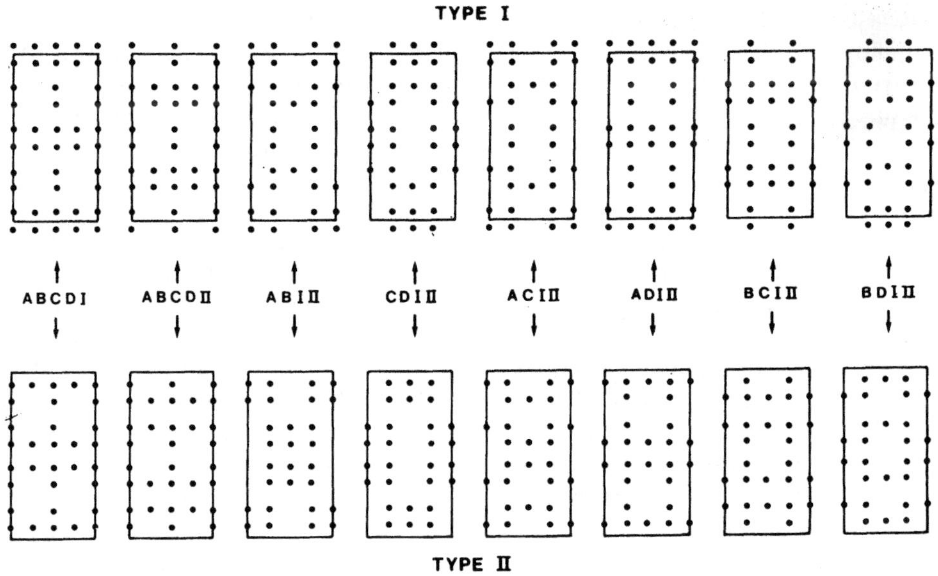

Fig. 22(a). The various combinations of tantalum positions for both types of TaS_3 (ortho) (see also Figure 22(b).

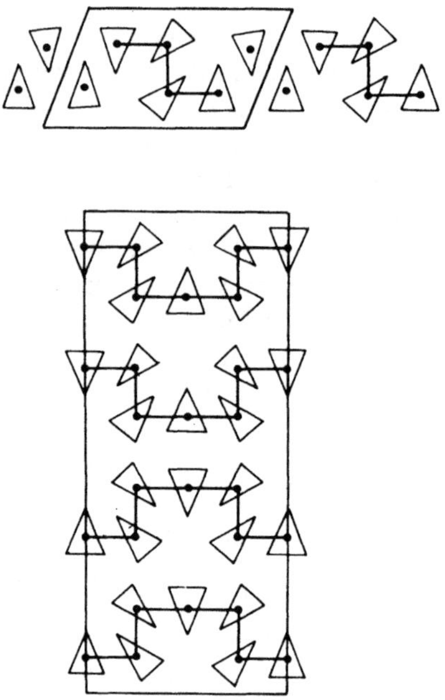

Fig. 22(b). Tantalum positions for both types of TaS_3 (see also Figure 22(a)).

Resistivity measurements performed on orthorhombic TaS_3 by Sambongi *et al.* [63] show a metal-insulator transition around 210 K due to a Peierls distortion. Electron diffraction studies [58, 63, 64] reveal diffuse scattering at room temperature and superlattice spots when the temperature is decreased below 210 K. X-ray studies of the superstructure indicate a new periodicity characterized by $a' = 2a$, $b' = 8b$, and $c' = 4c$ [65]. As the commensurability is one of the pinning mechanisms of a CDW, it was of great interest to look at the threshold field E_T for the onset of nonlinearity. Non-ohmicity was first reported by Takoshima *et al.* [66]. To ensure that the non-ohmicity was not due to sample heating, current—field (I—E) characteristics were measured using both DC and pulsed techniques at different temperatures below the transition temperature [66(a), (b)]. Nonlinear conductivity is observed when the applied electric field E is greater than a threshold field E_T which is estimated to be 0.6—2 V cm^{-1} [67(a), (b)]. E_T is nearly constant in the 215—130 K temperature range (incommensurate state); then, after a smooth decrease E_T increases sharply at lower temperature (commensurate state) [67(a)]. The threshold field E_T appears to be higher for TaS_3 (ortho) than for $NbSe_3$.

An important point is to evaluate the part of the pinning potential associated with the commensurability. Many experiments are currently underway, which should clarify the relative role of commensurability, impurities and defects. In this connection, Wang *et al.* [67] reported that the component of the distortion along the chains was weakly incommensurate below 215 K but locks to a commensurability of four c lattice parameters at $T_0' \sim 130$ K. Nevertheless, from a chemical point of view, one can say that both varieties of TaS_3 are more one-dimensional (1D) in character than $NbSe_3$. This is a consequence of weaker linking between adjacent chains probably because the Ta—S bonding is more ionic than the Nb—Se one. TaS_3 has far more anisotropic electronic properties, the conductivity anisotropy is of the order of 2×10^2 ($\sigma_\parallel$ along the chain/$\sigma_\perp$ transverse) [60] instead of ~10—20 in $NbSe_3$ [27(b)]. This anisotropy is also reflected in the CDW instability: for both TaS_3 precursor effects (diffuse lines) are visible well above the phase transition (up to 100 K) as it is typical of quasi-1D systems like the organic metals with CDW distortions. On the opposite, Roucau *et al.* [58] have not observed this pretransitional effects for $NbSe_3$. The superlattice spots appear in a temperature range less than 3 K. They are very sharp which indicates long-range order. XPS results [68] on the orthorhombic TaS_3 have shown evidence of the occurrence of two types of tantalum. It was not possible to observe such a situation in $NbSe_3$ by the same technique.

1.4. Amorphous MX_3 Chalcogenides

Amorphous trichalcogenides result from the chemical or thermal decomposition of ammonium thiosalts (for a review, see [69] and [70]). The method can be successfully used in the case of MoS_3, $MoSe_3$, and WS_3, which are obtained from the corresponding ammonium derivatives $(NH_4)_2MX_4$. For example, the decomposition of $(NH_4)_2MoS_4$ to MoS_3 is complete at 280 °C under nitrogen [71]. The MoS_3 thus formed is stable up to 335 °C beyond which it loses sulfur to form MoS_2. The same behavior is observed for the thio and seleno derivatives of molybdenum and tungsten [72], and also for the thio derivative of vanadium [73]. Instead of heating,

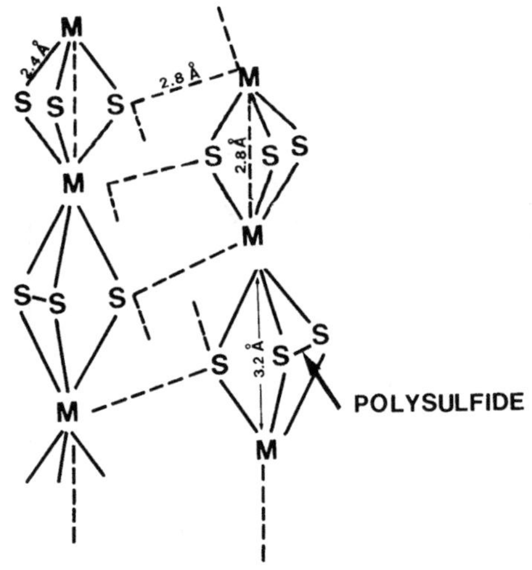

Fig. 23. The proposed structure of amorphous WS_3 and MoS_3. (After [76].)

an acid treatment may be used to prepare the trichalcogenides from the ammonium salts [74]. No significant deviation from stoichiometry has been observed.

Up to now the trichalcogenides of molybdenum and tungsten have been prepared only in the amorphous form. Besides the fact that these materials can be used as cathodes in new secondary advanced batteries, it was of interest to investigate the local structural arrangement and its possible relationship with that in crystalline MX_3. Some interesting structural information was deduced from various techniques such as EXAFS and XPS [75, 76]. It appears that amorphous trichalcogenides exhibit a chain-like structure as found in the crystalline materials of IVA and VA elements. XPS results indicate the presence of two types of chalcogen species, i.e. $(X_2)^{2-}$ and X^{2-} as in other trichalcogenides. The Mo—S and W—S distances are found to be close to 2.40 Å in the chains and 2.75 Å between them (Figure 23). An interesting feature is that the magnetic susceptibility measurements, along with the EXAFS results, indicate a dimerization of the metal atoms along the chains, giving rise to alternating short and long metal—metal distances (e.g. 2.8 and 3.2 Å in MoS_3).

1.5. Intercalation in Trichalcogenides

From the structural determinations it seems clear that transition metal trichalcogenides should be good candidates for intercalation. In fact the chemical behavior of these materials agrees with this prediction. However, the range of possibilities is not as wide as in the case of dichalcogenides due to the thermal, or chemical, instability of trichalcogenides.

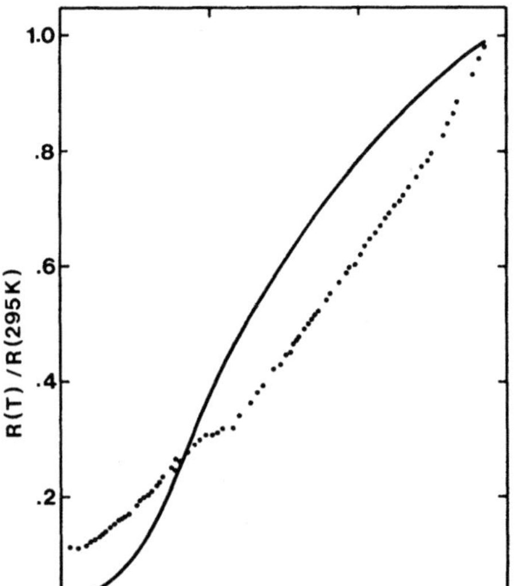

Fig. 24. The normalized resistance as a function of temperature in $TaSe_3(N_2H_4)_x$. (After [78].) Solid line refers to pure $TaSe_3$ and dotted line to intercalated materials.

When exposed to hydrazine vapor, tantalum triselenide forms a molecular intercalate $TaSe_3(N_2H_4)_x$ [77]. Within minutes of exposure to N_2H_4, $x = \frac{1}{3}$, but after 48 hours under 12 Torr of N_2H_4 the value of x is equal to 1. When the sample is evacuated, x is reduced to $\frac{2}{3}$ immediately, and after 48 hours under 10^{-5} Torr of N_2H_4, $x = \frac{1}{3}$. The composition $TaSe_3(N_2H_4)$ seems to represent a situation of particular stability. No structural investigations were undertaken due to the poor crystallization of the samples. The resistivity curves obtained with samples intercalated under a pressure of 12 Torr are of interest [78]. They show an anomaly near 100 K, which was not observed in pure $TaSe_3$. In addition, a positive curvature against temperature is observed above 120 K in contrast to the negative curvature observed in $TaSe_3$ (Figure 24). At the same time, the superconducting transition temperature T_c decreases from 2.2 to 1.5 K. The resistivity anomaly suggests that CDWs are produced, which in turn suppresses the superconductivity.

Alkali—metal intercalates are usually prepared by means of the following methods: use of alkali metal solutions in liquid ammonia or other solvents such as HMPT (hexamethylphosphotriamide), use of organometallics, electrochemical ways, and solid-state techniques. Due to the poor stability of the trichalcogenides, solid-state techniques (heating MX_3 with the alkali metal) are excluded.

Lithium solutions in liquid ammonia lead to a decomposition of the trichalcogenides with formation of Li_2X and an intercalated dichalcogenide. However, butyllithium and electrochemical systems were successfully used to intercalate lithium in trichalcogenides. Intercalation in crystalline trichalcogenides has been performed through the butyllithium technique by Chianelli and Dines [79]. Three lithium atoms per MX_3 were found to be incorporated into the host structure without disrupting that structure, according to a topochemical reaction:

$$3\text{Bu—Li} + MX_3 \longrightarrow Li_3MX_3 + \tfrac{3}{2}C_8H_{18}$$
$$(M = \text{Ti, Zr, Hf, Nb and } X = \text{S, Se.})$$

TABLE XIII
Lattice expansion of Li_3MX_3 (after [79])

	$2a$	b	c	β	Δa	Δb	Δc	$\Delta\beta$
TiS_3	9.916	3.4006	8.778	97.32				
Li_3TiS_3	11.1	3.46	9.12	98.0	1.18	0.06	0.34	0.7
ZrS_3	10.2486	3.6244	8.980	97.28				
Li_3ZrS_3	11.2	3.59	9.22	100.0	0.95	−0.03	0.24	2.7
$ZrSe_3$	10.8218	3.7488	9.444	97.48				
Li_3ZrSe_3	11.8	3.76	9.68	99.1	0.98	0.01	0.24	1.6
HfS_3	10.1846	3.5952	8.967	97.38				
Li_3HfS_3	12.1	3.52	9.24	100.0	1.92	−0.07	0.27	2.6
$HfSe_3$	10.776	3.7216	9.428	97.78				
Li_3HfSe_3	11.5	3.71	9.52	98.1	0.72	−0.12	0.09	0.32
$NbSe_3$	10.006	3.478	15.626	109.30				
Li_3NbSe_3	11.5	3.48	17.0	98.1	1.5	0.00	1.37	−11.2

Intercalation is followed by a lattice expansion. Table XIII gives the new lattice parameters for the Li_3MX_3 derivatives. It appears that the *b* axis does not expand, which would correspond to the fact that the infinite (MX_3) chains are not destroyed. A localization of lithium between these chains was proposed. However, the ribbons of fibers that these chains form are very likely to be preserved because of the strong lateral coupling between fibers. Lithium is certainly located in the van der Waals gaps between the zig-zag ribbons of fibers. A critical discussion of the possibility for the $NbSe_3$ structure to accommodate extra ions has been given by Hodeau *et al.* [26]. It appears at first that the Se—Se separations across the van der Waals gap are rather large when compared with the analogous distances in $NbSe_2$. In this compound the Se—Se distances across the van der Waals gap are equal to 3.55 Å whereas in $NbSe_3$ the corresponding distances range from 3.30 to 4.09 Å. In the van der Waals gap of $NbSe_3$ octahedral and tetrahedral sites are available

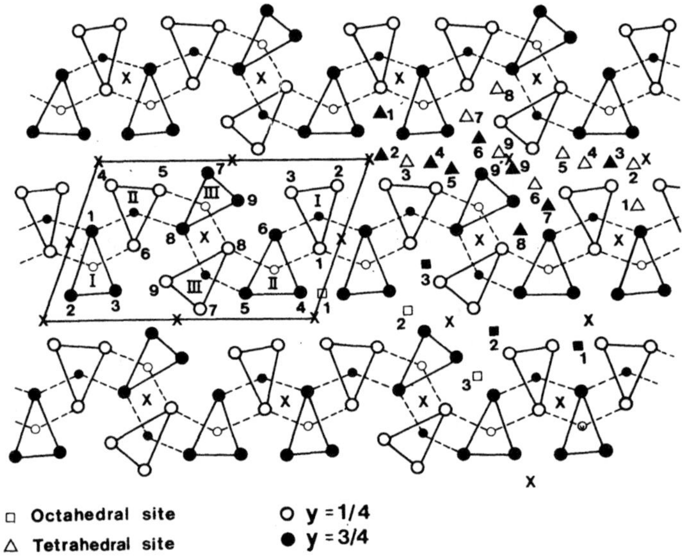

Fig. 25. Octahedral and tetrahedral sites available in the $NbSe_3$ structure. (After [26].)

TABLE XIV
The average distances (Å) from the center of the *Os* or *Ts* sites to the Se atoms (after [26])

Os 1—Se	2.43	*Ts* 4—Se	2.29
Os 2—Se	2.51	*Ts* 5—Se	2.18
Os 3—Se	2.37	*Ts* 6—Se	2.17
Ts 1—Se	2.04	*Ts* 7—Se	2.06
Ts 2—Se	2.20	*Ts* 8—Se	2.05
Ts 3—Se	2.26	*Ts* 9—Se	2.13

(Figure 25). Three types of octahedral sites (noted *Os* 1 to *Os* 3) and nine tetrahedral site (*Ts* 1 to *Ts* 9) can be recognized. The corresponding distances from the center of the site to the selenium atoms are given in Table XIV. It is clear that the three types of octahedral sites can easily accommodate lithium ions. Murphy and Trumbore [80] found that, up to the insertion of one lithium atom per niobium, the structure remains the same since the X-rays powder pattern of $LiNbSe_3$ is very similar to that of $NbSe_3$. As two lithium ions per niobium are inserted, the structure seems to change drastically. In the powder patterns of Li_2NbSe_3 a number of small peaks have disappeared and those which remain are broadened. The third lithium seems to break the structure as its powder pattern is totally different. On the contrary, Chianelli reported a distinct powder spectra for Li_3NbSe_3 which can be indexed with a monoclinic unit cell ~ 30% larger than that of $NbSe_3$.

It is of interest to know how the three electrons given by the lithium atoms are accommodated by the host structure. Infrared spectra show that the polysulfide peak at 560 cm^{-1} present in trisulfides such as TiS_3 or ZrS_3 disappears in the intercalated products [81]. Probably the first two electrons are used to break the X_2^{2-} bonds. The third electron may be delocalized along the metallic chains.

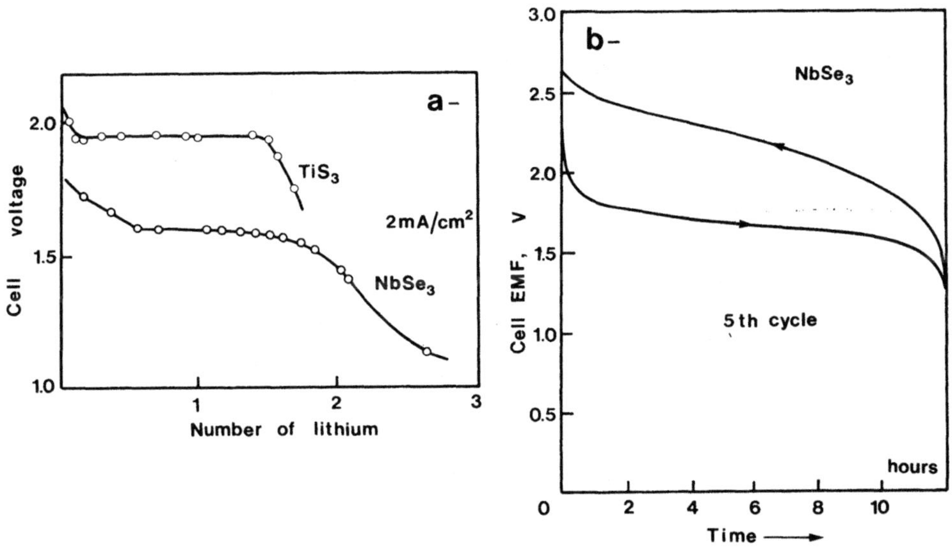

Fig. 26. Electrochemical behavior of TiS_3 and $NbSe_3$ structures. (After [83].)

Electrochemistry is a convenient way to intercalate transition metal trichalcogenides with lithium. Figure 26(a) shows first discharge curves for TiS_3 and $NbSe_3$ [82, 83]. In the case of TiS_3 [84] a real process of intercalation seems to occur only for very low lithium contents, where the reaction appears to be reversible. A disruption of the S_2^{2-} bond is soon observed corresponding to the formation of a ternary phase Li_2TiS_3. This reaction is irreversible and has been considered as due to a conversion of the titanium coordination from a trigonal prism to an octahedron leading possibly to a structure of the $BaTiS_3$ type. A third lithium would then react with Li_2TiS_3 leading to a nonstoichiometric phase $Li_{2+x}TiS_3$ with $0 < x < 1$. In the case of $NbSe_3$ the mechanism of the reaction seems to be different. It is possible to remove the three lithium electrochemically, and the cell shows some reversibility [80]. Figure 26(b) shows a cycling curve for $NbSe_3$. Cells discharged with pure powder $NbSe_3$ cathodes could be recharged to over 90% of the maximum capacity for several cycles, while thin film $NbSe_3$ cathodes have been cycled several hundred times [85].

The possible use of amorphous trichalcogenides as battery cathodes has been considered [86]. Amorphous MoS_3 reacts readily with *n*-butyllithium, *n*-butylsodium, or potassium naphthalides to give compounds A_xMoS_3 (A = Li, Na, K) with $0 < x < 4$. Compositions up to $Li_{3.8}MoS_3$ were reached electrochemically. The cathode was prepared by mixing MoS_3 with teflon and also graphite in some

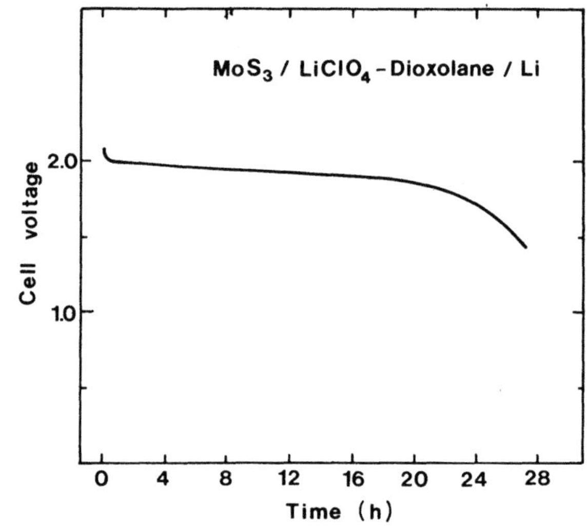

Fig. 27. Voltage composition profile for $MoS_3/LiClO_4$ — dioxolane/Li (1 lithium is equivalent to 7.44 h). (After [86].)

cases. The results are dependent on the crystallinity of MoS_3, which is manifested through experiments on a series of samples prepared at various temperatures. A very important feature of amorphous sulfides is the wide range of alkali-metal compositions, significantly higher than the values reported for crystalline sulfides. Figure 27 shows a voltage composition profile for a $MoS_3/LiClO_4$—dioxolane/Li cell.

2. $FeNb_3Se_{10}$: A New Structural Type

A way to modify the electronic density and the properties of the compounds is to change the chemical composition either by atomic substitutions in the chains or by intercalation between the chains. Only a very small amount of other atoms, particularly tantalum, has been substituted for niobium in $NbSe_3$. These atoms act as chemical defects in the course of the studies concerning the pinning of CDWs. This is different from transition metal dichalcogenides, which readily intercalate between the MX_2 slabs. Starting from a global composition $FeNb_3Se_{12}$ (or $Fe_{0.25}Nb_{0.75}Se_3$) Hillenius *et al.* [87] prepared a compound which was at first thought to have the same structural type as $NbSe_3$ with a substitution of niobium by iron in the trigonal prismatic [MX_3] chains. Crystals of this compound were also studied by Meerschaut *et al.* [88], who determined the structure thereby establishing the formulation $FeNb_3Se_{10}$. The same results were obtained independently by Cava *et al.* [89].

TABLE XV
$FeNb_3Se_{10}$ structure, monoclinic $P2_1/m$, $Z = 1$
All atoms in $2(e)$: $\pm(x, \frac{1}{4}, z)$
$a = 9.213$ Å, $b = 3.482$ Å, $c = 10.292$ Å, $\beta = 114.46$. (After [88])

	Fe	Nb II	Nb I	Se(1)	Se(2)	Se(3)	Se(4)	Se(5)
x	0.4480(3)	0.4480(3)	0.2750(1)	0.4887(1)	0.1608(2)	0.0388(2)	0.2430(2)	0.6605(2)
y	$\frac{3}{4}$	$\frac{3}{4}$	$\frac{1}{4}$	$\frac{3}{4}$	$\frac{3}{4}$	$\frac{3}{4}$	$\frac{1}{4}$	$\frac{1}{4}$
z	0.0922(3)	0.0922(3)	0.3639(1)	0.3642(1)	0.4927(1)	0.2421(1)	0.0893(1)	0.1654(2)
Occupancy (%)	50	50	100	100	100	100	100	100

Fibrous and needle-like crystals of $FeNb_3Se_{10}$ are obtained by heating the elements in silica tubes at about 650—700 °C. Vacuum is established before they are sealed. $FeNb_3Se_{10}$ presents a monoclinic unit cell. The atomic positions are listed in Table XV. Figure 28 represents a projection of the structure along the monoclinic *b* axis. Two types of chains run along this axis, and each chain is repeated twice. One of them corresponds to a trigonal prismatic frame of selenium atoms around Nb I, and the other develops an octahedral arrangement of selenium atoms around both Nb II and Fe atoms. Iron is octahedrally coordinated whereas niobium presents both types of coordination. In Table XVI the interatomic distances are summarized for each polyhedron.

The trigonal prismatic fiber exhibits a short Se—Se distance of 2.348 Å, which is almost identical to the MX_3 chain with the shortest Se—Se bond in $NbSe_3$. In the octahedral chain, one finds niobium and iron atoms both being statistically distributed along the chain axis, which explains the absence of a superstructure. A slight variation of the Fe/Nb ratio has been observed in some samples [89] and is very likely associated with the Fe—Nb distribution in the octahedral chain. Two octahedral chains bonded to each other by a common edge (Se—Se) form a rigid unit.

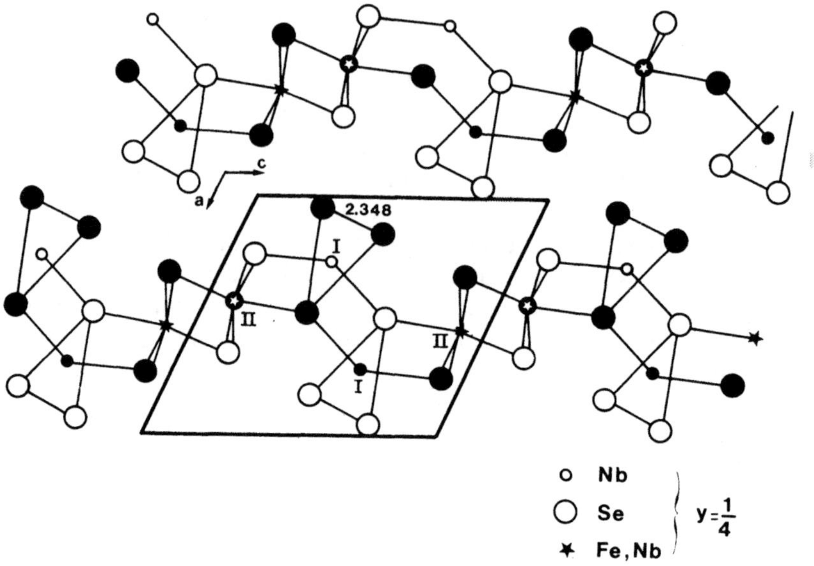

Fig. 28. The structure of $FeNb_3Se_{10}$ projected along [010]. (After [88].)

TABLE XVI
Interatomic distances in $FeNb_3Se_{10}$ (Å) (after [88])

Triangular base	Nb I—2Se(1)	2.628(1)	Se(1)—Se(2)	3.766(1)
	—2Se(2)	2.655(1)	Se(1)—Se(3)	3.801(1)
	—2Se(3)	2.660(1)	Se(2)—Se(3)	2.348(1)
'Caps'	Nb I—1Se(1*c*)	2.743(1)		
	—1Se(4)	2.718(1)		
	(Fe, Nb II)—2Se(4)	2.560(1)	Se(1)—Se(4)	3.295(1)
	—2Se(5)	2.492(1)	Se(1)—Se(5)	3.522(1)
	—1Se(1)	2.666(2)	Se(5c)—Se(4)	3.553(1)
	—1Se(5*c*)	2.416(1)	Se(5c)—Se(5)	3.877(2)
Intermetallic distances	Along the chains: 3.482			
	Across the chains: 3.011			
	(Fe, Nb II)—(Fe, Nb II*c*)			

The intermetallic distances across the common edge (3.011 Å) are shorter than intrametallic distances in each chain (3.482 Å). The separation of 3.011 Å implies probably a metal—metal bond, as in the case of $CrNb_2Se_4$ [90], where the Nb—Nb distance across adjacent octahedral chains is 2.93 Å whereas the Nb—Nb distance along the *b* axis (in each chain) is 3.481 Å. In the octahedral chains the Nb II or Fe coordination is truly six, whereas in the trigonal prismatic chains Nb I has two more nearest neighbors (Se 1*c*, Se 4) as in $NbSe_3$. Octahedral and trigonal prismatic chains form ribbons (parallel to the *b* axis) which are separated by a van der Waals gap.

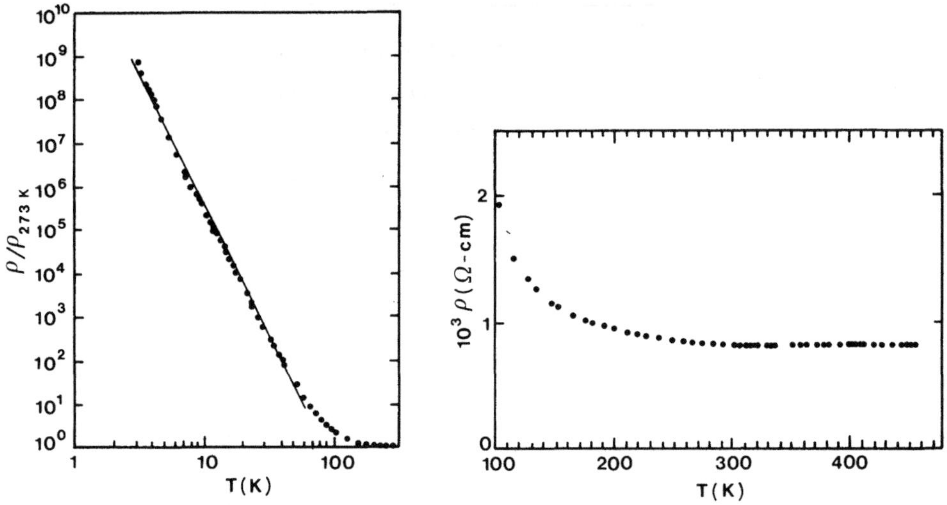

Fig. 29. Variation of the resistivity versus temperature for $FeNb_3Se_{10}$. (After [87].)

Resistivity measurements [87] indicate a metallic behavior down to 140 K. Then the resistivity rises by nine orders of magnitude as the temperature is lowered further (Figure 29). X-ray studies show the formation of superlattice peaks below 140 K. The components of the distortion wave vector are slightly temperature dependent, with values of **q** (0, 0.270(3), 0) near 140 K and **q** (0, 0.258(3), 0) near 6 K. The fact that the CDW has nearly the same wave vector components as those of the first CDW in $NbSe_3$ seems to be a proof that the CDW is related to the [$NbSe_3$] type of chain within the $FeNb_3Se_{10}$ compound. This suggests that the first CDW observed at 145 K for $NbSe_3$ occurs in the chain which has the lowest Se—Se distance. The disorder potential in the adjacent Nb—Fe octahedral chain may explain the fact that an insulating state is reached in the case of $FeNb_3Se_{10}$. It was suggested by Hillenius *et al.* [87] that the CDW-driven reduction in conduction electron density allows the observation of an Anderson type of metal-insulator transition (the disorder creating a random potential and then leading to the localization of the conduction electrons). The $FeNb_3Se_{10}$ structural type can accommodate some nonstoichiometry due to a change of the Fe/Nb ratio within the octahedral chain although the trigonal prismatic chain remains unchanged. Phase limits, even excluding the stoichiometric formulation, have been proposed referring to an expression $Fe_{1+x}Nb_{3-x}Se_{10}$ with $0.25 < x < 0.40$ [91]. However, no significant variation of the unit cell parameters is observed, which makes phase boundaries difficult to ascertain.

Ben Salem *et al.* [92] along with Cava *et al.* [93] have prepared isotypic compounds by replacing Nb atoms in octahedral sites with other transition metal elements such as V, Cr, Ta, etc. Resistivity measurements show that all these compounds exhibit a phase transition at nearly the same temperature $T \sim 140$ K (see Figure 30), but the resistance rise below the transition can differ drastically.

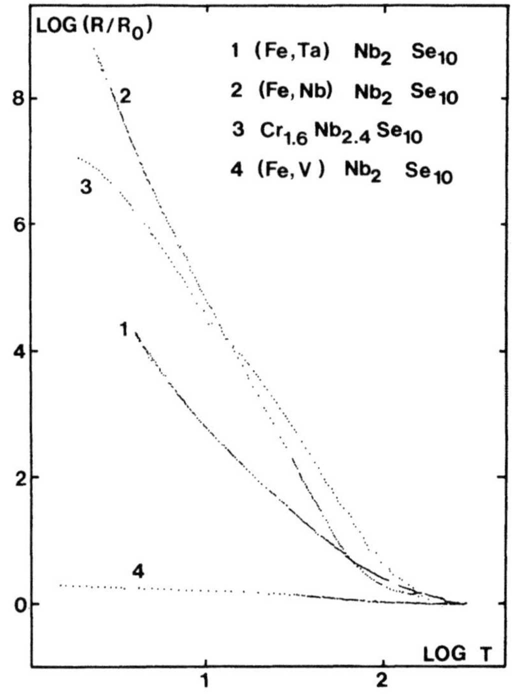

Fig. 30. Log of normalized resistance versus log T for $(Fe,Ta)Nb_2Se_{10}$, $(Fe,Nb)Nb_2Se_{10}$, $Cr_{1.6}Nb_{2.4}Se_{10}$, $(Fe,V)Nb_2Se_{10}$. (After [92].)

Moreover, CDW origin of the phase transition was only seen for the $FeNb_3Se_{10}$ case by X-ray diffuse scattering experiments. Explanation should be sought from the electron density of states at the Fermi level [91]. No nonlinear response was observed for electric field as high as 25 V cm^{-1}.

3. Tetrachalcogenides and Related Compounds

Transition metals tetrachalcogenides also belong to the family of compounds exhibiting a pseudo-1D character. VS_4 represents the structural archetype in this series. Indeed, it was the first tetrachalcogenide to be characterized [94, 95].

3.1. VS_4 STRUCTURAL TYPE

VS_4 crystallizes with the monoclinic symmetry, space group $I2/c$. Unit cell parameters and fractional coordinates are summarized in Table XVII. The structure consists of (VS_4) chains with the coordination polyhedron of the cation being a rectangular antiprism. Two anion pairs (S—S = 2.032 and 2.042 Å) form the short edges in each rectangular anionic plane (Figure 31). One electron from the vanadium atom is used for a V—V bond in agreement with the nonmetallic, diamagnetic behavior of VS_4. Thus, short (2.841 Å) and long (3.241 Å) V—V distances alternate along the c axis, parallel to the chains. Contrary to trichalcogenides, nearest-neighbor VS_4 chains are not shifted by $b/2$ along the chain axis and interchain V—V distances are large (more than 6 Å). Interatomic distances are given in Table XVIII.

3.2. $NbTe_4$ STRUCTURAL TYPE

A very similar structure has been proposed for the niobium tetratelluride. $NbTe_4$ crystallizes with the tetragonal symmetry, space group $P4cc$ or $P4/mcc$ [96]. The occurrence of a complicated superstructure makes the structural determination difficult. Only the subcell is characterized. It appears that Nb atoms are coordinated to eight Te atoms located on top of a slightly distorted square antiprism, as is

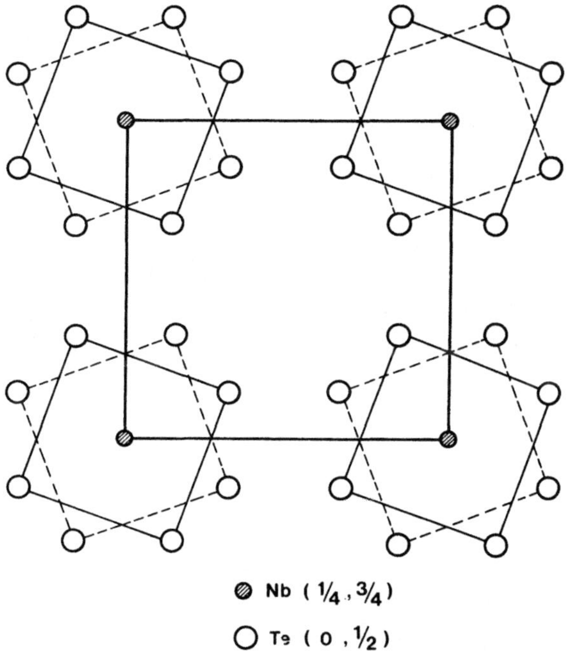

Fig. 32. The structure of the subcell of $NbTe_4$ projected along [001]. (After [96].)

shown in Figure 32. Unit cell parameters and fractional coordinates are listed in Table XIX. One finds eight identical Nb—Te distances equal to 2.902 Å. As Nb atoms occupy the $2a$ position ($\pm 0, 0, \frac{1}{4}$), this implies that Nb—Nb distances along the c axis are all the same. On this point $NbTe_4$ appears different from VS_4.

Electron diffraction studies were carried out by two groups [97, 98]. They both reported that $NbTe_4$ and $TaTe_4$ exhibited three independent CDWs at room

TABLE XVII
VS_4 structure, monoclinic $I2/c$, $Z = 8$
$a = 6.775$ Å, $b = 10.42$ Å, $c = 12.11$ Å, $\beta = 100.8°$. (After [95])

	V	S(1)	S(2)	S(3)	S(4)
x	−0.0002	−0.2369	−0.2365	0.2778	0.0531
y	0.0013	0.0980	−0.0961	0.0498	0.1832
z	0.1173	0.2180	0.2342	0.0286	0.0039

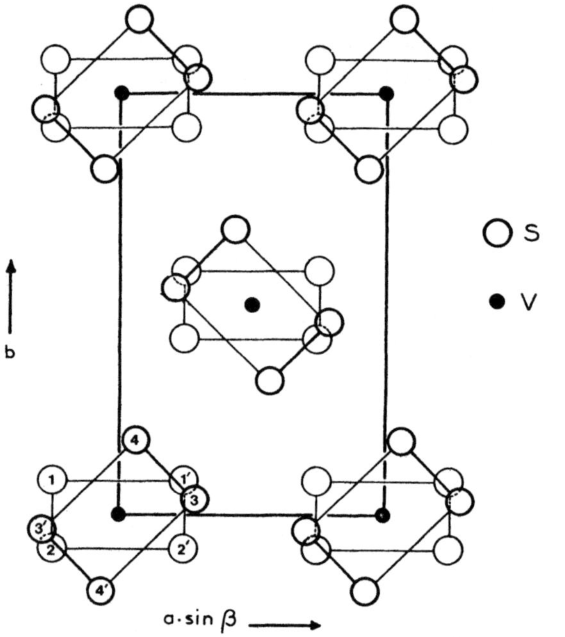

Fig. 31. The structure of VS_4 projected along [001]. (After [94, 95].)

TABLE XVIII
Interatomic distances in VS_4 (Å) (after [95])

Rectangular base	V—S(1)	2.411	V—S(3)	2.392	
	—S(1′)	2.526	—S(3′)	2.387	
	—S(2)	2.540	—S(4)	2.407	
	—S(2′)	2.400	—S(4′)	2.404	
Short S—S distances in rectangular bases					
	S(1)—S(2) and S(1′)—S(2′)			2.032	
	S(3)—S(4) and S(3′)—S(4′)			2.042	
Short V—V distance				2.841	along the c axis
Long V—V distance				3.214	along the c axis

TABLE XIX
$NbTe_4$ structure, tetragonal $P4/mcc$, $Z = 24$
$a = 2 \times 6.466$ Å, $c = 3 \times 6.837$ Å. (After [96])

Atomic arrangement for the subcell ($a' = 6.499$, $c' = 6.837$)
Nb in $2(a)$: $(0, 0, \frac{1}{4})$
Te in $8(m)$: $\pm(x, y, 0; \bar{y}, x, 0; \bar{x}, y, \frac{1}{2}; y, x, \frac{1}{2})$
with $x = 0.1425$ and $y = 0.3316$

temperature; Boswell *et al.* [97] interpreted the superlattice formation in terms of three incommensurate CDWs, for the $NbTe_4$ case only, for which the wave vector components are $\mathbf{q}_1$ $(0, 0, 0.311(2c^*))$, $\mathbf{q}_2$ $(\frac{1}{2}, \frac{1}{2}, 0.344(2c^*))$ and $\mathbf{q}_3$ $(\frac{1}{2}, 0, \frac{1}{3}(2c^*))$. These $\mathbf{q}$ vectors are shown in the reciprocal lattice diagram (see Figure 33). Therefore one has to consider three different types of $NbTe_4$ chains, namely one type I, two types II, and one type III as shown in Figure 34. The structural reason to consider three $NbTe_4$ different chains is not yet known. However, in a very recent short structural report (Böhm and von Schnering [99]), it is mentioned that the periodicity M along the c axis does not correspond to $M = 3c$ but rather to $M = 3.20c$, which generates a deformated modulated structure. As a consequence, a cation shift imposes slight modifications of interatomic Te—Te distances in the square tellurium planes. Whangbo [100] gives an attractive explanation. He con-

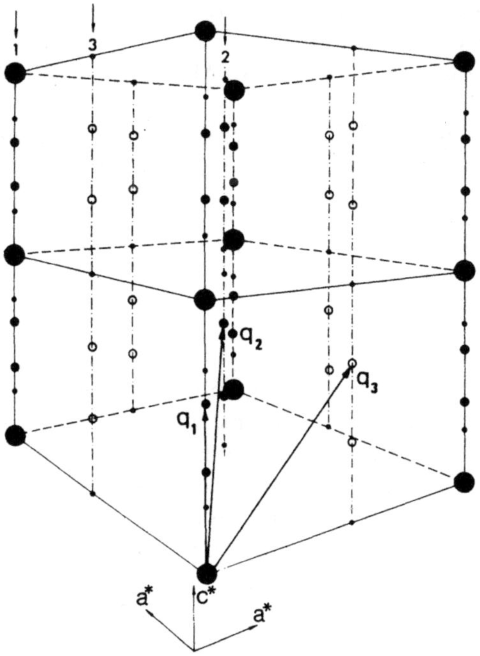

Fig. 33. The reciprocal lattice of $NbTe_4$. Large circles represent subcell reflections, smaller circles superlattice reflections. The three types of superlattice rows and the distortion wave vectors are labelled. (After [97].)

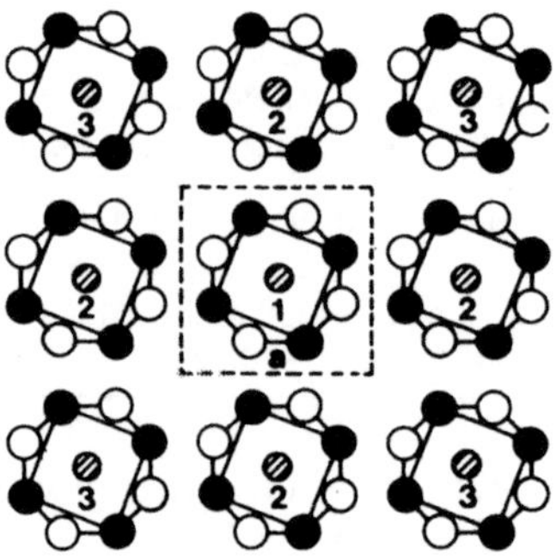

Fig. 34. Structure of $NbTe_4$ projected along [001] in order to identify the three kinds of $NbTe_4$ chains. (After [97].)

siders that the experimental value of $\simeq c^*/3$ (electron diffraction studies) means that the d_{z^2} bands are $\frac{1}{3}$ filled. Now, an ideal $NbTe_4$ structure made of undistorted Te square planes implies the formal oxidation state $Nb^{4+}(Te_2^{2-})_2$ indicating an $\frac{1}{2}$ filled d_{z^2} band. The discrepancy between the theoretical ($\frac{1}{2}$) and experimental ($\frac{1}{3}$) values of d_{z^2} band filling should be resolved through Te^{2-} species, which are produced by lengthening the Te—Te bonds. As the formula unit is represented by $Nb_{24}(Te_2)_{48}$ for the real unit cell ($2a$, $2a$, $3c$), we may assume an increase of 4 Te—Te bonds out of 48, corresponding to a change from $4(Te_2)^{2-}$ to 8 Te^{2-}. In this way 16 d electrons, instead of 24 d electrons, are available for 24 Nb atoms. This could explain the $\frac{1}{3}$ filled d_{z^2} band. More structural information is needed in this respect.

Three niobium selenides with an approximate $NbTe_4$ composition were reported [101]. However, it was recently proved that two of them were in fact Nb_2Se_9 and $(NbSe_4)_3I$. Discussion of Nb_2Se_9 will be given in the next section.

3.3. $(NbSe_4)_3I$ STRUCTURAL TYPE

3.3.1. *$(NbSe_4)_3I$*

$(NbSe_4)_3I$ was found to be the first member of a new series of compounds corresponding to the $(MX_4)_nY$ formulation (M = Nb, Ta; X = S, Se; Y = halogen). Single crystals of these phases were obtained by mixing starting elements in an evacuated sealed glass tube at temperatures ranging from 400 to 550 °C. Needle-shaped single crystals were grow from the reaction products. $(NbSe_4)_3I$ crystallizes with the tetragonal symmetry, space group $P4/mnc$. Fractional coordinates are listed in Table XX. Structural determination [102] revealed features similar to VS_4. Indeed, $(NbSe_4)_3I$ is constituted of $NbSe_4$ chains aligned along the direction parallel to the c axis. The iodine atoms are located in the channels between these chains (Figure 35). The niobium coordination is a rectangular antiprism made up of Se—Se pairs. The Se—Se distance in these pairs is 2.34 Å, corresponding typically to the $(Se_2)^{2-}$ pair. The antiprisms are stacked along the c axis in a screw-like

TABLE XX
$(NbSe_4)_3I$ structure, tetragonal $P4/mnc$, $Z = 4$
$a = 9.489$ Å, $c = 19.13$ Å. (After [102])

	I	Nb I	Nb II	Se(1)	Se(2)	Se(3)	Se(4)
Position	4(*e*)	8(*f*)	4(*d*)	8(*h*)	8(*h*)	16(*i*)	16(*i*)
x	0	0	0	0.2754(6)	0.3910(6)	0.1265(3)	0.7844(4)
y	0	$\frac{1}{2}$	$\frac{1}{2}$	0.0180(7)	0.8000(7)	0.6851(4)	0.5476(4)
z	0.6297(4)	0.4200(3)	$\frac{1}{4}$	0	0	0.3394(2)	0.3283(2)

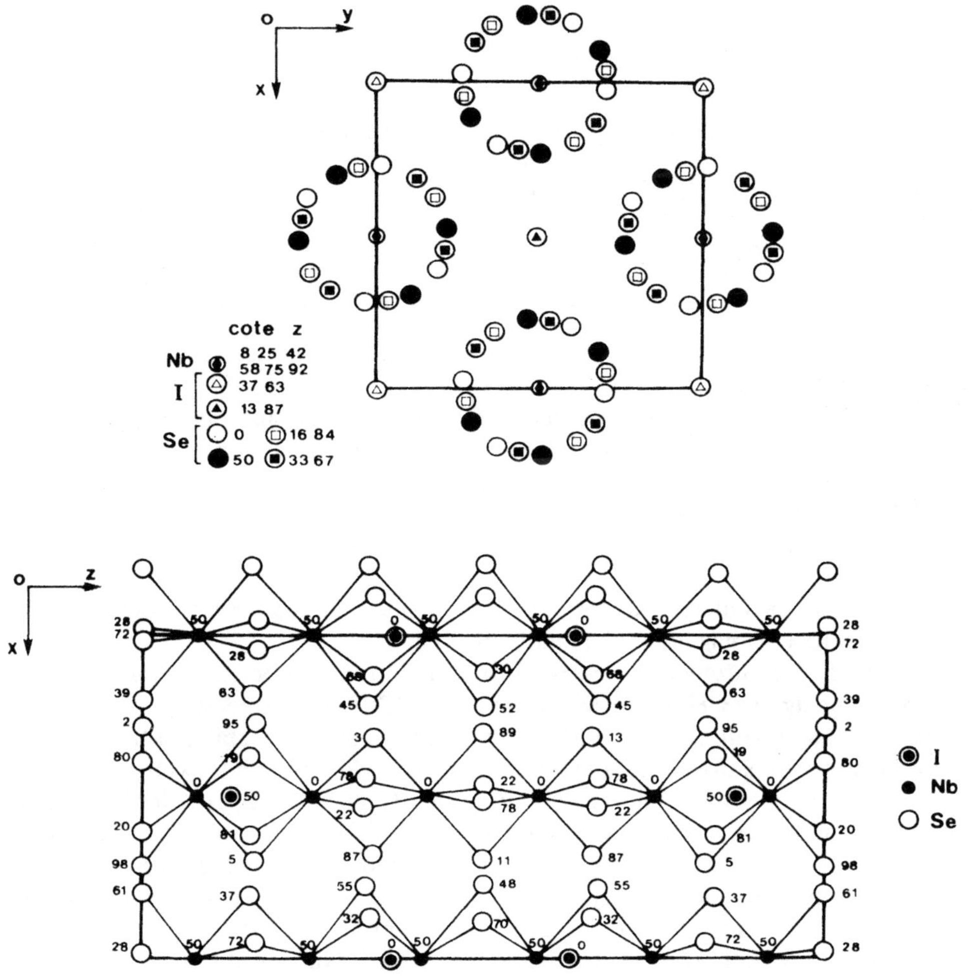

Fig. 35. The structure of $(NbSe_4)_3I$. (After [102].)

arrangement. Two different Nb—Nb distances along this axis are found to be 3.06 and 3.25 Å. Each short Nb—Nb bond is followed by two longer Nb—Nb distances (see Table XXI).

TABLE XXI
Interatomic distances $(NbSe_4)_3I$ (after [102])

Nb I (0, 0, 42)					
Rectangular base	Nb I—Se(31)	2.627(9)	Rectangular base	Nb I—Se(15)	2.63(1)
$Z \simeq 0.33$	—Se(32)	2.627(9)	$Z = 0.50$	—Se(16)	2.63(1)
	—Se(41)	2.73(1)		—Se(25)	2.65(1)
	—Se(42)	2.73(1)		—Se(26)	2.65(1)
	Se(32)—Se(41) Se(31)—Se(42)	2.37(1)		Se(15)—Se(25) Se(16)—Se(26)	2.34(2)
Nb II (0, 0, $\frac{1}{4}$)					
Rectangular base	Nb II—Se(316)	2.730(8)	Rectangular base	Nb II—Se(31)	2.730(8)
$Z \simeq 0.17$	—Se(315)	2.730(8)	$Z \simeq 0.33$	—Se(32)	2.730(8)
	—Se(415)	2.576(9)		—Se(41)	2.576(9)
	—Se(416)	2.576(9)		—Se(42)	2.576(9)
	Se(315)—Se(416) Se(316)—Se(415)	2.37(1)		Se(32)—Se(41) Se(31)—Se(42)	2.37(1)
Nb I—Nb I = 3.06 Å			Nb I—Nb II = 3.25 Å		

Simply writing down the most likely formal valences, one obtains the $(Nb^{4+})_2$ Nb^{5+} $(Se_2^{2-})_6I^-$ formulation, which could explain the succession of long and short Nb—Nb distances according to the sequence:

$$---Nb^{4+}—Nb^{4+}---Nb^{5+}---Nb^{4+}—Nb^{4+}---Nb^{5+}$$

This model postulates the formation of a bond between Nb^{4+} ions (d^1) and should lead to a diamagnetic semiconducting behavior.

Resistivity measurements were performed by a four-probe method using a low-frequency bridge (33 Hz) (see Monceau *et al.* [43] for the operating mode). Contacts were made with knotted gold wire around the single crystal and bonded with gold paint to minimize the contact resistances. Measurements were performed between 77 and 500 K. The resistivity curves (Figure 36(a), (b)) indicate semiconducting behavior above room temperature with a gap of 4400 K. The room-temperature resistivity is about 1 Ω cm. Around 275 K we measure a sharp decrease in the activation energy. At lower temperature two types of behavior have been observed. The corresponding crystals cannot be distinguished by X-ray techniques; they both present the characteristics of $(NbSe_4)_3I$. The measurements have been reproduced at least three times on different crystals of both types. Type I behavior is characterized by a very small semiconducting gap below 275 K. In type

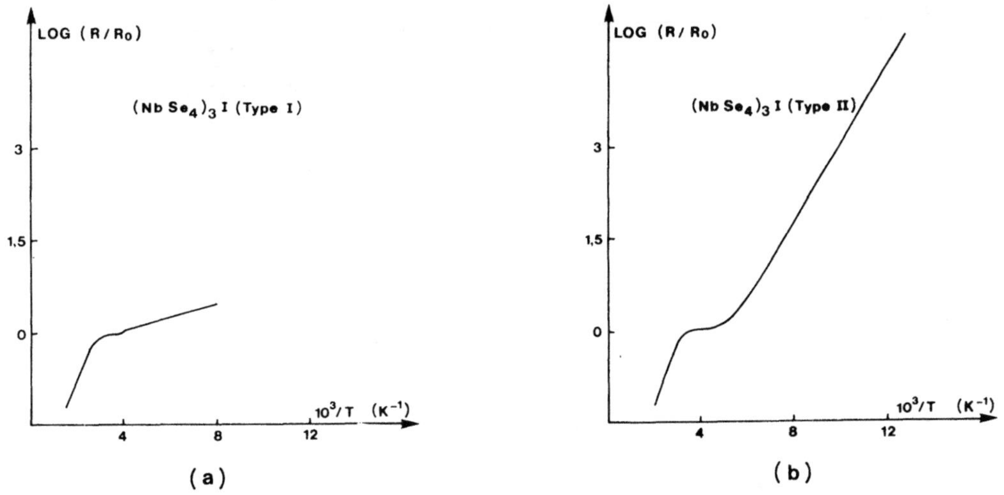

Fig. 36. Log R/R_0 versus $10^3/T$ for $(NbSe_4)_3I$ type I and type II compounds.

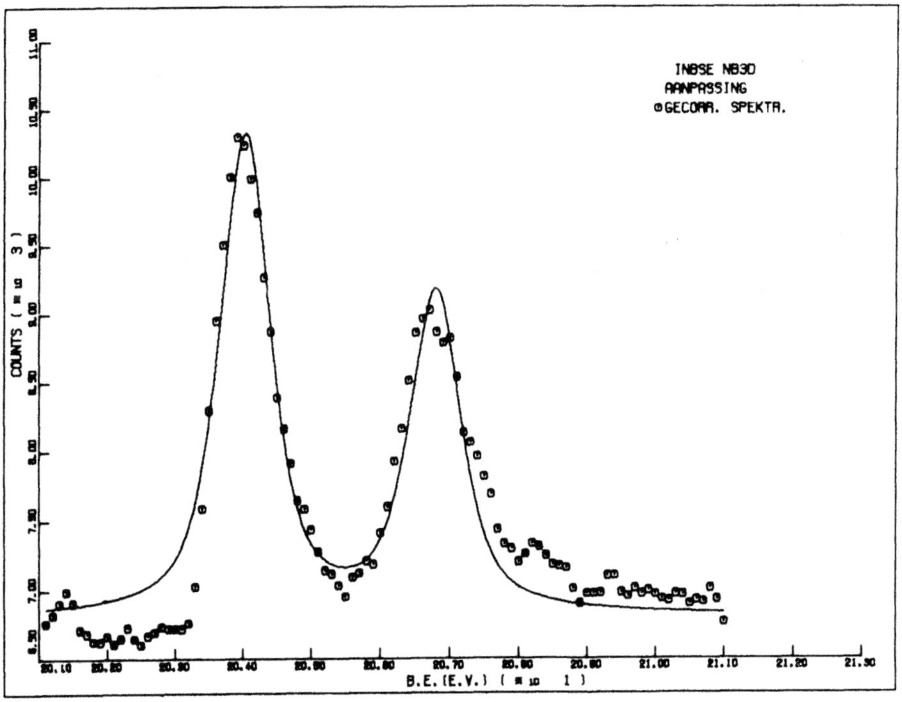

Fig. 37(a). XPS results on $(NbSe_4)_3I$: Nb 3d signal.

II crystals, after a decrease of the gap near 275 K, the resistivity increases again at lower temperatures, showing once more a typical semiconducting behavior, E_g = 3000 K [103].

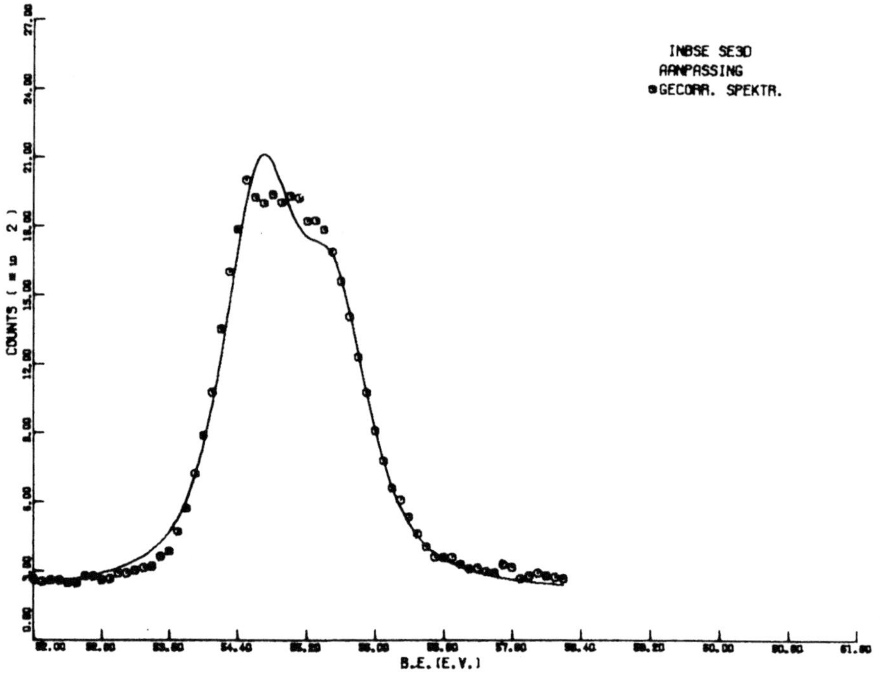

Fig. 37(b). XPS results on $(NbSe_4)_3I$: Se $3d$ signal fitted with one kind of Se.

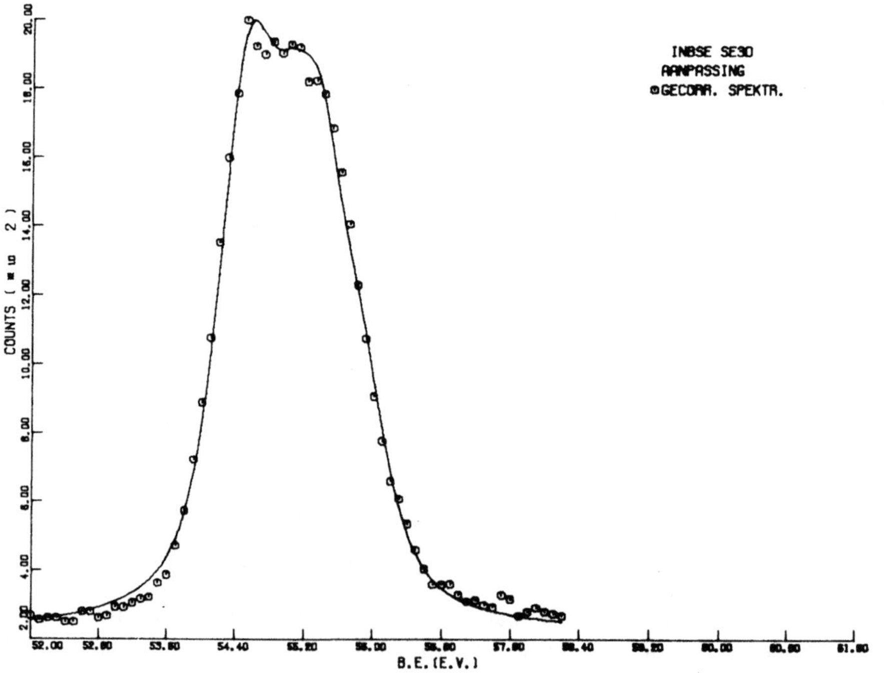

Fig. 37(c). XPS results on $(NbSe_4)_3I$: Se $3d$ signal fitted with two kinds of Se.

To test the model reported above, XPS experiments were done. Since there exist two kinds of Nb atoms, one would expect to see some mixed effects in the Nb $3d$ XPS spectrum, e.g. the occurrence of two signals in a 2 : 1 ratio. In fact, a quite normal narrow peak (0.88 eV) is observed [68] (Figure 37(a)), which means that either there is no mixed valence effect, or the Nb $3d$ level is too insensitive to such phenomena. Then, the Se $3d$ spectrum (Figure 37(b), (c)) does show an anomalous shape. The best fit of the XPS spectrum (Figure 37(c)) is obtained with two Se signals with a 2 : 1 ratio, as reported in Table XXII. The binding energy difference

TABLE XXII
Parameters used in the fit of Nb and Se $3d$ spectra
B.E. is the binding energy (in eV)
W is the width and I the intensity of the signal

		B.E.	W	I	
Se_2^{2-}	$3d$ 3/2	55.36	0.77	4	
	$3d$ 5/2	54.49	0.77	6	
$Se^{(-1+\delta)}$	$3d$ 3/2	55.86	0.77	2	
	$3d$ 5/2	54.99	0.77	3	
$Nb^{4+\delta'}$	$3d$ 3/2	206.79	0.88	2	
	$3d$ 5/2	204.02	0.88	3	
I^-	$3d$ 3/2	617.8	1.7	2	Nonmono
	$3d$ 5/2	629.3	1.7	3	chromated

between them is large (0.5 eV). One-third of the Se rectangular planes have a neighboring I atom at 3.27 Å, a distance shorter than the sum of the van der Waals radii. This may imply a charge transfer from these Se atoms to the iodine atoms, and then a formal valence state $Se^{-1+\delta}$ compared to Se^{-1} for the remaining Se atoms [104]. Thus, bridges are built between chains.

In order to get some information on the nature of the transition which is observed at 275 K in the resistivity curves, Roucau *et al.* [105] have performed an electron diffraction study. The anomaly in the transport properties is not related to a CDW phenomenon but to a change of the space group corresponding to the disappearing of the glide plane n (i.e. the hOl spots are visible at all values of the sum $h + l$).

Pulsed NMR studies of the ^{93}Nb have been performed on single crystals of $(NbSe_4)_3I$ [106]. The spectra observed above 275 K (Figure 38) are temperature independent. They can be easily interpreted according to the structure considering that only the paired Nb nuclei are observed, all of which experiencing a quadrupolar coupling $e^2qQ/h = 12.5$ MHz with an asymmetry parameter $\eta = 0.56$. At 275 K the onset of a structural phase transition is observed, corresponding to the softening (down to the NMR time scale) of an optical mode. Below 275 K, each resonance line splits into two (Figure 38), corresponding to the doubling of the number of inequivalent sites. The temperature dependence of this splitting gives

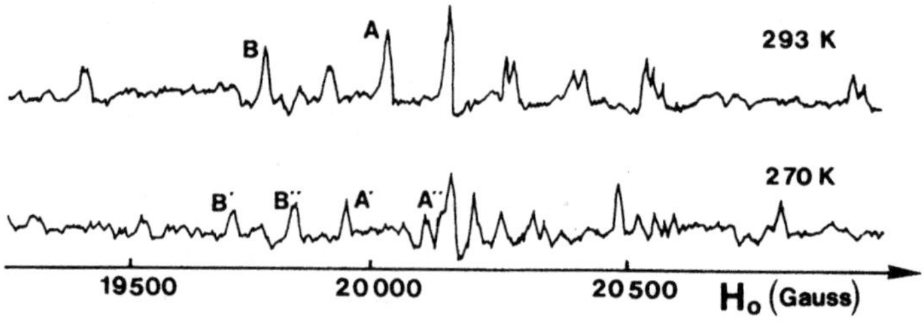

Fig. 38. Central part of the [93]Nb spectra recorded at the same orientation of H_0 in the a–b plane; peaks A and B correspond to the $(\frac{1}{2}, \frac{3}{2})$ transition of sites corresponding in a $\pi/2$ rotation. They respectively split into A', A'', B', and B''. (After [106].)

access to that of the order parameter of the transition [106]. Very recently, Izumi *et al.* [107] reported a phase transition in a new type of $(NbSe_4)_xI$ by resistivity and far-infrared reflectance measurements. Firstly, this compound was rather considered as $(NbSe_4)_2I$ and was later corrected as $(NbSe_4)_3I$ after re-examination of the chemical analyses and density measurement [108]. Resistivity measurements show a metal–insulator transition in the vicinity of 270 K. It seems that this compound is similar to that of type I mentioned earlier. Novel infrared-active phonon modes appear below 260 K with increasing intensities as the temperature decreases. Both features (resistivity anomaly and novel modes of infrared-active phonons) confirmed the existence of a structural phase transition below 270 K.

Investigation of this structural modification was completed by an X-ray diffraction study. As mentioned above in the electron diffraction part, condition limiting

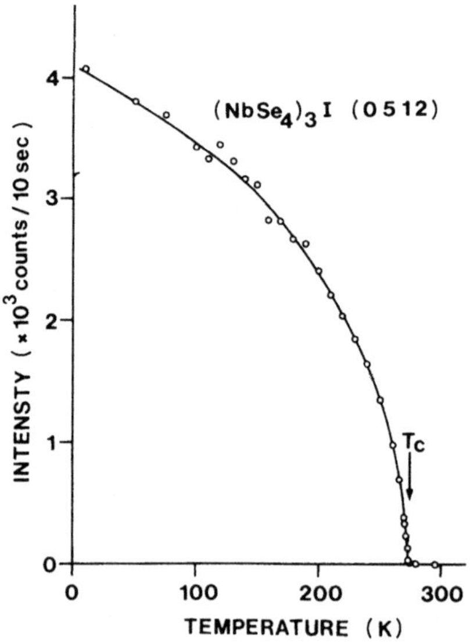

Fig 39. Temperature dependence of the peak intensity of (0, 5, 12) reflection for $(NbSe_4)_3I$ below the structural transition temperature (After [108].)

the possible reflection *Okl* with $k + l = 2n$ disappears in the low-temperature form. Thus, by following the variation of intensity of the reflection which is most sensitive to the phase transition (namely, type *Okl*, $k + l \neq 2n$; e.g. (0, 5, 12)) Izumi *et al.* [108] do not observe any hysteresis or jump of the peak intensity as a function of temperature (see Figure 39). Thus, it can be concluded that the phase transition is second order. The same structural model seems to be common to a large number of iodine derivatives $(MSe_4)_nI$ (with M = Nb, Ta; $n = 2, 3, \frac{10}{3}$). Among that series, three compounds are of particular interest, namely $(NbSe_4)_2I$, $(NbSe_4)_{10/3}I$, and $(TaSe_4)_2I$.

3.3.2. $(NbSe_4)_2I$

This is the most recently reported compound in the literature, and was mentioned by Fujishita *et al.* [109] at the International Symposium in Sapporo (Japan). Unit cell parameters ($a = 9.515$, $c = 12.80$ Å) and Weissenberg photographs ascertain that $(NbSe_4)_2I$ belongs to the $(TaSe_4)_2I$ structural type which will be described elsewhere.

Measurements of DC resistivity performed by a four-contacts technique showed a phase transition at about 210 K with a nonlinear conduction below T_c. Indeed, nonlinear conduction is observed when the applied electric field is greater than the threshold field $E_t(T)$. Far below T_c, $E_t(T)$ increases with decreasing temperature.

An X-ray diffraction study [110] revealed the presence of superlattice reflections around the Bragg spots in the low-temperature phase (below the transition temperature). This gives direct evidence of the existence of a CDW state. These superlattice reflections appear at the points $\mathbf{q}$ (± 0.065, ± 0.065, ± 0.159) away from the Bragg spots. Figure 40 shows the temperature dependence of peak intensity

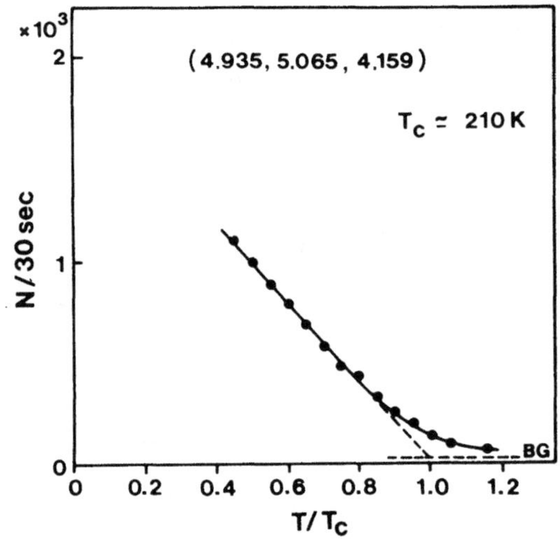

Fig. 40. Temperature dependence of peak intensity of superlattice reflection of $(NbSe_4)_2I$ at (4.935, 5.065, 4.159). $T_c \simeq 210$ K. (After [110].)

of the (4.935, 5.065, 4.159) satellite reflection. This linear variation of intensity proves the transition to be of second order. Though weak, the nonzero value component along the a^*, b^* ($\equiv a^*$) directions indicates a relative interaction between adjacent $(NbSe_4)_\infty$ chains. This interaction could be explained through the Nb—Se—I—Se—Nb bonds, as is demonstrated in the parent $(TaSe_4)_2I$ compound.

3.3.3. $(NbSe_4)_{10/3}I$

This compound crystallizes in tetragonal symmetry with the parameters $a = 9.464$, $c = 31.910$ Å, and space group $P4/mcc$ [111]. The fractional coordinates are listed in Table XXIII and the interatomic distances in Table XXIV. The a parameter is identical to the value found in other iodine derivatives, but the c parameter is $\frac{10}{6}$ times that of $(NbSe_4)_3I$, which implies 10 niobium atoms per unit cell

TABLE XXIII
Positional and thermal parameters and their estimated standard deviations in $(NbSe_4)_{10/3}I$ (after [111])

Atom	x	y	z	B_{eq} (Å^2)
Nb_1	0	$\frac{1}{2}$	$\frac{1}{4}$	1.88(7)
Nb_2	0	$\frac{1}{2}$	0.4506(1)	2.06(5)
Nb_3	0	$\frac{1}{2}$	0.3494(1)	2.13(5)
I_1	0	0	0.4205(1)	4.72(7)
I_2	$\frac{1}{2}$	$\frac{1}{2}$	$\frac{1}{4}$	7.1(2)
Se_1	0.3159(2)	0.1272(2)	0.40282(7)	2.15(4)
Se_2	0.8084(2)	0.3788(2)	0.29646(7)	2.25(4)
Se_3	0.7208(2)	0.0364(2)	0.30257(7)	2.35(4)
Se_4	0.7831(2)	0.5500(2)	0.39673(6)	2.28(4)
Se_5	0.8925(3)	0.7001(3)	$\frac{1}{2}$	2.44(6)
Se_6	0.4798(3)	0.2233(3)	$\frac{1}{2}$	2.37(6)

TABLE XXIV
Interatomic distances (Å) in $(NbSe_4)_{10/3}I$ (after [111])

Nb_1—$4Se_2$	2.608(3)	Nb_3—$2Se_2$	2.730(4)	Nb_2—$2Se_5$	2.665(4)
Nb_1—$4Se_3$	2.702(2)	Nb_3—$2Se_3$	2.591(3)	Nb_2—$2Se_6$	2.643(3)
		Nb_3—$2Se_1$	2.719(3)	Nb_2—$2Se_1$	2.610(3)
		Nb_3—$2Se_4$	2.593(3)	Nb_2—$2Se_4$	2.720(3)
Se_2—Se_3	2.362(4); 3.559(4)	Se_1—Se_4	2.381(3); 3.500(3)	Se_5—Se_6	2.356(6); 3.564(6)
Nb_1—Nb_3	3.171(3)	Nb_2—Nb_3	3.232(5)	Nb_2—Nb_2	3.150(7)

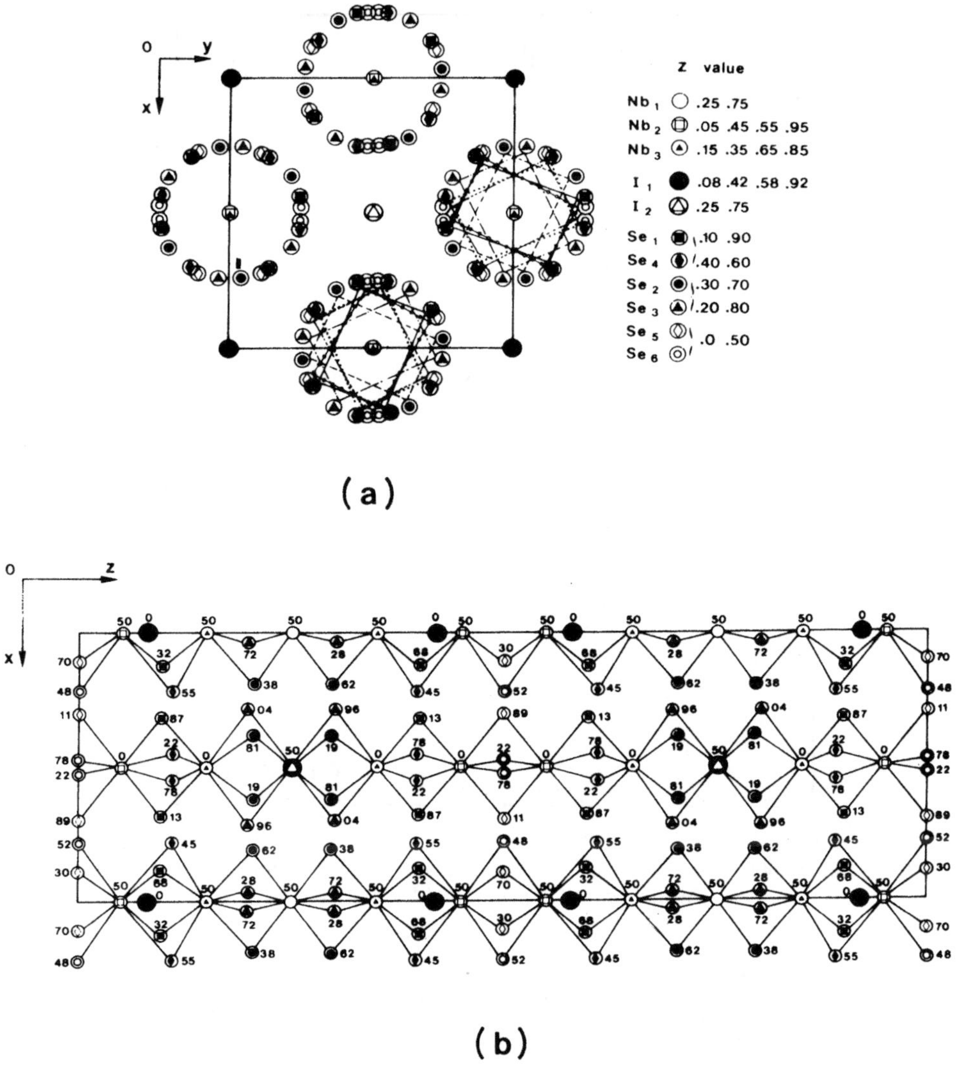

Fig. 41. The structure of $(NbSe_4)_{10/3}I$. (After [111].)

along each $[NbSe_4]$ chain (Figure 41(a), (b)). Nb—Nb distances along this chain can take different values, as represented by the following sequence:

$$\begin{array}{l} \text{——Nb2} \overset{3.232}{\text{——}} \text{Nb3} \overset{3.171}{\text{——}} \text{Nb1} \overset{3.171}{\text{——}} \text{Nb3} \overset{3.232}{\text{——}} \text{Nb2} \overset{3.150}{\text{——}} \text{Nb2——} \\ \quad\downarrow \qquad\qquad\qquad\qquad\quad \downarrow \qquad\qquad\qquad\qquad\quad \downarrow \\ z \text{ values } 0 \qquad\qquad\qquad \tfrac{1}{4} \qquad\qquad\qquad\qquad \tfrac{1}{2} \end{array}$$

The occupancy of channels between $(NbSe_4)$ chains is different; four iodines atoms are located along the $00z$ axis but only two are along the $\frac{1}{2}\frac{1}{2}z$ axis. In the

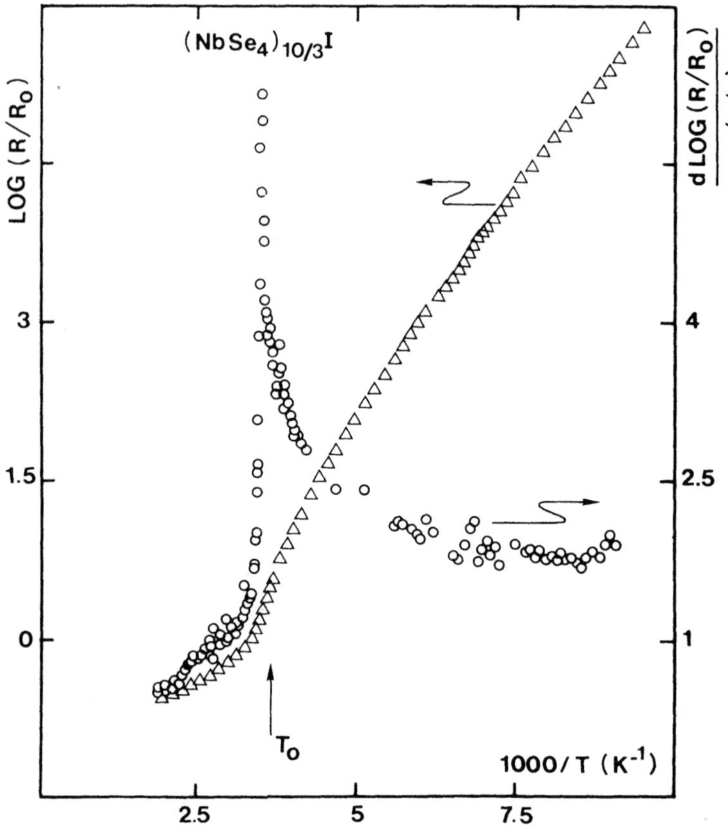

Fig. 42. Variation of the resistance and of its logarithmic derivative of $(NbSe_4)_{10/3}I$ as a function of the inverse of the temperature $10^3/T$. (After [112].)

first case, iodine is close to four selenium atoms (I—4Se = 3.272 Å), while in the second case iodine is surrounded by eight selenium in a square antiprismatic arrangement at a somewhat longer distance (I—8Se = 3.469 Å).

The room temperature resistivity along the needle axis is 1.5×10^{-2} Ω cm. A phase transition is observed at 285 K [112] as is clearly shown (Figure 42) by a peak in the logarithmic derivative d $\log(R/R_0)/d(1/T)$ resistance curve.

Electron diffraction studies [105] show that this transition originates from a Peierls instability. Satellite spots are observed on diffraction patterns of the (a^*, c^*) plane. They appear below T_c and are located along the c^* axis with no component along the a^* or b^* axes (in fact, satellite spots are elongated along a^* in the a^*—c^* plane). The reported wave vector components are (0, 0, 0.487), in good agreement with the predictions based upon band calculations [113].

For any temperature below the transition, nonlinear transport properties can once more be observed if the applied field exceeds a threshold value. An interesting point is that a hysteresis phenomenon takes place in the 285—220 K tempera-

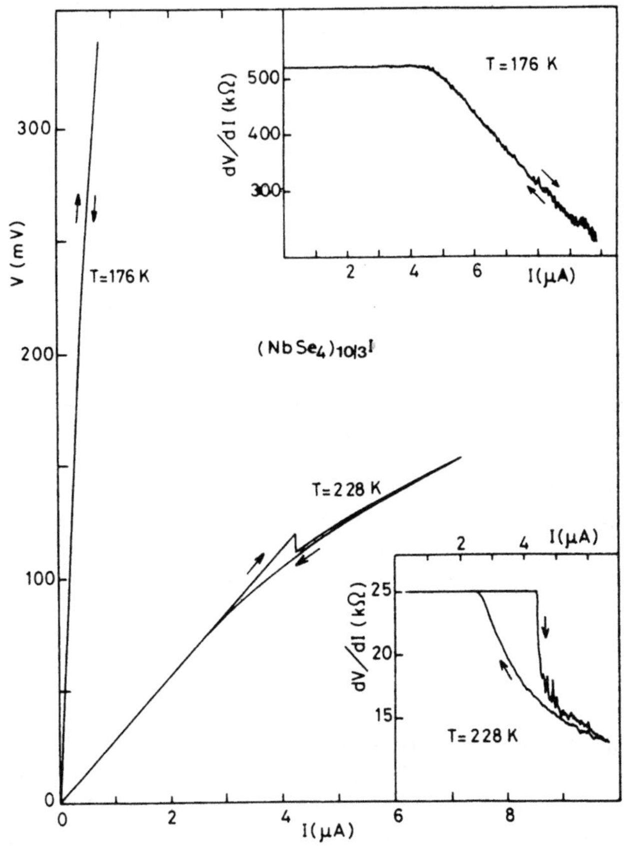

Fig. 43. Variation of the characteristics $V(I)$ and of the differential resistance dV/dI as a function of the applied current for $T = 176$ K and $T = 228$ K. The beginning of the decrease of dV/dI defines the value of the threshold field. Between 285 and 220 K, switching and hysteresis are observed at the onset of nonlinearity. (After [112].)

ture range [112] (see Figure 43). This type of memory effect, also observed in both orthorhombic TaS_3 [114] and $NbSe_3$ [115], and in the blue bronze $K_{0.30}MoO_3$ [116], depends on the electrical history of the sample. It implies that, probably, closely related metastable states of comparable energy are involved in the CDW pinning effects. We have a threshold field when the field is increased and a different one when it is decreased. As in $(TaSe_4)_2I$, E_c increases approximately exponentially when the temperature is reduced below T_c. Near T_c it goes through a minimum. Below the transition the fundamental frequency of the noise spectra varies linearly as a function of J_{CDW}.

Pulsed NMR studies of the ^{93}Nb have also been achieved on single crystals of $(NbSe_4)_{10/3}I$ [117]. The onset of the incommensurate CDW below 285 K has been followed on the $(\frac{1}{2}, -\frac{1}{2})$ transition resonance lines. The temperature dependence of the line shape is in agreement with that expected from a purely plane wave ICDW;

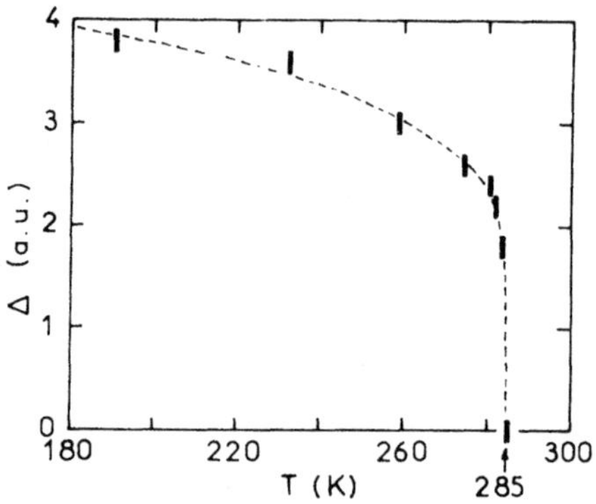

Fig. 44. Temperature dependence of the order parameter Δ of the transition in $(NbSe_4)_{10/3}I$, assuming the amplitude of the frequency distribution associated with the ICDW linear with Δ. (After [117].)

this allows the determination of the temperature dependence of the CDW which is the order parameter of the transition (see Figure 44). Furthermore, comparisons of the spin-lattice relaxation rates in both compounds $(NbSe_4)_3I$ and $(NbSe_4)_{10/3}I$ indicate that for $(NbSe_4)_{10/3}I$, the fluctuations of the pinned CDW are the dominant mechanism.

3.3.4. $(TaSe_4)_2I$

$(TaSe_4)_2I$ exhibits the simplest structure of the series. It crystallizes with tetragonal symmetry with parameters $a = 9.531$ and $c = 12.824$ Å, and space group $I422$ [118]. The fractional coordinates are listed in Table XXV. The c parameter, which is two-thirds that of $(NbSe_4)_3I$, means that there are four Ta atoms per unit cell along the chain (see Figure 45(a), (b)). Tantalum atoms are equally spaced along the chain axis (Ta—Ta = 3.206 Å) (Table XXVI).

When the metal—metal distances are equal, a high conductivity is expected. That is the case for $(TaSe_4)_2I$, which has a room temperature resistivity of 1.5×10^{-3} Ω cm. As shown in Figure 46, this compound undergoes a transition towards a semiconducting state at 263 K [104]. The nonlinear conductivity observed below this temperature suggests that $(TaSe_4)_2I$ is a new example of a compound exhibiting CDW transport phenomena [119, 120].

This was indeed later confirmed by electron diffraction studies [105]. Patterns corresponding to the (a^*, c^*) plane reveal, at low temperature (90 K), satellite spots very close to the main Bragg spots. The components of the distortion vector q are found equal to (1, 0, 0.94) [105]. Band structure calculations led to a Fermi wave vector $\mathbf{q} = 2k_F \simeq 0.88c^*$, in reasonable agreement with this value.

X-ray diffraction studies in the low-temperature phase reveal superlattice reflections which are observed at points (0.05, 0.05, 0.085) on both sides of Bragg spots (Fujishita *et al.* [109]). Figure 47 shows the temperature dependence of the peak intensity which clearly indicates a second-order transition. The same comments as those given for $(NbSe_4)_3I$ can be put forward.

TABLE XXV
$(TaSe_4)_2I$ structure, tetragonal $I422$, $Z = 4$
$a = 9.531$ Å, $c = 12.824$ Å. (After [118])

	I	Ta I	Ta II	Se(1)	Se(2)
Position	4(*e*)	4(*c*)	4(*d*)	16(*k*)	16(*k*)
x	0	0	0	0.1212(2)	0.2160(2)
y	0	$\frac{1}{2}$	$\frac{1}{2}$	0.3120(2)	0.5449(2)
z	0.1553(3)	0	$\frac{1}{4}$	0.8809(2)	0.8693(2)

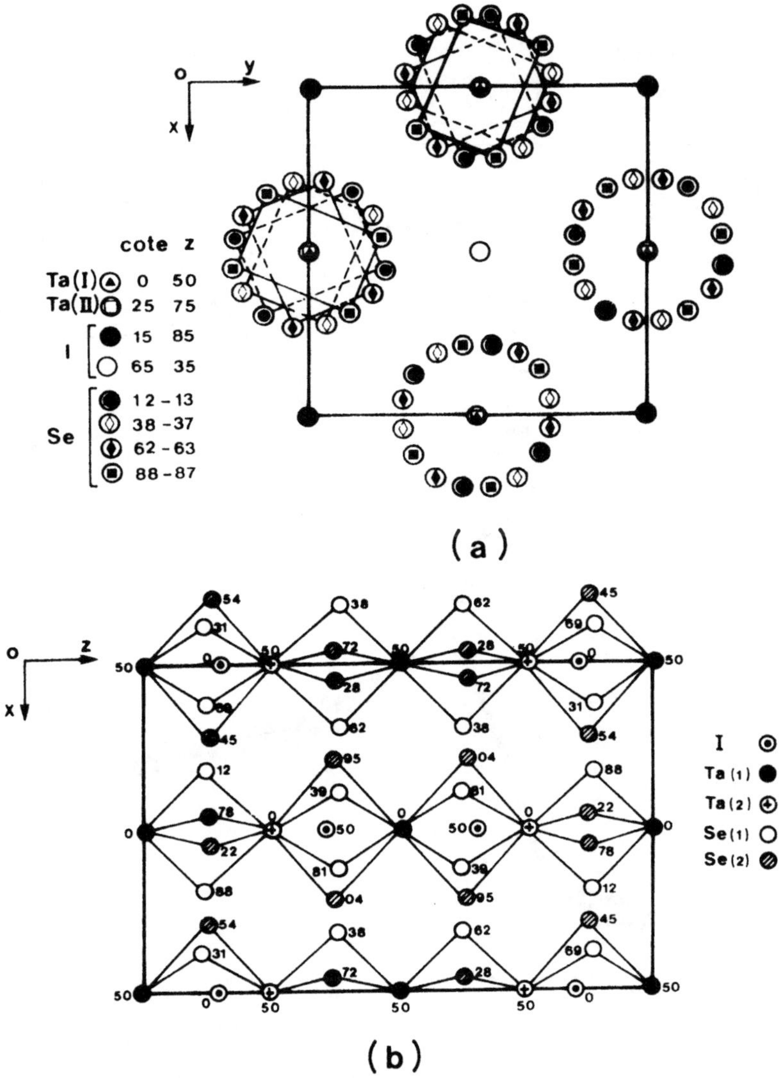

Fig. 45. The structure of $(TaSe_4)_2$. (After [118].)

TABLE XXVI
Interatomic distances in $(TaSe_4)_2I$ (after [118])

Ta I—Se(11) —Se(12) —Se(13) —Se(14)	2.623(1)	Ta II—Se(23) —Se(24) —Se(215) —Se(216)	2.600(1)
Ta I—Se(21) —Se(22) —Se(23) —Se(24)	2.689(1)	Ta II—Se(13) —Se(14) —Se(115) —Se(116)	2.713(1)
Se(14)—Se(24) Se(11)—Se(21)	2.401(1)		
Ta I—Ta II	3.206(1)		

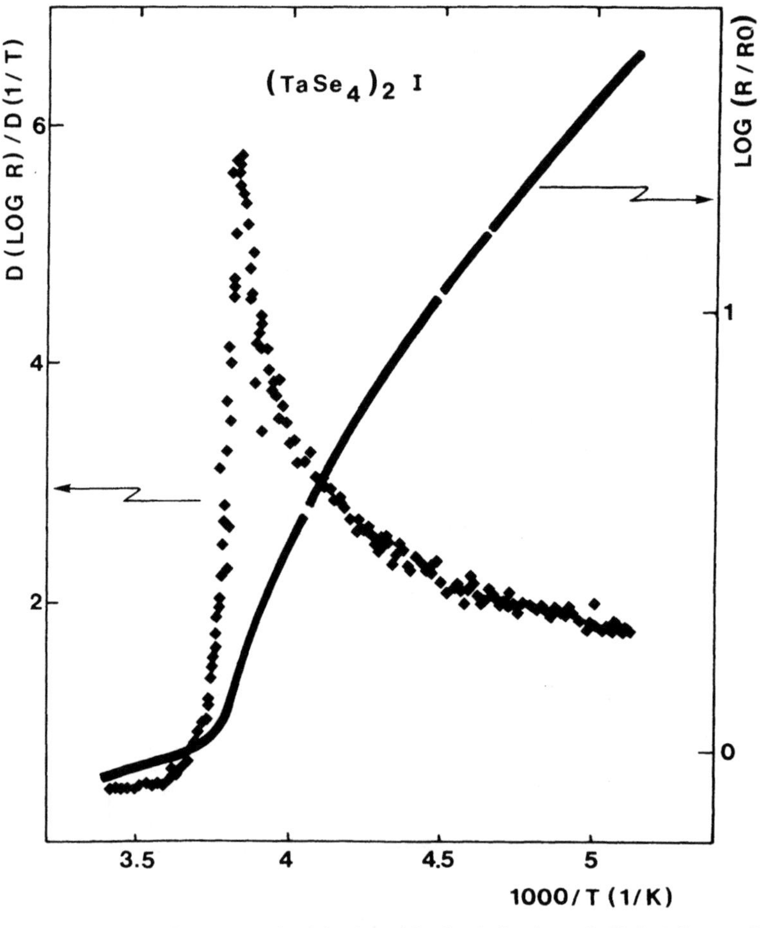

Fig. 46. Variation of the resistance and of its logarithmic derivative of $(TaSe_4)_2I$ as a function of $10^3/T$. (After [119].)

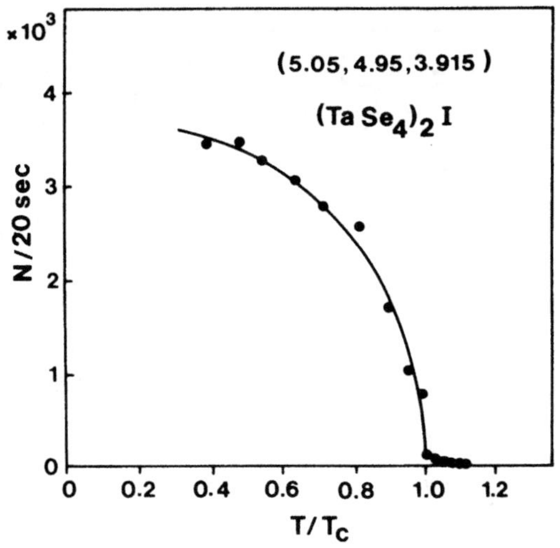

Fig. 47. Temperature dependence of peak intensity of superlattice reflection at (5.05, 4.95, 3.915) for $(TaSe_4)_2I$, $T_c \simeq 260$ K. (After [109].)

3.3.5. *$(TaSe_4)_3I$*

$(TaSe_4)_3I$ is very comparable to the $(NbSe_4)_3I$ compound. It presents similar structural and electrical transition when the temperature is decreased, although the transition takes place at a lower temperature ($T \sim 200$ K) [103].

3.3.6. *Discussion*

The many $(MSe_4)_nI$ phases form one of the best series by which to study CDW phenomena. They possess all the striking properties observed for trichalcogenides. In addition, they lead to a variety of slightly different species by simply changing the metal and halogen atoms. Interchain coupling can change according to the size and electronegativity of the counter ions (iodine, bromine, or chlorine). The amount of counter-ion also changes the electronic density available along the chains, and from that viewpoint the tetrachalcogenides are reminiscent of Krogmann salts. Structurally they also remind us of some organic conductors containing iodine, such as tetraselenotetracene I_x (see Figure 48) [121]. The electronic properties of a MSe_4 chain are largely determined by the interactions of the metal d_{z^2} orbitals along the chain. This is illustrated in the resistivity values of the compounds, the highest one referring to the greatest difference between long and short M—M distances. If the role of the halogen is neglected, each $[MSe_4]$ chain has an $M^{4+}(d^1)$ metal ion at the center of every Se_8 rectangular antiprism. Iodine atoms, and other halogens too, are isolated from one another (e.g. the I—I interatomic distance is 4.76 Å in $(NbSe_4)_3I$ which is very large when compared to 2.68 Å in solid I_2 and a maximum of 3.07 Å in I_3^-). Each iodine takes one electron. Thus, for $(MSe_4)_nI$ the average number of d electrons on each M would be $(n-1)/n$. The

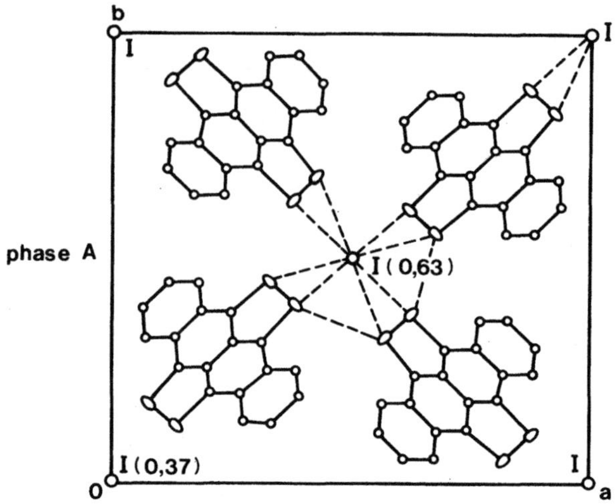

Fig. 48. The structure of TSeT—I_x (phase A) projected along [001]. (After [121].)

corresponding filling of the d_{z^2} band is $f = (n - 1)/2n$, which leads to a distortion increasing the repeat distance by a factor of $1/f$. This suggests that $(NbSe_4)_3I$, $(NbSe_4)_{10/3}I$, and $(TaSe_4)_2I$ would have $\frac{1}{3}$, $\frac{7}{20}$, and $\frac{1}{4}$ filled d_{z^2} bands, respectively. As n increases, the d_{z^2} band of $(MSe_4)_nI$ becomes closer to $\frac{1}{2}$-filled. Vanadium tetrasulfide finally constitutes an example of the $n \rightarrow \infty$ limit. It has a $2c$ insulating Peierls distortion. More precisely, tight-binding band structure calculations are in rather good agreement with the observed distortions [113].

Like transition metal trichalcogenides, the $(MX_4)_nY$ phases are obtained in sealed tubes, i.e. under thermodynamic conditions that are not well defined. A small range of temperature is associated with particular starting conditions (quantities of elements, volume of the tube) in order to get a given compound. In addition, the structural similarity of all the compounds with only slight changes in the iodine distribution and in the metallic sequence along the chains makes it possible to observe intergrowth phenomena. This is what was probably observed in the case of some $(NbSe_4)_3I$ crystals, which present at low temperature a particular behavior (Figure 36(a), (b)). The corresponding crystals were not distinguished by X-ray techniques. Both types of crystals have a room temperature resistivity of about 1 Ω cm. Type I behavior is characterized by a very small semiconducting gap below 275 K. In type II crystals, after the same decrease of the gap at 275 K, the resistivity increases again at lower temperatures showing a behavior very similar to that of $(NbSe_4)_{10/3}I$ at these temperatures (E = 3000 K instead of 3900 K for $(NbSe_4)_{10/3}I$). This suggests that type II crystals could present microdomains of the $(NbSe_4)_{10/3}I$ composition, these domains acting as local defects perpendicular to the chain axes. This looks like a stacking fault. Without any modification of the structural framework this would essentially result in different local arrangements of iodine ions. This is an important point concerning chemistry, i.e. the first illustration of a one-dimensional phase intergrowth. On the other hand, it may explain some unexpected results obtained recently in another group [122]. It may also shed some light on the real nature of defects pinning charge density waves. As well as in trichalcogenides there is a high probability that such defects are not only represented by chemical impurities but also by chemical defects such as a local variation of the Se—Se bond lengths or structural defects due to intergrowth mechanisms.

A last point to discuss concerns the decrease of the semiconducting energy gap in both types of $(NbSe_4)_3I$ at 275 K. This unusual behavior leads to a higher conducting state when lowering the temperature. It implies a modification of the electronic density along the chains. Thus, this structural transition will result in a change of the metal—metal bond length sequences. X-ray investigation of the low-temperature form has just been started. Preliminary results [123] indicate that the low-temperature form belongs to the $P\bar{4}2_1c$ space group (which is a subgroup of $P4/mnc$). This means that the inversion center is missing. The Nb—Nb bond sequence is slightly different from that of the high-temperature state. Indeed three Nb—Nb bond lengths are observed.

4. Nb_2Se_9 and Pentachalcogenides

Chalcogen pairs appear in chalcogen-rich chalcogenides MX_n when $n \geqslant 3$. For trichalcogenides it is typical to find one chalcogen pair per metal atom. With $n = 4$, two chalcogen pairs are present. Nb_2Se_9 and the pentachalcogenides MX_5 represent particular situations in which chalcogen pairs exist along with more complex arrangements of atoms.

4.1. Nb_2Se_9

Nb_2Se_9 is prepared by direct combination of the elements in evacuated sealed silica tubes at temperatures ranging from 500 to 600 °C [124]. It crystallizes in a needle-

TABLE XXVII
Nb_2Se_9 structure, triclinic $P\bar{1}$, $Z = 2$
$a = 8.1829$ Å, $b = 8.3230$ Å, $c = 13.084$ Å, $\alpha = 120.893°$,
$\beta = 124.017°$, $\gamma = 92.274°$, V = 562.9 Å^3. (After [124])

	x	y	z
Nb(1)	0.049(2)	0.034(2)	0.136(2)
Nb(2)	0.982(2)	0.036(2)	0.398(2)
Se(1)	0.449(3)	0.215(3)	0.417(3)
Se(2)	0.361(3)	0.204(3)	0.687(3)
Se(3)	0.309(2)	0.123(3)	0.914(2)
Se(4)	0.194(3)	0.340(3)	0.152(3)
Se(5)	0.156(3)	0.337(3)	0.417(3)
Se(6)	0.113(3)	0.333(3)	0.676(3)
Se(7)	0.823(3)	0.188(3)	0.006(3)
Se(8)	0.796(2)	0.108(2)	0.700(2)
Se(9)	0.736(3)	0.194(3)	0.281(2)

TABLE XXVIII
Interatomic distances (Å) in Nb_2Se_9 (after [124])

	Nb(1)—Se(4)	2.62(4)	Nb(2)—Se(2)	2.64(2)
	—Se(7)	2.63(3)	—Se(6)	2.58(2)
	—Se(4*c*)	2.63(2)	—Se(2*c*)	2.62(3)
	—Se(7*c*)	2.60(3)	—Se(6*c*)	2.63(3)
	—Se(3*c*)	2.72(3)	—Se(3*c*)	2.72(2)
	—Se(5)	2.70(3)	—Se(9)	2.64(3)
	—Se(1)	2.64(2)	—Se(5)	2.66(4)
	—Se(8*c*)	2.70(3)	—Se(8*c*)	2.76(3)
Isolated pairs	Se(4)—Se(7)	2.28(3)	Se(2)—Se(6)	2.30(3)
Pentanion	Se(3*c*)—Se(9)	2.36(3)		
	Se(9)—Se(5)	2.66(3)		
	Se(5)—Se(1)	2.64(4)		
	Se(1)—Se(8*c*)	2.34(3)		
	Nb(1)—Nb(2)	3.76(3)	Nb(1)—Nb(1*c*) Nb(2)—Nb(2*c*)	2.89(3)

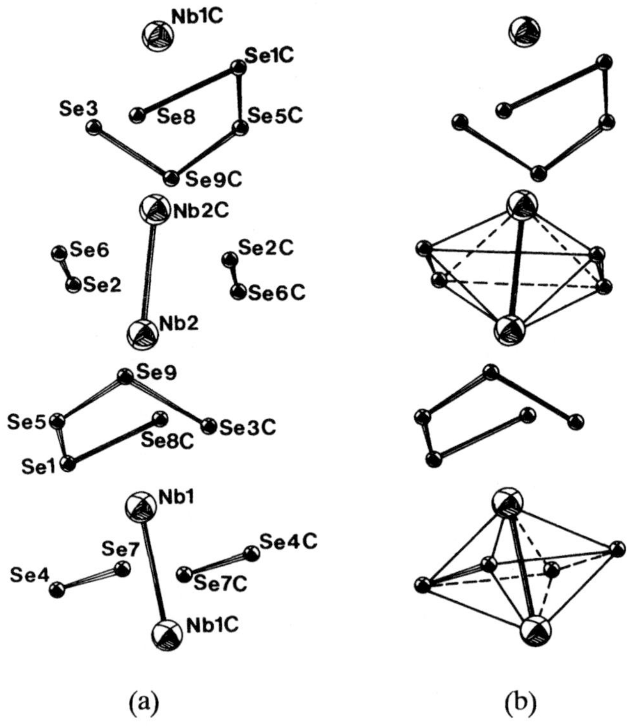

Fig. 49. The structure of Nb_2Se_9 developed along the *C* axis. (After [124].)

shaped specimen with triclinic symmetry, space group $P\bar{1}$. Only twinned crystals were available for the structural determination, which had to be made by a photographic film technique [124]. Fractional coordinates and unit cell parameters are listed in Table XXVII, and the interatomic distances in Table XXVIII. The structure can be described on the basis of linear Nb_2Se_9 chains running along the *c* axis. Within these chains, the Nb atoms are not exactly aligned as shown in Figure 49(a), and long (3.76 Å) and short (2.89 Å) Nb—Nb distances alternate. The short Nb—Nb distance is representative of an octahedral $[Nb_2Se_4]$ cage (Figure 49(b)), while the long Nb—Nb distance occurs through a Se_5 unit. Thus, stacking along a chain may be considered as:

$$---Nb_2Se_4—Se_5—Nb_2Se_4—Se_5---.$$

Then, we can observe a common feature with other compounds such as NbX_2Y_2

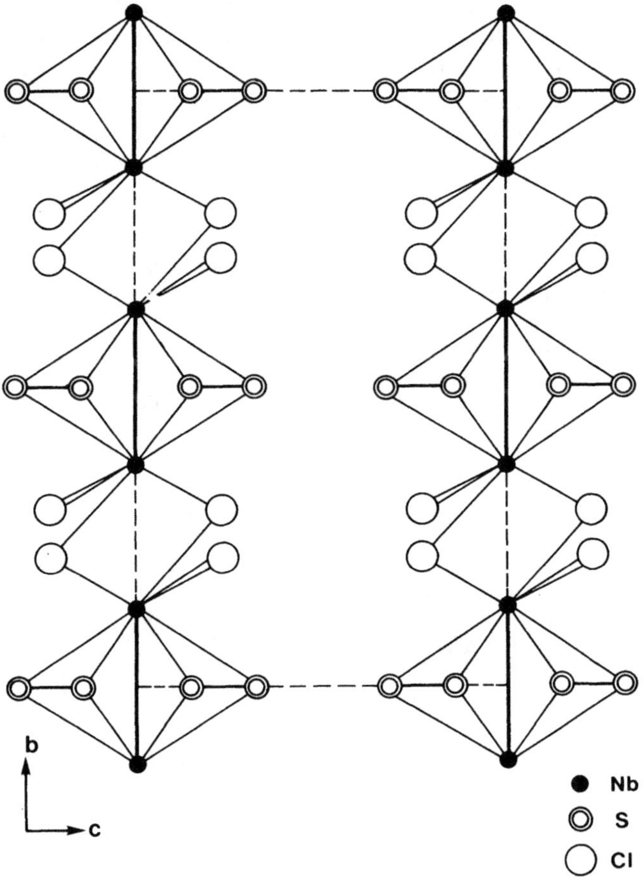

Fig. 50. The structure of NbS_2Cl_2 (monoclinic form) projected along [001]. (After [8].)

(X = S, Se; Y = Cl, Br, I) and $Nb_3Se_5Cl_7$ [125]. Indeed, these compounds (Figure 50) exhibit similar Nb_2X_4 cages along their chain axes. For example, NbS_2Cl_2 can be viewed as:

$$\text{---}Nb_2S_4\text{—}Cl_4\text{—}Nb_2S_4\text{—}Cl_4\text{---}.$$

Thus, the occurrence of Nb^{4+} (d^1) in pairs would explain the diamagnetic and semiconducting properties of these materials. The activation energy calculated for Nb_2Se_9 (E_g = 0.85 eV) is greater than that of $(NbSe_4)_3I$. This is related to a larger difference between short and long Nb—Nb distances in Nb_2Se_9. The Se_5 unit corresponds to an open ring with two short terminal Se—Se distances and two elongated distances in the middle. This agrees with the formal oxidation state $(Nb^{4+})_2$ $(Se_2^{2-})_2$ Se_5^{4-}.

4.2. Pentachalcogenides of Group IVA Elements

The pentatellurides of Zr and Hf exhibit a structure which is related to the MX_3 type. One important structural feature of this compound is a tellurium bridge which connects the trigonal prismatic chains (see Figure 51). The pentatellurides are prepared from the elements (Zr or Hf + 5Te) at temperatures ranging from 450 to 500 °C. Needle-shaped single crystals were obtained when using iodine as a transport agent. The crystal can be easily cleaved parallel to the *a—c* plane.

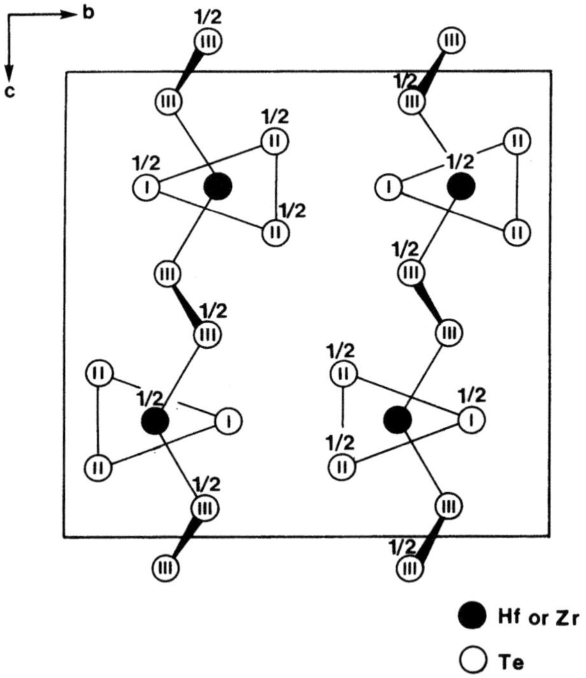

Fig. 51. The structure of $HfTe_5$ projected along [100]. (After [126].)

TABLE XXIX

$HfTe_5$ structure, orthorhombic *Cmcm*, $Z = 4$ (after [126])

$ZrTe_5$: $a = 3.988$ Å, $b = 14.502$ Å, $c = 13.727$Å

$HfTe_5$: $a = 3.974$ Å, $b = 14.492$ Å, $c = 13.730$ Å

Coordinates of equivalent positions: $(0, 0, 0; \frac{1}{2}, \frac{1}{2}, 0)+$

Hf and Te I in position 4(*c*): $\pm(0, y, \frac{1}{4})$

Te II and Te III in position 8(*f*): $\pm(0, y, z; 0, y; \frac{1}{2} - z)$

	Hf	Te I	Te II	Te III
y	0.3143(1)	0.6635(2)	0.9299(2)	0.2099(2)
z	$\frac{1}{4}$	$\frac{1}{4}$	0.1494(2)	0.4353(2)

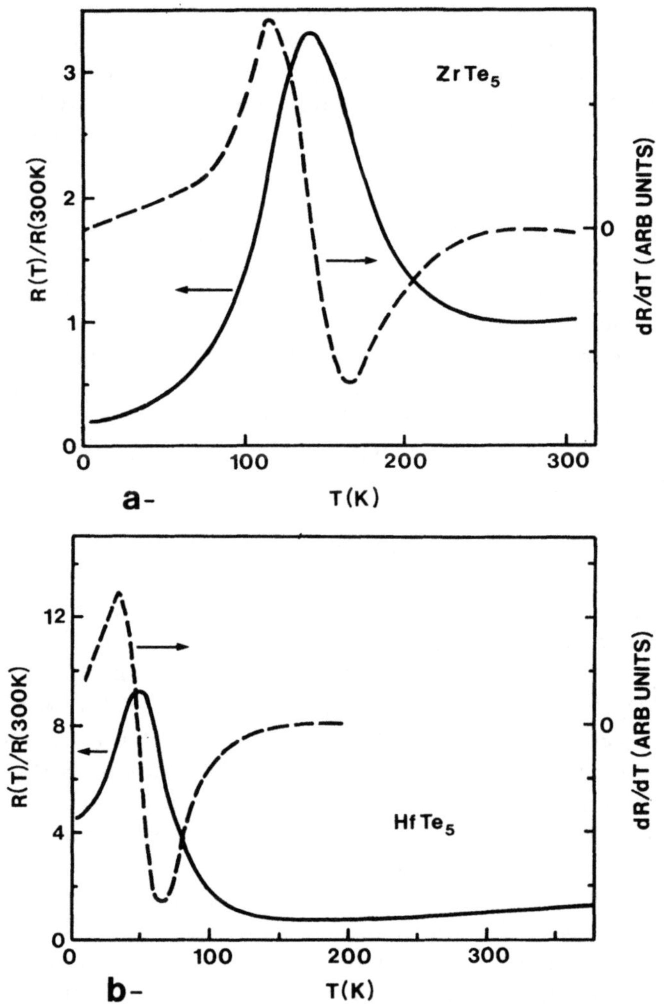

Fig. 52. Variation of the resistance and of its derivative (dashed line) of (a) $ZrTe_5$ and (b) $HfTe_5$. (After [129].)

$ZrTe_5$ and $HfTe_5$ crystallize with orthorhombic symmetry, space group *Cmcm*. The structures were determined by Furuseth *et al.* [126]. The fractional coordinates are summarized in Table XXIX. We find again the same polyhedron around the metal atoms as observed in the trichalcogenides, but the trigonal prism is more regular in the case of pentachalcogenides. Indeed, the six M—Te distances are almost identical (e.g. for $HfTe_5$, Hf—4Te II at 2.94 Å, Hf—2Te I at 2.95 Å and the two radial Hf—Te III distances at 2.96 Å). A short Te—Te bond length in the triangular base (Te II—Te II = 2.763 Å) is reminiscent of the short X—X bond length in the MX_3. Pauling [127] proposed a value of 2.74 Å for the Te—Te bond. On the other hand, the Te bridge between chains shows a longer distance (Te III—Te III at 2.908 Å).

Resistivity measurements on these compounds have been reported [128, 129]. The normalized resistance versus temperature between 4 and 350 K (Figure 52) reveals a metallic behavior and shows a large peak around 150 K in $ZrTe_5$ and 50 K in $HfTe_5$. Studies have been carried out to characterize nonlinear effects in the case of $HfTe_5$, which presents a larger peak in resistivity than $ZrTe_5$ [129]. Physical developments will be largely discussed in another section by Sambongi.

5. Discussion and Concluding Remarks

Various approaches to the electronic structure of the trichalcogenides have been attempted in order to explain their physical properties. From a chemical viewpoint an important feature is the variable length of the chalcogen pairs in these materials. A lengthening of the X—X pair corresponds to a weakening of the X—X bond. The X—X pair behaves as an electron reservoir, which governs indirectly the electronic density available along the metallic chains. The origin of the distortion of the trigonal prisms has been discussed by Hoffmann *et al.* [130]. Using a tight binding scheme based on the extended Hückel method, calculations were performed on mononuclear $[NbSe_6^{6-}]$ and binuclear $[Nb_2Se_9^{6-}]$ building blocks. Each is examined in a D_{3h} symmetry corresponding to a regular frame with an equilateral triangular base and also in a C_{2v} symmetry, which is the actual symmetry of the chains. The d-block orbitals of these building blocks are represented in Figure 53. With the D_{3h} symmetry, we have the classical splitting with $a_1'(z^2)$ and $e'(x^2 - y^2, xy)$ below $e''(xz, yz)$ [131], the relative position of a_1' being below e' for $NbSe_3$. The highest orbital set, e'', derived from the six selenium atoms is above a block of filled Se levels. In this scheme an $[NbSe_6^{6-}]$ complex corresponds to Nb^{5+}, with empty d levels, and a hole in the Se e'' levels. When distortion occurs, a Se—Se bond is formed with a breaking of the degeneracy of the $(e'')^3$ configuration. Two Se ligands are transformed into a coordinated Se_2 ligand. One electron occupies the lowest d level, i.e. $a_1'(z^2)$. The scheme represents the formal oxidation state $Nb^{4+}(Se_2)^{2-}Se^{2-}$. It is important to notice that the calculations have shown the splitting of the $(e'')^3$ configuration not to be stabilizing by itself. This was taken as proof of the importance of lateral interchain bonds to stabilize the distorted chain. This 'chemical approach' also allows a band structure to be derived by applying successively a

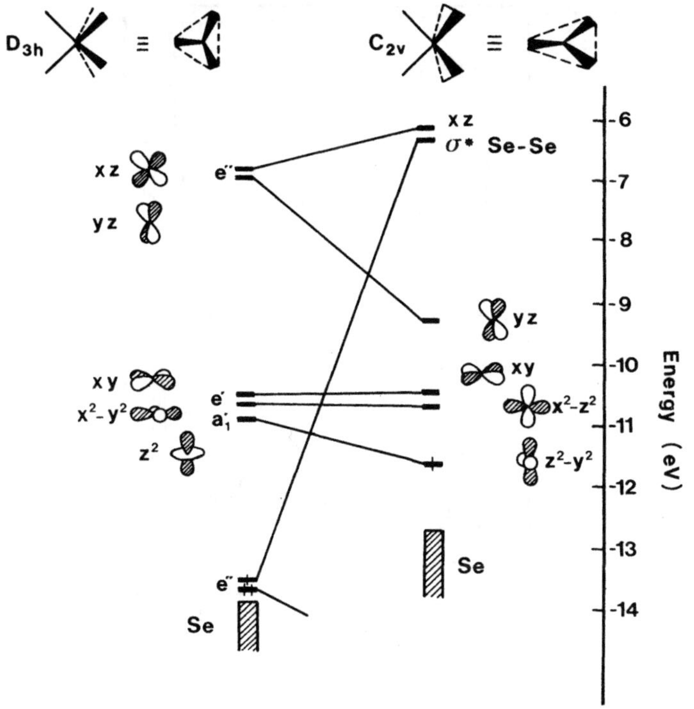

Fig. 53. The electronic structure of [$NbSe_6^{6-}$] in D_{3h} and C_{2v} geometries. (After [130].)

translational symmetry along b, which builds the chains, then moving towards construction of the slabs along c, and finally considering interactions between slabs along the a axis. In the construction of the slabs the xy and $x^2 - y^2$ monomer orbitals play an essential role because they are directed towards the Se atoms of neighboring chains. Calculations show that the energy bands rise rapidly along the chain axis but very slowly along the directions perpendicular to the chain axis, which reveals the 1D character. In addition, the energy bands along the chains are only slightly affected by the introduction of interactions between the slabs, which means that the interactions are weak. Considering the lower bands it is found that they are mainly z^2 in character with a 1D nature. Higher energy bands based on xy and $x^2 - y^2$ naturally show substantial dispersion in the plane perpendicular to the chain axis. Their bandwidth of about 0.2 eV explains the low conductivity in the a—c plane. The much larger conductivity along b is associated with the wider d_{z^2} bands. Tight binding calculations [132] show that the presence of three nonequivalent pairs of chains give rise to four roughly $\frac{1}{4}$-filled d-block bands. This is in complete agreement with Devreux's suggestion [133]. The calculated $2k_F$ values along the chain direction suggest that high- and low-temperature CDW transitions occur primarily in type III (Se—Se = 2.374 Å) and type I (Se—Se = 2.485 Å) chains, respectively.

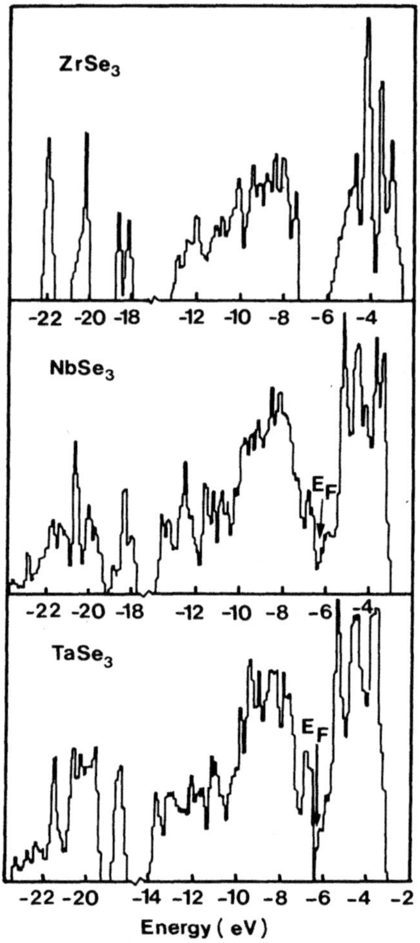

Fig. 54. Electron densities of states in the valence and conduction bands of $ZrSe_3$, $NbSe_3$, and $TaSe_3$. (After [134].)

Bullett [134] has made extended band calculations for various trichalcogenides. This allows some comparison of the properties of these materials. The resulting densities of states for $ZrSe_3$, $NbSe_3$, and $TaSe_3$ are shown in Figure 54. $ZrSe_3$ was found to have a semiconducting direct gap of 1.6 eV and an indirect gap of 1.5 eV, to be compared with the experimental value of 1.8 eV [135, 6]. The main valence bandwidth (~ 6 eV) in $ZrSe_3$ is in agreement with X-ray photoemission spectra [6]. An examination of the energy bands at the Fermi level (Figure 55) explains why phase transitions occur in $NbSe_3$ but not in $TaSe_3$. In $NbSe_3$, with $(NbSe_3)_6$ per unit cell, 69 energy bands are exactly filled, but there is a sharp separation between the 68th, relatively flat, and the 69th band that belongs to a group of four, almost dispersionless in the *a—c* plane but rising rapidly as *k* moves away in the chain direction. This level has a strong d_{z^2} component along the *b* axis. The result

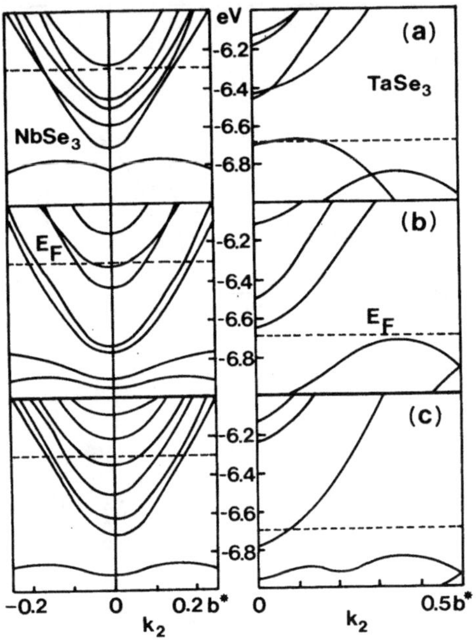

Fig. 55. Dispersion parallel to the chain direction for bands near the Fermi energy in $NbSe_3$ and $TaSe_3$ along (a) $(0, k_2, 0)$; (b) $(\pi/a, k_2, 0)$; (c) $(0, k_2, \pi/c)$. (After [134].)

is a pseudo-1D metal with four energy levels crossing the Fermi energy. A Peierls transition associated with $\frac{1}{4}b^*$ is expected. In the case of $TaSe_3$, 46 energy levels are filled but there is a slight semimetallic overlap with the next higher band. No CDW is expected because of the much stronger dispersion in the a^*—c^* plane of the relevant bands.

A recent study by Endo *et al.* [136] shows very good agreement between the density of state calculations and the corresponding XPS spectra (Figure 56) for $NbSe_3$ and $TaSe_3$. The XPS spectra of $NbSe_3$ and $TaSe_3$ also show an increased 2D character of these materials with respect to NbS_3 and TaS_3. This is reflected in the CDW transition temperatures. Interchain interactions, investigated by Levy *et al.* [137] in the course of Raman and infrared studies of chemical bonding in the IVA and VA trichalcogenides suggest that the strength of the interchain M—X bond is a substantial fraction (0.15—0.28) of that of the intrachain M—X bond.

The most striking properties of trichalcogenides are undoubtedly the nonlinear effects associated with CDW phenomena in $NbSe_3$ and TaS_3. An important problem in this field is to understand the nature of the pinning of CDWs by defects, impurities, or commensurability. From this viewpoint, the preparation of two types of TaS_3, with noncommensurate (TaS_3 mono) and commensurate (TaS_3 ortho) CDWs, should help to understand the role of commensurability.

Due to the nonlinear electrical properties, observed for the first time in $NbSe_3$, transition metal trichalcogenides have become important in the field of the physics and chemistry of low-dimensional conductors. Theoretical works are in progress to

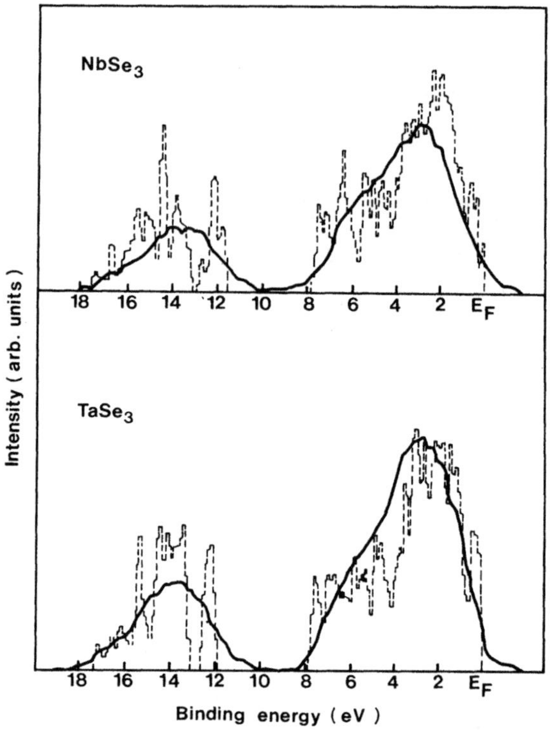

Fig. 56. XPS valence band spectra (solid lines) of $NbSe_3$ and $TaSe_3$ compared with the theoretical density of states (dashed lines). (After [136].)

understand the depinning phenomena of CDWs. The chemistry of these materials is probably still very rich. For example, polytypism relating various types of chains according to new arrangements is to be foreseen. As we may be dealing with metastable phases, it is possible that, under pressure and at higher temperatures, the very symmetrical $ZrSe_3$ structural type might be obtained. The $(MX_4)_nY$ type materials show a great originality because of the different mixing of long and short bondings in the metallic chain. In this bond density wave system, chemistry will allow various possibilities with respect to the nature of the counter ion Y and its concentration.

Acknowledgements

We are indebted to our colleagues of the laboratory at Nantes — Mrs. L. Guemas, P. Gressier, and A. Ben Salem — who are working with us on new low-dimensional conductors. Indeed, this work would not have raised such interest if the Grenoble group had not performed transport measurements of very high quality leading to the discovery of new intriguing properties associated with this class of materials. Professor N. P. Ong has also been much involved in the field. We thus take advantage of this review article to acknowledge all our colleagues working in the CRTBT—CNRS center at Grenoble and particularly Dr P. Monceau (who has initiated most of the physical studies reported here) and Professors M. Renard, J. Richard, Z. Z. Wang, H. Salva, and M. C. St Lager. We would also like to thank C. Roucau and R. Ayroles for their remarkable contribution through their electron diffraction studies. X-ray diffuse scattering were performed by R. Moret and P. Pouget, the NMR studies by C. Berthier and P. Butaud, and the irradiation experiments by L. Zuppiroli, G. Mihaly, and H. Mutka at the C.E.N. (Fontenay aux Roses). The present article was kindly revised by Professor M. H. Whangbo during his stay in the laboratory, and by Professor R. Brec. The authors would like to acknowledge numerous discussions with Professor T. Sambongi, S. Kikkawa, M. Izumi and many other colleagues, Professor Lederer, Professor C. Grüner, and Dr F. Di Salvo.

Finally, the authors are indebted to Mrs M. P. Guilbaud for typing this manuscript with great care and to Mr J. Y. Guilbaud for drawing all the figures.

References

1. J. A. Wilson, F. J. Di Salvo, and S. Mahajan, *Advances in physics* **24** (2), 117 (1975).
2. A. Meerschaut and J. Rouxel, *J. Less Common Met.* **39**, 197 (1975).
3. J. A. Wilson, *Phys. Rev. B: Condens. Matter* **19** (12), 6456 (1979).
4. W. Kröniert and K. Plieth, *Z. Anorg. allg. Chem.* **336**, 207 (1965).
5. S. Furuseth, L. Brattås, and A. Kjekshus, *Acta Chem. Scand.* **A29**, 623 (1975).
6. F. Jellinek, R. A. Pollack, and M. W. Shafer, *Mat. Res. Bull.* **9**, 845 (1974).
7. J. Rijnsdorp and F. Jellinek, *J. Solid State Chem.* **25**, 325 (1978).
8. J. Rijnsdorp, Thesis, Univ. of Gröningen, 1978.
9. K. Endo, H. Ihara, K. Watanabe, and S. I. Gonda, *J. Solid State Chem.* **39**, 219 (1981).
10. T. Cornelissens, G. Tendeloo, J. Landuyt, and S. Amelinckx, *Phys. Stat. Sol.* **A48**, K5 (1978).
11. F. W. Boswell and A. Prodan, *Physica* **99B**, 361 (1980).
12. C. Roucau, T. Granier, and R. Ayroles, *J. Phys. C,* **3, 44**, 1725 (1983).
13. P. Haen, J. M. Mignot, P. Monceau, M. Nunez-Regueiro, and J. Richard, *Lecture Notes in Physics* **95**, 323 (1978), International Conference, Dubrovnik.
14. D. W. Bullett, *J. Solid State Chem.* **33**, 13 (1980).
15. S. Kikkawa, N. Ogawa, M. Koizumi, and Y. Onuki, *J. Solid State Chem.* **41**, 315 (1982).
16. M. Izumi, T. Nakayama, and R. Yoshizaki, *Proc. Int. Symp. on Nonlinear Transport and Related Phenomena in Quasi-1D Inorganic Conductors*, Sapporo 20—22 October 1983 (Japan).
17. A. Zettl, C. M. Jackson, A. Janossy, G. Grüner, A. Jacobsen, and A. H. Thompson, *Solid State Commun.* **43** (5), 345 (1982).
18. E. Bjerkelund, J. H. Fermor, and A. Kjekshus, *Acta Chem. Scand.* **20**, 1836 (1966).
19. T. Sambongi, M. Yamamoto, K. Tsutsumi, Y. Shiozaki, K. Yamaya, and Y. Abe, *J. Phys. Soc. Japan* **42** (4), 1421 (1977).
20. P. Haen, F. Lapierre, P. Monceau, M. Nunez-Regueiro, and J. Richard, *Solid State Commun.* **26**, 725 (1978).
21. R. M. Fleming, J. A. Polo, and R. V. Coleman, *Phys. Rev.* **B17** (4), 1634 (1978).
22. K. Yamaya, T. H. Geballe, J. F. Kwack, and R. L. Greene, *Solid State Commun.* **31**, 627 (1979).

23. Y. Tajima and K. Yamaya, *J. Phys. Soc. Japan* **53** (2), 495 (1984).
24. S. Kikkawa, K. Shinya, and M. Koizumi, *J. Solid State Chem.* **41**, 323 (1982).
25. K. Yamaya and Y. Abe, *Mol. Cryst. Liq. Cryst.* **81**, 133 (1982).
26. J. L. Hodeau, M. Marezio, C. Roucau, R. Ayroles, A. Meerschaut, J. Rouxel, and P. Monceau, *J. Phys. C: Solid State Phys.* **11**, 4117 (1978).
27. (a) P. Haen, P. Monceau, B. Tissier, G. Waysand, A. Meerschaut, P. Molinié, and J. Rouxel, *Proc. 14th Int. Conf. on Low Temperature Physics*, Otaniemi, Finland, Vol. 5, p. 445 (1975); (b) N. P. Ong and J. W. Brill, *Phys. Rev.* **B18** (10), 5265 (1978).
28. J. Chaussy, P. Haen, J. L. Lasjaunias, P. Monceau, G. Waysand, A. Waintal, A. Meerschaut, P. Molinié, and J. Rouxel, *Solid State Commun.* **20**, 759 (1976).
29. R. Delaplace, P. Molinié, and D. Jérome, *J. Physique Lett.* **37**, L13 (1976).
30. A. Briggs, P. Monceau, M. Nunez-Regueiro, J. Peyrard, M. Ribault, and J. Richard, *J. Phys. C: Solid State Phys.* **13**, 2117 (1980).
31. K. Tsutsumi, T. Tagagaki, M. Yamamoto, Y. Shiozaki, M. Ido, T. Sambongi, K. Yamaya, and Y. Abe, *Phys. Rev. Lett.* **39**, 1675 (1977).
32. M. H. Whangbo and P. Gressier, *Inorg Chem.* **23**, 1305 (1984).
33. R. M. Fleming, D. E. Moncton, and D. B. McWhan, *Phys. Rev.* **B18** (10), 5560 (1978).
34. J. P. Pouget, R. Moret, A. Meerschaut, L. Guemas, and J. Rouxel, *J. Physique Colloq. C3*, supplément au n° **6** (44), (1983), Les Arcs (France).
35. P. Monceau, N. P. Ong, A. M. Portis, A. Meerschaut, and J. Rouxel, *Phys. Rev. Lett.* **37**, 602 (1976).
36. N. P. Ong and P. Monceau, *Phys. Rev.* **B16**, 3443 (1977).
37. A. Larkin and P. A. Lee, *Phys. Rev.* **B17**, 1596 (1978).
38. K. Maki, *Phys. Rev. Lett.* **39**, 46 (1977).
39. J. Bardeen, *Highly Conducting One-Dimensional Solids* (ed. J. T. Devreese), Plenum, New York.
40. R. M. Fleming, *Phys. Rev. Lett.* **42**, 1423 (1979).
41. P. A. Lee and T. M. Rice, *Phys. Rev.* **B19**, 3970 (1979).
42. R. M. Fleming, *Phys. Rev.* **B22** (12), 5606 (1980).
43. P. Monceau, J. Richard, and M. Renard, *Phys. Rev.* **B25** (2), 931 (1982); J. Richard, P. Monceau, and M. Renard, *Phys. Rev.* **B25** (2), 948 (1982).
44. J. Richard, Thesis, Univ. of Grenoble (1981); see also J. Richard, P. Monceau, and M. Renard, *Phys. Rev.* **B25**, 948 (1982).
45. J. W. Brill, N. P. Ong, J. C. Eckert, J. W. Savage, S. K. Khanna, and R. B. Somoano, *Phys. Rev.* **B23** (4), 1517 (1981).
46. W. W. Fuller, P. M. Chaikin, and N. P. Ong, *Solid State Commun.* **39**, 547 (1981).
47. P. Monceau, J. Richard, and R. Lagnier, *J. Physique C,* **14**, 2995 (1981).
48. M. Nunez-Regueiro, Thesis, Univ. of Grenoble, France (1979).
49. W. W. Fuller, G. Grüner, P. M. Chaikin, and N. P. Ong, *Phys. Rev.* **B23** (12), 6259 (1981).
50. A. Briggs, P. Monceau, M. Nunez-Regueiro, M. Ribault, and J. Richard, *J. Physique* **42**, 1453 (1981).
51. P. A. Lee and T. M. Rice, *Phys. Rev.* **B19**, 3970 (1979).
52. J. Bardeen, *Phys. Rev. Lett.* **42** (22), 1498 (1979).
53. J. Bardeen, *Proc. Conf. on Low Dimensional Conductors, Boulder, 1981: Molecular Crystals and Liquid Crystals*, Vol. 81, Part C, p. 719 (1982).
54. A. Meerschaut, J. Rouxel, P. Haen, P. Monceau, and M. Nunez-Regueiro, *J. Phys. Lett.* **40**, L157 (1979).
55. E. Bjerkelund and A. Kjekshus, *Z. Anorg. allg. Chem.* **B328**, 235 (1964).
56. S. Kikkawa, M. Koizumi, S. Yamanaka, Y. Onuki, R. Inada, and S. Tanuma, *J. Solid State Chem.* **40**, 28 (1981).
57. A. Meerschaut, L. Guemas, and J. Rouxel, *J. Solid State Chem.* **36**, 118 (1981).
58. C. Roucau, R. Ayroles, P. Monceau, L. Guemas, A. Meerschaut, and J. Rouxel, *Phys. Stat. Sol.* **62**, 483 (1980).
59. K. Hasegawa, A. Maeda, S. Uchida, and S. Tanaka, *Physica* **117–118B**, 599 (1983).

60. G. Mihaly, N. Housseau, H. Mutka, L. Zuppiroli, J. Pelissier, P. Gressier, A. Meerschaut, and J. Rouxel, *J. Physique Lett.* **42**, L263 (1981).
61. L. Zuppiroli, *Proc. Conf. on Low Dimensional Conductors, Boulder 1981: Molecular Crystal and Liquid Crystals*, Vol. 81, Part D (1982).
62. A. Meerschaut and P. Grenouilleau (unpublished results).
63. T. Sambongi, K. Tsutsumi, Y. Shiozaki, M. Yamamoto, K. Yamaya, and Y. Abe, *Solid State Commun.* **22**, 729 (1977).
64. G. Van Tendeloo, J. Van Landuyt and S. Amelinckx, *Phys. Stat. Sol.* **A43**, K137 (1977).
65. K. Tsutsumi, T. Sambongi, S. Kagoshima, and T. Ishiguro, *J. Phys. Soc. Japan Lett.* **44** (5), 1735 (1978).
66. (a) T. Takoshima, M. Ido, K. Tsutsumi, T. Sambongi, S. Honna, K. Yamaya, and Y. Abe, *Solid State Commun.* **35**, 911 (1980); (b) A. H. Thompson, A. Zettle, and G. Grüner, *Phys. Rev. Lett.* **47** (1), 64 (1980).
67. (a) Z. Z. Wang, H. Salva, P. Monceau, M. Renard, C. Roucau, R. Ayroles, F. Lévy, L. Guemas, and A. Meerschaut, *J. Physique Lett.* **44** (8), 311 (1983); (b) A. Zettl, G. Grüner, and A. H. Thompson, *Phys. Rev.* **B26** (10), 5760 (1982).
68. (a) A. Meerschaut, J. Rijnsdorp, and F. Jellinek, *Int. Congress of Low Dimensional Solids*, Hawaïi (1979); (b) A. Meerschaut, J. Rijnsdorp, J. C. W. Folmer, and F. Jellinek, *Solid State Chemistry, Proc. Second European Conf., Veldhoven, The Netherlands, June* 1982, Vol. 3, p. 777 (1982).
69. E. Dieman and A. Müller, *Coord. Chem. Rev.* **10**, 79 (1973).
70. T. P. Prasad, E. Dieman, and A. Müller, *J. Inorg. Nucl. Chem.* **35**, 1895 (1973).
71. A. J. Jacobson, R. R. Chianelli, S. M. Rich, and M. S. Whittingham, *Mat. Res. Bull.* **14** (11), 1437 (1979).
72. R. J. H. Voorhoeve and H. M. B. Wolters, *Z. Anorg. allg. Chem.* **376**, 165 (1970).
73. A. Buisine and G. Tridot, *Bull. Soc. Chim. Fr.* **00**, 1383 (1961).
74. J. C. Wilderwanck and F. Jellinek, *Z. Anorg. allg. Chem.* **328**, 309 (1964).
75. E. Dieman, *Z. Anorg. allg. Chem.* **432**, 127 (1977).
76. K. S. Liang, S. P. Cramer, D. C. Johnston, C. H. Chang, A. J. Jacobson, J. P. Neufville, and R. R. Chianelli, *J. Noncryst. Solids* **42**, 345 (1980).
77. J. V. Acrivos, C. Delios, N. Y. Topsoe, and J. R. Salem, *J. Phys. Chem.* **79**, 3003 (1975).
78. K. Yamaya, T. H. Geballe, J. V. Acrivos, and J. Code, *Physica* **105B**, 444 (1981).
79. R. R. Chianelli and M. B. Dines, *Inorg. Chem.* **14** (10), 2419 (1975).
80. D. W. Murphy and F. A. Trumbore, *J. Electrochem. Soc.* **123**, 960 (1976).
81. C. Perrin, A. Perrin, and J. Prigent, *Bull. Soc. Chim. Fr.* **8**, 3086 (1972).
82. F. A. Trumbore and J. Broodhead, *The Electrochemical Society Extended Abstracts*, p. 445, Chicago Meeting, May (1973).
83. M. S. Whittingham in *Electrode Materials and Processes for Energy Conservation and Storage* (ed. J. D. MacIntyre), Princeton, N.J. (1977).
84. M. S. Whittingham, *J. Electrochem. Soc.* **122**, 526 (1975).
85. F. A. Trumbore, J. Broadhead, and T. Pudvinski (unpublished results).
86. A. J. Jacobson, R. R. Chianelli, S. M. Rich, and M. S. Whittingham, *Mat. Res. Bull.* **14** (11), 1439 (1979).
87. S. J. Hillenius, R. V. Coleman, R. M. Fleming, and R. J. Cava, *Phys. Rev.* **B23** (4), 1567 (1981).
88. A. Meerschaut, P. Gressier, L. Guemas, and J. Rouxel, *Mat. Res. Bull.* **16**, 1035 (1981).
89. R. J. Cava, V. L. Hines, A. D. Mighell, and R. S. Roth, *Phys. Rev.* **B24** (6), 3634 (1981).
90. A. Meerschaut, M. Spiesser, J. Rouxel, and O. Gorochov, *J. Solid State Chem.* **31**, 31 (1980).
91. M. H. Whangbo, R. J. Cava, F. J. Di Salvo, and R. M. Fleming, *Solid State Commun.* **43**, 277 (1982).
92. A. Ben Salem, A. Meerschaut, H. Salva, Z. Z. Wang, and T. Sambongi, *J. Physique* **45**, 771 (1984).
93. R. J. Cava, F. J. Di Salvo, M. Eibschutz, and J. V. Waszczak, *Phys. Rev.* **B27** (12), 7412 (1983).
94. R. Allman, L. Bauman, A. Kutoglu, H. Rösch, and E. Hellner, *Naturwiss.* **51**, 263 (1964).
95. A. Kutoglu and R. Allman, *Neues Jahrb. Miner. Monat.* **H8**, 339 (1972).

96. K. Selte and A. Kjekshus, *Acta Chem. Scand.* **18**, 690 (1964).
97. (a) F. W. Boswell, A. Prodan, and J. K. Brandon, *J. Phys. C: Solid State Phys.* **16**, 1067 (1983); (b) F. W. Boswell and A. Prodan, *Mat. Res. Bull.* **19**, 93 (1984).
98. (a) J. Mahy, J. Van Landuyt, and S. Amelinckx, *Phys. Stat. Sol.* **A77**, K1 (1983); (b) J. Mahy, G. A. Wiegers, J. Van Landuyt, and S. Amelynckx, *Mat. Res. Soc. Symp. Proc.* **21**, 181 (1984).
99. H. Böhm and H. G. von Schnering, *Z. Kristallogr.* **162**, 26 (1983).
100. M. H. Whangbo (private communication).
101. K. Selte and A. Kjekshus, *Acta Chem. Scand.* **19**, 1002 (1965).
102. A. Meerschaut, P. Palvadeau, and J. Rouxel, *J. Solid State Chem.* **20**, 21 (1977).
103. P. Gressier, A. Meerschaut, L. Guemas, J. Rouxel, and P. Monceau, *J. Solid State Chem.* **51**, 141 (1984).
104. P. Gressier, A. Meerschaut, L. Guemas, J. Rouxel, and P. Monceau, *J. Physique* **44**, C3-1741 (1983).
105. C. Roucau, R. Ayroles, P. Gressier, and A. Meerschaut, *J. Phys. C: Solid State Phys.* **17**, 2993 (1984).
106. P. Butaud, P. Segransan, C. Berthier, and A. Meerschaut, (to be published in *Proceedings of the XXII Congress Ampere*).
107. M. Izumi, T. Iwazumi, T. Seino, K. Uchinokura, R. Yoshizaki, and E. Matsuura, *Solid State Commun.* **49** (5), 423 (1984).
108. M. Izumi, T. Iwazumi, K. Uchinokura, R. Yoskizaki, and E. Matsuura, *Solid State Commun.* **51**, 191 (1984).
109. H. Fujishita, M. Sato, and S. Hoshino, *Solid State Commun.* **49** (4), 313 (1984).
110. H. Fujishita, M. Sato, S. Sato, and S. Hoshino (to be published in *J. Phys. C*).
111. A. Meerschaut, P. Gressier, L. Guemas, and J. Rouxel, *J. Solid State Chem.* **51** (3), 307 (1984).
112. Z. Z. Wang, P. Monceau, M. Renard, P. Gressier, L. Guemas, and A. Meerschaut, *Solid State Commun.* **47** (6), 439 (1983).
113. P. Gressier, M. H. Whangbo, A. Meerschaut, and J. Rouxel, *Inorg. Chem.* **23**, 1221 (1984).
114. G. Mihaly and L. Mihaly, *Solid State Commun.* **48**, 449 (1983).
115. J. C. Gill, *Solid State Commun.* **39**, 1203 (1981).
116. R. M. Fleming and L. F. Schneemeyer, *Phys. Rev.* **B28** (12), 6996 (1983).
117. P. Butaud, P. Segransan, C. Berthier, and A. Meerschaut, *Proc. Conf. on CDW in Solids, Budapest, 1984* (to be published).
118. P. Gressier, L. Guemas, and A. Meerschaut, *Acta Cryst.* **B38**, 2877 (1982).
119. Z. Z. Wang, M. C. Saint Lager, P. Monceau, M. Renard, P. Gressier, A. Meerschaut, L. Guemas, and J. Rouxel, *Solid State Commun.* **16**, 325 (1983).
120. M. Maki, M. Kaiser, A. Zettl, and G. Grüner, *Solid State Commun.* **46**, 497 (1983).
121. P. Delhaes, C. Coulon, S. Flandrois, B. Hilti, C. W. Mayer, C. Rihs, and J. Rivory, *J. Chem. Phys.* **73**, 1452 (1980).
122. M. Izumi, T. Iwazumi, K. Uchinokura, R. Yoshizaki, and E. Matsunura (to be published in *Molecular Crystals and Liquids Crystals, Conference, Abano, Italy, June 1984*).
123. P. Gressier, A. Meerschaut, L. Guemas, and J. Rouxel (unpublished results).
124. A. Meerschaut, L. Guemas, R. Berger, and J. Rouxel, *Acta. Cryst.* **B35**, 1747 (1979).
125. (a) J. Rijnsdorp, G. J. de Lange, and G. A. Wiegers, *J. Solid State Chem.* **30**, 365 (1979); (b) J. Rijnsdorp and F. Jellinek, *J. Solid State Chem.* **28**, 149 (1979).
126. S. Furuseth, L. Brattås, and A. Kjekshus, *Acta Chem. Scand.* **27**, 2367 (1973).
127. L. Pauling, *The Nature of the Chemical Bond*, Cornell Univ. Press, Ithaca (1960).
128. S. Okada, T. Sambongi, and M. Ido, *J. Phys. Soc. Japan* **49** (2), 839 (1980).
129. F. J. Di Salvo, R. M. Fleming, and J. V. Waszczak, *Phys. Rev.* **B24** (6), 2935 (1981).
130. R. Hoffmann, S. Shaik, J. C. Scott, M. H. Whangbo, and M. J. Foshee, *J. Solid State Chem.* **34**, 263 (1980).
131. R. Huisman, R. Dejonge, C. Haas, and F. Jellinek, *J. Solid State Chem.* **3**, 56 (1971).
132. M. H. Whangbo and P. Gressier, *Inorg. Chem.* **23**, 1305 (1984).
133. F. Devreux, *J. Phys.* **43**, 1489 (1982).
134. D. W. Bullett, *J. Phys. C*, **12**, 277 (1979).

135. W. Schairer and M. W. Shafer, *Phys. Stat. Sol.* **A17**, 181 (1973).
136. K. Endo, H. Ihara, S. Gonda, and K. Watanabe, *Physica* **B105**, 159 (1981).
137. (a) T. J. Wieting, A. Grisel, F. Lévy, and P. H. Schinid, *Lecture Noters in Physics* **95**, 354 (1979); (b) A. Grisel, Thesis, Univ. of Lausanne, Switzerland (1981); (c) J. Wieting, A. Grisel, and F. Lévy, Boulder Meeting August 1981, *Molecular Crystals and Liquid Crystals*, Vol. 81, Part C, p. 835 (1982).

PENTACHALCOGENIDES OF TRANSITION METALS

TAKASHI SAMBONGI

Department of Physics, Faculty of Science,
Hokkaido University, Sapporo 060, Japan

1. Introduction

The existence of pentatellurides was established in 1971 by Brattås and Kjekshus [1]. Through the systematic investigation of Hf—Te system, they found that $HfTe_5$ is the compound with the highest Te concentration. The composition was determined by the neutron activation technique. No homogeneity range was detected. Two years later the same group synthesized $ZrTe_5$ and determined their crystal structure [2]. Since large peaks of electrical resistivity were found in both $ZrTe_5$ and $HfTe_5$, rapidly growing interest has been devoted to ascertain if the anomalies in transport properties are associated with a structural phase transition, e.g. the formation of a charge density wave (CDW). Research on MTe_5 has been stimulated by the discovery of the CDW states and their unique properties in transition-metal trichalcogenides, because the crystal structure of MTe_5 is similar to that of trichalcogenides and the resistivity peaks of MTe_5 are more remarkable than those observed in $NbSe_3$. A unified explanation has not been given, however, to the whole class of the accumulated experimental data. Difficulty also lies in that various types of structural modulation have been found but cannot always be reproduced; some researchers failed to confirm the findings of others. Therefore, the role of the possible structural modulation, which should affect various physical properties significantly, has not been ascertained.

2. Preparation

Powder crystals are prepared by sealing weighted amounts of starting materials in an evacuated quartz tube and heating at 450—800 °C for 1—2 weeks. Single crystals have been prepared by the iodine vapour transport method. The transport conditions are:

(i) iodine concentrations of ~ 5 mg cm^{-3} of quartz tube [1];
(ii) a temperature gradient of 500—450 °C for $HfTe_5$ [1, 2, 5], 580—430 °C for $ZrTe_5$ [2], 500—420 °C for both compounds [6], and 500—300 °C for both [10, 11].

Other transport agents have not been examined, presumably because iodine works satisfactorily; iodine inclusion has not been detected so far.

J. Rouxel (ed.), Crystal Chemistry and Properties of Materials with Quasi-One-Dimensional Structures, 281—313.

The products are long parallelepiped shiny crystals with the longest dimension parallel to the orthorhombic *a* axis and the shortest parallel to the *b* axis. Good crystals can be easily cleaved parallel to the *a—c* plane.

$TiTe_5$ has not been synthesized in powder or in single crystalline form [2].

It has been noted that another kind of needle-shaped crystals is often obtained within the tube, especially when $HfTe_5$ is to be grown. They are elemental Te [5, 6]. No appreciable concentration of iodine or metal is incorporated in the by-product Te.

$ZrTe_5$ can be also prepared by the transport method from $ZrTe_2$ or $ZrTe_3$ in powder form with the addition of a suitable amount of Te [7].

In general, a noticeable amount of Hf is found as impurity in metallic Zr and more Zr in Hf. Therefore, it should be noted that 'pristine' MTe_5 is not always free from an isoelectronic metallic impurity, though its level is different from that in the starting material due to the different vapor transport coefficients.

3. Crystal Structure

3.1. BASIC STRUCTURE [3]

The fundamental crystal structure of $HfTe_5$ was determined by Furuseth *et al.* [2]. Though the structure of $ZrTe_5$ has not been determined independently, it is isostructural with $HfTe_5$ because a close relationship is found in the lattice dimensions, symmetry, and X-ray reflection intensities of the two compounds. The lattice parameters were determined by Furuseth *et al.* from Guinier photographs with monochromatic Cu $K\alpha_1$ radiation using KCl as the internal standard. The atom positions of $HfTe_5$ were determined using reflection intensities collected from integrating Weissenberg photographs with Zr-filtered Mo $K\alpha$ radiation. Values of density at room temperature [2] determined pycnometrically (ρ_p) are in good agreement with ρ_x calculated from the lattice constants. Structural and density data are summarized in Table I.

TABLE I
Crystal system: orthorhombic
Space group: *Cmcm*
Lattice constants (Å) and density (g cm^{-3})

	$ZrTe_5$	$HfTe_5$
a	3.9876(11)	3.9743(5)
b	14.502(4)	14.492(2)
c	13.727(3)	13.730(2)
ρ_p	6.079	6.806
ρ_x	6.102	6.858

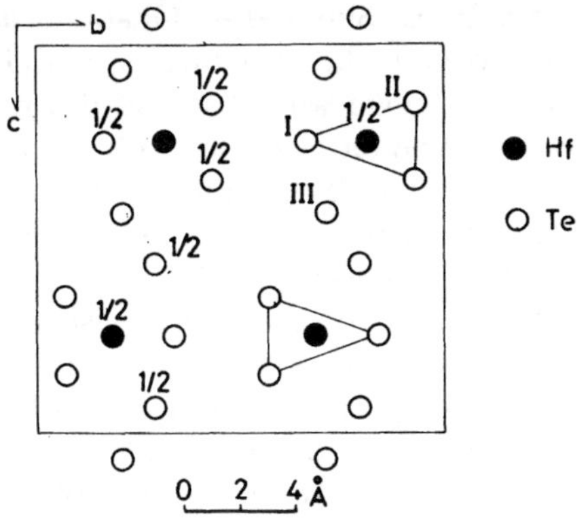

Fig. 1. Crystal structure of $HfTe_5$. (After Furuseth *et al.* [2].)

The structure given by Furuseth *et al.* is illustrated in Figure 1. There are four MTe_5 chemical formulae within the unit cell. Four MTe_3 chains, with the cross section of an isosceles triangle, are running parallel to the *a* axis. The metel atom is positioned at the center of an MTe_6 'bicapped trigonal' prism. MTe_3 chains are linked together by $—Te_2—$ chains along the *c* axis. The interatomic distances between the metal atom and Te atoms are approximately the same:

$Hf—Te_I$	$Hf—Te_{II}$	$Hf—Te_{III}$
2.954(3) Å	2.944(2) Å	2.960(2) Å

The coordination number of the metal atom should be 8 but not 6; the structure can be viewed as being composed of MTe_5 layers stacked along the *b* axis. The distance between Te_{II} atoms within the chain is short (2.763(4) Å); they are paired. The interlayer Te—Te distances ($\gtrsim$4.161(3) Å) are comparable to the van der Waals distance of 4.4 Å for Te—Te; a two-dimensional character is expected in properties of $ZrTe_5$ and $HfTe_5$.

The structures of transition-metal trichalcogenides are based on the unit of the bicapped MX_6 prism but they are directly connected to each other, as are fully discussed by Rouxel and Meerschaut in this volume.

3.2. Structure Modulation and Polytype

Various kinds of modulation of the fundamental structure have been observed at room temperature. Di Salvo *et al.* [6] observed half-integer reflections at and below room temperature. By scanning within the $(\xi, \eta, 0)$ plane, they found several $(h + \frac{1}{2}, k + \frac{1}{2}, 0)$ reflections but no extra reflection along the [100] or [010] direction in both $ZrTe_5$ and $HfTe_5$. Intensities of these half-integer reflections are weak; $< 4 \times 10^{-2}$ of the Bragg (200). Because these extra reflections are observed even at room temperature, they proposed that the actual lattice constants are twice (along the *a* and *b* axes) those given by Furuseth *et al.* [2]. No extra reflection in (ξ,

η, $\zeta \neq 0$) was looked for by Di Salvo *et al.* Okada *et al.* [7] examined $ZrTe_5$ by electron diffraction to discover if the structure is modulated at low temperature. The result was negative; no change was observed down to ~ 12 K. The use of a high-voltage electron microscope (accerelation voltage up to 1 MeV) made it possible to use relatively thick crystals. The sample stage was cooled by liquid helium down to ~ 12 K, though the temperature of the sample was somewhat higher. Because $ZrTe_5$ is cleaved most easily parallel to the a—c plane, the net planes approximately parallel to the a^*—c^* plane are observed. No satellite or diffuse streak, which would correspond to a precursor of structural modulation, was observed even at the lowest temperature. Elongation of the Bragg spots along the b^* axis is inferred from the regular array of diffractions spots over a large area without dark gaps between the Laue zone; it is improbable that an extra spot with nonzero η was missed in the electron diffraction experiment.

Skelton *et al.* [11] examined both $ZrTe_5$ and $HfTe_5$ by a X-ray counter method below room temperature. The half-integer reflections found by Di Salvo *et al.* were looked for but not observed. No extra reflection was observed along the a^* and b^* axes and long the $[0, \eta, \zeta = 3\eta]$ and $[0, \eta = 3\zeta, \zeta]$ directions. They reported that weak (0, 0, odd integer) reflections, which are forbidden by the space group *Cmcm*, were observed only by the counter method but not by camera techniques. The intensities of $(0, 0, 2n + 1)$ reflections with low n show maxima at the temperature corresponding to the resistivity maxima (cf. Section 4.1.1), as shown in Figure 2. Although they concluded recently that the forbidden reflections are

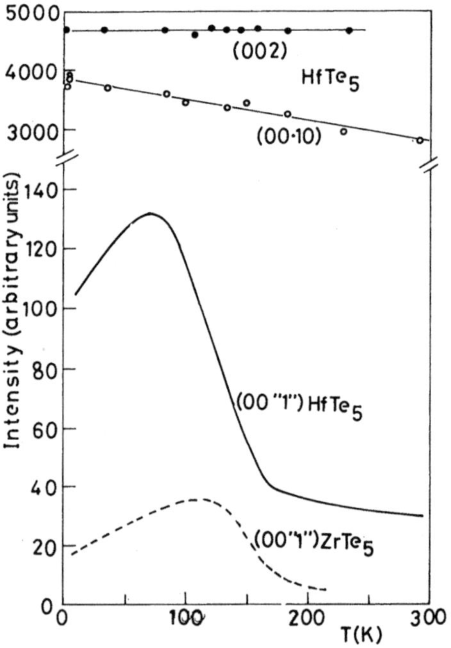

Fig. 2. Temperature dependence of the 'extra' reflections. (After Skelton *et al.* [11].)

not indicative of the presence of superlattice but due to multiple scattering (T. J. Wieting, private communication), the peculiar temperature dependence shown in Figure 2 may tell us something about the nature of the resistivity maxima.[1]

It was found [24] that the orthorhombic lattice is accompanied with another set of reflections with the periods incommensurate to the host lattice, by camera techniques at room temperature. The periods of the modulation are ~26.5 Å, ~31.6 Å, and ~30.1 Å along the *a*, *b*, and *c* axes, respectively. The extinction rule is the same as of the host lattice. Half-integer reflections [6] were not observed. However, some of the extra reflections observed by Di Salvo *et al.* [6] and by Skelton *et al.* [11] can be indexed by the above long period modulation, if some uncertainty is allowed in their indexing. Absence of a structural phase transition was confirmed by powder X-ray and neutron diffraction experiments below room temperature [22].

A monoclinic polytype has been found in $ZrTe_5$ [7, 24]. The lattice parameters are $a = 14.59$ Å, $b = 3.98$ Å, $c = 31.29$ Å, and $\beta = 96.7°$. From the extinction rule, the most probable space group is $C2/c$. Note that $b_m = a_o$ and $a_m = b_o$, where the subscripts m and o denote monoclinic and orthorhombic structures, respectively. When crystals of the two polytypes are stuck together the a_o axis is parallel to the b_m axis and b_o to a_m, in agreement with the above relations. Since $|b_m \cdot (a_m \times c_m)| \cong 2.3\,|a_o \cdot (b_o \times c_o)|$, the intralayer distance between chains must be longer in the monoclinic structure than in the orthorhombic if the —MTe_3—Te_2— arrangement is retained in the former. Stoichiometry of the monoclinic polytype has not been examined. Electrical resistivity of the monoclinic $ZrTe_5$ shows a small hump at ~85 K [24]. In this temperature range an anomaly in the Young's modulus [21] and a broad peak in the heat capacity [24] have been found (cf. Section 8). These observations, together with the unit cell size ~15% larger, lead to the expectation that the monoclinic $ZrTe_5$ is a quasi-one-dimensional (quasi-1D) conductor and shows the lattice instability at ~85 K. However, no good crystal has been separated from the orthorhombic matrix, therefore its physical properties have not been made clear.

4. Transport Properties and Related Phenomena

4.1. ELECTRICAL RESISTIVITY

4.1.1. *DC Resistivity*

Electrical resistivity has been measured by many workers. Because single crystals have the largest dimension along the *a* axis, the resistivity along that direction, ρ_a, is most reliable.

Electrical contacts were made by thin metal wires attached by indium solder [6, 7] or by direct painting of silver paste [5, 6]. It was found that contacts of silver paste can be improved by painting on evaporated pads of gold with tellurium [7].

[1] By reproducing the misleading result, it is not the present author's intention to show its inaccuracy, but rather to emphasize the importance of repeated measurements in this field.

Resistivities at room temperature are:

	ρ_a (10^{-4} Ω cm)
$ZrTe_5$	7 [4, 7], 5.5 [6], 7.2 [11]
$HfTe_5$	4.0 [6], 6.7 [11].

The scatter of the above values is not serious; accuracy of the geometrical factor is limited because of the small sample size.

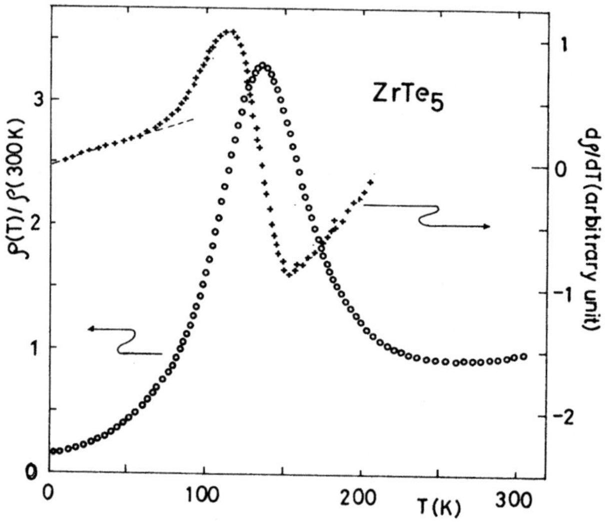

Fig. 3. Electrical resistivity of $ZrTe_5$. (After Okada *et al.* [4].)

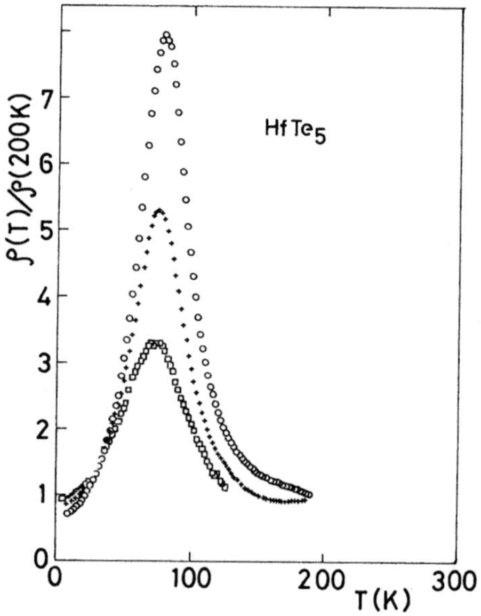

Fig. 4. Electrical resistivity of $HfTe_5$. (After Izumi *et al.* [5].)

Temperature dependence of the resistivity is shown in Figure 3 and 4. For both compounds the resistivity is metallic at high and low temperatures and increases with increasing temperature.

In the intermediate temperature range large peaks are found at ~140 K in $ZrTe_5$ and ~80 K in $HfTe_5$. The temperature where the resistivity is maximum, T_m, is sample dependent:

	T_m
$ZrTe_5$	~150 K [4, 7], 143 K [6], 141 ± 3 K [11]
$HfTe_5$	76 K ($Hf_{0.92 \pm 0.06}Te_5$) [5], 50 K [6], 80 K [10], 83 ± 3 K [11], 70 K ($Hf_{1.00}Zr_{0.004}Te_5$) [16]

The magnitude of the peak shows a larger scatter; the maximum resistivities normalized by the room temperature values, $\rho(\max)/\rho(\sim 300\text{ K})$, are:

	$\rho(\max)/\rho(\sim 300\text{ K})$
$ZrTe_5$	<4 [4], 3.3 [7], 2.8—3.5 [11]
$HfTe_5$	3—8 [5], 8—10 [6], ~5 [10], 5—6 [11].

It seems that the scatter in the data of the relative peak height and of T_m is related to the purity and/or the stoichiometry of samples.

If the resistivity peak corresponds to a phase transition — e.g. into the CDW state — the transition temperature is not equal to T_m but to the temperature where the logarithmic temperature derivative $|\mathrm{d}\ln\rho/\mathrm{d}\ln T|$ is maximum (above T_m), as was verified in many quasi-1D and quasi-2D conductors. In low-dimensional conductors, the transition curve is broadened by fluctuation but the transition temperature can be determined from the transition curve; $|\mathrm{d}\ln\rho/\mathrm{d}\ln T|$ becomes very large (it diverges in the ideal case) at the transition temperature. With increasing impurity concentration the structure near the transition temperature becomes smeared. As for MTe_5, the maxima of $|\mathrm{d}\ln\rho/\mathrm{d}\ln T|$ are not large as shown in Figures 3 and 4; it seems that the resistivity peaks in MTe_5 are not indicative of a phase transition. Referring to the small anisotropy of the resistivity in the *a—c* plane (see Section 4.1.3), strong one-dimensionality is not expected in MTe_5.

Qualitative information on the sample quality can be obtained from the residual resistance ratio, RRR = $\rho(\sim 300\text{ K})/\rho(\sim 4\text{ K})$. In normal metals the impurity scattering resistivity or residual resistivity, $\rho_i = \rho(T \to 0)$ which is approximately given by $\rho(T \cong 4\text{ K})$, is determined by impurities and/or structural imperfections. The resistivity at finite temperature is given approximately by the sum of ρ_i and the phonon resistivity ρ_{ph}; $\rho(T) = \rho_i + \rho_{\text{ph}}(T)$, as long as impurities or imperfections do not alter the electronic and phonon structures significantly (Mattheissen's rule). When RRR >> 1, RRR $= 1 + (\rho_{\text{ph}}(\sim 300\text{ K})/\rho_i) \cong \rho_{\text{ph}}(\sim 300\text{ K})/\rho_i$ is inversely proportional to ρ_i and hence to, for example, impurity concentration. Okada *et al.* [4] used $(\text{RRR-1})^{-1}$ as a measure of the sample quality; it is equal to $\rho_i/\rho_{\text{ph}}(\sim 300$

K), the relative residual resistivity free from geometrical uncertainty. The usefulness of this parameter is somewhat obscured, however, by the presence of the resistivity peak. The Mattheissen's rule is based on the assumption that the impurity resistivity at finite temperature is equal to the residual resistivity measured at low temperature. In MTe_5 the carrier concentration well above T_m is different from that below T_m, $\rho_i(T > T_m) \neq \rho_i(T \sim 0)$, regardless of the mechanism of the resistivity maximum. Nevertheless, RRR can be used as a qualitative guide; hopefully both $\rho_i(T > T_m)$ and $\rho_i(T \sim 0)$ are proportional to impurity/defect concentration and

$$\mathrm{RRR} = \frac{\rho_i(T > T_m) + \rho_{\mathrm{ph}}(\sim 300\ \mathrm{K})}{\rho_i(T \sim 0)} = \mathrm{constant} + \frac{\rho_{\mathrm{ph}}(\sim 300\ \mathrm{K})}{\rho_i(T \sim 0)} .$$

Okada *et al.* [4] found that the relative height of the peak increases and T_m is lowered with decreasing RRR for $ZrTe_5$. The former correlation can be understood

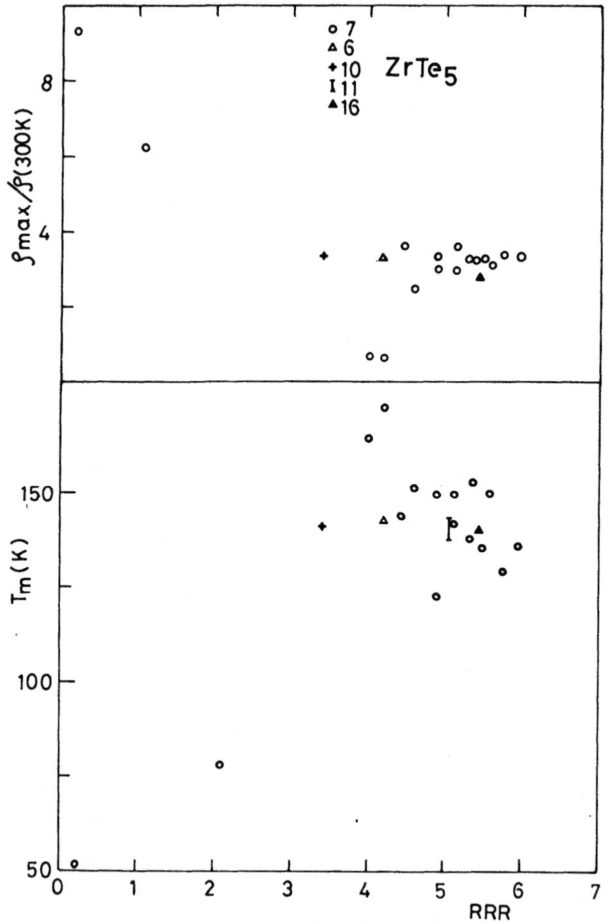

Fig. 5. Correlation between the temperature of resistivity peak, T_m, resistivity peak normalized by the room temperature value, $\rho(\max)/\rho(\sim 300\ \mathrm{K})$ and the residual resistance ratio (RRR) of $ZrTe_5$.

qualitatively as follows. Assume that (i) $\rho(T) = \rho_i + AT$ (A constant) well above T_m and (ii) the resistivity peaks at a transition (not necessarily a phase transition) from a state with a small number of carriers to one with a large carrier density $\rho(\max) = \alpha(\rho_i + AT_m)$, where $\alpha > 1$ is the ratio of the carrier density at high temperature to that at low temperature. No change in the mobility is assumed here. The relative peak height $\rho(\max)/\rho(\sim 300\text{ K}) = \alpha(\rho_i + AT_m)/(\rho_i + A \times 300)$ is an increasing function of ρ_i, as long as the shift of T_m with impurity concentration is smaller. If data points for samples that are too impure are excluded, however, such a correlation becomes obscure.

Di Salvo *et al.* [6] observed that, for both compounds, T_m and the peak height are sample dependent and that T_m for $HfTe_5$ is between 46 and 55 K. On the other hand, Izumi *et al.* [5] found that T_m for $HfTe_5$ shows only a small variation near 76 K (considerably higher than that given by Di Salvo *et al.*) whereas the relative peak height, $\rho(T_m)/\rho(T = 200\text{ K})$, is sample dependent (3.4— ~ 8). The composition of the samples used in [5] was determined by the inductively coupled plasma optical emission method, as $Hf_{0.92 \pm 0.06}Te_5$. In a later paper they found that T_m = 70 K for $Hf_{1.00}Zr_{0.004}Te_5$ [16]. Izumi *et al.* suggested that $HfTe_5$ synthesized by Di Salvo *et al.* [6] is also off-stoichiometric; RRR is considerably smaller than 1 and the Curie—Weiss tail has been observed in the magnetic susceptibility (cf. Section 6).

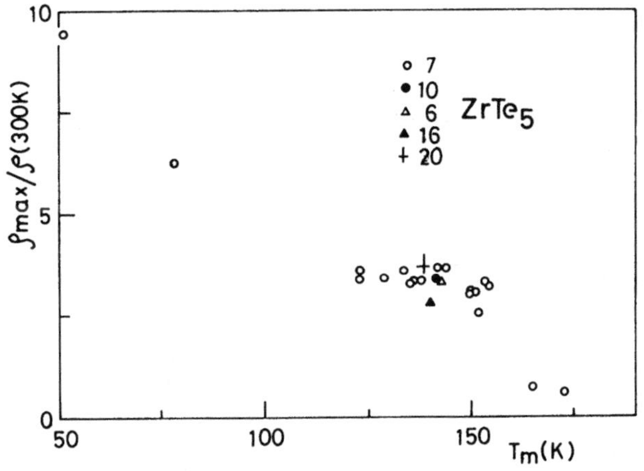

Fig. 6. Correlation between T_m and $\rho(\max)/\rho(\sim 300\text{ K})$.

All the available data are summarized in Figures 5 and 6 for $ZrTe_5$ and in Figures 7 and 8 for $HfTe_5$. No clear correlation can be found between RRR, T_m, and the relative peak height, if data for samples with too small RRR are excluded. Even if these quantities are functions of the impurity concentration in stoichiometric crystals or of the degree of off-stoichiometry in chemically pure samples, the influence of these two types of irregularity has not yet been separated.

As shown in Figure 9, the temperature-dependent part of resistivity of $ZrTe_5$ is proportional to T^2 below 60 K, which was explained as arising from electron—

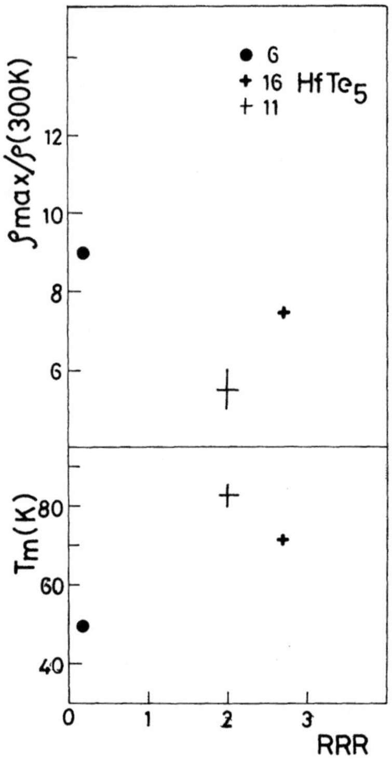

Fig. 7. Correlation between T_m, $\rho(\mathrm{max})/\rho(\sim 300\ \mathrm{K})$ and RRR of $HfTe_5$.

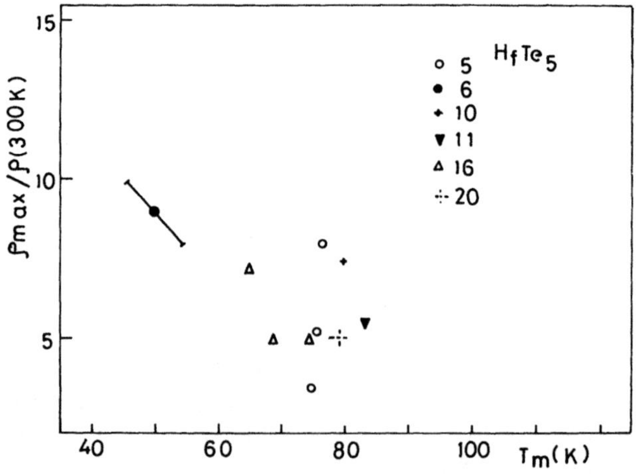

Fig. 8. Correlation between T_m and $\rho(\mathrm{max})/\rho(\sim 300\ \mathrm{K})$ of $HfTe_5$.

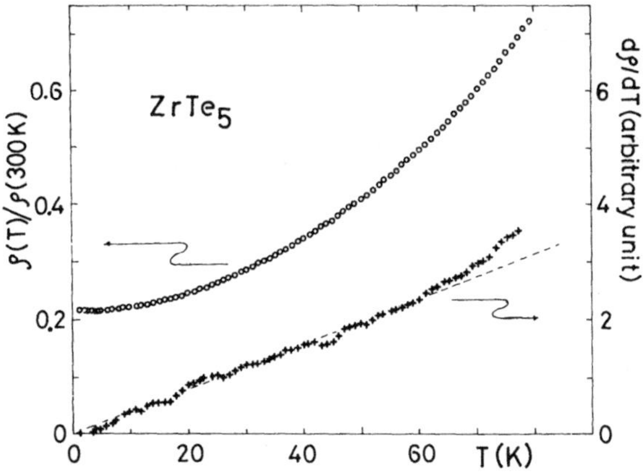

Fig. 9. Resistivity of $ZrTe_5$ at low temperature [7].

electron scattering [6, 7, 11]. In three-dimensional metals, electron—electron Umklapp scattering leads to a resistivity proportional to T^2 [25]. Oshiyama *et al.* [26] calculated the resistivity of quasi-1D metals with the Fermi surface warped but open. They found that it is proportional to T^n, where $2 \leqslant n \leqslant 3$. With increasing curvature of the Fermi surface, n approaches 2. It is not clear why such a large electron—electron scattering resistivity is observed in MTe_5 (cf. Sections 5.1 and 5.2).

In $ZrTe_5$ superconductivity has not been observed down to 1.2 K [4]. It is quite probable that neither $ZrTe_5$ nor $HfTe_5$ is superconducting at least down to 30 mK [11].

4.1.2. *High Electric Field*

Resistivity was measured in a high electric field in an attempt to discover if MTe_5 shows non-ohmic conductivity. In the case of MTe_5, an increase of conductivity can be expected for the conductivity along the *a* axis. In $NbSe_3$, TaS_3 and other quasi-1D inorganic conductors the increase of conductivity with a well-defined threshold electric field has been observed below the transition temperature into the CDW state and is attributed to the sliding motion of the CDW. On the other hand, this type of non-ohmic conductivity has not been observed in transition-metal dichalcogenides, which show triple-q CDW states, presumably because of stronger effects of impurities.

Di Salvo *et al.* observed that the resistivity of $HfTe_5$ is ohmic up to 1.5 V cm^{-1} near $T_m \sim 50$ K [6]. Okada *et al.* [7] measured the voltage response to pulsed current in $ZrTe_5$ between 115 and 291 K. Up to 50 V cm^{-1} the resistivity ρ_a is ohmic within the experimental uncertainty, as shown in Figure 10. Skelton *et al.* [11] confirmed the ohmic resistivity up to 50 V cm^{-1} in $ZrTe_5$ and up to 20 V cm^{-1} in $HfTe_5$. The temperature range examined was not specified.

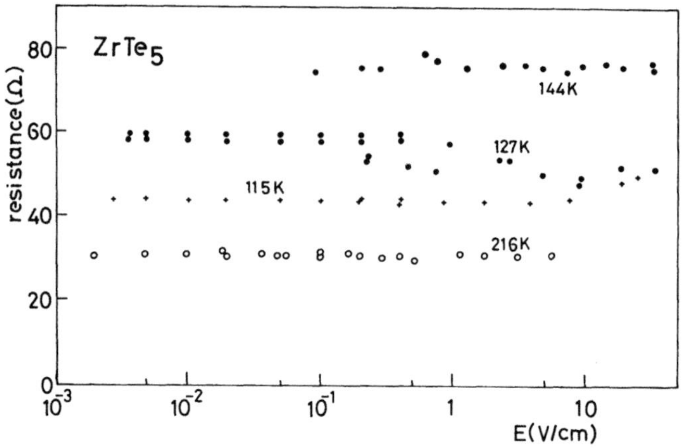

Fig. 10. Electric field dependence of the resistivity of $ZrTe_5$. (After Okada *et al.* [7].)

4.1.3. *Anisotropy*

Anisotropy of the resistivity within the cleverage plane, ρ_a/ρ_c, was measured for $ZrTe_5$ [7]. The Montgomery technique [27] was used. As shown in Figure 11, the anisotropy is not large; it is between 2 and 3 and slightly dependent on temperature with no peak near T_m. Presumably, the peak is indicative of a change in the density of carriers rather than in their mobility. Both ρ_a and ρ_c show the peak at the same temperature. The anisotropy shows an apparent increase at low temperature, because the residual resistivity ratio is larger along the *a* axis. Because of the anisotropic crystal structure, imperfections are expected to show themselves in the transverse resistivities. A small anisotropy was confirmed for $HfTe_5$ by Izumi *et al.* [9]. The resistivity along the *b* axis has not been measured. MTe_5 is not likely to be one-dimensional, though the possibility of two-dimensionality has not been excluded.

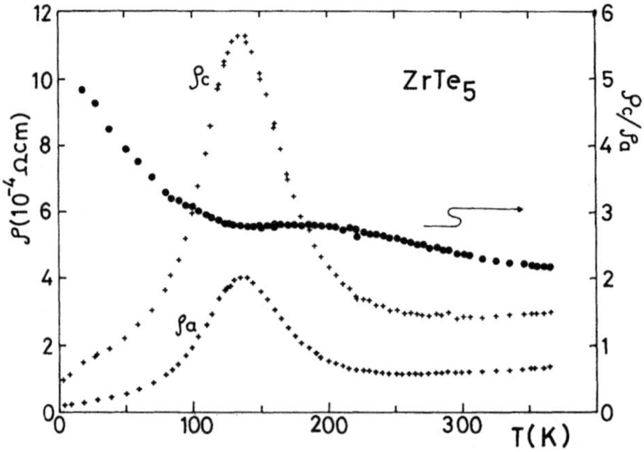

Fig. 11. Anisotropy of the resistivity of $ZrTe_5$ in the *a*—*c* plane. (After Okada *et al.* [7].)

4.1.4. *AC Resistivity*

One of the most puzzling features of MTe_5 is that the resistivity is strongly dependent on frequency. Skelton *et al.* [11] reported that at 9.3 GHz the resistivity anomaly is completely suppressed for $HfTe_5$ while the effect is much weaker for $ZrTe_5$. Qualitatively identical but clearer data were given by Jackson *et al.* [17]. For $HfTe_5$ the peak vanishes almost completely at 9.1 GHz (Figure 12(a)), while for $ZrTe_5$ (Figure 12(b)) it is considerably suppressed. For $Zr_{0.5}Hf_{0.5}Te_5$ the peak is suppressed only weakly; the difference from the dc resistivity is within the experimental error though an apparent difference can be found in the figure. Samples used in [17] were supplied from the same stocks as used in [6]; the composition of the mixed compound is nominal and $HfTe_5$ is presumably off-stoichiometric.

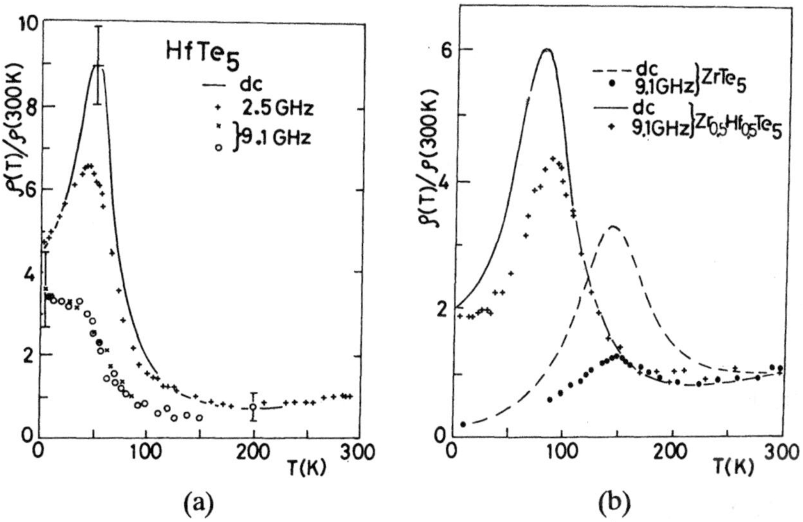

Fig. 12. Resistivity at various frequencies, of (a) $HfTe_5$ and (b) $ZrTe_5$ and $Zr_{0.5}Hf_{0.5}Te_5$. (After Jackson *et al.* [17].)

The strong dependence on frequency suggests that the peak of the dc resistivity is attributed to the condensation of a collective mode. The energy of $\hbar\omega$ for 9.3 GHz is equivalent to 0.3 K; the transport process involving such a small energy should be wiped out by thermal broadening as long as a single electron process is concerned. In $NbSe_3$, the resistivity anomaly associated with the onset of the CDW is scarcely observable at 9.3 GHz. The ac conductivity corresponds to the oscillation of the CDW, as a rigid body, in the pinning potential. Electrons participate coherently to the CDW so the energy $\hbar\omega \times$ (number of electrons participating to the CDW) should be compared with $k_B T$.

Similarity with $TiSe_2$ has been noted by Jackson *et al.* [17]. It shows a large resistivity peak at 165 K and the superlattice $2a_0 \times 2a_0 \times 2c_0$ is observed below 202 K. In the ac resistivity at 8.9 GHz the peak is significantly suppressed. As to the origin, both the formation of the CDW of the excitonic insulating state and a band Jahn—Teller mechanism have been proposed, but which is more reasonable is an open question.

Although most of the available data show that the transport anomalies of MTe_5 are not likely to be associated with a structural phase transition, the strong dependence of the resistivity on frequency remains unexplained.

4.2. Hall Effect

The Hall coefficient R_H of $HfTe_5$ (assigned as $Hf_{0.88}Zr_{0.12}Te_5$) was measured [8] with the current and magnetic field parallel to the a and b axes, respectively. With lowering temperature, the positive R_H increases and shows a maximum just above T_m and then crosses zero near T_m as shown in Figure 13. Such a change near T_m might be assigned as evidence for a (structural) phase transition, but a change in the

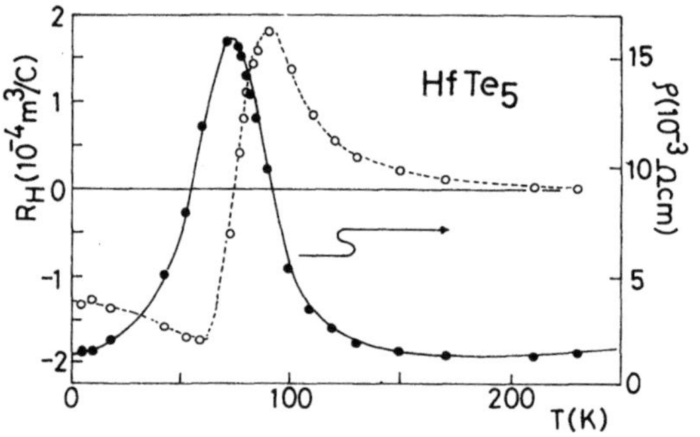

Fig. 13. Hall coefficient of $HfTe_5$ in low-magnetic field (0.17 T). (After Izumi *et al.* [8].)

electronic population or mobility that is not also associated with a phase transition can induce such a change in transport properties. With increasing magnetic field T_m shifts higher and the negative R_H at lower temperature decreases in magnitude. In the high-field limit, R_H shows the same negative value at 1.5, 4.2, and 80 K ($> T_m$).

From the simplest assumption of a single band, the carrier is assigned as holes with a density of 1.5×10^{18} cm^{-3} and mobility of 3×10^3 (cm/sec)(V/cm)$^{-1}$ at room temperature, from ρ_a and the positive R_H. These values should be used only as a guide; the single-band model leads to the unreasonable result that the density of holes decreases and their mobility increases with increasing temperature above ~150 K, because ρ_a is dependent on temperature only weakly.

If a two-band model of electron and hole bands is applicable, R_H in low field is written as

$$R_H(B) = -\frac{n\mu^2 - p\nu^2}{|e|(n\mu + p\nu)^2} - \frac{np\mu^2\nu^2(n-p)(\mu^2+\nu^2)}{|e|(n\mu+p\nu)^4}B^2 \tag{1}$$

while in the high-field limit

$$R_H(B) = -\frac{1}{|e|(n-p)} + \frac{np(\mu+\nu)^2}{|e|\mu^2\nu^2(n-p)^3}\frac{1}{B^2} \tag{2}$$

where n, p, μ, and ν are, respectively, the density of electrons and holes, and the mobility of electrons and holes. Anisotropy of the mobilities within the a—c plane is neglected.

Above T_m the first term of (1) is positive while the second is negative [9]; it is inferred that $n > p$ and $\mu < \nu$, different from the assignment by the single-band model. Below T_m, $R_H(0)$ is negative but an inconsistency is found in the high-field data. $R_H(\infty)$ is negative ($n > p$) while the second term of (2) is also negative ($n < p$); the two-band model of electrons and holes is not applicable.

Izumi *et al.* [9] proposed later that a two-band model, with heavy and light electron bands, can reconcile with the data below T_m. In this case $R_H(\infty)$ gives the total electron density of $\sim 6 \times 10^{16}$ cm^{-3}.

As shown in Figure 14, T_m shifts higher in a magnetic field. If the resistivity peak is associated with the charge or spin density wave, the polarization of conduction electrons would lower the transition temperature, as was observed in the A-15 compound V_3Si [28].

The transverse magnetoresistance in a weak magnetic field [9] is positive and proportional to B^2. The coefficient of the B^2 term is shown in Figure 15. It is related to the mobility of carriers in the two-band model or to the nonsphericity of the Fermi surface [25]. It is independent of temperature below 90 K, while it

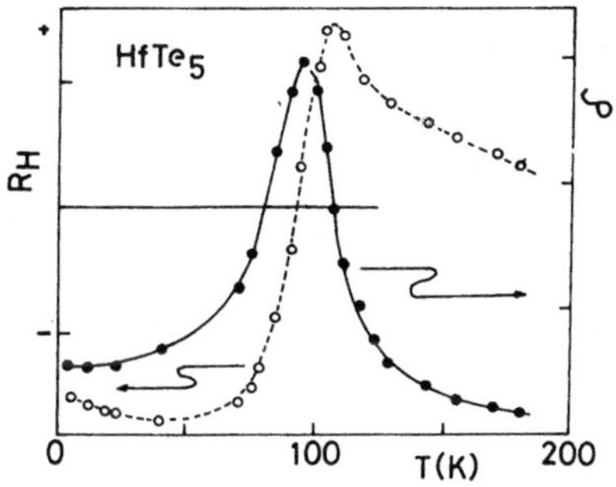

Fig. 14. Resistivity and Hall coefficient of $HfTe_5$ in high-magnetic field (11.7 T). (After Izumi *et al.* [9].)

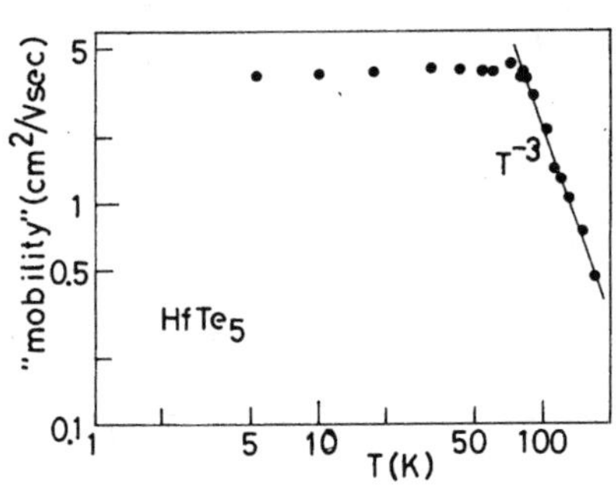

Fig. 15. 'Magnetoresistance mobility' of $HfTe_5$. (After Izumi *et al.* [9].)

decreases as T^{-3} at higher temperature. Again, this result demonstrates a change in the scattering process and/or population of conduction electrons near T_m. No explanation has been given to the T^{-3} dependence.

4.3. Thermoelectric Power

The thermoelectric power $S(T)$ of $ZrTe_5$ and $HfTe_5$ was measured by Jones *et al.* [10]. A low-frequency ac temperature modulation technique was used. The results are shown in Figures 16 and 17 compared with ρ_a. In both compounds $S(T)$ is positive above T_m but changes its sign near T_m.

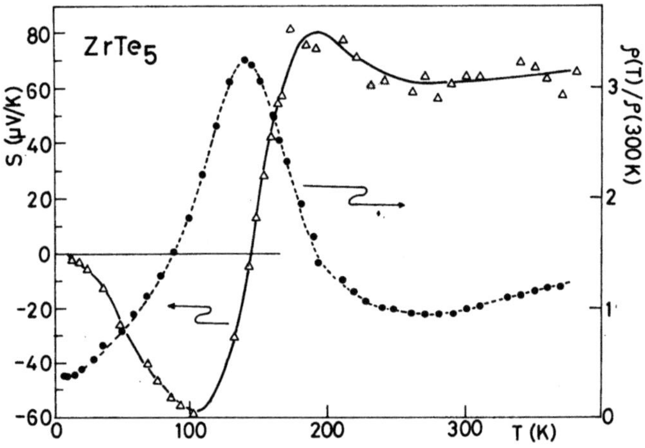

Fig. 16. Thermoelectric power of $ZrTe_5$. (After Jones *et al.* [10].)

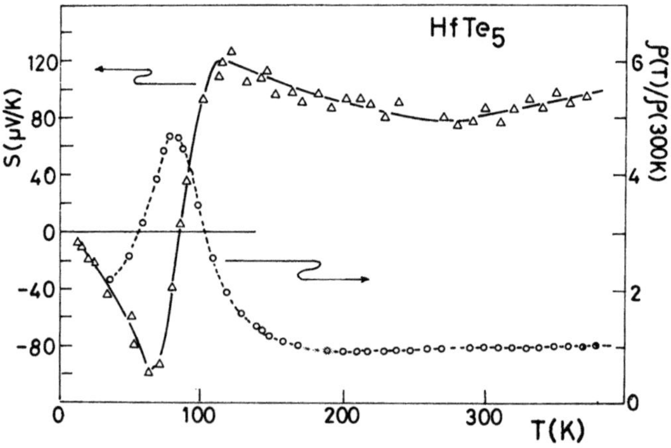

Fig. 17. Thermoelectric power of $HfTe_5$. (After Jones *et al.* [10].)

In the intermediate temperature range the analysis is complicated because the phonon-drag contribution is usually important. Nevertheless, the anomaly is clearly observable, which corresponds to the resistivity peak. Absolute magnitude of $S(T)$ is somewhat sample dependent. The diffusion thermopower is given by

$$S(T) = -\frac{\pi^2}{3}\frac{k_B^2 T}{|e|}\left[\frac{\partial \ln \sigma(E)}{\partial E}\right]_{E=E_F}. \qquad (3)$$

If the effective mass m^* in the conductivity $\sigma = ne^2\tau/m^*$ is independent of energy E,

$$S(T) = -\frac{\pi^2}{3}\frac{k_B^2 T}{|e|}\left[\frac{\partial \ln n}{\partial E} + \frac{\partial \ln \tau}{\partial E}\right]_{E=E_F}, \qquad (4)$$

where n and τ are the density and relaxation time of conduction electrons, respectively. Jones *et al.* [10] assumed the 1D band, $n(E) \propto E^{1/2}$, and obtained that $E_F \sim 0.1$ eV as long as the last term of (4) is neglected. They pointed out that E_F is not much larger than the thermal energy, 300 k_B, therefore the relaxation time τ cannot be approximated as independent of energy E.

This difficulty cannot be removed even if a 2D free electron system is assumed. It must be remarked that near room temperature $S(T)$ is not proportional to T and the Hall coefficient R_H is still dependent on temperature though only weakly; the gradual change in the carrier density cannot be neglected even in this temperature range.

The sign of $S(T)$ is the same as that of the Hall coefficient both above and below T_m; the charge of major carriers is positive above T_m and negative at low temperature. Again a change in the electronic structure is undoubtedly occurring near T_m, though not necessarily corresponding to a phase transition.

4.4. Mechanism of Transport Anomalies

Two possible mechanisms have been proposed of the transport anomalies.

The first is that density and/or mobility of conduction electrons is modified near T_m by a phase transition, e.g. formation of the charge or spin density wave or the Jahn—Teller like distortion or transition into the excitonic insulator state. Because the family compounds, $NbSe_3$ and TaS_3, show the CDWs at low temperature, the phase transitions characteristic of low-dimensional conductors have been looked for. The peaks of resistivity, changes in sign of the Hall coefficient and of the thermoelectric power have been interpreted as indicating the formation of the CDW. However, changes of these properties are rather sluggish even in samples of good quality. No new modulation of the crystal structure has been observed by X-ray and electron diffraction experiments below T_m. Several types of modulation, explained in Section 3.2, have been observed but even at room temperature. Therefore, they are not related to the transport anomalies. Referring to the band structure calculated by Whangbo *et al.* [14], the wave vector of the modulation associated with the Fermi surface nesting would be very small. Extra reflection associated with it would be observable in electron diffraction and powder diffraction methods. A change in the diffraction patterns related to the band Jahn—Teller distortion or expected in an excitonic insulator, if any, could be detected more easily.

The second possibility is that the transport anomalies are not associated with a phase transition but a semimetallic band structure, with the characteristic energy comparable with $k_B T$, is responsible for these anomalies. For example, if the conduction electron density increases with temperature by thermal excitation across a small gap and their mobility is limited by phonon scattering, conductivity shows a maximum (resistivity minimum) at an intermediate temperature. Because the conductivity is metallic and the Schubnikov de Haas effect (cf. Section 5.2) can be observed, both $ZrTe_5$ and $HfTe_5$ are expected to be semimetallic and a finite carrier concentration remains at low temperature.

Assume that the band structure near the Fermi level consists of an electron band and a pair of semiconducting ones separated by a gap E_g. The conductivity σ is a sum of the contribution from the electron band (σ_1) and from the semiconducting region (σ_2); if the interband scattering is neglected,

$$\begin{aligned}\sigma &= \sigma_1 + \sigma_2 \\ &= |e|^2 \{n_1(T^\alpha + \text{const})^{-1} + n_2 T^{-\beta} \exp(-E_g/2k_B T)\}\end{aligned} \quad (5)$$

where n_1, n_2, α, and β are constants. An example of model calculations is shown in Figure 18 compared with the measured ρ_a. Agreement with the experimental result can be improved by, for example, allowing a slight temperature dependence to α and β in (5). The experimental finding, that $\rho(\max)/\rho(\sim 300\ \text{K})$ is sample dependent, suggests that it is sensitive to the carrier concentration and/or the magnitude of E_g; a slight modification by impurities or off-stoichiometry would change the band parameters and the second term of (5). Temperature dependences of the Hall

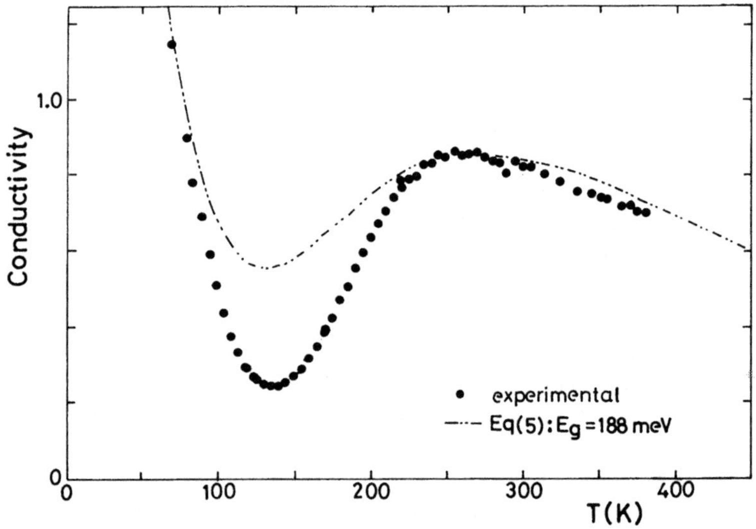

Fig. 18. Parameter fitting for the conductivity of $ZrTe_5$.

coefficient and the thermoelectric power can be reproduced, at least qualitatively, by the same model. A variant of this model can reproduce the broad peak of the heat capacity at T_m (cf. Section 8.2). The remaining difficulty is that in the band structure calculated by Whangbo *et al.* [14] no semiconducting pair can be found.

From the magnetic susceptibility (cf. Section 6) a pair of bands separated by a small gap is expected along the b^* and/or c^* direction(s). It may be possible that the band crossing at some point(s) in the Brillouin zone is removed by the spin-orbit coupling, leaving a small gap. In such a case, the Fermi surface may become open along some direction(s).

It must be stressed that the possibility of a phase transition is not completely ruled out. Especially the strong frequency dependence (cf. Section 4.1.4) of the resistivity is difficult to understand within the single particle picture. The question, then, is: "What type of collective condensation is occurring?"

5. Band Structure and Fermi Surface

5.1. BAND STRUCTURE CALCULATION

The electronic band structure of $ZrTe_5$ was calculated independently by Bullett [13] and Whangbo *et al.* [14].

Bullett [13] obtained a semiquantitative band picture by a nonempirical atomic orbital method. In most parts of the first Brillouin zone, the empty conduction band (mainly Zr *d* states) is separated from the filled valence band (mainly Te *p* states). The semiconducting gap is ~1 eV. Band crossing is observed along the [0, *k*, 0] and [1, *k*, 0] directions, which results from coupling between neighboring $ZrTe_5$ layers. The density of states near the Fermi level is very small; a semimetallic picture is appropriate. From these results the CDW instability is not likely to be expected in $ZrTe_5$ and, presumably, in $HfTe_5$ also; the transport anomalies should be explained in terms of the strong temperature dependence of carrier concentration and/or mobility.

Whangbo *et al.* [14] used the tight-binding approach based on the extended Hückel method. In the band structure for a 2D $ZrTe_5$ layer, two branches are located near the Fermi level. They are separated by a small band gap of ~0.05 eV. The hole-like band is mainly Te $5p_y$ on the trigonal prism, while the higher is Te_{III} (in the Te_2 chain) $5p_y$ at the zone center. The Zr *d* bands overlap with the latter but are located ~1 eV higher than the Fermi level. The 3D structure is shown in Figure 19. Two bands are crossing the Fermi level; one is electron-like and another hole-like. The Fermi surface has two pockets with approximately equal sizes. Both are cigar-shaped but more flattened along the a^* axis; the axial dimensional are: $k_a \sim 10^{-2}(a^*/2)$, $k_b \sim 0.35(b^*/2)$, and $k_c \sim 10^{-1}(c^*/2)$. These values are in good agreement with those determined by the Schubnikov de Haas measurements [16] (next section). The band masses of these two bands are not significantly different; the question is still open why only one pocket has been observed experimentally. Another point to be mentioned is that the band structure calculation is based on

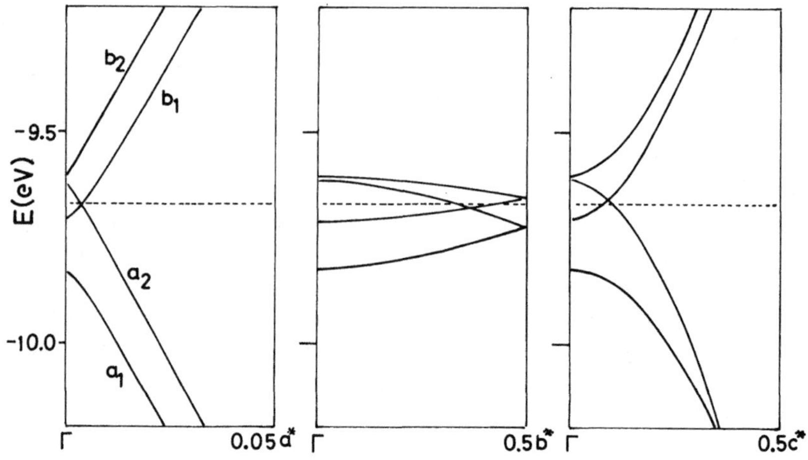

Fig. 19. Band structure of $ZrTe_5$. (After Whangbo *et al.* [14].)

the crystal structure determined by Furuseth *et al.* [2]. It is not clear whether the structure modulation (cf. Section 3.2) has no effect on the geometry of the Fermi surface. For example, the radial dimension k_b may be altered by the band folding in the superstructure observed by Di Salvo *et al.* [6]. The effect of the magnetic breakdown was not found in the Schubnikov de Haas data by Izumi *et al.* [16], however.

In spite of quantitative differences, both calculations agree in: (1) $ZrTe_5$ is a semimetal, (2) wave functions near the Fermi level are mainly of Te *p* states, and (3) the CDW instability is not expected. Whangbo *et al.* predicted a stronger effect of chalcogen substitution in contrast with the smooth variation of the resistivity with substitution of the metal atom. Such a study has not yet been tried, but careful examination of the stoichiometry is indispensable for this kind of experiment. A detailed discussion on band structure will be found in the article by Whangbo in this volume.

5.2. Schubnikov de Haas Effect

At low temperature the electrical resistance of a pure metal changes oscillatorily in a high-magnetic field (B). Its period in the reciprocal of the field $\Delta(B^{-1})$ is related to the extremal cross-sectional area A_0 of the Fermi surface by

$$\Delta(B^{-1}) = (2\pi e/\hbar)A_0^{-1}. \tag{6}$$

Izumi *et al.* observed the Schubnikov de Haas oscillation superposed on the positive transverse magnetoresistance in both $ZrTe_5$ and $HfTe_5$ [15, 16]. The modulation technique, with a 33 Hz ac magnetic field of 2.2×10^{-2} T, was used and the result was compared with the dc data.

In both $ZrTe_5$ and $HfTe_5$ only one oscillation period is observed. Note that at least two different periods are expected from the calculated band structure [14], or if the two-band model corresponds to the presence of plural pieces of the Fermi surface.

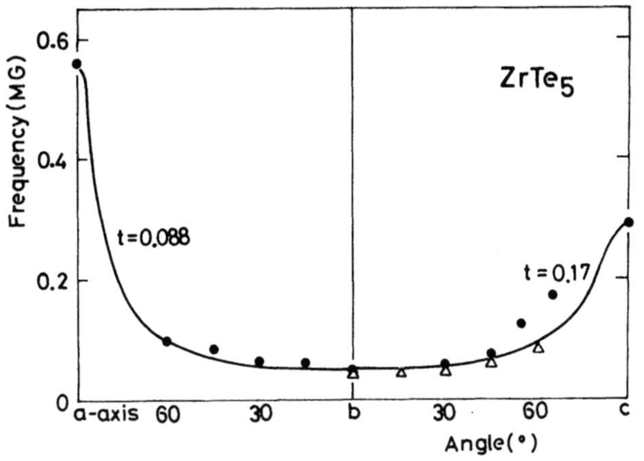

Fig. 20. Schubnikov de Haas oscillation frequency of $ZrTe_5$. (After Izumi *et al.* [16].)

From the oscillation frequency in (B^{-1}) for a field parallel to the b axis, A_0 parallel to the a^*—c^* plane was obtained:

	$ZrTe_5$	$HfTe_5$
$A_0/(a^*c^*)$	6×10^{-4}	2×10^{-4}

Figure 20 shows the angular dependence of the oscillation frequency in (B^{-1}) for $ZrTe_5$ in the a^*—c^* and a^*—b^* planes. The observed Fermi surface is approximately ellipsoidal. Solid lines in the figure are the best fits to an ellipsoidal:

$$\text{Frequency} \propto (\cos^2 \theta + t^2 \sin^2 \theta)^{-1/2},$$

where t^2 is equal to m_c^*/m_b^* or m_a^*/m_b^*. From the temperature-dependent oscillation amplitude when the magnetic field B is parallel to the b axis, the cyclotron mass was found for $ZrTe_5$ as $(m_a^* m_c^*)^{1/2} = (0.04 + 0.01)m_0$, where m_0 is the free mass. Combined with the values of t^2, the band masses were obtained. The axial lengths of the ellipsoidal Fermi surface were calculated from the band masses and the relation

$$k_a^2/m_{a^*} = k_b^2/m_{b^*} = k_c^2/m_{c^*}.$$

The shape of the Fermi surface is shown schematically in Figure 21. For both $ZrTe_5$ and $HfTe_5$, the Fermi surface has cigar-like shape; the longest dimension is parallel to the b^* axis as expected from the crystal structure. The axial ratio is less than 10; these compounds are not strongly 2D. Band parameters obtained as above are summarized in Table II. The carrier density belonging to the observed Fermi surface shown in Figure 21, calculated from its volume, is in good agreement with the total carrier densities estimated from the high-field Hall data (assuming two electron bands) as shown in Table II.

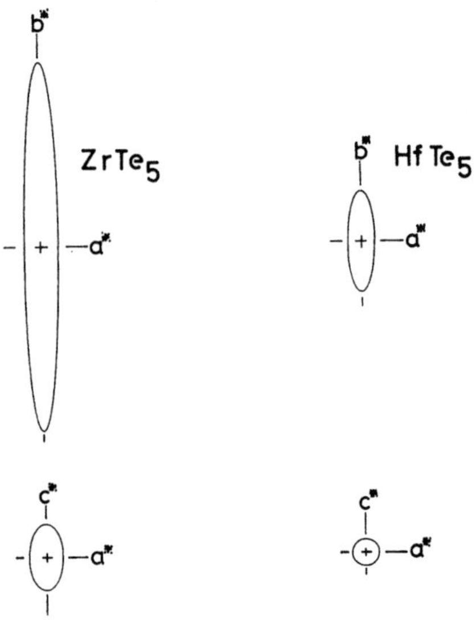

Fig. 21. Fermi surface of $ZrTe_5$ and $HfTe_5$ (schematic). (After Izumi *et al.* [16].)

TABLE II
Band parameters of $ZrTe_5$ and $HfTe_5$ determined from the Schubnikov de Haas experiments [15, 16]

	$ZrTe_5$	$HfTe_5$
Band mass (m_0 units)		
m_{a^*}	0.020	
m_{b^*}	2.7	
m_{c^*}	0.078	
Axial lengths of the Fermi surface (10^6 cm^{-1})		
k_a	0.87	0.68
k_b	9.5	2.6
k_c	1.7	0.68
Carrier density n (10^{16} cm^{-3})	51	4.0
Carrier density from the high-field Hall coefficient	65	5.9

It should be stressed that, while the shape of the observed Fermi surface is in good agreement with that calculated by Whangbo *et al.* [14], only one piece has been observed against the theoretical expectation that two sheets should appear.

6. Magnetic Susceptibility

Magnetic susceptibility, $\varkappa$, was measured first by Furuseth *et al.* [2]. Orientation of the samples was not specified in their paper. Both $ZrTe_5$ and $HfTe_5$ are diamagnetic; $\varkappa(HfTe_5) = -3.8 \times 10^{-7}$ emu g^{-1} and $\varkappa(ZrTe_5) = -4.7 \times 10^{-7}$ emu g^{-1} at 300 K. Above ~100 K the absolute values decrease with temperature and are temperature independent above ~500 K. The temperature range was extended down to 4 K by Di Salvo *et al.* [6]. Polycrystalline samples were used. At 300 K, $\varkappa(HfTe_5) = -2.9 \times 10^{-7}$ emu g^{-1} and $\varkappa(ZrTe_5) = -2.9 \times 10^{-7}$ emu g^{-1}. The Curie tail was observed at low temperature, which is presumably due to excess metal atoms occupying the irregular sites. Absolute values of the negative susceptibilities show maxima near T_m; for $ZrTe_5$ the maximum of $|\varkappa|$ is at T_m while such a coincidence was not found in $HfTe_5$.

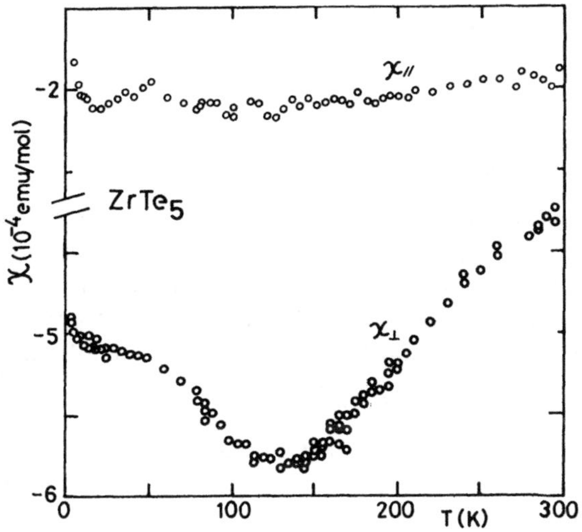

Fig. 22. Magnetic susceptibility of $ZrTe_5$. (After Okada *et al.* [7].)

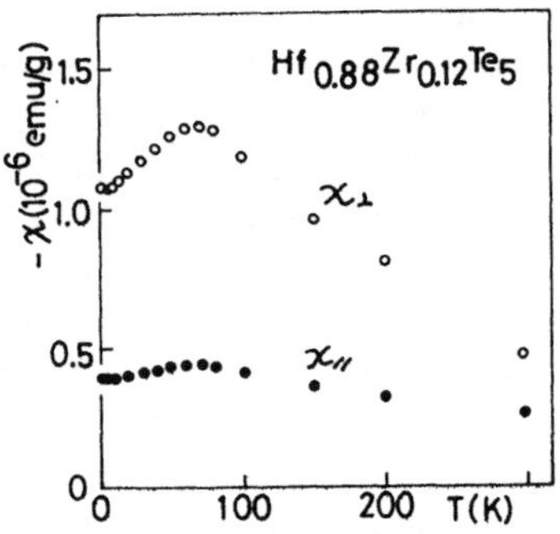

Fig. 23. Magnetic susceptibility of $HfTe_5$. (After Izumi *et al.* [8].)

Single crystal data [7, 8] are shown in Figures 22 and 23. For both compounds the susceptibility $\chi_{\parallel}$ (magnetic field B parallel to the a axis) is weakly dependent on temperature, while the absolute values of the transverse components $\chi_{\perp}$ (B parallel to the b axis for $HfTe_5$ [8], an average of B parallel to the b and c axes for $ZrTe_5$ [7]) are larger and show broad maxima centered at T_m. Room temperature values are in a reasonable agreement with the powder data, if $\chi(\text{powder}) \cong \frac{1}{3}\chi_{\parallel} + \frac{2}{3}\chi_{\perp}$. Coincidence of T_m with the temperature of the maximum $|\chi|$ was confirmed for $HfTe_5$ [8].

Izumi *et al.* [8] tried to evaluate the Landau—Peierls contribution, χ_{LP}. To estimate the ion—core diamagnetic susceptibility, χ_{ion}, they assumed the metal atom as M^{4+}, Te_I and Te_{II} as Te^{6-}, and Te_{III} as Te^{4-}; $\chi_{ion} = -9.8 \times 10^{-8}$ emu g^{-1} for $HfTe_5$. Neglecting the Pauli and Van Vleck contributions, the ratio of the effective masses of conduction electrons, m_b/m_a, was estimated as ~ 2.8 at 1.5 K from the relation

$$\chi_j \propto \left\{ \frac{(\alpha_{i \perp j}\alpha_{k \perp j})^2}{\alpha_j} \right\}^{1/3} \tag{7}$$

for the parabolic band

$$E(k) = (\hbar^2/2m_0)(\alpha_1 k_x^2 + \alpha_2 k_y^2 + \alpha_3 k_z^2). \quad [30]$$

The above value is much smaller than ~ 14.6 obtained from the Schubnikov de Haas effect (cf. Section 5.2). In fact, the difference becomes even larger when a finite value of the positive Pauli susceptibility χ_P is taken into account. In a 3D system χ_P is of the same order as $|\chi_{LP}|$. In the free electron approximation $\chi_{LP} = -(\frac{1}{3})\chi_P$. When the anisotropy of the effective mass is taken into account explicitly,

$$\chi = (m_{3D}^*/m_0 - (\tfrac{1}{3})m_0/m_{2D}^*)\chi_P^0 \tag{8}$$

where $m_{3D}^* = (\alpha_1\alpha_2\alpha_3)^{-1/3}m_0$ is the effective mass, m_{2D}^* given by (7) is the average perpendicular to the magnetic field, and χ_P^0 corresponds to the free electron system with equal electron density.

As long as the additivity of χ_{LP} and χ_P is valid, the negative value of χ comes from the small effective masses, and its anisotropy from the anisotropic m_{2D}^*. Serious difficulty is found, however, when the measured susceptibility is compared with the Schubnikov de Haas data (Table II). From the carrier density n and the size of the observed Fermi surface, $\chi_P^0 = 2.3 \times 10^{-6}$ emu mole^{-1}, $m_{3D}^*/m_0 = 0.16$, $(\alpha_2^2\alpha_3^2/\alpha_1)^{1/3} = 0.77$ and $(\alpha_1^2\alpha_3^2/\alpha_2)^{1/3} = 103$ for $ZrTe_5$. In spite of the unreliability of the ion-core diamagnetic susceptibility, the calculated value of $|\chi| = 7.9 \times 10^{-5}$ emu mole^{-1} is much smaller than the measured $|\chi_{\perp}| = 5.4 \times 10^{-4}$ emu mole^{-1} and $|\chi_{\perp} - \chi_{\parallel}| = 3 \times 10^{-4}$ emu mole^{-1}. Note that when the measured susceptibility is corrected for ion-core diamagnetism the difference becomes even larger. The decrease of $|\chi_{\perp}|$ with increasing temperature should be assigned to the increase of m_{2D}^* as a result of, for example, thermal excitation of electrons into a band of heavier mass. In this case the band structure along the a axis must be more complicated than that calculated by Whangbo *et al.*, because such an effect is not observed in $\chi_{\parallel}$ which is constant down to low temperature; variation of χ_{LP} with temperature is partially cancelled by that of $\varkappa_P$. Another possibility is that a larger piece of the Fermi surface is missed in the Schubnikov de Haas measurement and, as the result, comparison with the susceptibility data is not feasible.

It is well known that Bi and its alloys exhibit anisotropic and temperature-dependent diamagnetic susceptibility. Pure Bi is a semimetal; its Fermi surface consists of a hole pocket at the point T (along the trigonal axis) and electron pockets at points L. With increasing electron concentration the hole pocket shrinks and $|\varkappa_{\parallel}|$ (parallel to the trigonal axis) increases and shows a maximum when the hole pocket vanishes. With decreasing electron concentration by alloying Pb, $|\varkappa_{\perp}|$ shows a maximum when electron pockets vanish. Fukuyama and Kubo [29] pointed out that in heavy element materials $\varkappa_{LP}$ and $\varkappa_P$ are not separable because the spin-orbit coupling is important. They calculated the electronic susceptibility for a pair of two bands separated by a small direct gap and found a diamagnetic term which shows logarithmic divergence if the Fermi level lies within the gap. Temperature dependence comes from a shift of the Fermi level by redistribution of electrons with temperature.

Comparing MTe_5 with Bi, the similarity is apparent. The susceptibility $\varkappa_{\parallel}$ of MTe_5 is temperature independent; in fact the small change in the measured $\varkappa_{\parallel}$ may be due to a slight misorientation of samples. Band structure along the a^* axis has no special feature. On the other hand, $\varkappa_{\perp}$ for both $ZrTe_5$ and $HfTe_5$ is strongly temperature dependent. Since the constituent elements are heavy, the spin-orbit coupling is expected to play an important role; separation of conduction electron susceptibility into $\varkappa_P$ and $\varkappa_{LP}$ is not plausible. Although we must be content, at present, with the expectation that the complicated temperature dependence will be explained in terms of a semimetallic band structure, it must be stressed that an interpretation by assuming a phase transition is more difficult.

7. Raman Scattering

Raman scattering was studied for $ZrTe_5$ by Zwick *et al.* [12], Taguchi *et al.* [19], and Landa *et al.* [22, 23] and for $HfTe_5$ by Taguchi *et al.* [18] and Landa *et al.* [23], over a wide temperature range. The Toulouse group used back-scattering geometry with the nearly normal incident beams onto a freshly cleaved (010) [12] or (001) [22, 23] surface. With this geometry $k = 0$ phonons are observed. Taguchi *et al.* [18, 19] used the forward-scattering geometry to detect $k \neq 0$ phonons. Partly because of the differences in the geometry, some differences are found in the data from the two groups. Moreover, some peaks are assigned differently.

Nevertheless, there is no ambiguity in that no new peak appears near T_m in $ZrTe_5$ or $HfTe_5$; the transport anomalies are not associated with a change in the structure. As shown in Figure 24 for $ZrTe_5$, several peaks exhibit strong temperature dependence in their intensity and profile, which can be explained as due to a gradual change in the band structure with temperature [22]. For example, intensities of $\sim 147\ \mathrm{cm}^{-1}\ A_g$ and $\sim 86\ \mathrm{cm}^{-1}\ B_{2g}$ peaks decrease rapidly with increasing temperature. These lines correspond to vibrational modes of Te_{III} chains, which are responsible for the metallic nature of MTe_5 (cf. Section 5.1).

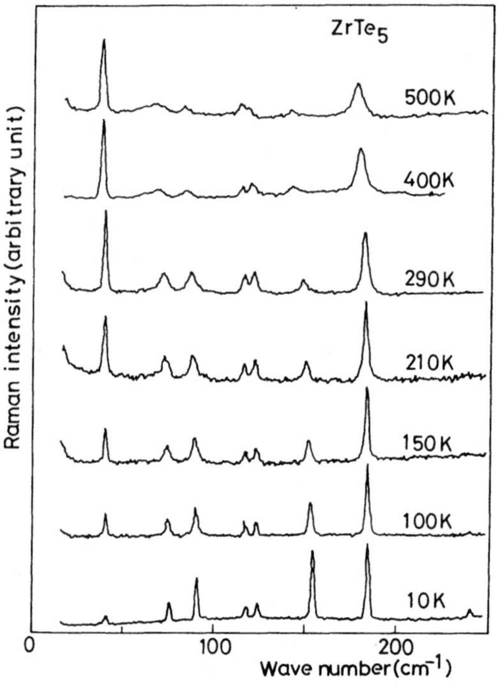

Fig. 24. Raman scattering spectra of $ZrTe_5$. (After Landa *et al.* [22].)

8. Elastic Anomaly and Specific Heat

8.1. Elastic Anomaly

The Young's modulus of $ZrTe_5$ was measured by a vibrating reed technique [21]. Single crystals were mounted as cantilevers by gluing one end to a rod and the fundamental flexural resonant frequency $f \propto (E/\rho)^{1/2}$ was measured. By a refined electronic circuit, high sensitivity $\Delta f/f \sim 2 \times 10^{-5}$ was attained.

At room temperature, the Young's modulus E is of the order of 1.1×10^{12} dyne cm^{-2}, which is comparable with that of $NbSe_3$. As shown in Figure 25, the Young's modulus shows a smooth variation near T_m without any singular behavior; the upper bound, if any, of $\Delta E/E$ is estimated to be less than 3×10^{-4}. On the other hand, a sharp decrease is observed at 84 K; $\Delta E/E \sim 4 \times 10^{-3}$. No change in internal friction is observable in this temperature range. Though the samples used were confirmed as orthorhombic by X-ray, the change at 84 K is presumably associated with the monoclinic crystallites stuck to the host crystal. A small resistivity anomaly has been observed in the monoclinic crystal near that temperature (cf. Section 3.2).

In general, this technique is quite powerful for detecting a structural change; therefore, the absence of elastic anomaly near T_m provides further evidence that the transport anomalies are not associated with a phase transition.

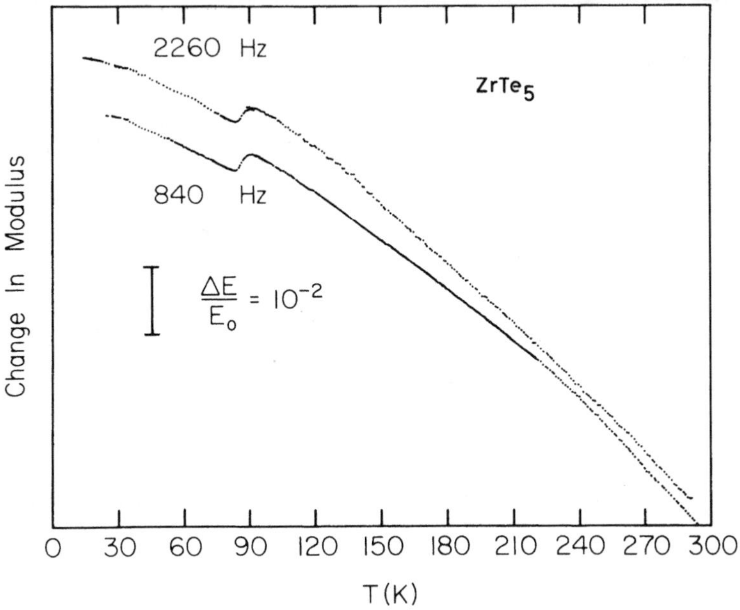

Fig. 25. Young's modulus of $ZrTe_5$. (After Brill and Sambongi [21].)

8.2. Specific Heat

The specific heat of $ZrTe_5$ was measured [24] between 60 and 190 K, by a high-resolution dc relaxation method. Large single crystals, assigned as orthorhombic by X-ray, were used.

There appear two broad maxima at ~135 and ~90 K, as shown in Figure 26(a). The magnitudes of these peaks are small; both are ~2% of the background. No latent heat is associated with these peaks. The temperature width of the high-temperature peak is ~40 K, comparable with the widths of transport anomalies.

Since $ZrTe_5$ is a semimetal, the background is mainly of the lattice heat capacity. As the Young's modulus is continuous and does not show any singularity near T_m, the heat capacity peak near T_m is not attributable to a change in the phonon system. The semimetallic band model, explained in Section 4.4, shows a structure at $k_B T \sim E_g$ but its shape is sensitive to the density of the electronic states curve near the gap. For example, if the density of states, $D(E)$, is constant outside the gap, the heat capacity shows a step near $k_B T \sim E_g$ (Figure 26(b)). A heat capacity peak is retained if the density of states curve has peak(s) near the gap, as in a two-level system (Figure 26(c)). Such a structure can be realized if mixing of two bands occurs at the band crossing, as illustrated in Figure 26(d)).

The peak at ~90 K is attributable to the monoclinic phase. Although the sample was assigned as orthorhombic by X-ray, an inclusion of the monoclinic polytype cannot be ruled out. Because the X-ray absorption coefficients of the constituent

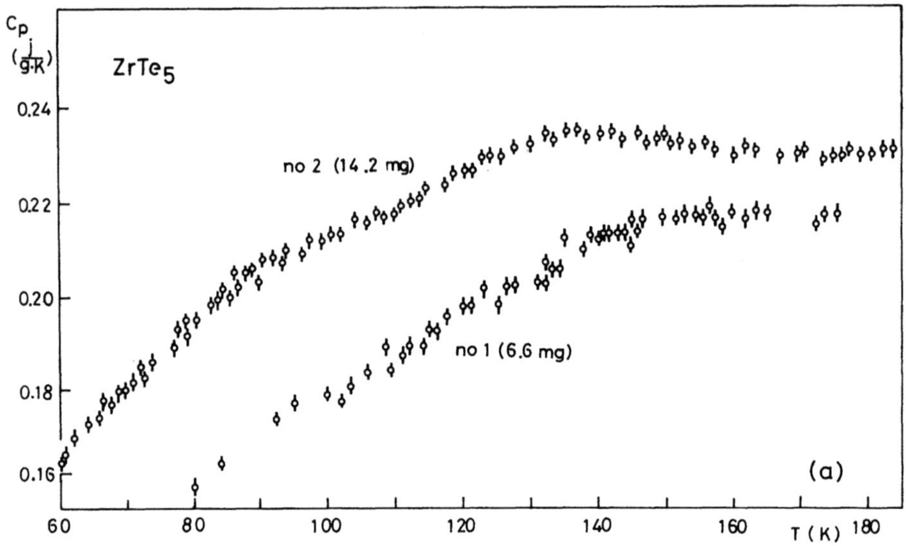

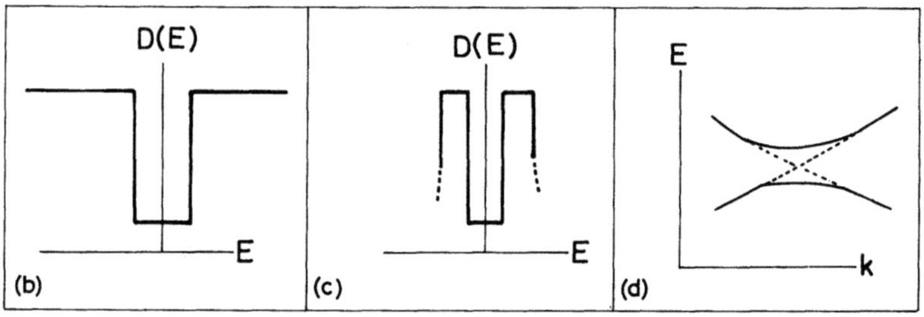

Fig. 26. (a) Specific heat of $ZrTe_5$ for two samples (after Biljakovic and Smontara [24]); (b)—(d) see text.

elements are large, the diffracted beam comes mainly from near the edges or surfaces of the sample. Therefore, an inclusion of other phase(s) may be missed by an X-ray technique.

9. Pressure Effect

Resistivity and thermoelectric power were measured under pressure by Fuller *et al.* [20]. A clamp cell was used to pressurize the sample up to 17 kbar. The pressure was calibrated against the superconducting transition temperature of Sn. Effects of the pressure were confirmed to be reversible; no permanent degradation was found in repeated measurements with decreasing pressure.

The thermoelectric power does not show a large pressure effect. For both $ZrTe_5$ and $HfTe_5$ the zero-crossing temperature ($\sim T_m$) and the absolute values increase only slightly under pressure. On the other hand, the change in resistivity is remarkable. For both compounds T_m shifts higher under pressure. The room temperature

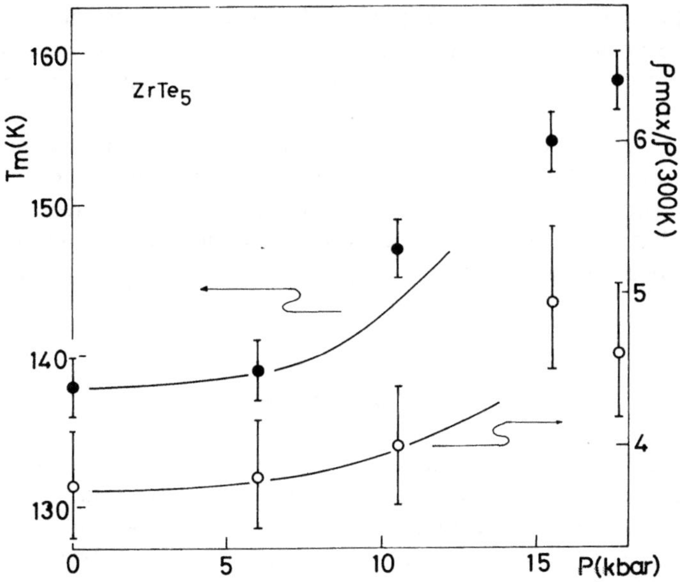

Fig. 27. Pressure dependences of T_m and $\rho(\max)/\rho(\sim 300\ K)$ for $ZrTe_5$. (Reproduced from Fuller *et al.* [20].)

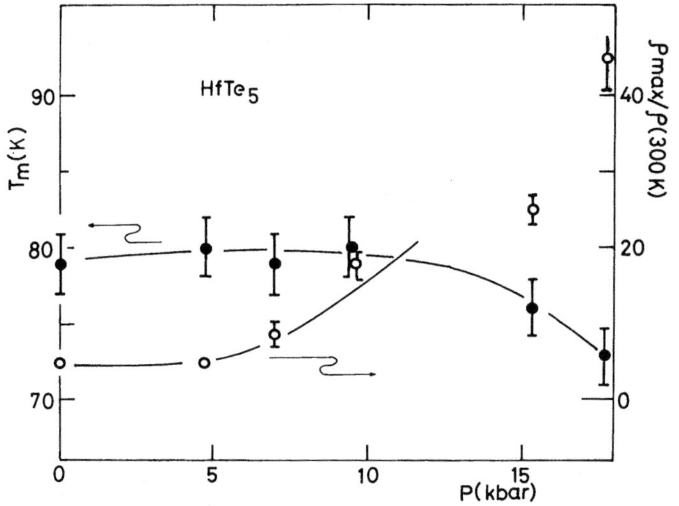

Fig. 28. Pressure dependences of T_m and $\rho(\max)/\rho(\sim 300\ K)$ for $HfTe_5$. (Reproduced from Fuller *et al.* [20].)

resistivity of $ZrTe_5$ decreases while it remains constant for $HfTe_5$. In normal metals the impurity resistivity increases and the phonon resistivity decreases under pressure, but the decrease of the latter in $ZrTe_5$, by a factor of 8, is quite remarkable. Both the residual resistivity and $\rho(\max)/\rho(\sim 300\ K)$ increase in $HfTe_5$ under pressure. In Figures 27 and 28, T_m and $\rho(\max)/\rho(\sim 300\ K)$ are reproduced from the tabulated values in their paper [20]. Though not conclusive, the pressure coefficients seem to be nonlinear; for $ZrTe_5$ both T_m and $\rho(\max)/\rho(\sim 300\ K)$ show a clear increase above ~ 5 kbar while for $HfTe_5$ both are essentially constant below 10 kbar. As is well known, the actual pressure within the usual clamping cell is temperature dependent due to the differential thermal expansion of the cell and the pressure-transmitting fluid. Therefore, it is not clear whether the nonlinear pressure effect is intrinsic or due to the incomplete pressure scale. Moreover, because the compressibility has not been measured in either compound, the effects of possible nonlinear and anisotropic changes in the lattice dimensions have not been separated.

10. Mixed Systems

A mixed system $Zr_{1-x}Hf_xTe_5$ was examined by Di Salvo *et al.* [6]. As shown in Figure 29, the resistance varies smoothly with Hf concentration; the peak shifts to lower temperature and the residual resistance increases with increasing x ($x <$ 0.5). The residual resistivity of $HfTe_5$ is larger; $HfTe_5$ synthesized by [6] is presumably off-stoichiometric. In Figure 30, T_m is plotted against x. Data by Izumi *et al.* [8] are also shown in the figure. The poor agreement comes from a large deviation of the actual Hf concentration from the nominal ones; the values of x by [6] are nominal while those of Izumi *et al.* [8] were determined by the inductively coupled plasma optical emission technique. It is not surprising that the actual composition of mixed crystals deviates from the nominal value especially in single

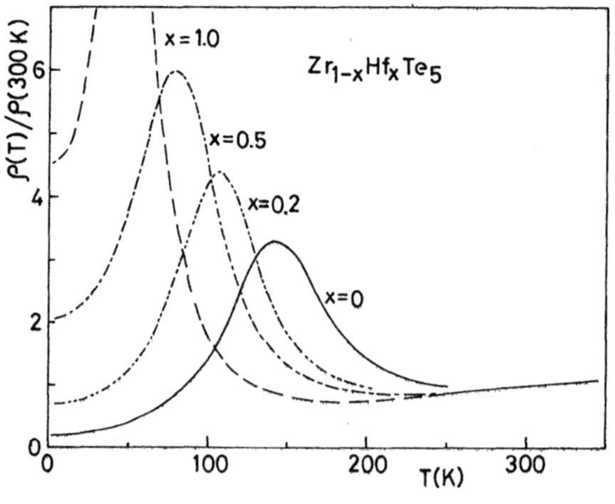

Fig. 29. Resistivity of $Zr_{1-x}Hf_xTe_5$. (After Di Salvo *et al.* [6].)

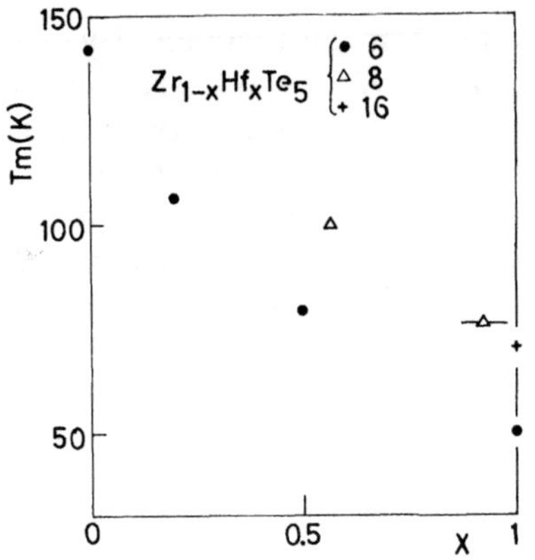

Fig. 30. Concentration dependence of T_m for $Zr_{1-x}Hf_xTe_5$.

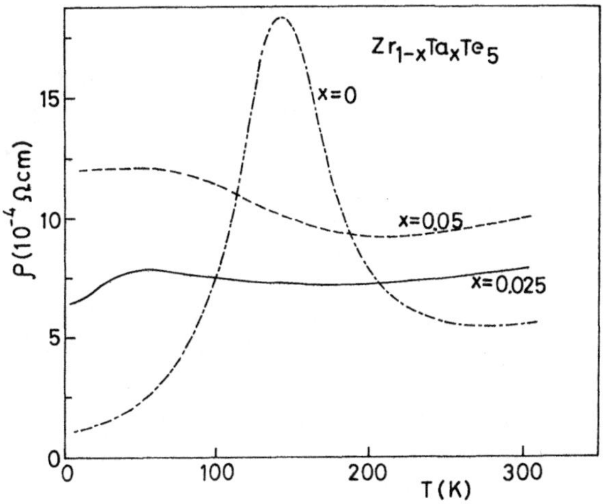

Fig. 31. Resistivity of $Zr_{1-x}Ta_xTe_5$. (After Di Salvo *et al.* [6].)

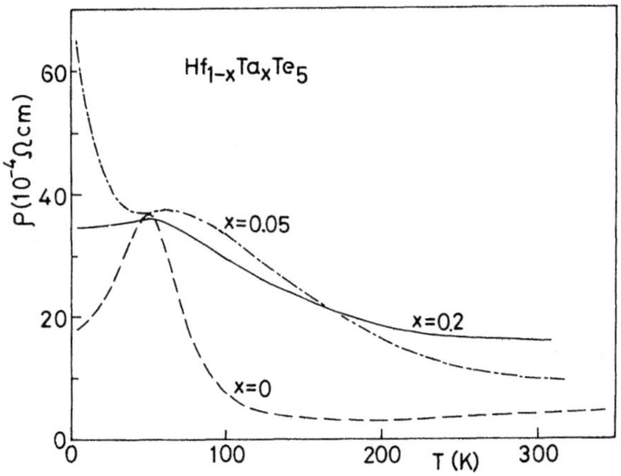

Fig. 32. Resistivity of $Hf_{1-x}Ta_xTe_5$. (After Di Salvo *et al.* [6].)

crystals prepared by the vapor transport method. By extrapolating T_m by [8] to $x = 1$, T_m of $HfTe_5$ is obtained as 70 K, in good agreement with that of $Hf_{1.00}Zr_{0.004}Te_5$ [16]. In the $Zr_{1-x}Hf_xTe_5$ system the effect of alloying is not so remarkable in contrast with the observed effects in typical 1D systems. In the latter even a low level of isoelectronic impurity atoms is sufficient to modify the CDW transition and the associated changes in physical properties. The gradual change in the resistivity of $Zr_{1-x}Hf_xTe_5$ can presumably be explained by shifts of electronic band parameters with alloying.

The effect of alloying Ta in $ZrTe_5$ and in $HfTe_5$ is quite remarkable [6]. As shown in Figures 31 and 32, the resistivity peak is suppressed by a small concentration of Ta, not inconsistent with the above idea that modification of the band structure is the main effect of alloying.

Ackowledgements

The author is grateful to: Acta Chemica Scandinavia for kind permission to reproduce the data given in Figure 1 from [2]; The American Institute of Physics for kind permission to reproduce the data given in Figure 19 from [14] and in Figures 29—32 from [6]; The Physical Society of Japan for kind permission to reproduce the data given in Figure 3 from [4], in Figures 10, 11, and 22 from [7], and in Figure 25 from [21]; Gordon & Breach Science Publishers Ltd for kind permission to reproduce the data given in Figures 13 and 23 from [8]; Les Editions de Physique for kind permission to reproduce the data given in Figures 20 and 21 from [16], and in Figure 27 and 28 from [20]; Pergamon Press for kind permission to reproduce the data given in Figure 2 from [11], in Figure 4 from [5], in Figures 12(a) and (b) from [17], in Figures 14 and 15 from [9], in Figure 16 and 17 from [10], and in Figure 24 from [22]; and K. Biljakovic and A. Smontara for kind permission to reproduce their data in Figure 26(a).

References

1. L. Brattås and A. Kjekshus, *Acta Chem. Scand.* **25**, 2783 (1971).
2. S. Furuseth, L. Brattås, and A. Kjekshus, *Acta Chem. Scand.* **27**, 2367 (1973).
3. F. Hulliger, *Structural Chemistry of Layer-Type Phases* (ed. F. Lévy), D. Reidel, Dordrecht (1976).
4. S. Okada, T. Sambongi, and M. Ido, *J. Phys. Soc. Japan* **49**, 839 (1980).
5. M. Izumi, K. Uchinokura, and E. Matsuura, *Solid State Commun.* **37**, 641 (1981).
6. F. J. Di Salvo, R. M. Fleming, and J. V. Waszczak, *Phys. Rev.* **B24**, 2935 (1981).
7. S. Okada, T. Sambongi, M. Ido, Y. Tazuke, R. Aoki, and O. Fujita, *J. Phys. Soc. Japan* **51**, 460 (1982); S. Okada, M.Sc. Thesis, Hokkaido Univ. (Sapporo) (1981).
8. M. Izumi, K. Uchinokura, S. Harada, R. Yoshizaki, and E. Matsuura, *Mol. Cryst. Liq. Cryst.* **81**, 141 (1982).
9. M. Izumi, K. Uchinokura, E. Matsuura, and S. Harada, *Solid State Commun.* **42**, 773 (1982).
10. T. E. Jones, W. W. Fuller, T. J. Wieting, and F. Lévy, *Solid State Commun.* **42**, 793 (1982).
11. E. F. Skelton, T. J. Wieting, S. A. Wolf, W. W. Fuller, D. U. Gubser, T. L. Francavilla, and F. Lévy, *Solid State Commun.* **41**, 1 (1982).

12. A. Zwick, G. Landa, R. Carles, M. A. Renucci, and A. Kjekshus, *Solid State Commun.* **44**, 89 (1982).
13. D. W. Bullett, *Solid State Commun.* **42**, 691 (1982).
14. M. H. Whangbo, F. J. Di Salvo, R. M. Fleming, *Phys. Rev.* **B26**, 687 (1982).
15. M. Izumi, S. Harada, R. Yoshizaki, K. Uchinokura, and E. Matsuura, *Physica* **117—118B**, 605 (1983).
16. M. Izumi, K. Uchinokura, R. Yoshizaki, S. Harada, T. Nakayama, A. Yamada, and E. Matsuura, *J. Phys. C,* **3**, 1705 (1983).
17. C. M. Jackson, A. Zettl, G. Grüner, and F. J. Di Salvo, *Solid State Commun.* **45**, 247 (1983).
18. I. Taguchi, A. Grisel, and F. Lévy, *Solid State Commun.* **45**, 541 (1983).
19. I. Taguchi, A. Grisel, and F. Lévy, *Solid State Commun.* **46**, 299 (1983).
20. W. W. Fuller, S. A. Wolf, T. J. Wieting, R. C. Lacoe, P. M. Chaikin, and C. Y. Huang, *J. Phys. C,* **3**, 1709 (1983).
21. J. W. Brill, and T. Sambongi, *J. Phys. Soc. Japan* **53**, 20 (1984).
22. G. Landa, A. Zwick, R. Carles, M. A. Renucci, and S. Kjekshus, *Solid State Commun.* **49**, 1095 (1984).
23. G. Landa, A. Zwick, R. Carles, M. A. Renuccil, and A. Kjekshus, *Solid State Commun.* **50**, 297 (1984).
24. K. Biljakovic and A. Smontara, private communication; T. Sambongi, K. Biljakovic, A. Smontara, and L. Guemas, (to appear in *Synthetic Metals*).
25. J. Ziman, *Electrons and Phonons*, Clarendon Press, Oxford (1962).
26. A. Oshiyama, K. Nakao, and H. Kamimura, *J. Phys. Soc. Japan* **45**, 1136 (1978).
27. H. C. Montgomery, *J. Appl. Phys.* **42**, 2971 (1971).
28. J. P. Maita and E. Bucher, *Phys. Rev. Lett.* **29**, 93 (1972).
29. H. Fukuyama and R. Kubo, *J. Phys. Soc. Japan* **27**, 604 (1969); **28**, 570 (1970).
30. N. F. Mott and H. Jones, *The Theory of the Properties of Metals and Alloys*, Clarendon Press, Oxford (1936).

FROM THREE-DIMENSIONAL TO ONE-DIMENSIONAL CLUSTER Mo_6 CHALCOGENIDES

R. CHEVREL and M. SERGENT

Laboratoire de Chimie Minérale B, Associé au CNRS, n⁰ 254, Avenue du Géneral Leclerc, Université de Rennes Beaulieu, 35042 Rennes Cédex, France

1. Introduction

Many compounds of low-oxidation state transition metal atoms contain discrete M_6L_8 and M_6L_{12} units [1—4] with octahedral M_6 clusters such as, for instance, the $|Mo_6Cl_8|^{4+}$ unit in the $MoCl_2$ compound and the $|Nb_6Cl_{12}|^{2+}$ unit in the Nb_6Cl_{14} compound. The M_6L_8 type displays a M_6 octahedron face capped by eight L^i ligands while the M_6 octahedron in the M_6L_{12} type is coordinated by twelve L^i ligands bridging the edges of the octahedron and forming a cuboctahedron. An example of each cluster type is shown in Figure 1. Generally, each unit is surrounded by six (maximum) other nonmetal ligands, L^a, which can be linked to the octahedral apices either in the form of further ligands or internal ligands of adjacents units (see Figure 2).

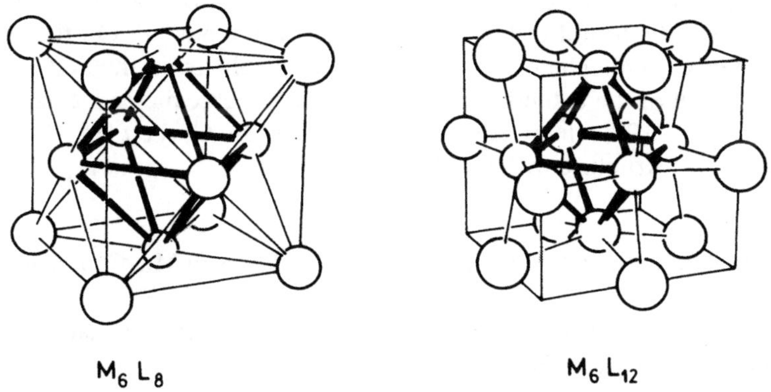

Fig. 1. Two types of environment of the octahedral M_6 clusters in the M_6L_8 and M_6L_{12} units.

The sharing of these six external ligands (L^a) (change of terminal ligands by bridging ligands) allows us to lower the ratio L/Metal, for instance: MMo_6Cl_{14}, MMo_6Cl_{13}, Mo_6Cl_{12}, $Mo_6Cl_{10}Se$ [5—7]. It is interesting to note that one of the easiest ways to obtain low-dimensional materials is the possible condensation of these discrete units. A further condensation is the sharing of L^i internal ligands along the pseudo-three-fold axis, for instance, in the Mo_6L_8 units. The corresponding compounds exhibit a certain anisotropic feature as in the one-dimensional (1D)

J. Rouxel (ed.), Crystal Chemistry and Properties of Materials with Quasi-One-Dimensional Structures, 315—373.

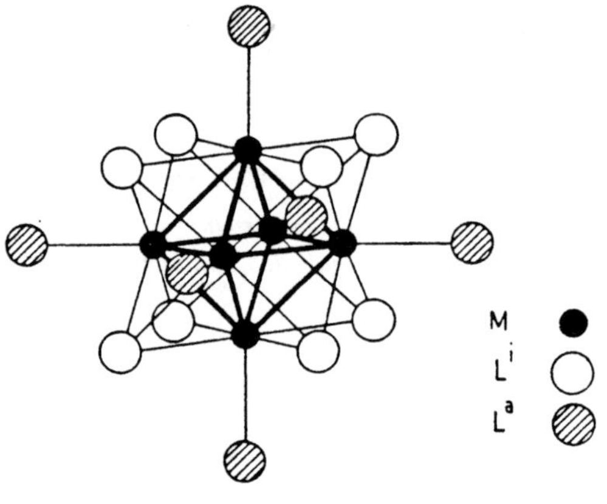

Fig. 2. Environment of the $M_6L_8^i$ unit by the six L^a ligands.

$Mo_6Br_8S_2$ chalcogenhalide and the new $Ba_2Mo_6S_7$ and $Cs_{0.6}Mo_6S_9$ chalcogenides [8—10]. Beyond the condensation of M_6L_8 units, many types of octahedral M_6 cluster condensations are reported in the metal—metal bonding literature. First, some examples show a tridimensional apex-sharing cluster condensation in M_6L_{12} units as in the well-known reduced NbO oxide, or a 1D *trans*-edge-sharing Mo_6 cluster condensation, forming linear chains, as in the new $NaMo_4O_6$ oxide [11, 12]. Furthermore, the M_6 octahedral cluster condensation in the M_6L_8 unit can be seen as a formation of chains, or 1D columns of vertex-sharing M_6 clusters in Ti_5Te_4 type and *trans*-edge-sharing M_6 clusters in the Gd_2Cl_3 type [4, 13]. These two kinds of mixed chains can be found in Nb_2Se and Ti_2S compounds [4]. On the other hand, the pairs of Gd_4Cl_6 type chains sharing edges are displayed in the Sc_7Cl_{10} type [14]. The starting point of the extended condensation of double metal sheets in ZrCl and derived types [15—17] is the result of very nice work by Corbett [18—20]. The ZrCl structure is a layer structure with graphitic properties.

2. Principle of the Cluster Condensation in Ternary Mo-Chalcogenides

In solid-state chalcogenide chemistry, two opposite methods have made it possible to get low-dimensional materials:

(1) one method is to increase the chalcogen/metal ratio in order to condense NbX_6 trigonal-prism units, for instance in $NbSe_3$ and $I_{0.33}NbSe_4$ [21], MTe_5 [22];
(2) the second method is to decrease the chalcogen/metal ratio in order to generate new extended face-sharing Mo_6 octahedral clusters. This method is the purpose of the paper.

The Mo_6X_8 basic unit of the very large family of ternary molybdenum chalcogenides with extended clusters is found in the first $M_xMo_6X_8$ compounds, the so-called Chevrel phases [23] (X = chalcogen). The Mo_6X_8 unit has only $\overline{3}$ symmetry;

the Mo_6 octahedron (two types of Mo—Mo intracluster distances: $d\overline{\text{Mo—Mo}}$ ~ 2.70 Å) is surrounded by a X_8 cube (dMo—S intraunit = 2.50 Å and dMo—Se ~ 2.60 Å).

The originality of these materials arises from the 'interconnection' of these different Mo_6X_8 units giving very striking electronic properties and making these Chevrel phases into high-temperature and high-field ternary superconductors. The tridimensional stacking of these Mo_6X_8 building blocks is illustrated in Figure 3.

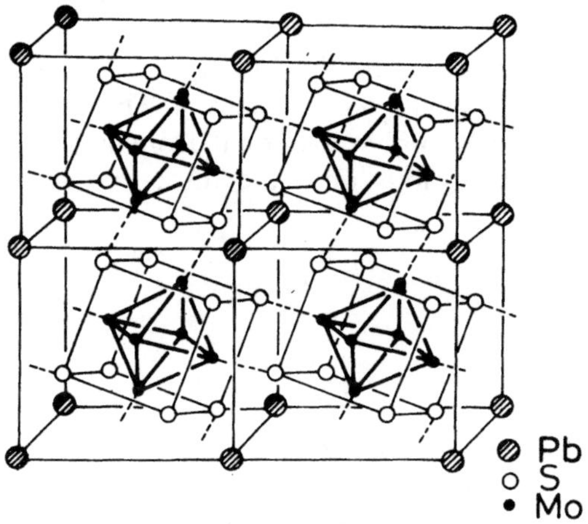

Fig. 3. View of the rhombohedral structure of the $PbMo_6S_8$ compound.

Each of the Mo_6X_8 building blocks is rotated by 27° about the only three-fold axis of the Pb cube due to the intercluster chalcogen—chalcogen repulsions in order to minimize this repulsion energy. The Pb atoms are located at the origin of the rhombohedral unit cell, thus retaining one Mo_6X_8 unit block per unit cell and forming a simple cubic array.

The M atoms do not necessarily locate at the origin of the unit cell but are often completely delocalized (in position and stoichiometry) in the sites between the Mo_6X_8 building blocks [24]. A large variety of M cations may be intercalated, often reversibly, into the Mo_6X_8 host lattice [25—27]. Such intercalations were found to exert a pronounced effect on electronic properties (charge and content of the M^{n+} counterions) but changed only little the structural arrangement (steric effect). Yet, the intercluster arrangement does not change; the clusters pack in such a way the Mo atoms, in a square-planar chalcogen coordination within a unit, 'try' to connect strongly with one chalcogen on neighboring units thus becoming base-squared pyramidal coordinated. In other words, some of the inner (i) chalcogen atoms (6/8) also function as outer apical (a) chalcogen atoms to vertices in other clusters [11].

The situation "the exo-chalcogens on neighboring block units place over all the Mo atoms in apical positions" looks much more general since it may be extended to the terminal Mo atoms on new cluster chalcogenides described below. This new type of stacking of the M_6L_8 units in Chevrel phases allows exactly the same number of ligands per formula unit as the number of internal ligands of the units: the ratio X/Mo corresponds to the ratio 8/6 = 1.33. The arrangement of the Mo_6X_8 cubic units, by means of shifted square faces, is responsible for the weak intercluster Mo—Mo coupling between the Mo_6 clusters and corresponds to a new way to chain the clusters. The Mo—Mo intercluster distances depend on the nature of the chalcogen atoms and the nature and/or the amount of M elements: they are about 3.10—3.30 Å in the sulfides, 3.30—3.50 Å in the selenides, and are 20—30% larger than the average metal—metal distance 'into' the Mo_6 cluster or in the elemental Mo metal (2.72 Å).

The simplest manner to describe the Mo_6X_8 unit (X = chalcogen) is to see it along the three-fold axis as illustrated in Figure 4. This view allows one fragmental decomposition of the Mo_6X_8 units: it is built from two Mo_3X_3 staggered planes capped by two chalcogen ligands onto the three-fold axis getting a discrete or molecular unit.

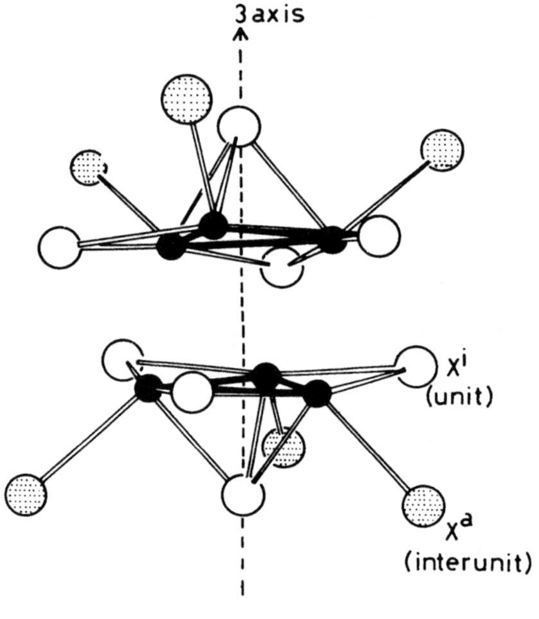

Fig. 4. Fragmental view of the $Mo_6X_8^i$ cluster unit along the three-fold axis with the six X^a ligands.

Now, it is possible to visualize the stacking of three staggered planes capped by two chalcogen atoms so as to get the new Mo_9X_{11} unit. This reduced cluster unit corresponds to a lowering of the X/Mo ratio from 8/6 = 1.33 to 11/9 = 1.22. So, it can be illustrated as arising from the 'fusion' of two Mo_6X_8 block units by sharing a Mo_3X_3 face and loss of two capped chalcogen atoms. The Mo atoms form a bioctahedron and the X_{11} polyhedron is composed of six square faces (as in the Mo_6X_8 unit) and six triangular faces in Figure 5(a). In this condensation there appears another type of environment of chalcogen ligands. The three medium

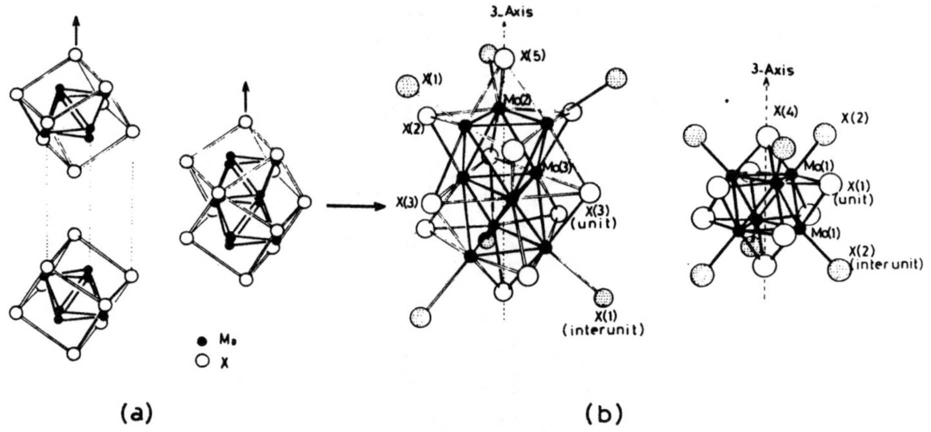

Fig. 5. View of (a) the Mo_9X_{11} unit from the 'fusion' of two Mo_6X_8 units and (b) the Mo_9X_{11} unit with the six neighboring interunit ligands comparatively to the Mo_6X_8 unit with the six X neighboring ligands.

chalcogens bridge two butterfly-shaped and edge-sharing faces of Mo atoms when the six extremal chalcogens bridge only the Mo_3 triangular faces as in the Mo_6X_8 type. The three medium molybdenum atoms have changed their number of coordinates: they are coordinated by only four butterfly-shaped chalcogen atoms when the six external Mo atoms have a base-squared pyramidal coordination. This squared base is composed of the four inner X^i chalcogens and the apex of the pyramid is constituted from one of the six exo (outer) X^a chalcogens belonging to the adjacent units. This coordination hardly looks like the Mo environment in the Chevrel phases (see Figure 5(b)). Furthermore, the constitution of six extremal square faces of chalcogen atoms allows six extremal Mo atoms in the middle of these square faces to link to six other Mo atoms of six adjacent units. Let us note that this Mo extremal atom is linked to four inner Mo atoms of the cluster and one exo Mo atom of an adjacent cluster while the medium Mo atom is linked to six inner Mo atoms of the cluster. The metallic environment of the Mo atom becomes richer and richer.

The addition of a new Mo_3X_3 plane takes away these latter intercluster Mo—Mo bonds and adds only one new central plane. This addition is illustrated in the new $Mo_{12}X_{14}$ unit in Figure 6 and results from the condensation of three Mo_6X_8 units along the three-fold axis leading to a trioctahedron. This condensation displays a unit more and more elongated along the three-fold axis and is a stepwise condensation process with the general formula $Mo_{6n}X_{6n+2}$.

The further reduction of the X/Mo ratio, by addition of new star-shaped Mo_3X_3 planes, leads to new, even more extended cluster units, $Mo_{12}X_{14}$ (X/Mo = 14/12 = 1.167) ($n = 2$), $Mo_{18}X_{20}$ ($n = 3$, X/Mo = 1.111), $Mo_{24}X_{26}$ ($n = 4$, X/Mo 3 1.083), $Mo_{30}X_{32}$ ($n = 5$, X/Mo = 1.066) where each cluster (Mo_{6n}) is a multiple of the primary Mo_6 octahedron (Figure 7).

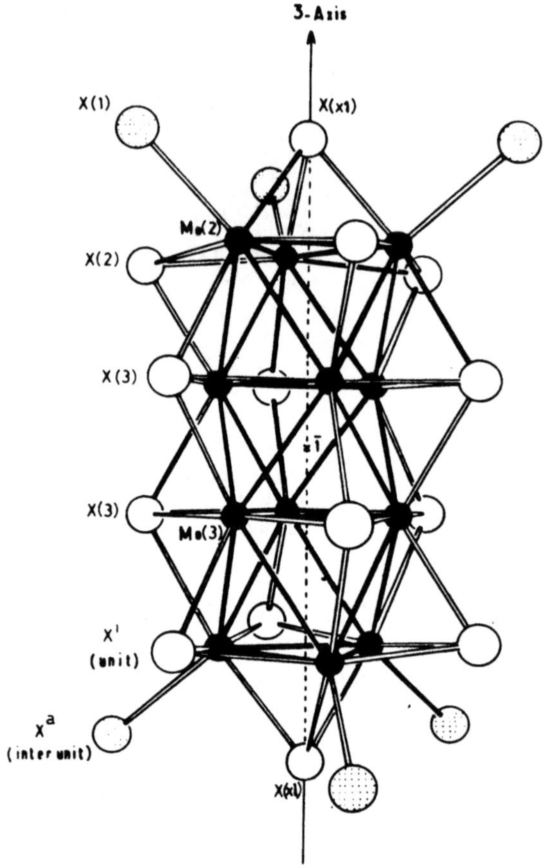

Fig. 6. The $Mo_{12}X_{14}$ unit and the six X^a further ligands.

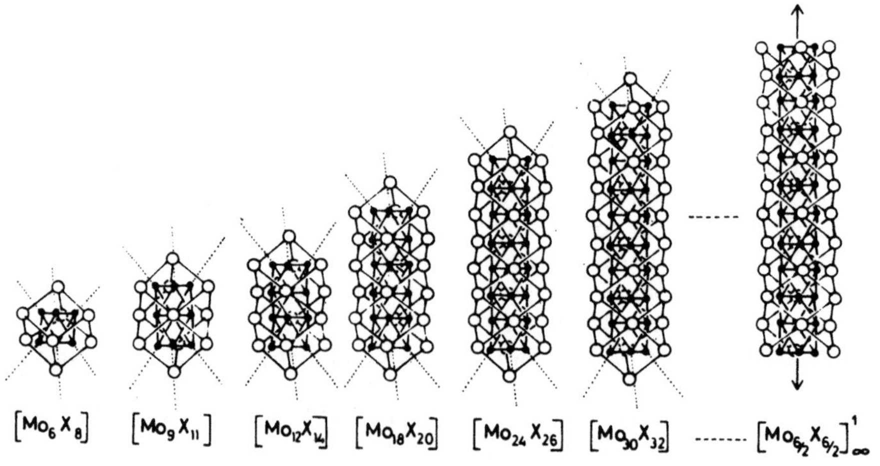

Fig. 7. The principle of face-sharing Mo_6 cluster condensation.

The Mo_6X_8, Mo_9X_{11}, $Mo_{12}X_{14}$, . . . , $Mo_{30}X_{32}$ units form a linear progression which has, as its limit $n = \infty$, the one-dimensional $(Mo_3X_3)^1_\infty$ chain (X/Mo = 1) (Figure 7). There are no extremal Mo atoms because there is no capped-chalcogen ligand on the three-fold axis and thus no face-squared ligand, and so, there are no close contacts between Mo atoms of different chains. Actually, only one type of Mo coordination remains (butterfly-shaped coordination).

3. Metal—Metal Bonding Approach in the Mo-Cluster Compounds

It is well known in the literature [28, 29] that, according to molecular orbital and band structure calculations, the M_6L_8 cluster unit — for instance in the $(Mo_6Cl_8)^{4+}$ aqueous chemistry of Mo_6Cl_{12} — provides 12 bonding states for 24 d electrons (i.e. it is then possible to assign an electron-pair bond to each Mo—Mo pair (12 edges of the octahedron)) against the M_6L_{12} cluster unit which provides 8 bonding states for 16 d electrons. The latter assigns a fractional bond order to, for instance, the Nb_6 cluster on the $(Nb_6Cl_{12})^{2+}$ cation (2/3); the former provides a single bond model (one electron pair per Mo—Mo bond). Especially interesting are the $M_xMo_6X_8$ compounds where not all of the bonding cluster states are occupied.

A way to understand the Mo—Mo bonding in these compounds has been emphasized by Corbett [30]. Really, a simple approach to the metal—metal bond order is given by the empirical Pauling expression:

$$D_{n'} = D_1 - 0.6 \log n',$$

where $D_{n'}$ is the interatomic distance for a bond with bond order n' (<1) and e^- deficient compounds. D_1 is the bond distance for $n' = 1$. This equation correlates the bond order of Me—Me bonds with the bond length. Firstly, we calculate the D_1 value from the $2R_{12}$ values of Teatum *et al.* [31] (coordinance 12) of the metal (D_1 (Mo) = 2.619 Å). The application of this formula to the Mo_6S_8-reduced Mo chalcogenide [25] proceeds as follows: each molybdenum atom has two pairs of neighbors at 2.698 Å ($(Mo—Mo)_\Delta$) and 2.862 Å ($Mo_\Delta—Mo_\Delta$) — the intracluster distances and the Mo—Mo intercluster distance at 3.084 Å, which, by the bond order equation, yield the sum of 2(0.738) + 2(0.3935) + 1(0.168) = 2.432. It is the sum of the bond orders per Mo atom. Furthermore, a simple calculation of the number of valence electrons per Mo atom (VEC: Valence Electron Concentration) yields 3.33 electrons per molybdenum atom (6 valence electrons per Mo metal: $6 \times 6 = 36e^- - 16e^-$ (S^{2-}): as there are $20e^-$ per Mo_6 cluster, we have $20/6 = 3.33e^-$ per Mo atom). With $3.33e^-$ per Mo atom left for the metal—metal bonding, the Pauling bond order per electron (PBO/e^-) is 2.432/3.33 = 0.73. Whereas the value is close to 1 in discrete clusters with extra exo atoms (as $MoCl_2$), the reduced value 0.73 of the Pauling bond order per electron results from the matrix effect due to the mixing of particularly strong Mo—chalcogen intercluster bonds and closed-shell chalcogen—chalcogen repulsions [30]. This steric effect forces an elongation of the Mo_6 trigonal antiprism and reduces the Mo—Mo bonding with respect to only the electronic effect. Another relation used by Honle *et al.*

[32] ($n' = (d_1/d_{n'})^5$) allows the bond order per electron (BO/e^-) to fit well in the $M_xMo_6S_8$ compounds ($0.95 \leqslant BO/e^- \leqslant 1.10$ as compared to $0.73 \leqslant PBO/e^- \leqslant 0.85$).

The $M_xMo_6X_8$ chalcogenides are electron deficient Mo—Mo bonding compounds where, in function of the n charge and the x content of the counterions M^{n+}, the high VEC per molybdenum atom can vary from 3.33 to $4e^-$, leading to unusual and striking physical properties [23].

This method of counting electrons on the cluster, and the study of the metal—metal bond order, will appear again in the new extended cluster compounds described below. So, it looks very interesting to show how, from the simple pseudo-molecular Mo_6X_8 unit forming a tridimensional framework, one can get new extended clusters with more and more chain-like properties.

Of course, the condensation of the Mo_6X_8 isolated clusters will be related to the lowering of the X/Mo, chalcogen/molybdenum ratio and to the low oxidation state of the molybdenum atom or the high VEC. Actually, it was known that compounds with M_6L_8 isolated clusters formed the boundary at low values of X/Mo [4]. It was presumed, therefore, that clusters of the Mo_6X_8 type would be retained as structural units on further reduction of X/Mo ratio < 1.33.

4. New Ternary Mo—Chalcogenides with Discrete Mo_{6n} Clusters (n finite)

All of these different cluster units are found in a very large family of new ternary molybdenum chalcogenides. The counterions are generally large cations with different charges and x content, especially alkaline, alkaline earth, In, Tl, Ag elements. While the Mo_9X_{11}, $Mo_{12}X_{14}$ larger cluster units co-crystallized with the Mo_6X_8 units in the first compounds $In_2Mo_{15}Se_{19}$, $In_{\sim 3}Mo_{15}Se_{19}$, $Tl_2Mo_9S_{11}$ types; more recently these large units have been found exclusively in new materials, as well as new and more extended cluster materials, satisfying the general formula $M_{2n-2}Mo_{6n}X_{6n+2}$.

4.1. Co-crystallization of Mo_9X_{11} and $Mo_{12}X_{14}$ Units with Mo_6X_8 Units

The first examples of condensed (M_6X_8) clusters in molybdenum chalcogenides were found with the Mo_9X_{11} unit which coexists with the Mo_6X_8 unit in $In_{\sim 3}Mo_{15}Se_{19}$ and $In_2Mo_{15}Se_{19}$ compounds [33—38]. The tridimensional stacking of both types of Mo_6X_8 and Mo_9X_{11} units is illustrated in Figure 8. The different stackings of units create different types of channels: zig-zag tunnels in $In_{\sim 3}Mo_{15}Se_{19}$ and tridimensional chains of channels in $In_2Mo_{15}Se_{19}$ type, like that found in the Chevrel phases. Each unit is always surrounded by six units of a different type, leading to six intercluster Mo—Mo interactions as shown for the environment of the Mo_9X_{11} unit in Figure 9.

The VEC is very interesting to study in these compounds, since it is easy to change the content of indium in $In_{\sim 3}Mo_{15}Se_{19}$ solid solution, to intercalate some

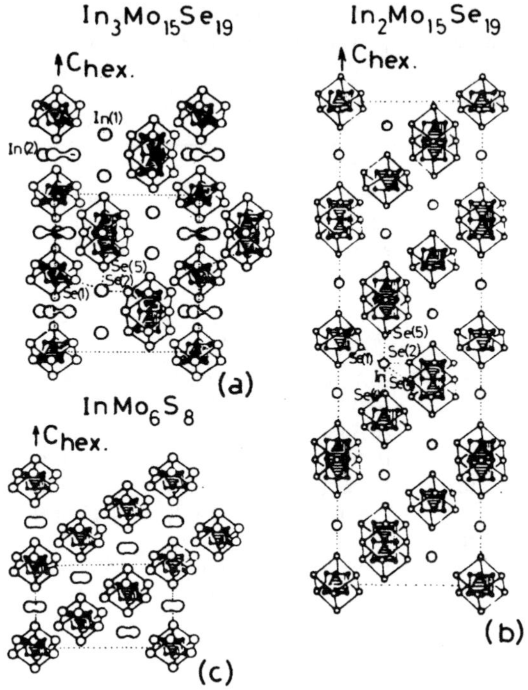

Fig. 8. Projection onto the $(11\bar{2}0)$ hexagonal plane: (a) $In_{\sim 3}Mo_{15}Se_{19}$; (b) $In_2Mo_{15}Se_{19}$; (c) $InMo_6S_8$.

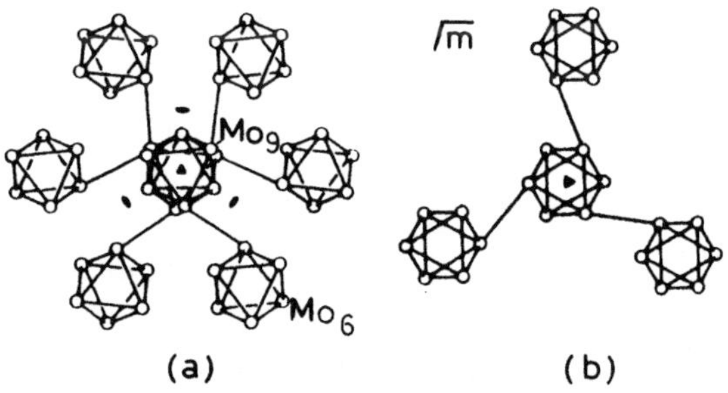

Fig. 9 Mo—Mo intercluster bonds from the Mo_9 cluster to the six Mo_6 octahedral clusters: (a) in $In_2Mo_{15}Se_{19}$ (symmetry 32); (b) in $In_3Mo_{15}Se_{19}$ (symmetry $3/m$). The chalcogen atoms are omitted for clarity.

quantities of other counterions, or to remove completely the indium ions in $In_2Mo_{15}Se_{19}$ and $In_{\sim 3}Mo_{15}Se_{19}$ getting new metastable binaries by 'soft chemistry' and then to reintercalate another type of ion [27, 39] (see Table I).

It is well known in the $M_xMo_6X_8$ compounds that the higher the VEC (i.e. the charge transfer from M to Mo increases), the shorter the Mo—Mo intracluster distance [40]. The same resuts are found in the Mo_6 clusters of the new co-crystal-

TABLE I
Lattice parameters for $In_{\sim 3}Mo_{15}Se_{19}$-type compounds ($P6_3/m$) and $In_2Mo_{15}Se_{19}$-type compounds ($R\bar{3}c$)

(a)	a (Å)	c (Å)	V (Å^3)
α-$Mo_{15}Se_{19}$	9.47	19.63	1524
$In_{2.9}Mo_{15}Se_{19}$	9.80	19.49	1622
$In_{3.33}Mo_{15}Se_{19}$	9.83	19.40	1624
$Sn_xMo_{15}Se_{19}$ ($x \sim 2.7$)	9.73	19.46	1595
$Pb_xMo_{15}Se_{19}$ ($x \sim 2.2$)	9.62	19.61	1572

(b)	a_R (Å)	α_R (°)	V_R (Å^3)	a_h (Å)	c_h (Å)
$In_2Mo_{15}Se_{19}$	20.159(3)	27.808(7)	1575.4	9.688	58.10
$Ba_2Mo_{15}Se_{19}$	20.00	28.42	1602	9.82	57.54
β-$Mo_{15}Se_{19}$	20.348(3)	26.983(2)	1529.7(3)	9.494(2)	58.786(6)
$Ag_3In_2Mo_{15}Se_{19}$	19.944(6)	28.938(6)	1642.7(7)	9.966	57.288
$Ag_3Mo_{15}Se_{19}$ (low temperature)	19.865(3)	28.887(3)	1617.9(3)	9.910	57.070
$Ag_{2.4}In_2Mo_{15}Se_{19}$	19.959(3)	28.754(2)	1626.8(3)	9.912	57.363
$Ag_1In_1Mo_{15}Se_{19}$ (low temperature)	19.997(5)	28.336(5)	1592.0(6)	9.789	57.545
$Sn_xMo_{15}Se_{19}$ ($x \sim 2.7$)	19.99	28.31	1589	9.78	57.54
$Pb_xMo_{15}Se_{19}$ ($x \sim 2$)	20.02	28.01	1560	9.68	57.66

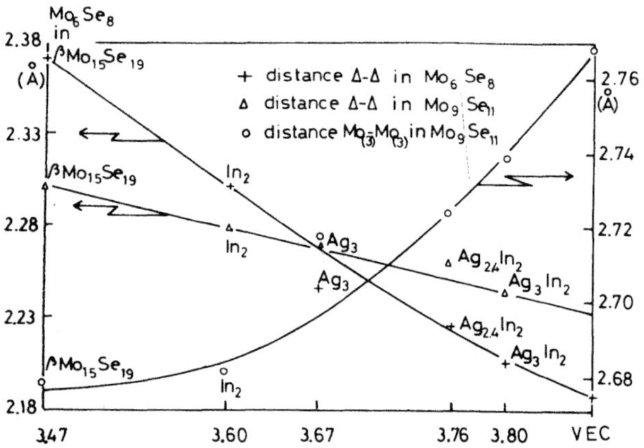

Fig. 10. Variation of interplane Mo_Δ—Mo_Δ so-called Δ—Δ distances and the $(Mo(3)—Mo(3))_\Delta$ (cf. Fig. 5) distance of the central plane of the Mo_9Se_{11} unit versus the VEC.

lized phases whereas the $(Mo—Mo)_\Delta$ distance in the central plane of the Mo_9 cluster increases when the charge transfer or the VEC increases (see Table II and Figure 10). Actually contrary to the Mo_6 cluster, for which 12 bonding orbitals correspond to the 12 localized bonds (equal to the number of edges of the Mo_6 polyhedron), the 18—19 bonding orbitals of the Mo_9 cluster do not correspond

TABLE II

The Mo—Mo intracluster and intercluster distances into the Mo_6 and Mo_9 clusters in In_2-type and $In_{\sim 3}Mo_{15}Se_{19}$-type compounds

VEC	3.80	3.76	3.67	3.60	3.47	3.70	3.87
	Ag_3In_2	$Ag_{2.4}In_2$	Ag_3	In_2	β-$Mo_{15}Se_{19}$	$In_{2.9}Mo_{15}Se_{19}$	$In_{3.3}Mo_{15}Se_1$
Mo_6Se_8							
$(Mo(1)—Mo(1))_\Delta$	2 × 2.678	2 × 2.684	2 × 2.695	2 × 2.686	2 × 2.675	2 × 2.684	2 × 2.688
$Mo(1)_\Delta—Mo(1)_\Delta$	2 × 2.704	2 × 2.715	2 × 2.733	2 × 2.772	2 × 2.827	2 × 2.712	2 × 2.683
$\Delta(1)—\Delta(1)$	2.207	2.226	2.245	2.302	2.371	2.23	2.19
Mo_9Se_{11}							
$(Mo(2)—Mo(2))_\Delta$	2 × 2.645	2 × 2.648	2 × 2.656	2 × 2.674	2 × 2.697	2 × 2.654	2 × 2.655
$(Mo(3)—Mo(3))_\Delta$	2 × 2.739	2 × 2.724	2 × 2.717	2 × 2.680	2 × 2.677	2 × 2.705	2 × 2.768
$Mo(2)_\Delta—Mo(3)_\Delta$	2.684	2.676	2.704	2.712	2.719	2.729	2.725
	2.775	2.772	2.799	2.808	2.830	2.770	2.771
$\Delta(2)—\Delta(3)$	2.244	2.261	2.271	2.280	2.300	2.27	2.26
Mo(1)—Mo(2) inter	3.568	3.530	3.478	3.389	3.275	3.498	3.512

*For Molybdenum notation (1), (2), (3), see Fig. 5.

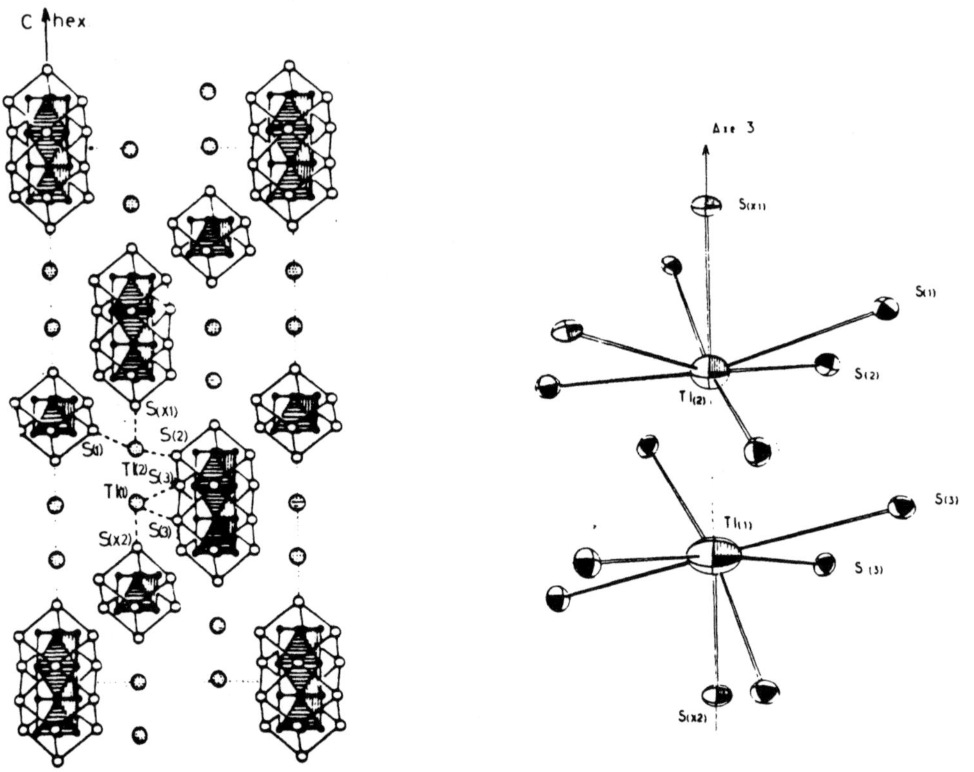

Fig. 11. Projection onto the $(11\overline{2}0)$ hexagonal plane of the $Tl_2Mo_9S_{11}$ compound and the environment of the thallium ions.

TABLE III
The Mo—Mo intracluster and intercluster distances into both Mo_6 and Mo_{12} clusters in the $Tl_2Mo_9S_{11}$ compound

Distances in Mo_6S_8 cluster		Distances in $Mo_{12}S_{14}$ cluster*	
$(Mo(1)—Mo(1))_\Delta$	2 × 2.693(1)	$(Mo(2)—Mo(2))_\Delta$	2 × 2.658(1)
$Mo(1)_\Delta—Mo(1)_\Delta$	2 × 2.780(5)	$(Mo(3)—Mo(3))_\Delta$	2 × 2.688(1)
$\Delta(1)—\Delta(1)$	2.30	$Mo(2)_\Delta—Mo(3)_\Delta$	2.743(1)
			2.771(1)
		$Mo(3)_\Delta—Mo(3)_\Delta$	2 × 2.669(5)
		$\Delta(2)—\Delta(3)$	2 × 2.28
		$\Delta(3)—\Delta(3)$	2.17
Intercluster distance		Mo(1)—Mo(2)	3.217

*For Mo notation, see Fig. 6.

to the 21 edges of the Mo_9 cluster (two-e^--single covalent bondings) that involves a delocalized character of the bonding. This new situation will still occur in the new condensed clusters, and will be explained in more detail below. Likewise, the $Mo_{12}X_{14}$ unit has been found to co-crystallize with the Mo_6X_8 unit in the $Tl_2Mo_9X_{11}$

compound. Thus, the latter does not contain Mo_9X_{11} clusters, but a 1—1 mixture of Mo_6X_8 and $Mo_{12}X_{14}$ clusters. The symmetry of the $Mo_{12}X_{14}$ unit is exactly the same ($\bar{3}$) as that of the Mo_6X_8 unit. Thus, in fact, the stacking of the units in the $Tl_2Mo_9S_{11}$ compound is like the one found in $M_xMo_6X_8$ phases: every second Mo_6X_8 unit has been replaced by every other $Mo_{12}X_{14}$ unit in Figure 11. This substitution by a larger unit creates a larger site of 14 anions, composed of two fused-cubic sites, sum of two cubic sites of the $M_xMo_6X_8$ compounds containing two monovalent thallium ions [41, 42]. The nonequivalence of the Mo_Δ—Mo_Δ intertriangle distances turns out to be an important remark: the Mo_Δ—Mo_Δ distance of the central octahedron (2.67 Å) is much shorter than the corresponding Mo_Δ—Mo_Δ distances of the both terminal octahedra (2.74 and 2.77 Å) (Table III).

4.2. Crystallization of Only Condensed Units: $Mo_{6n}X_{6n+2}$ Units

The *trans*-face-sharing octahedral clusters have been isolated recently in new ternary molybdenum chalcogenides [9] of general formula $M_xMo_{6n}X_{6n+2}$ units.

The Mo_9X_{11}, $Mo_{12}X_{14}$, $Mo_{18}X_{20}$, $Mo_{24}X_{26}$, $Mo_{30}X_{32}$ only units are illustrated in Figure 7 and observed in the $M_xMo_{6n}X_{6n+2}$ chalcogenides. These compounds are the intermediate steps of the 1D limit and lead inevitably to the linear chain $(Mo_3X_3)^1_\infty$.

4.2.1. *The Mo_9X_{11} Units* ($n = 3/2$)

The molybdenum chalcogenides, with only Mo_9X_{11} units, crystallize in two different structures: $Ag_{2.3}CsMo_9Se_{11}$ (space group $P6_3/m$) and $Ag_{3.6}Mo_9Se_{11}$ (space group

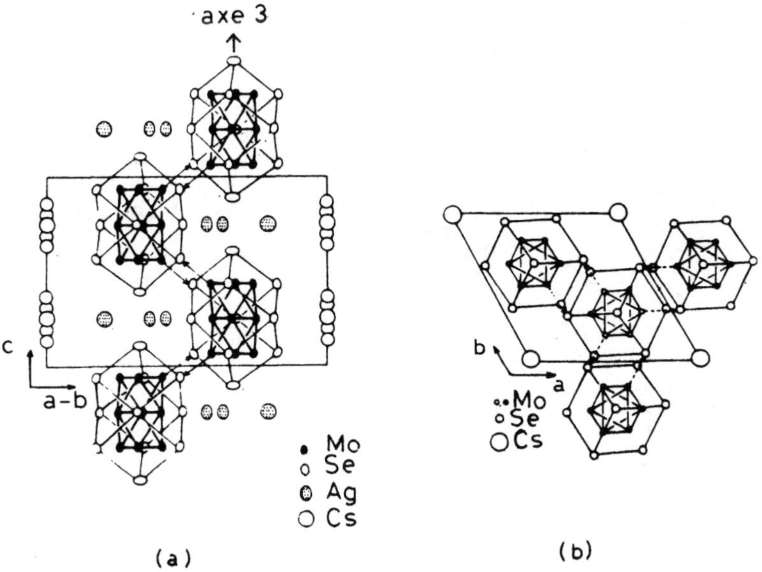

Fig. 12. Projection of the $Ag_{2.3}CsMo_9Se_{11}$ hexagonal structure onto: (a) the $(11\bar{2}0)$ hexagonal plane; (b) the (0001) base hexagonal plane.

Cmcm) [9, 43, 44]. In the former hexagonal compound, the Mo_9Se_{11} unit has six neighboring Mo_9Se_{11} units in eclipsed positions with the classical intercluster Mo—Mo interactions (the Mo_9Se_{11} unit is surrounded by six Mo_6Se_8 units in $In_{\sim 3}Mo_{15}Se_{19}$) between the square-faced chalcogens in Figure 12, while, in the latter compound, the Mo_9Se_{11} unit is surrounded by eight Mo_9Se_{11} units, but is only bonded to four units by Mo—Mo intercluster interaction, yielding a bidimensional character to this compound (see Figure 13).

Let us compare the structures of the $Ag_{2.3}CsMo_9Se_{11}$ and $In_{\sim 3}Mo_{15}Se_{19}$ compounds. One can imagine a chain of Mo_6Se_8 units along the |001| row (Mo_6Se_8—In(2)—Mo_6Se_8) in the $In_{\sim 3}Mo_{15}Se_{19}$ structure being replaced by a 1D column of cesium ions. The contribution of the condensed Mo_9X_{11} clusters trends to give a low-dimensional feature to these materials.

Note: an isoelectronic and isostructural $Ag_{4.4}ClMo_9Se_{11}$ compound with the $Ag_{2.3}CsMo_9Se_{11}$ compound has been so isolated [9] (the same VEC per Mo atom); some Ag^+ ions and Cl^- anions are located along the channel *c* axis (see Figure 12(a)).

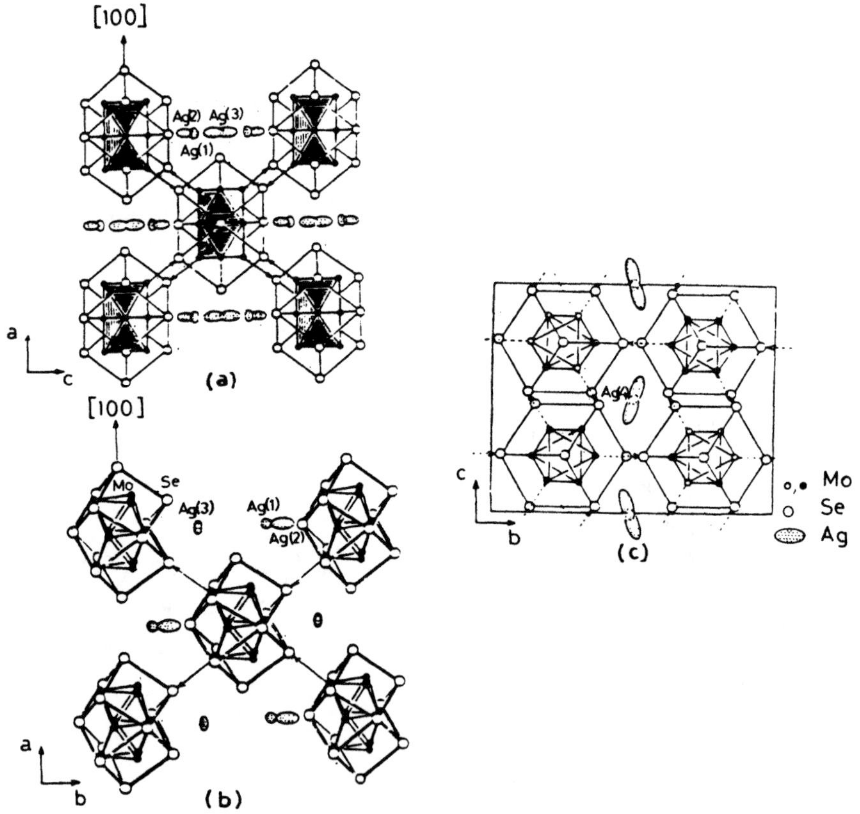

Fig. 13. Projection of the $Ag_{3.6}Mo_9Se_{11}$ compound onto: (a) the (010) plane; (b) the (001) plane; (c) the (100) plane of the orthorhombic unit cell (space group: *Cmcm*).

4.2.2. *The Mo_{6n} Condensed Clusters in $M_{2n-2}Mo_{6n}X_{6n+2}$ Compounds (n = 2, 3, 4, 5)*

The *trans*-face-sharing Mo_6 cluster condensation occurs in a stepwise process in a very nice family of compounds: the new $M_{2n-2}Mo_{6n}X_{6n+2}$ ternary chalcogenides, where the new $Mo_{6n}X_{6n+2}$ units contain the extended clusters Mo_{12} ($n = 2$), Mo_{18} ($n = 3$), Mo_{24} ($n = 4$), Mo_{30} ($n = 5$), and where M stands for the cesium or rubidium ions and X = S, Se [9, 45, 46].

The compounds are prepared from mixtures in appropriate amounts of the M_2MoS_4 ternary, the MoS_2 disulfide and the molybdenum element for the sulfur compounds, and the $M_2Mo_6Se_6$ ternary with the binary Mo_6Se_8 selenide for the selenide compounds. These mixtures were heated up to 1200—1500 °C into sealed molybdenum crucibles. Single crystals of all these phases are obtained by annealing and are black shaped rhomboids as shown in Figure 14. It is remarkable to note

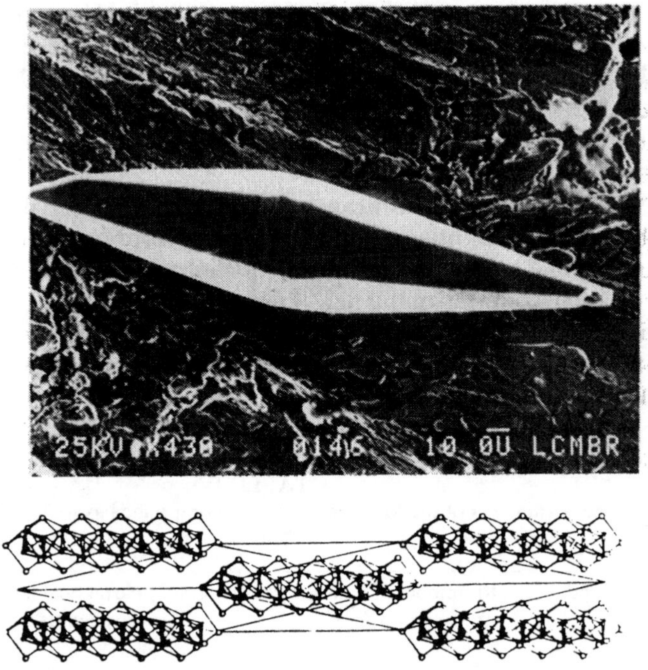

Fig. 14. SEM micrograph exhibiting the morphology of a $Cs_8Mo_{30}Se_{32}$ single crystal and the corresponding view of the $Mo_{30}Se_{32}$ unit in the rhombohedral unit cell.

that this rhombohedral morphology of the single crystals corresponds exactly to the shape of the rhombohedral unit cell and demonstrates marvelously well the pseudo-molecular aspect of these building blocks.

The $M_{2n-2}Mo_{6n}X_{6n+2}$ chalcogenides crystallize in a hexagonal-rhombohedral unit cell, space group $R\bar{3}$, isotypical with the Mo_6X_8 binary compound, type-structure of the Chevrel phases. The Mo_6X_8 unit is the first member of this series ($n = 1$). Actually, the extended $Mo_{6n}X_{6n+2}$ unit has exactly the same $\bar{3}$ point symmetry as the Mo_6X_8 unit, and replaces this latter one to the center of the unit cell (see Figure 14) (one block $Mo_{6n}X_{6n+2}$ unit by primitive rhombohedral cell). The list of these compounds and the crystalline rhombohedral and hexagonal parameters, together with the theoretical densities, are summarized in Table IV. One may notice that the density decreases towards a limit when n tends towards infinity: the maximum density is equal to 6.80 for $n = 1$ (Mo_6Se_8) and decreases to 6.15 for $n = \infty$ in the $Cs_2Mo_6Se_6$ compound for a series with the same ion (M = Cs).

The tridimensional packing of these condensed cluster units is illustrated in Figure 15. The cohesion of these structures is always ensured by strong extremal Mo—chalcogen interunit bonds through the six square faces. The alkaline cations sit onto the larger and larger holes in columns of 2, 4, 6, and 8 ions along the ternary axis located between the extended clusters. These cationic sites form intersecting channels which run along the three rhombohedral axes.

The lengthening of the unit occurs by the increasing of the a rhombohedral axis and the lowering of the α rhombohedral angle that involve an ever-stretching rhomboid along the three-fold axis. The visible outcome of the lengthening of the rhomboid will be an increase in the c hexagonal parameter (|111| rhombohedral row), the a hexagonal axis being almost invariant. So, if we plot the c hexagonal parameter versus n, the order of the condensed $Mo_{6n}X_{6n+2}$ cluster, we find a linear law (see Figure 16): $c_{hex} = 13.5n - 2.3$.

In order to understand the exact meaning of this law, it is possible to change the set of axes: substitution of another physical x variable for the n variable ($x = 6n - 3$) (x is the many-fold Mo—Mo interplane distance, so called Δ—Δ, repeated along the c hexagonal axis). Then, the new linear law is written $c'_{hex} = 2.25x + 4.45$ (a = 2.25 Å is the average Mo_Δ—Mo_Δ interplane distance into the cluster; b = 4.45 Å expresses three times the interplane distance between two extremal average Mo_Δ—Mo_Δ planes of two adjacent units).

TABLE IV
Lattice parameters of the $M_{2n-2}Mo_{6n}X_{6n+2}$ compounds as well as their theoretical densities

Compounds	Rhombohedral lattice parameters (space group $R\overline{3}$)			Hexagonal lattice parameters		Density
	a_R (Å)	α_R (°)	V_R (Å^3)	a_h (Å)	c_h (Å)	
Mo_6S_8	6.432	91.34	265.8	9.20	10.88	5.20
$Cs_6Mo_{24}S_{26}$	17.705	30.62	1276.8	9.351	50.59	5.12
$Cs_8Mo_{30}S_{32}$	21.960	24.56	1608.7	9.423	64.83	5.13
Mo_6Se_8	6.658	91.58	294.7	9.54	11.21	6.80
$Cs_2Mo_{12}Se_{14}$	9.905	58.38	661.6	9.661	24.55	6.33
$Rb_4Mo_{18}Se_{20}$	13.896	39.94	998.9	9.492	38.31	6.07
$Rb_6Mo_{24}Se_{26}$	18.110	30.23	1333.9	9.444	51.81	6.06
$Cs_6Mo_{24}Se_{26}$	18.087	30.91	1384.7	9.639	51.63	6.18
$Cs_8Mo_{30}Se_{32}$	22.409	24.78	1738.4	9.616	65.13	6.18

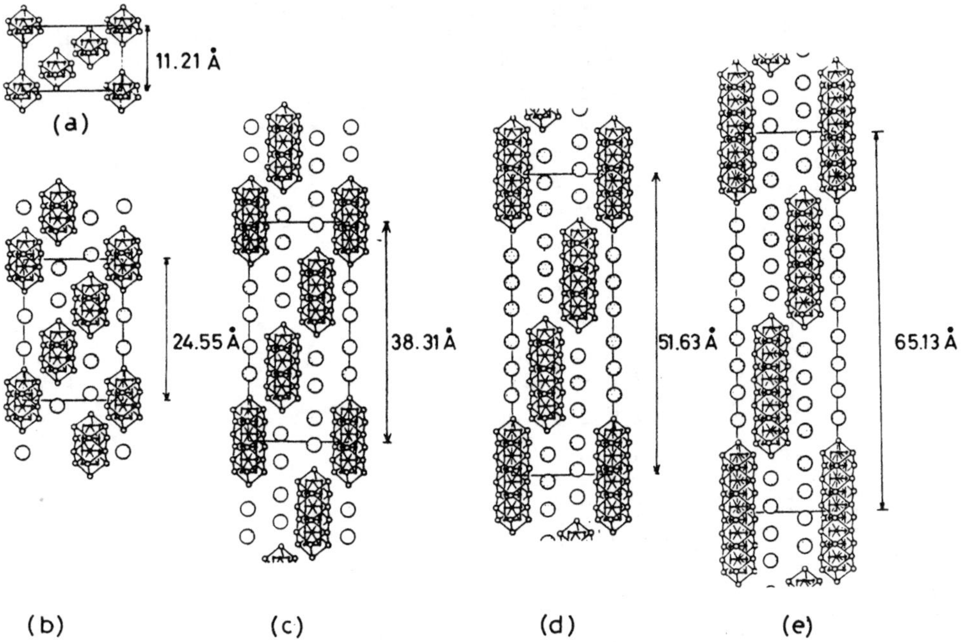

Fig. 15. Projection onto the $(11\bar{2}0)$ hexagonal plane of the $M_{2n-2}Mo_{6n}X_{6n+2}$ series (space group $R\bar{3}$): (a) Mo_6Se_8; (b) $Cs_2Mo_{12}Se_{14}$; (c) $Rb_4Mo_{18}Se_{20}$; (d) $Cs_6Mo_{24}Se_{26}$; (e) $Cs_8Mo_{30}Se_{32}$.

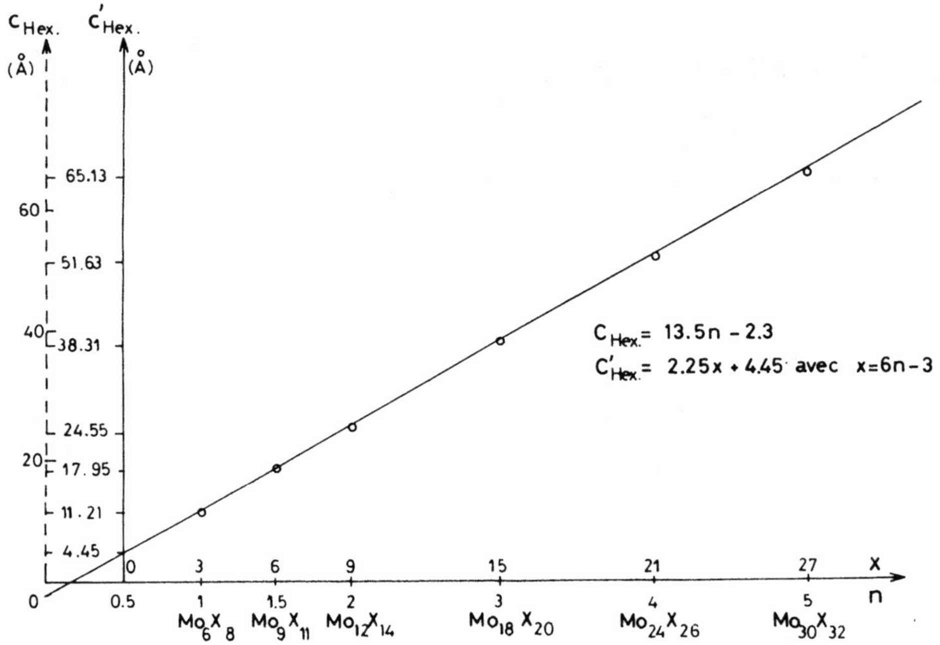

Fig. 16. Variations of the c hexagonal axis of the $M_{2n-2}Mo_{6n}X_{6n+2}$ compounds versus n, order of the condensed $Mo_{6n}X_{6n+2}$ clusters, or versus x, the many-fold Mo—Mo interplane distance (Δ—Δ) along the c hexagonal axis.

TABLE V

The Mo—Mo intratriangle $(Mo—Mo)_\Delta$ and intertriangle $(Mo_\Delta—Mo_\Delta)$ distances and the interplane distances $(\Delta—\Delta)$ in the clusters Mo_{6n} of the $M_{2n-2}Mo_{6n}X_{6n+2}$ compounds
[Note: the notation of indices of the Mo triangles begins from the terminal Mo planes]

	$Cs_2Mo_{12}Se_{14}$		$Rb_4Mo_{18}Se_{20}$		$Rb_6Mo_{24}Se_{26}$		$Cs_6Mo_{24}S_{26}$		$Cs_6Mo_{24}Se_{26}$		$Cs_8Mo_{30}S_{32}$		$Cs_8Mo_{30}Se_{32}$	
Mo—Mo intratriangle distance	$(Mo(1)—Mo(1))_\Delta$	2 × 2.656	$(Mo(1)—Mo(1))_\Delta$	2 × 2.643	$(Mo(1)—Mo(1))_\Delta$	2 × 2.646	$(Mo(1)—Mo(1))_\Delta$	2 × 2.649	$(Mo(1)—Mo(1))_\Delta$	2 × 2.645	$(Mo(1)—Mo(1))_\Delta$	2 × 2.650	$(Mo(1)—Mo(1))_\Delta$	2 × 2.644
	$(Mo(2)—Mo(2))_\Delta$	2 × 2.682	$(Mo(2)—Mo(2))_\Delta$	2 × 2.674	$(Mo(2)—Mo(2))_\Delta$	2 × 2.672	$(Mo(2)—Mo(2))_\Delta$	2 × 2.670	$(Mo(2)—Mo(2))_\Delta$	2 × 2.673	$(Mo(2)—Mo(2))_\Delta$	2 × 2.669	$(Mo(2)—Mo(2))_\Delta$	2 x 2.680
			$(Mo(3)—Mo(3))_\Delta$	2 × 2.664	$(Mo(3)—Mo(3))_\Delta$	2 × 2.655	$(Mo(3)—Mo(3))_\Delta$	2 × 2.663	$(Mo(3)—Mo(3))_\Delta$	2 × 2.656	$(Mo(3)—Mo(3))_\Delta$	2 × 2.663	$(Mo(3)—Mo(3))_\Delta$	2 × 2.655
					$(Mo(4)—Mo(4))_\Delta$	2 × 2.655	$(Mo(4)—Mo(4))_\Delta$	2 × 2.659	$(Mo(4)—Mo(4))_\Delta$	2 × 2.660	$(Mo(4)—Mo(4))_\Delta$	2 × 2.657	$(Mo(4)—Mo(4))_\Delta$	2 × 2.637
											$(Mo(5)—Mo(5))_\Delta$	2 × 2.651	$(Mo(5)—Mo(5))_\Delta$	2 × 2.674
Mo—Mo intertriangle distances	$Mo(1)_\Delta—Mo(2)_\Delta$	2.758	$Mo(1)_\Delta—Mo(2)_\Delta$	2.745	$Mo(1)_\Delta—Mo(2)_\Delta$	2.739	$Mo(1)_\Delta—Mo(2)_\Delta$	2.732	$Mo(1)_\Delta—Mo(2)_\Delta$	2.739	$Mo(1)_\Delta—Mo(2)_\Delta$	2.728	$Mo(1)_\Delta—Mo(2)_\Delta$	2.745
		2.796		2.782		2.770		2.746		2.759		2.738		2.772
	$Mo(2)_\Delta—Mo(2)_\Delta$	2 × 2.682	$Mo(2)_\Delta—Mo(3)_\Delta$	2.690	$Mo(2)_\Delta—Mo(3)_\Delta$	2.694	$Mo(2)_\Delta—Mo(3)_\Delta$	2.663	$Mo(2)_\Delta—Mo(3)_\Delta$	2.694	$Mo(2)_\Delta—Mo(3)_\Delta$	2.660	$Mo(2)_\Delta—Mo(3)_\Delta$	2.706
				2.709		2.715		2.685		2.716		2.689		2.724
			$Mo(3)_\Delta—Mo(3)_\Delta$	2 × 2.745	$Mo(3)_\Delta—Mo(4)_\Delta$	2.727	$Mo(3)_\Delta—Mo(4)_\Delta$	2.702	$Mo(3)_\Delta—Mo(4)_\Delta$	2.734	$Mo(3)_\Delta—Mo(4)_\Delta$	2.701	$Mo(3)_\Delta—Mo(4)_\Delta$	2.702
						2.739		2.709		2.743		2.709		2.724
					$Mo(4)_\Delta—Mo(4)_\Delta$	2 × 2.695	$Mo(4)_\Delta—Mo(4)_\Delta$	2 × 2.673	$Mo(4)_\Delta—Mo(4)_\Delta$	2 × 2.700	$Mo(4)_\Delta—Mo(5)_\Delta$	2.669	$Mo(4)_\Delta—Mo(5)_\Delta$	2.702
												2.675		2.704
											$Mo(5)_\Delta—Mo(5)_\Delta$	2 × 2.705	$Mo(5)_\Delta—Mo(5)_\Delta$	2 × 2.750
Mo—Mo interplane Mo_3	Δ(1)—Δ(2)	2.309	Δ(1)—Δ(2)	2.295	Δ(1)—Δ(2)	2.286	Δ(1)—Δ(2)	2.268	Δ(1)—Δ(2)	2.286	Δ(1)—Δ(2)	2.264	Δ(1)—Δ(2)	2.282
	Δ(2)—Δ(2)	2.192	Δ(2)—Δ(3)	2.214	Δ(2)—Δ(3)	2.224	Δ(2)—Δ(3)	2.185	Δ(2)—Δ(3)	2.224	Δ(2)—Δ(3)	2.181	Δ(2)—Δ(3)	2.239
			Δ(3)—Δ(3)	2.268	Δ(3)—Δ(4)	2.262	Δ(3)—Δ(4)	2.228	Δ(3)—Δ(4)	2.262	Δ(3)—Δ(4)	2.229	Δ(3)—Δ(4)	2.247
					Δ(4)—Δ(4)	2.218	Δ(4)—Δ(4)	2.190	Δ(4)—Δ(4)	2.218	Δ(4)—Δ(5)	2.189	Δ(4)—Δ(5)	2.227
											Δ(5)—Δ(5)	2.221	Δ(5)—Δ(5)	2.271
Mo—Mo interunit	Mo(1)—Mo(1)	3.317	Mo(1)—Mo(1)	3.295	Mo(1)—Mo(1)	3.274	Mo(1)—Mo(1)	3.181	Mo(1)—Mo(1)	3.347	Mo(1)—Mo(1)	3.198	Mo(1)—Mo(1)	3.336

This law can be extended in order to calculate the packing axis of the units in the $M_xMo_{6n}X_{6n+2}$ compounds. For instance, the $Ag_{2.3}CsMo_9Se_{11}$ compound, with n equal to 3/2, leads to a parameter equal to 17.95 Å in a $3R$-type stacking, or 11.9 Å in a $2H$-type stacking (the experimental c parameter is 11.95 Å).

Before examining the electronic properties of these compounds it is interesting to study the Mo—Mo bondings in these different clusters.

The different Mo—Mo intracluster distances and the Mo—Mo intercluster distances are reported in Table V. How 'chain-like' are the $Mo_{6n}X_{6n+2}$ clusters? The $(Mo—Mo)_\Delta$ intratriangle distances are all close to 2.66 Å, whereas the $(Mo_\Delta—Mo_\Delta)$ intertriangle distances vary. These latter distances seem to obey to a general rule: the $Mo_\Delta—Mo_\Delta$ intertriangle distances of both terminal octahedra are always seen to be long about 2.75 Å.

Then, the sequence along the three-fold axis from the terminal octahedra is: long $(Mo_\Delta—Mo_\Delta)$ distance (2.75 Å), then short $(Mo_\Delta—Mo_\Delta)$ distance (2.7 Å), and so on. The $Mo_\Delta—Mo_\Delta$ distance of the central octahedron is short if n is an even number or long if n is odd. Thus, we encounter between planar Mo_3X_3 units a succession of *trans*-face-sharing Mo_6 octahedra with alternate short and long $Mo_\Delta—Mo_\Delta$ distances. This distortion of the one-column chain looks like a Peierls distortion of a 1D chain of atoms where, alternatively, long and short distances are repeated between the atoms.

5. Electronic Properties and Bond Order

5.1. THEORETICAL ELECTRONIC PROPERTIES

Many theoretical studies have been carried on the electronic structure of Mo_6X_8 clusters but very few have investigated the electronic structure of $Mo_{6n}X_{6n+2}$ clusters and the infinite chain $(Mo_3X_3)^1_\infty$ [47, 48].

We have seen, for instance, that the intercluster linkages with extended cluster compounds are all structurally similar to the Chevrel phases. So, we shall try to analyze from band structure and molecular orbital calculations how this series of compounds offers a method of investigating the evolution from molecules or discrete clusters to extended clusters up to infinite chains. Before establishing the relationships between the energy diagrams of the different extended clusters, we shall discuss in some detail the M—O diagram calculated for the Mo_6S_8 cluster (Figure 17).

A very important result on this M—O diagram is that the levels displayed are all primarily of Mo d character. Actually, the sulfur-blocks levels are just below the lowest a_{1g} level of the Mo—Mo molecular orbitals. Now, the model of the VEC can easily be understood if we considered the $(Mo_6S_8)^{4-}$ charged unit. Really, we see that the 12 Mo d levels are filled and consistent with the fact of 24 cluster-Mo electrons based upon an oxidation state 2^- for sulfur. The purely ionic assumption (S^{2-}) is not very realistic because a charge integration result provides a very strongly pronounced covalent character of the Mo—chalcogen bonding, but that in no way changes the VEC (VEC = 24 Mo electrons per cluster). Actually, the

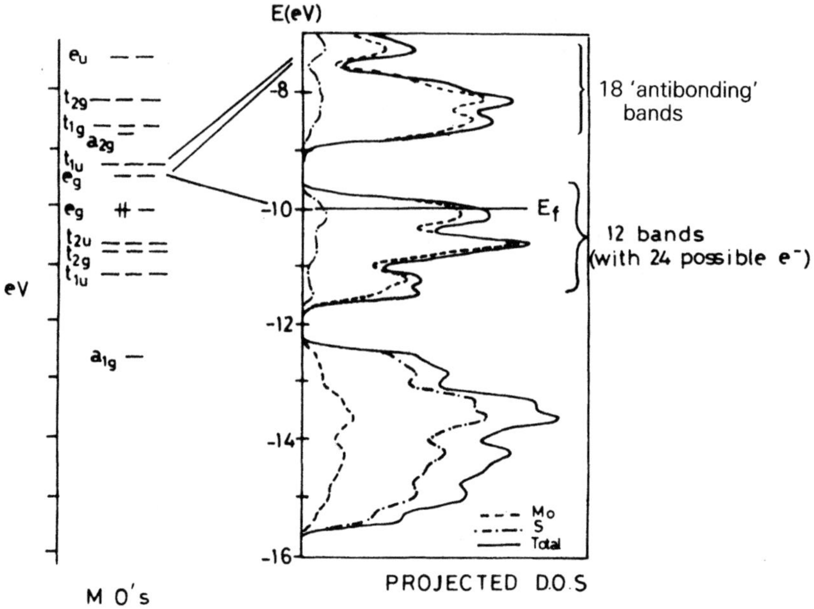

Fig. 17. Mo-based MOs for the $Mo_6S_8^{2-}$ cluster and Mo-d and S-p contributions to the total density of state (DOS).

$Mo_2Re_4S_8$ and $Mo_4Ru_2Se_8$ compounds with $24e^-$ as VEC are found semiconducting [49, 50], in agreement with a closed-shell configuration.

The symmetry of the Mo_6S_8 cluster is so high that the level diagram can be derived from group theory (symmetry *Oh* of Mo_6 octahedron). A convention has been adopted: every Mo metal atom has a local coordinate system such as the 'z axes' are normal to the square faces of the cubes and the d_{xy} orbitals are directed to the sulfur atoms. Thus, with this convention, the xy-type orbitals may be neglected because they are directed to the sulfur atoms and then are pushed up. In this M—O diagram we display only the high-lying e_u orbital being the lowest of the xy orbitals. The 24 remaining levels corresponding to the Mo—Mo molecular orbitals of the $d_{x^2-y^2}$, d_{z^2}, d_{xz}, and d_{yz} atomic orbitals are shown in the diagram in Figure 17.

(1) The highest occupied e_g orbitals (HOMOs) are of very large import for superconductivity: this e_g orbitals is of local δ symmetry proceeding from the $x^2 - y^2$ character and thus has very little interaction with neighboring cluster. This e_g HOMO of the cluster can thus explain the high critical fields of these materials (the mean-free path being very short).

(2) The lowest unoccupied molecular orbitals (LUMOs), are reported to be the t_{1u} and e_g orbitals. These orbitals consist of mainly 'z^2' character and point directly towards the apical sulfur atoms on neighboring clusters. The effect of this perturbation is to strongly push up the LUMOs with a small part which contributes down to the HOMO level (see Figure 18).

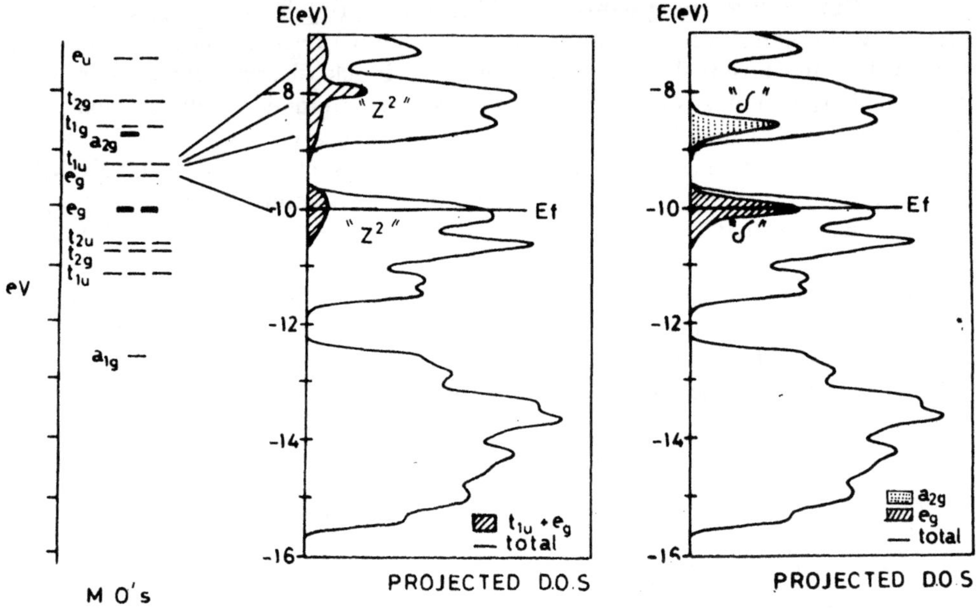

Fig. 18. Projected DOS curves for the 'z^2' type orbitals and for 'δ pseudo-symmetry' orbitals of the Mo_6S_8 unit are displayed. The MO diagrams of the Mo_6 cluster and the total DOS are included for reference.

This suggests that the principal structure determining interaction between clusters in the Chevrel phases $M_xMo_6X_8$ is the donor—acceptor interaction between sulfur atoms on neighboring clusters and the LUMO e_g and t_{1u} 'z^2-type' orbitals of the Mo atoms located at the middle of the square faces of the cluster. This situation will be repeated in larger clusters and a study of the electronic structure will shown how these frontier orbitals evolve.

The pushing up of both $2e_g$ and $2t_{1u}$ hybrid orbitals opens up a HOMO—LUMO gap about 1 eV. Then, the 'new' LUMO is an a_{2g} orbital of $d_{x^2-y^2}$ character and thus has a δ pseudo-symmetry as the $1e_g$ HOMO. This δ symmetry leads to narrow bands in both LUMOs and HOMOs, as shown in Figure 18. This a_{2g} orbital will be investigated anew in the molecular orbitals (MOs) of the extended clusters.

A M—O calculation to understanding the mode of packing of these condensed clusters can be explained in terms of the cluster frontier orbitals. Indeed, we have seen the mode of structural arrangement of the Mo_6X_8, Mo_9X_{11}, $Mo_{12}X_{14}$, . . . , extended units by fragmental decomposition of staggered Mo_3X_3 planes. Hoffman and Lissillour have treated these new clusters in the D_{3d} or (C_{3v}) symmetry by changing the referential axes: the three-fold axis is the z general axis (not local axis) [47, 51]. The M—O diagrams display a very large analogy between the simple cluster Mo_6X_8 and the other extended clusters. We find again the orbitals of the same symmetry at the Fermi level. The frontier orbitals are very similar in all the clusters of these different systems: the LUMOs are orbitals of z^2 character and, as seen previously, are destabilized up and slightly down because they are located at the level of the terminal square faces of these block discrete units. Intercluster Mo—chalcogen bonds are seen to result from an interaction between chalcogen donor orbitals and the cluster LUMOs (z^2 type) that are localized on Mo atoms residing on the middle of the 'square faces' of the clusters. Likewise, the HOMOs, which are very important for the explanation of the superconducting properties, are again orbitals located on the Mo atoms of the 'square faces' and are of pseudo-symmetry δ or $x^2 - y^2$ character, showing that the orbitals at the Fermi level reside in narrow bands as in the previous Mo_6X_8 cluster (see Figure 19). The LUMOs (z^2 type) are so mainly pushed up in the larger clusters and for the Mo_9S_{11} unit allow a a_2' nonbonding orbital between the bonding and antibonding gap. Likewise, an a_{2g} orbital lies just above the HOMOs δ in the $Mo_{12}S_{14}$ unit.

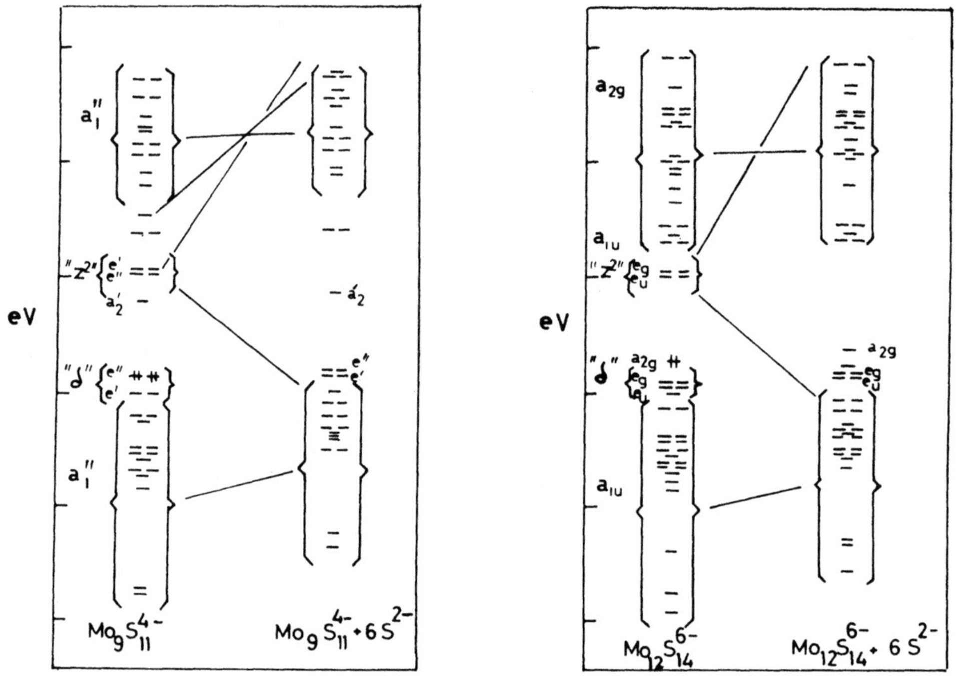

Fig. 19. MO diagrams of the $Mo_9S_{11}^{4-}$ and $Mo_{12}S_{14}^{6-}$ units and perturbation of these levels by six capping S^- ions (apical sulfur ions).

What is the nature of these orbitals?

All these a_2 orbitals (C_{3v} symmetry) (a_2', a_{2g}, ...) are seen arising from MOs of Mo_3X_3 fragments. The $1a_1''$ orbital on the simple Mo_3X_3 fragment is of d_{xz} or d_{yz} character, as seen in Figure 20. Then, the interaction of two staggered Mo_3X_3 planes 'splits' the $1a_1''$ orbital into two orbitals: one $1a_{1u}$ bonding orbital and one

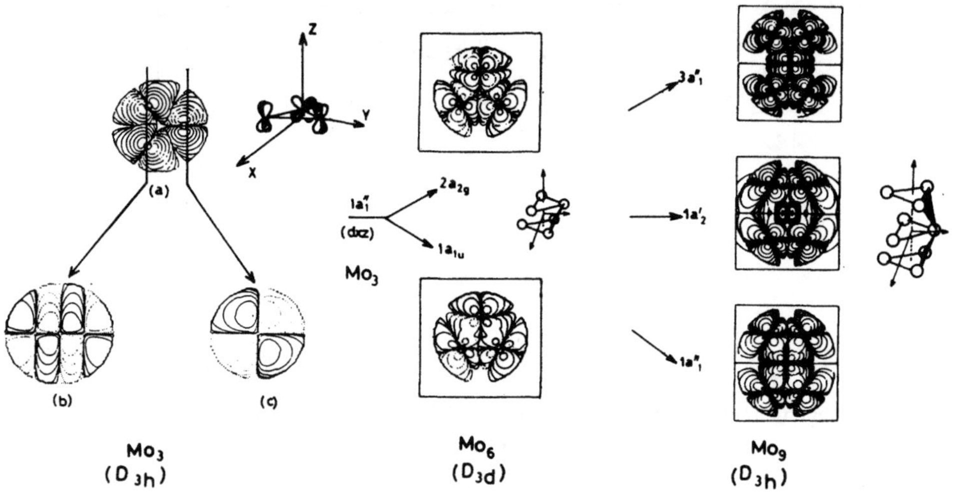

Fig. 20. Wave function contours of the $1a_1''$ MO arising from d_{xz} or d_{yz} AOs of the Mo_3 plane ((a) onto the XY plane; (b) and (c) perpendicular to the XY plane), of the $1a_{1u}$ and $2a_{2g}$ MOs for both staggered Mo_3 planes (Mo_6 cluster) and of the $1a_1''$, $1a_2'$, and $3a_1''$ MOs for three staggered Mo_3 planes (Mo_9 cluster). The MOs come from the same atomic orbitals (d_{xz} or d_{yz} symmetry). (After Reference [51].)

$2a_{2g}$ antibonding orbital in the Mo_6X_8 unit (this $2a_{2g}$ has alread been seen in the M—O treatment of the Mo_6X_8 unit as a new LUMO). The interaction of three Mo_3X_3 planes leads to three orbitals: one $1a_1''$ bonding orbital, one $1a_2'$ nonbonding orbital, and one $3a_1''$ antibonding orbital in the Mo_9X_{11} unit. The interation of four Mo_3X_3 planes give four orbitals: two bonding and two antibonding orbitals in the $Mo_{12}X_{14}$ unit, as shown in Figure 21. The interaction of $n Mo_3X_3$ planes leads to a wide band a_2 (~ 4 eV) which is the only one crossing over the gap. This band is generally half-filled for the major part of these compounds synthesized so far (13 electrons Mo per Mo_3X_3). Thus, the presence of the single half-filled a_2 band crossing the Fermi level implies that it may be possible to undergo a Peierls transition for the $(Mo_3X_3)_\infty$ chain. The $1a_2'$ orbital seen in the bonding—antibonding gap of the Mo_9X_{11} unit arises from the nonbonding combination of the Mo_3X_3 a_1'' fragment orbitals (see Figures 20 and 21), and will be expected in every 'like-chain' composed of an odd number of Mo_3X_3 fragments. This result can explain a certain thermodynamic instability in these kinds of compounds and that, so far, only compounds with Mo_9X_{11} units have been prepared [9]. As against this, 'like-chains' with an even number of Mo_3X_3 fragments (up to 10) have been prepared. Actually, the contribution of the a_{2g} orbital of d_{xz} or d_{yz} internal character at the Fermi level becomes more and more important with respect to the δ-type frontier orbitals (of $d_{x^2-y^2}$ character) (for instance, in the $Mo_{12}X_{14}$ unit) so exhibiting a more and more pronounced pseudo 1D character since, in fact, there remains only the a_2 band in the Mo_3X_3 infinite chain. Thus, the occurrence of these a_2 orbitals at the Fermi level in these new larger extended clusters (Mo_{18}, Mo_{24}, Mo_{30}, . . .) ought to lead to very striking electronic properties in these materials and help us to understand the one-dimensional $M_2Mo_6X_6$ phases.

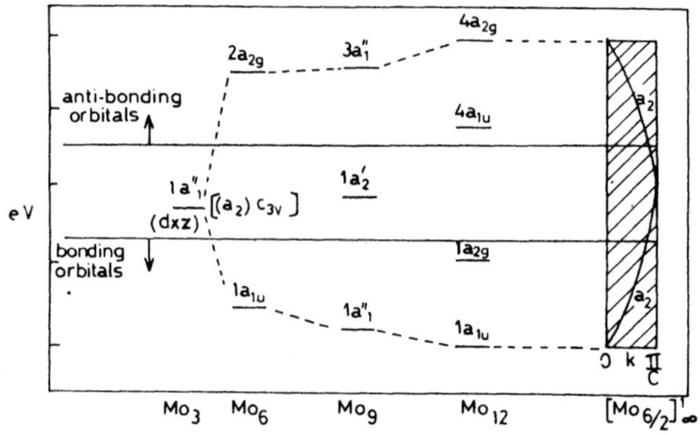

Fig. 21. Schematic representation of orbital energy levels derived from the $1a''_1$ fragment orbital of the Mo_3 plane.

From the M—O results, $(13n - 1)$ bonding orbitals can be deduced: thus, the maximum number of valence electrons leading to a closed-shell configuration is $(26n - 2)$ electrons for the $Mo_{6n}X_{6n+2}$ extended clusters. We find again $24e^-$ per Mo_6 cluster for the semiconducting $M_xMo_6X_8$ compounds ($n = 1$, $Mo_6X_8^{4-}$ units), 50 electron-cluster system for a $Mo_{12}X_{14}$ closed-shell cluster ($n = 2$), and so on . . . (see Table VI), but $37e^-$ per Mo_9 cluster ($n = 3/2$) — a cluster with an 'odd' number of Mo_3X_3 planes. This result is due to the presence of one a'_2 nonbonding orbital. That is a very curious situation, but, after all, semiconducting behavior is observed in compounds with only Mo_9X_{11} units containing $36e^-$ or $38e^-$ per Mo_9 cluster (if in the latter case, the a'_2 orbital is occupied).

Now, we remark that when a Mo_3X_3 plane adds to the previous unit, $13e^-$ are added, a number which is very consistent with the 13 electrons per Mo_3X_3 unit found for the $(Mo_3X_3)^{-1}_\infty$ chain. The different $Mo_{6n}X_{6n+2}$ extended clusters are reported in Table VI. The number of VEC per cluster is calculated taking into

TABLE VI

Number of 'delocalized' and 'localized' bonds and evolution of the valence electron concentration (VEC), bond order comparatively with the Pauling bond order versus n, nuclearity of the Mo_{6n} cluster in the new $M_{2n-2}Mo_{6n}X_{6n+2}$ chalcogenides

n	Units	'Delocalized' or bonding orbitals $13n - 1$	Localized orbitals $18n - 6$	Closed-shell configuration	VEC/ cluster	VEC/ Mo atom	VEC/ bonding orbital	Bond order	PBO/e
1	Mo_6X_8	12	12	24	20	3.333	1.667	0.833	0.73—0.84
1.5	$\|Mo_9X_{11}\|^{1-}$	18—19	21	36—38	33	3.666	1.783	0.891	0.85—0.96
2	$\|Mo_{12}X_{14}\|^{2-}$	25	30	50	46	3.833	1.84	0.94	0.92—0.96
3	$\|Mo_{18}X_{20}\|^{4-}$	38	48	76	72	4	1.894	0.947	
4	$\|Mo_{24}X_{26}\|^{6-}$	51	60	102	98	4.083	1.921	0.96	
5	$\|Mo_{30}X_{32}\|^{8-}$	64	84	128	124	4.133	1.937	0.968	
⋮	⋮	⋮	⋮	⋮	⋮	⋮	⋮	⋮	
∞	$\|Mo_{6/2}X_{6/2}\|^{1-}_\infty$	6.5	9	12—14	13	4.333	2	1	1.05

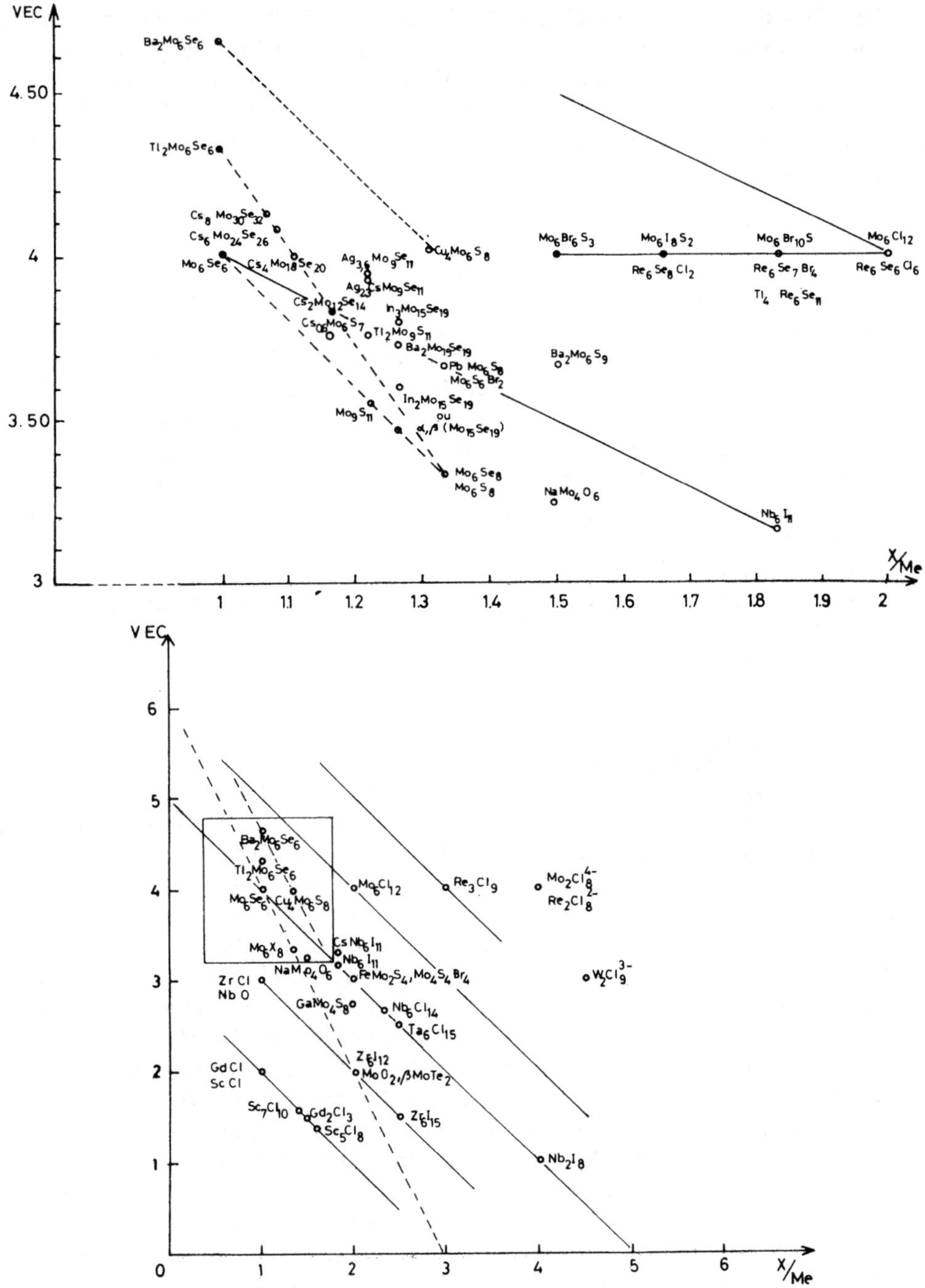

Fig. 22. Influence of valence electron concentration (VEC) of the Me metal atom and nonmetal—metal ratio (X/Me) on the size of metal cluster in the neutral or charged units of halogen and chalcogen compounds (and some oxides). The Me atoms are 'early transition' metals or rare earth metals. In the inset are illustrated the Mo chalcogenides described in this paper. The chalcohalides of Mo or Re cluster atoms are reported in References [6] and [7].

account the experimental unit charge, for instance, in the $TlMo_3Se_3$ compound, the $(Mo_3Se_3)^{-1}$ unit is $6 \times 3 = 18e^- - (3 \times 2) + 1 = 13e^-$ per Mo_3 plane. It is interesting to show the influence of the VEC per Mo atom (VEC/atom) versus the ratio of the nonmetal over the metal atoms (X/Me) in these new extended cluster chalcogenides with respect to the other clusters in solid-state chemistry. The VEC can be calculated from the law:

$$\mathrm{VEC} = \left(\mathrm{VEC}_0 - \frac{m}{\mathrm{Me}}\right) + \frac{n''\mathrm{X}}{\mathrm{Me}},$$

where VEC is the number of the valence electrons on the metal atom which are available for Me—Me bondings, VEC_0 is the valence electron concentration of the elemental metal, m is the charge of the unit (for instance $(Mo_6S_8)^{4-}$: $m = -4$), n'' is the charge of the ligand X (−2 for chalcogen in a purely ionic model and −1 for halogen) and is the slope of the full or dashed line, X is the number of ligands surrounding the metal cluster, and Me is the number of metal atoms of the cluster.

From the view in both parts of Figure 22, the following conclusions can be drawn: when the number of X ligands decreases (i.e. the X/Me ratio of nonmetal over metal decreases), the Me—Me bonding dominates. So, the formal oxidation state of the metal lowers and the compounds become more and more reduced. An increase of the VEC per metal connected with a lowering of the X/Me ratio allows new clusters of higher and higher nuclearity to be formed. Actually, the lower the X/Mo ratio and the higher the VEC, the larger the size of the cluster.

5.2. BOND ORDER

In the $M_xMo_6X_8$ compounds, we have seen that these compounds are models for electron-deficient bondings with a calculated bond order less than 1. The Mo—Mo bonding of the Mo_6 octahedral cluster in the 'Chevrel phases' decreases when the VEC increases, i.e. the bond order n' increases (when $n' = 1$, d Mo—Mo = 2.619 Å, the Mo_6 octahedron is almost regular as in Mo_6Cl_{12}). Thus, Figure 23 exhibits the average Mo_Δ—Mo_Δ distance versus n in the $Mo_{6n}X_{6n+2}$ extended clusters. We remark that this distance becomes shorter and shorter and converges towards the Mo_Δ—Mo_Δ distance of the infinite chain. These distances vary as the Mo_Δ—Mo_Δ distance in the $M_xMo_6X_8$ compounds when the Mo_6 cluster is reduced and thus contains more electrons [52]. How do the bond order and the electronic properties evolve in the extended clusters and the limit chain? What is the significance of lowering the Mo_Δ—Mo_Δ distance with respect to the bond order?

Firstly, lowering the Mo—Mo distances, although weak for high n values, leads to a greater cohesion of building blocks when the n degree of condensation increases. In Table VI, we have reported the VEC per cluster for the different extended clusters (in taking into account the ionic model: the monovalent M cation and the divalent chalcogen); it can be calculated according to the formula $26n - 6$ electrons. What is the average electronic repartition versus the number of localized Mo—Mo bands? This last number is equal to the edges of the polyhedra, i.e. $18n - 6$. Now, if we divide the VEC over this number, we find the number of electrons

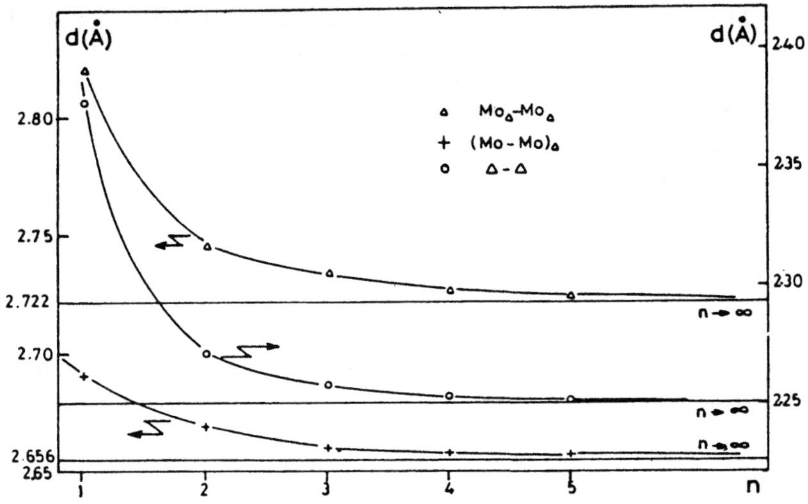

Fig. 23. Variation of the average Mo—Mo distances as a function of the nuclearity of the Mo_{6n} cluster in the $M_{2n-2}Mo_{6n}X_{6n+2}$ compounds.

per localized bond: this ratio varies from 1.667 ($n = 1$) to 1.444 ($n = \infty$). Thus, the bond order should decrease; that is exactly opposite to the largest cohesion of these extended clusters. So, it is necessary to introduce a new parameter: the number of Mo—Mo bonding orbitals. It is no longer equal to the number of localized bonds, but to the number of delocalized bonds. This amount is, as seen previously, equal to $13n - 1$ bonding orbitals. We can notice that, only in the Mo_6X_8 units, the number of bonding orbitals is equal to the number of localized orbitals of the Mo_6 cluster, and shifts more and more in the condensed clusters. The ratio of the number of valence electrons (VEC) to the number of bonding orbitals can be interpreted as the new 'electron number per bond', which varies from 1.667 to 2. Thus, the new 'bond order' varies from 0.83 to 1. This number of e^- per bond increases and varies as the VEC per Mo atom in the more and more extended and reduced clusters, as shown in Table VI. These values fit relatively well with the results found by Corbett [30] for the Pauling bond order per electron (PBO/e^-). This trend to fill more and more the Mo—Mo bonding orbitals (i.e. to fill the conduction band) is reflected by the electronic physical properties of these compounds.

5.3. Electronic Physical Properties

The new extended cluster compounds, owing to their crystalline and electronic structure, have an electronic behavior close to the first $M_xMo_6X_8$ compounds. They exhibit narrow Mo—Mo d bands at the Fermi level due to the pseudo-molecular nature of these compounds and properties related to frontier orbitals as seen above, with a contribution more and more important of the a_2 internal orbitals as soon as the Mo_6 clusters condense.

The $M_2Mo_{15}X_{19}$ phases are new superconducting materials, very similar to the $M_xMo_6X_8$ phases as seen in the structural packing of the block units. The critical temperatures are low, but the critical fields are comparable with those found in the $M_xMo_6X_8$ compounds: dHc_2/dT for $K_2Mo_{15}Se_{19}$ and $Ba_2Mo_{15}Se_{19}$ yields 34 kG K^{-1} and 32 kG K^{-1}, respectively [37].

The $In_{\sim 3}Mo_{15}Se_{19}$ phase is a superconducting phase with a nonstoichiometry in indium ions (from 2.9 to ~3.5). The superconducting critical temperature varies from 4.24 to 1.8 K. The zig-zag packing of the intercluster interactions involves a short mean-free path and consequently a high normal resistivity and a high critical field. Actually, the slopes (dHc_2/dT) of the critical field are between 68 kG K^{-1} and 78 kG K^{-1} [53]. The χ_g magnetic susceptibility (mainly Pauli susceptibility) at room temperature decreases versus the increasing content of indium ions and goes towards a diamagnetic susceptibility in Figure 24. Measurements of

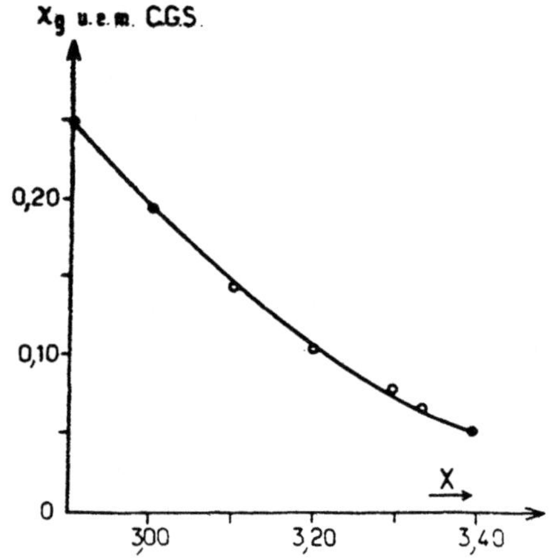

Fig. 24. Variation of the χ_g susceptibility versus the x content of indium ions in $In_xMo_{15}Se_{19}$ compounds at room temperature.

resistivity versus temperature carried out on a single crystal of limit composition $In_{\sim 3.5}Mo_{15}Se_{19}$ exhibit a semiconducting behavior with a very low gap (10^{-4}—10^{-5} eV) [54]. This behavior can be explained by the filling of the HOMO band of δ symmetry ($d_{x^2-y^2}$ character) up to $36e^-$ for the Mo_9 cluster attempting the $24e^-$ closed-shell Mo_6 cluster.

The filling of the HOMO band of the Mo_9X_{11} unit can be so found in the semiconducting $Ag_{2.3}CsMo_9Se_{11}$ compound (~$36e^-$) [9]. The calculated gap energy is very low ($E_g = 6.5 \times 10^{-3}$ eV) (Figure 25). On the other hand, the compounds $Ag_{4.4}ClMo_9Se_{11}$ and $Ag_{3.6}Mo_9Se_{11}$ both have a semimetallic behavior [43, 9]. The resistivities carried on single crystals are 2.43×10^{-3} Ω cm and 6.88×10^{-3} Ω cm, respectively, at room temperature and vary very little with temperature except at low temperature (< 80 K) for $Ag_{3.6}Mo_9Se_{11}$ (Figure 26). This behavior may be due to the filling of the nonbonding orbital $1a_2'$, the VEC being slightly higher than that of the silver semiconducting compound.

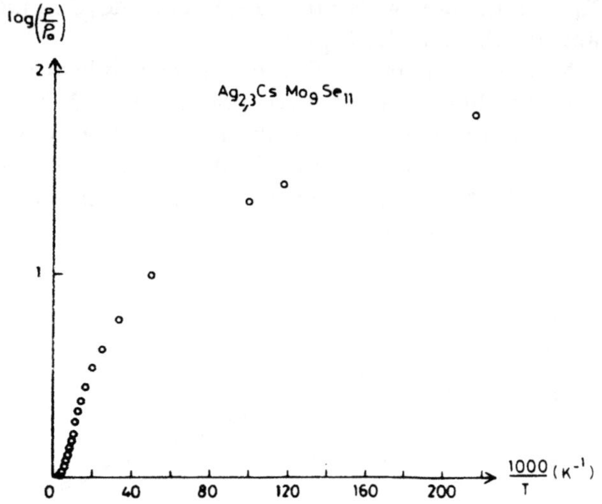

Fig. 25. Normalized resistance (log scale) as a function of reciprocal temperature in a $Ag_{2.3}CsMo_9Se_{11}$ single crystal.

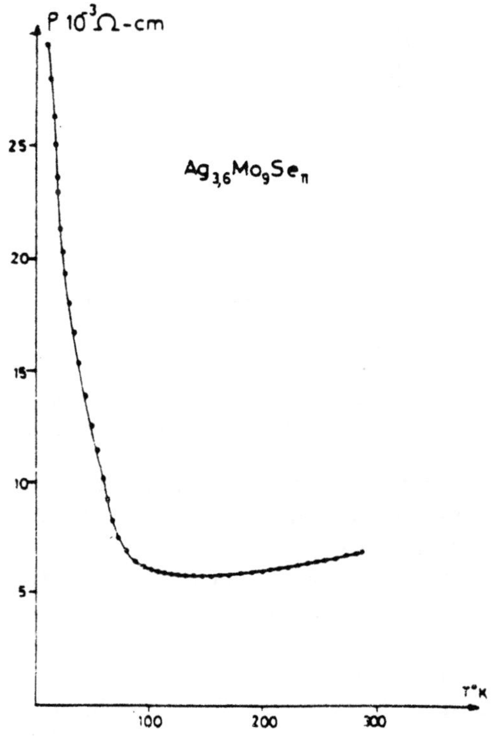

Fig. 26. Resistivity versus temperature in a $Ag_{3.6}Mo_9Se_{11}$ single crystal.

The $M_2Mo_9X_{11}$ compounds, where the Mo_6X_8 and $Mo_{12}X_{14}$ units co-crystallize, are metallic conductors down to 1.8 K [42].

All the $M_{2n-2}Mo_{6n}X_{6n+2}$ compounds are metallic conductors at room temperature [9, 45, 46]. So far, these additional compounds are only the subject of partial physical studies. The curious behavior of the resistivity study of the $Cs_2Mo_{12}Se_{14}$ compound performed on a single crystal is shown in Figure 27. Resistivity increases slightly when the temperature lowers and decreases roughtly with a poorly

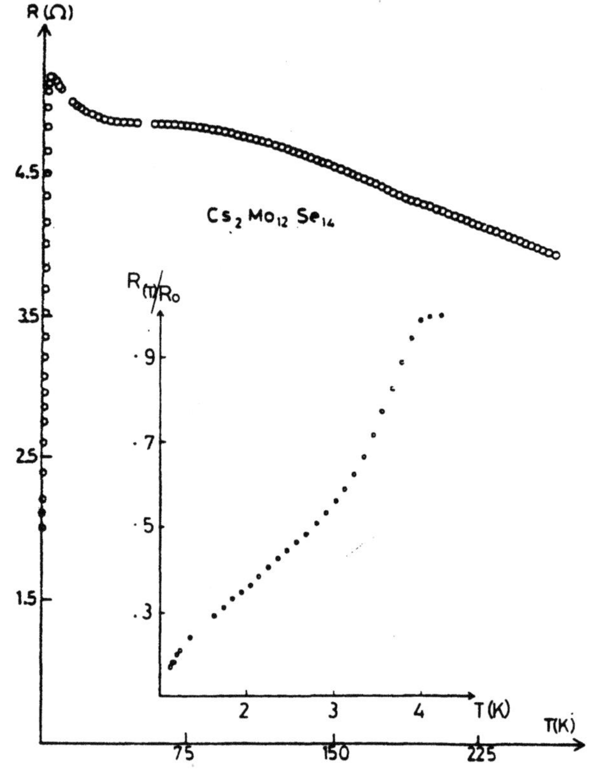

Fig. 27. Resistance versus temperature in a $Cs_2Mo_{12}Se_{14}$ single crystal. Inset: normalized resistance versus temperature displays the wide superconducting transition.

defined superconducting transition (transition-tail). This aspect is strongly suggestive of the one-dimensional organic superconductors [55] and the one-dimensional $Tl_2Mo_6Se_6$ compound as shown below [56]. The curve of resistivity carried on a single crystal of the $Cs_6Mo_{24}Se_{26}$ chalcogenide exhibits a superconducting transition temperature starting from 3 K and a behavior analogous to the previous phase [45]. The $Cs_6Mo_{24}S_{26}$ isotypical compound displays in a single crystal, a metallic-type behavior of resistivity from room temperature to 120 K, and down to this temperature the resistivity increases abruptly to reach, at 10 K, a value 10^5-fold higher than that at room temperature (see Figure 28). The nature of this metal-insulator transition is as yet unknown. The temperature is more accurately given on the derivative curve shown in the inset of Figure 28 ($T = 112$ K). Such a rough

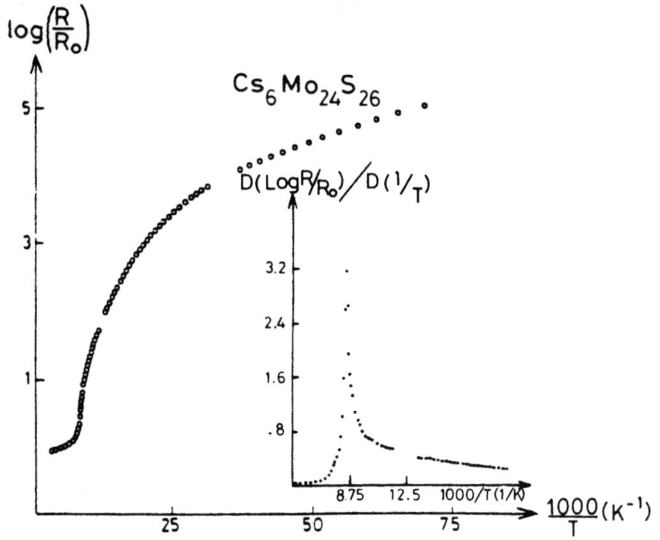

Fig. 28. Normalized resistance (log scale) as a function of temperature in a $Cs_6Mo_{24}S_{26}$ single crystal (after Reference [9]).

transition is generally observed in the 1D $M_2Mo_6X_6$ compounds reported below. This trend towards a behavior that looks more and more like the behavior of 1D compounds can be explained, as stated above in Section 5.1, by the more dominating character of the a_2 internal orbital (d_{xz} or d_{yz} character) with respect to the z^2-type frontier orbitals at the Fermi level. Actually, the role of this a_2 band will be of major importance in the $M_2Mo_6X_6$ compounds.

6. New Pseudo-One-Dimensional $(Mo_{6/2}X_{6/2})^1_\infty$ Infinite Chain

The new 1D cluster $(Mo_{6/2})^1_\infty$ may be viewed as the result of a linear condensation of an infinite number of face-sharing Mo_6 octahedral clusters. The $(Mo_{6/2}X_{6/2})^1_\infty$ chain is therefore the last member of the discrete condensed cluster units: Mo_6X_8, Mo_9X_{11}, $Mo_{12}X_{14}$, ... These striking compounds of stoichiometry $M_2Mo_6X_6$ (M = K, Rb, Cs with X = S; M = Li, Na, K, Rb, Cs, Ba, In, Tl with X = Se; M = Na, K, Rb, Cs, Ba, In, Tl with X = Te) have been synthesized first of all by Potel *et al.* [33, 54, 57], then by Gougeon *et al.* [9, 58], Honle *et al.* [59], and Huster *et al.* [60].

6.1. Preparation

The new $M_2Mo_6Se_6$ and $M_2Mo_6Te_6$ compounds can be made from the elements Na, In, K, Tl in the form of lumps and Mo in the form of powder and Se and Te in the form of grains in evacuated silica tubes. The mixture is put in alumina or molybdenum crucibles. The $M_2Mo_6S_6$ and $M_2Mo_6Se_6$ compounds (M = K, Rb, Cs, and Ba) can be synthesized by another method: the sulfides have been made up at 900 °C either by reduction of M_xMoS_2 under H_2 flow [61] or from stoichiometric

TABLE VII

Lattice parameters and volume for hexagonal $M_2Mo_6X_6$ compounds

$M_2Mo_6X_6$ Hexagonal symmetry Space group : $P6_3/m$ Laue group : $6/m$ $Z = 1$

	Ionic radius of cation (Å)	a_H (Å)	c_H (Å)	V_H (Å^3)	Ref.
$K_2Mo_6S_6$	1.55	8.76	4.42	293.73	54, 33
		8.72	4.41	290.39	9, 27
		8.720	4.407	290.20	60
$Rb_2Mo_6S_6$	1.63	8.96	4.41	306.60	9, 27
		8.958	4.411	306.53	60
$Cs_2Mo_6S_6$	1.78	9.26	4.42	328.21	9, 27
		9.269	4.419	328.78	60
$Li_2Mo_6Se_6$	—	8.513	4.48	281.16	62
$Na_2Mo_6Se_6$	1.24	8.741	4.421	292.52	57
		8.65	4.49	290.93	9, 24, 27
		8.629	4.479	288.81	62
$K_2Mo_6Se_6$	1.55	9.011	4.501	316.49	57
		9.022	4.481	315.86	62
$Rb_2Mo_6Se_6$	1.63	9.26	4.49	333.41	9, 24, 27
		9.243	4.483	331.67	62
$Cs_2Mo_6Se_6$	1.78	9.54	4.50	354.67	9, 24, 27
		9.533	4.484	352.89	62
$Ba_2Mo_6Se_6$	1.47	9.07	4.31	307.05	24, 54
$In_2Mo_6Se_6$	1.40	8.85	4.50	305.22	9, 27
		8.881	4.491	306.75	57
		8.851	4.497	305.09	62
		8.835	4.492	303.65	59
$Tl_2Mo_6Se_6$	1.60	8.9398(7)	4.4997(4)	311.44	27, 34, 57
		8.918	4.482	308.69	59, 63
		8.933	4.478	309.45	62
$Na_2Mo_6Te_6$	1.24	9.23	4.60	339.37	9, 27
		9.237	4.613	340.85	62
$K_2Mo_6Te_6$	1.55	9.60	4.60	367.13	9, 27
		9.597	4.600	366.90	62
$Rb_2Mo_6Te_6$	1.63	9.76	4.60	379.47	9, 27
		9.812	4.599	383.44	62
$Cs_2Mo_6Te_6$	1.78	10.09	4.60	405.56	9, 27
		10.09	4.596	405.21	62
$Ba_2Mo_6Te_6$	1.47	9.41	4.57	350.44	9, 24
$In_2Mo_6Te_6$	1.40	9.35	4.59	347.50	9, 27
		9.381	4.581	349.12	57
		9.342	4.591	347.04	62
		9.326	4.590	345.71	59
$Tl_2Mo_6Te_6$	1.60	9.441	4.591	354.37	9, 27, 57
		9.434	4.594	354.07	62
		9.428	4.583	352.78	59

TABLE VIII
Lattice parameters of metastable binary, ternary, or solid solution phases of $M_2Mo_6X_6$-type compounds

$M_2Mo_6X_6$		a_H (Å)	c_H (Å)	V_H (Å^3)	Ref.
$Ag_2Mo_6Se_6$		8.56	4.50	285.54	54, 9
		8.51	4.49	281.60	62
$Cu_xMo_6Te_6$		No defined parameters			62
$Pb_xMo_6Se_6$	$(x \sim 1)$	8.52	4.49	282.26	27
$Sn_xMo_6Se_6$	$(x \sim 1)$	8.44	4.49	276.98	27
$In_{2-x}Li_xMo_6Se_6$	$(x = 0.5)$	8.737	4.50	297.48	62
	$(x = 1.2)$	8.619	4.493	289.05	
	$(x = 1.5)$	8.57	4.49	285.58	
$In_{2-x}Mo_6Se_6$	$(x = 0.5)$	8.705	4.497	295.10	62
	$(x = 1.2)$	8.565	4.493	285.43	
	$(x = 1.5)$	8.466	4.495	278.99	
	$(x = 1.8)$	8.399	4.478	273.56	
Mo_6Se_6		8.35	4.44	268.08	54, 9, 24, 27
Mo_6S_6, Mo_6Te_6		Not well defined			61, 54, 27

amounts of tetrathiomolybdate, molybdenum, and sulfur [9, 60]. High-temperature synthesis (1000 °C) has been generally used to prepare the ternary selenides or tellurides. Recently, ion exchange reactions as well as extraction or insertion reactions of the ternary element at low temperatures have been carried on as for the synthesis of new Chevrel phases, nonstable at high temperatures [9, 24, 27, 54, 62]. The compounds with alkali or alkali earth metal are synthesized according to the following reaction from $In_2Mo_6X_6$ (X = Se, Te):

$$In_2Mo_6X_6 + 2MY \longrightarrow M_2Mo_6X_6 + 2InY^{\uparrow}, \qquad Y = Cl, Br, I.$$

Appropriate amounts of $In_2Mo_6X_6$ and salts (10% in excess) were mixed together and pressed into a pellet, and then placed in a silica tube under vacuum. The samples were rapidly heated up to 100 °C below the melting point of the salt, then the temperature was raised very slowly. During the heating process, the end tube opposite to the sample is kept close to room temperature in order to condense the vapors of the In halide. The samples were finally heated to about 10 °C above the melting point of the salt and maintained at this temperature for several hours. Metastable phases are obtained by this exchange reaction for M = Ag, Cu, Pb, Sn; X = Se. The new binaries Mo_6X_6 are synthesized by the reaction of HCl flow at 400 °C from $M_2Mo_6X_6$ ternaries according to the following reactions:

$$K_2Mo_6S_6 + 2HCl \longrightarrow 2KCl + Mo_6S_6 + H_2^{\uparrow} \quad [61]$$

or

$$In_2Mo_6X_6 + 6HCl \longrightarrow 2InCl_3^{\uparrow} + Mo_6X_6 + 3H_2^{\uparrow} \quad (X = S, Se, Te) \quad [27]$$

All the $M_2Mo_6X_6$ chalcogenides are reported in Table VII and VIII with lattice parameters and volume of the hexagonal unit cell.

It is thus possible to prepare mixed solid solutions $(M, M')_2Mo_6X_6$ by exchange reactions [62]. In the $Li_xIn_{2-x}Mo_6Se_6$ phase, the lithium ion can be removed to form $In_{2-x}Mo_6Se_6$ ($0 < x < 1.8$) by treatment with iodine in acetonitrile at room temperature (see Table VIII). The VEC per Mo atom can be so modified continuously from 4.03 to 4.33 when the x content decreases from 1.8 to 0.

All these compounds are black microcrystalline powders stable in the air atmosphere except $Li_2Mo_6Se_6$. Single crystals of these compounds are obtained by prolonged reaction between 1300 and 1600 °C in sealed molybdenum crucibles under low argon pressure. Needle-like crystals up to 1 cm long are obtained (see Figure 29).

Fig. 29. SEM picture of $Cs_2Mo_6Te_6$ single crystals.

6.2. Structural Description

X-ray diffraction powder patterns reveal the existence of a single phase material for the $M_2Mo_6X_6$ composition. All these compounds have a hexagonal symmetry, space group $P6_3/m$ with hexagonal parameters $a_{hex} \sim 9$ Å and $c_{hex} \sim 4.5$ Å.

The X-ray powder diffraction diagram of the $Tl_2Mo_6Se_6$ type is shown on Figure 30. The lattice parameters of these compounds are reported in Tables VII and VIII. According to the authors, such a difference in the reported parameters may result from a weak nonstoichiometry in either the M elements or the chalcogens. The first complete structural determinations of the $Tl_2Mo_6Se_6$ and $In_2Mo_6Se_6$ compounds have been established by Potel *et al.* [64] and Honle *et al.* [59]. The crystalline structure consists of a 1D condensation of an infinite number of Mo_6Se_6 units running along the c hexagonal axis and separated by cations M^+ or M^{2+}. It is interesting to note that the environments of molybdenum and the chalcogens are geometrically

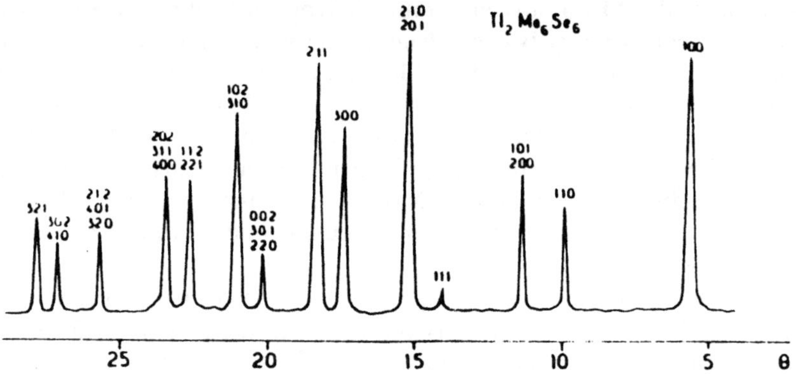

Fig. 30. X-ray diffraction powder pattern of $Tl_2Mo_6Se_6$-type compound (λ Cu K$\bar{\alpha}$).

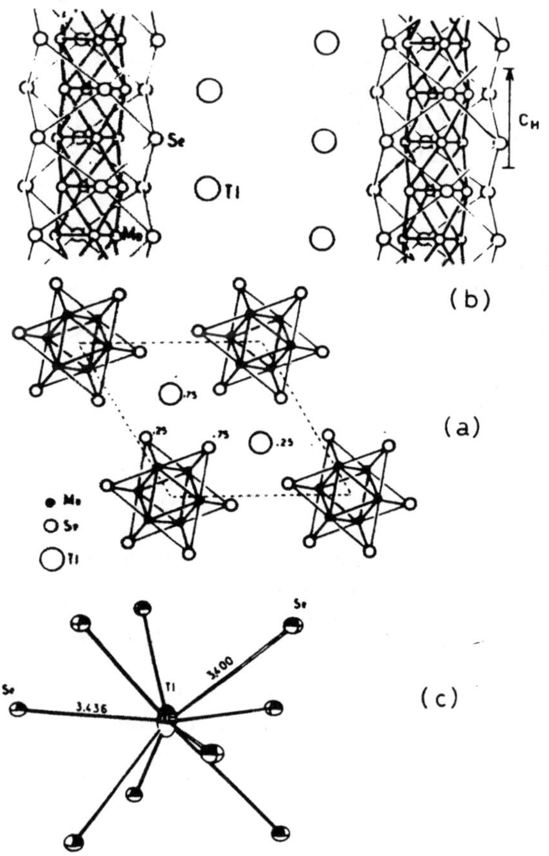

Fig. 31. Projection of the $Tl_2Mo_6Se_6$ structure onto: (a) the (0001) hexagonal plane; (b) the $(11\bar{2}0)$ hexagonal plane; (c) the thallium counterions in 9-prismatic site.

similar. In fact, the Mo atoms face four chalcogens forming two triangles with a common edge, and conversely each chalcogen atom faces two Mo triangles having a common edge. The intrachain Mo—Mo bonds are very close to the intracluster bond observed in the other ternary molybdenum chalcogenides, and also very close to the nearest neighbor distance in the Mo metal element (Mo—Mo = 2.723 Å). In $Tl_2Mo_6Se_6$, the Mo—Mo intratriangle and intertriangle distances are 2.663 and 2.725 Å, respectively. The interplane Mo—Mo distance is 2.25 Å. The M cation is located in a 9-prismatic site forming linear chains between the $(Mo_3X_3)^1_\infty$ units, as shown in Figure 31.

In these compounds, the Mo first has a valence lower than 2 in chalcogenides. In fact, the formal number of electrons on each Mo atom is 4.33 (VEC). Another difference with the other chalcogenides comes from the fact that there is no chalcogen capping on the 'ternary' axis in $M_2Mo_6X_6$ compounds. In the other chalcogenides, the chalcogens located on the ternary axis capped the Mo_3 staggered planes and created six opposite square chalcogen faces from which the intercluster bonding to the adjacent cluster was realized. In the new linear-chain materials, these latter bondings are absent and the shortest Mo—Mo interchain distance is greater than 6 Å (6.34 Å in $Tl_2Mo_6Se_6$), in contrast to the distance of about 3.40 Å commonly observed in the other cluster chalcogenides. Thus, there is almost no direct interchain $4d$—$4d$ overlap, and the electronic interchain coupling has to pass through chalcogen—chalcogen van der Waals bonds (Figure 32). From structural considerations, we therefore expect these materials to be highly anisotropic.

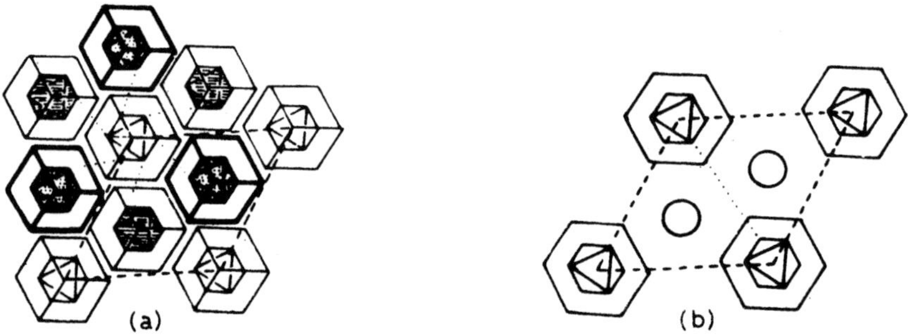

Fig. 32. Schematic projection of the units onto the (0001) hexagonal plane. (a) The intercluster Mo—Mo bonds are illustrated as well as in MMo_6X_8, $M_2Mo_{15}X_{19}$, $M_2Mo_9X_{11}$, and $M_{2n-2}Mo_{6n}X_{6n+2}$ compounds. For clarity, the M counterions are omitted. (b) The dotted line shows the shortest Mo—Mo intercluster distance in $M_2Mo_6X_6$ compounds.

Tables IX and X summarize the structural parameters of the $M_2Mo_6X_6$ chalcogenides with the $Tl_2Mo_6Se_6$-type structure. These results show that with monovalent cations the intratriangle distance in the Mo chain remains constant in these chalcogenides (2.65 Å). However, the intertriangle distance increases from sulfides (2.68 Å) to selenides (2.72 Å) and from selenides to tellurides (2.76 Å). That seems to arise from the dominating influence of the chalcogen size related to the peculiar environment of the chalcogen atom (two butterfly-shaped one-edge-sharing Mo_3 faces). The Mo—Mo interplane distance here is equal to the X—X interplane

TABLE IX

Structural parameters of $M_2Mo_6X_6$ compounds (X = S, Se, Te), space group $P6_3/m$; hexagonal set: Mo and X in $6h$, M atoms in $2d$ ($\frac{2}{3}$, $\frac{1}{3}$, $\frac{1}{4}$); $Z = 1$

		a_H (Å)	c_H (Å)	V_H (Å^3)	Mo x	Mo y	Mo z	X x	X y	X z	Ref.
$Tl_2Mo_6Se_6$		8.9398	4.4997	311.44	0.1861	0.1530	$\frac{1}{4}$	0.0663	0.3661	$\frac{1}{4}$	64
	Single crystal	8.918	4.482	308.69	0.1857	0.1530	$\frac{1}{4}$	0.0656	0.3658	$\frac{1}{4}$	59
$K_2Mo_6S_6$	Powder	8.7203	4.4076	290.25	0.1881	0.1559	$\frac{1}{4}$	0.059	0.349	$\frac{1}{4}$	60
$Rb_2Mo_6S_6$	Powder	8.9589	4.4114	306.63	0.1802	0.1548	$\frac{1}{4}$	0.056	0.346	$\frac{1}{4}$	60
$Cs_2Mo_6S_6$	Single crystal	9.2603	4.4251	328.61	0.1764	0.1520	$\frac{1}{4}$	0.0474	0.3307	$\frac{1}{4}$	9
	Powder	9.2698	4.4191	328.86	0.1741	0.1528	$\frac{1}{4}$	0.043	0.336	$\frac{1}{4}$	60
$Na_2Mo_6Se_6$	Single crystal	8.6552	4.4852	291.01	0.1949	0.1514	$\frac{1}{4}$	0.0849	0.3836	$\frac{1}{4}$	9
$K_2Mo_6Se_6$	Single crystal	9.0494	4.4932	318.62	0.1831	0.1517	$\frac{1}{4}$	0.0628	0.3612	$\frac{1}{4}$	9
$Rb_2Mo_6Se_6$	Single crystal	9.2642	4.4911	333.81	0.1776	0.1499	$\frac{1}{4}$	0.0556	0.3506	$\frac{1}{4}$	9
$Cs_2Mo_6Se_6$	Single crystal	9.5443	4.4995	354.94	0.1713	0.1474	$\frac{1}{4}$	0.0483	0.3376	$\frac{1}{4}$	9
$In_2Mo_6Se_6$	Single crystal	8.835	4.492	303.65	0.1886	0.1540	$\frac{1}{4}$	0.0698	0.3708	$\frac{1}{4}$	59
$Cs_2Mo_6Te_6$	Single crystal	10.086	4.6015	405.37	0.1625	0.1399	$\frac{1}{4}$	0.0479	0.3424	$\frac{1}{4}$	9
$In_2Mo_6Te_6$	Powder	9.326	4.590	345.71	0.1792	0.1412	$\frac{1}{4}$	0.0711	0.3711	$\frac{1}{4}$	59
$Tl_2Mo_6Te_6$	Single crystal	9.428	4.583	352.78	0.1763	0.1384	$\frac{1}{4}$	0.0683	0.369	$\frac{1}{4}$	59

TABLE X

Interatomic distances (Å) in $M_2Mo_6X_6$ compounds

		$K_2Mo_6S_6$ [60]	$Rb_2Mo_6S_6$ [60]	$Cs_2Mo_6S_6$ [9]	$Cs_2Mo_6S_6$ [60]	$Tl_2Mo_6Se_6$ [64]	$Tl_2Mo_6Se_6$ [59]	$Na_2Mo_6Se_6$ [9]	$K_2Mo_6Se_6$ [9]	$Rb_2Mo_6Se_6$ [9]	$Cs_2Mo_6Se_6$ [9]	$In_2Mo_6Se_6$ [59]	$Cs_2Mo_6Te_6$ [9]	$In_2Mo_6Te_6$ [59]	$Tl_2Mo_6Te_6$ [9]	$Tl_2Mo_6Te_6$ [59]
(a) $(Mo-Mo)_\Delta$	(×2)	2.631(8)	2.640(10)	2.654(1)	2.641(10)	2.663(3)	2.652(4)	2.658(1)	2.653(1)	2.656(1)	2.650(1)	2.661(3)	2.661(1)	2.64(3)	2.653(1)	2.65(2)
(b) $Mo_\Delta-Mo_\Delta$	(×4)	2.677(3)	2.681(3)	2.689(1)	2.685(3)	2.725(1)	2.724(3)	2.715(1)	2.720(1)	2.719(1)	2.722(1)	2.721(2)	2.763(1)	2.76(1)	2.757(1)	2.75(2)
(c) Mo—Mo interchain	(×2)	6.150(3)	6.365(3)	6.610(1)		6.343(6)		6.122(1)	6.434(1)	6.637(1)	6.877(1)	6.250(2)	7.403(1)	6.772(3)	6.828(2)	
(d) Mo—X	(×2)	2.56(1)	2.59(2)	2.582(1)	2.62(2)	2.695(5)	2.686(2)	2.689(1)	2.695(1)	2.695(1)	2.695(1)	2.689(2)	2.857(1)	2.83(3)	2.859(1)	2.85(3)
(e) Mo—X	(×1)	2.44(2)	2.48(4)	2.477(1)	2.55(2)	2.611(5)	2.604(6)	2.614(1)	2.612(1)	2.614(1)	2.601(1)	2.604(3)	2.802(1)	2.72(3)	2.806(1)	2.73(3)
(f) Mo—X	(×1)	2.45(2)	2.52(2)	2.493(1)	2.54(3)	2.622(5)	2.614(4)	2.620(1)	2.618(1)	2.623(1)	2.613(1)	2.619(3)	2.804(1)	2.79(3)	2.807(1)	2.83(3)
(g) X—X interchain	(×2)	3.95(4)	4.03(6)	4.270(2)	4.14(5)	3.877(6)	3.863(5)	3.769(1)	3.932(1)	4.076(1)	4.281(1)	3.820(3)	4.398(1)	4.00(3)	3.951(1)	4.02(3)
(h) X—X	(×4)	3.58(2)	3.63(2)	3.620(1)	3.68(2)	3.767(2)	3.755(4)	3.762(1)	3.764(1)	3.766(1)	3.759(1)	3.760(2)	3.968(1)	3.92(3)	3.969(1)	3.94(3)
(i) M—X	(×3)	3.36(2)	3.44(3)	3.537(1)	3.47(3)	3.436(2)	3.422(4)	3.424(1)	3.459(1)	3.525(1)	3.614(1)	3.411(3)	3.795(1)	3.61(1)	3.586(1)	3.63(3)
(j) M—X	(×6)	3.41(1)	3.48(2)	3.652(1)	3.64(2)	3.400(2)	3.394(2)	3.222(1)	3.448(1)	3.561(1)	3.708(1)	3.349(2)	3.849(1)	3.48(2)	3.523(1)	3.52(3)
		Powder	Powder	Monocryst.	Powder	Monocryst.	Monocryst.	Monocryst.	Monocryst.	Monocryst.	Monocryst.	Monocryst.	Monocryst.	Powder	Monocryst.	Monocryst.

distance. This steric effect of the chalcogen atom has already a major role in the previous Mo_{6n} clusters; whereas this role is minor with respect to the electronic effect in the Chevrel phases. It suggests that the Mo—Mo bond length is determined by a balance between Mo—Mo and Mo—X bond energies. Firstly, it remains constant in a similar chalcogen series and, secondly, is independent of the size of the intercalated monovalent ternary element. On the contrary, this ion size strongly affects interchain distances. Now, if we compare the X—X interchain distance with the van der Waals chalcogen distances, the relation $X—X_{inter}/X—X_{vdW}$ is close to 1 for the sulfurs, 0.98 for the selenides, and 0.90 for the tellurides. This divergence measures the covalency character of the chalcogen. Figure 33 shows the variation of the $X—X_{inter}/X—X_{vdW}$ ratio versus the ionic radius of the M cation. It is interesting to note that the dependence of this ratio on the ionic radius is similar for all the chalcogenides, except for M = Tl. This change is probably due to the more covalent nature of the Tl ion. An increase in the covalent character of the Tl—X bond yields a shorter distance between interchain chalcogen atoms.

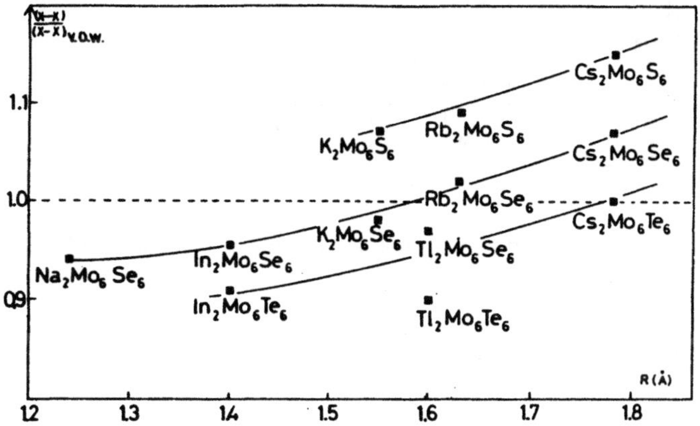

Fig. 33. The shortest chalcogen—chalcogen interchain distance over the twice chalcogen van der Waals radius $X—X_{inter}/X—X_{vdW}$ ratio versus the ionic radius of the M cation in $M_2Mo_6X_6$ compounds.

The same silver telluride cannot be prepared with this type of structure. Actually, the X-ray diffraction pattern shown in Figure 34 reveals the existence of a new single phase material for the composition $AgMo_6Te_6$. This compound is prepared at 900 °C from the elements (Ag, Mo, Te) in adequate proportions. A final reaction at 1200 °C leads to black microcrystalline powder stable in air atmosphere [9]. The single crystals are obtained by a chemical vapor reaction with the transport agent, silver chloride, in a gradient temperature (1050—1060 °C). This compound has a monoclinic unit cell (space group $C2/m$) with parameter a = 17.450 Å, b = 4.585 Å, c = 9.129 Å, β = 118°75, V = 640.4 Å^3, and Z = 2. The crystalline structure (Tables XI and XII) consists of an infinite number of Mo_6Se_6 units running along the b axis; each line is surrounded, as seen in the (010) plane (Figure 35), by six neighbor-lines according to the |100|, |001|, and |102| directions. According to

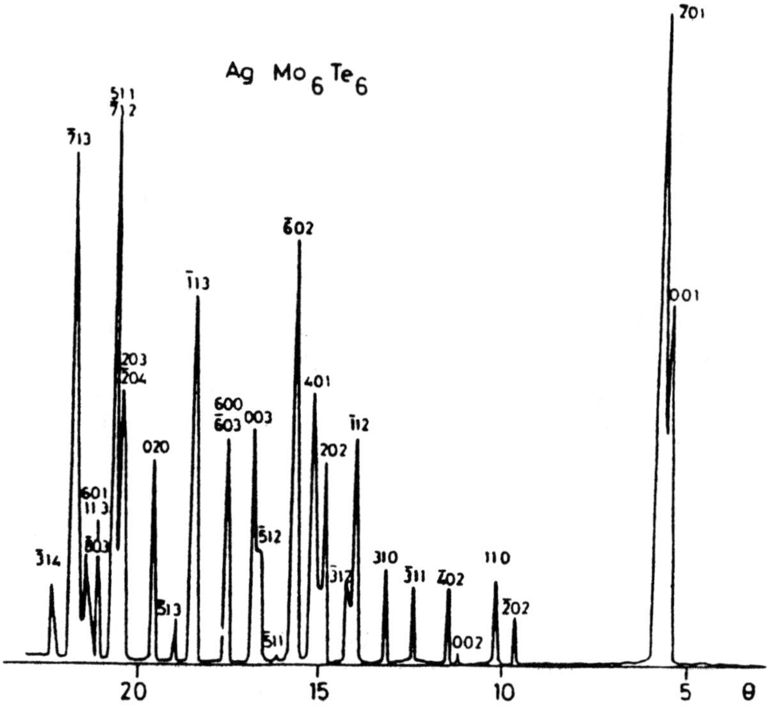

Fig. 34. X-ray diffraction pattern of $AgMo_6Te_6$ compound (λ Cu K$\bar{\alpha}$).

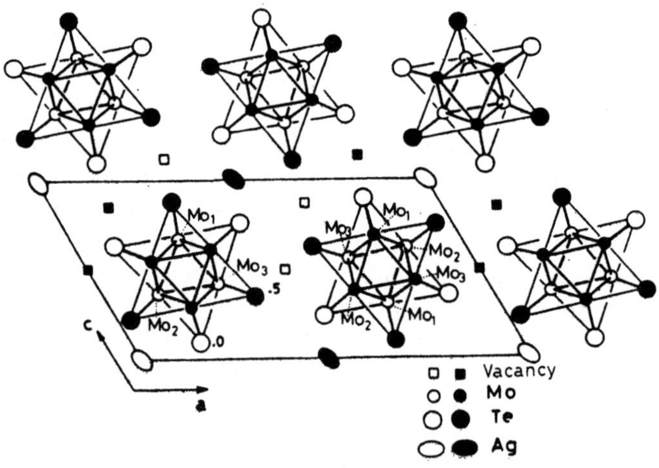

Fig. 35. Projection of the $AgMo_6Te_6$ structure onto the (010) monoclinic plane.

TABLE XI
Positional and thermal parameters (Å^2) of the monoclinic $AgMo_6Te_6$ compound

Atoms	Mo(1) (4*i*)	Mo(2) (4*i*)	Mo(3) (4*i*)	Te(1) (4*i*)	Te(2) (4*i*)	Te(3) (4*i*)	Ag (2*a*)
x	0.71559(4)	0.84841(4)	0.68426(4)	0.88622(3)	0.53957(3)	0.81636(4)	0
y	0	0	0	0	0	0	0
z	0.31020(8)	0.62319(8)	0.56850(8)	0.35805(7)	0.25262(7)	0.90128(7)	0
B_{eq} (Å^2)	0.49(1)	0.49(1)	0.49(1)	0.85(1)	0.95(1)	0.95(1)	2.97(3)

TABLE XII
Interatomic distances (Å) in $AgMo_6Te_6$ compound

Intratriangles distances Mo—Mo		Intertriangles distances Mo—Mo	
Mo(1)—Mo(2)	2.677(1)	Mo(1)—Mo(2)	2.743(1)
Mo(1)—Mo(3)	2.664(1)	Mo(1)—Mo(3)	2.770(1)
Mo(2)—Mo(3)	2.661(1)	Mo(2)—Mo(3)	2.772(1)
Intertriangle distance Δ—Δ	2.293		
Interchain distances Mo—Mo (<7 Å)			
Parallel at the direction |100|		Parallel at the direction |102|	
Mo(3)—Mo(3)	5.931(2)	Mo(2)—Mo(2)	6.349(2)
Mo(2)—Mo(3)	2 × 6.532(2)	Mo(1)—Mo(2)	2 × 6.855(1)
Mo(2)—Mo(2)	6.670	Mo(1)—Mo(1)	6.906(2)
Parallel at the direction |001|			
Mo(1)—Mo(3)	6.526(1)		
Mo(1)—Mo(1)	2 × 6.732(2)		
Intrachain Mo—Te distances			
Mo(1)—Te(1)	2.894(1)	Mo(3)—Te(2)	2.772(1)
Mo(1)—Te(2)	2.856(1)	Mo(3)—Te(3)	2.794(2)
Mo(1)—Te(3)	2 × 2.874(1)	Mo(3)—Te(1)	2 × 2.830(1)
Mo(2)—Te(1)	2.798(1)		
Mo(2)—Te(3)	2.850(1)		
Mo(2)—Te(2)	2 × 2.869(1)		
Interchain Te—Te distances			
Parallel at the direction |100|		Parallel at the direction |102|	
Te(1)—Te(1)	3.547(1)	Te(2)—Te(2)	4.130(1)
Te(1)—Te(2)	2 × 3.977(1)	Te(2)—Te(3)	2 × 4.324(1)
Parallel at the direction |001|			
Te(1)—Te(3)	2 × 3.740(1)		
Te(3)—Te(3)	2 × 4.229(1)		
Other distances			
Ag—Te(3)	2 × 2.881(1)	(1)—Te(3)	2.70
Ag—Te(2)	4 × 3.085(1)	(1)—Te(2)	3.00
		(1)—Te(1)	2 × 2.96
		(1)—Te(3)	2 × 2.96
		(2)—Te(2)	2 × 2.66
		(2)—Te(1)	4 × 2.90

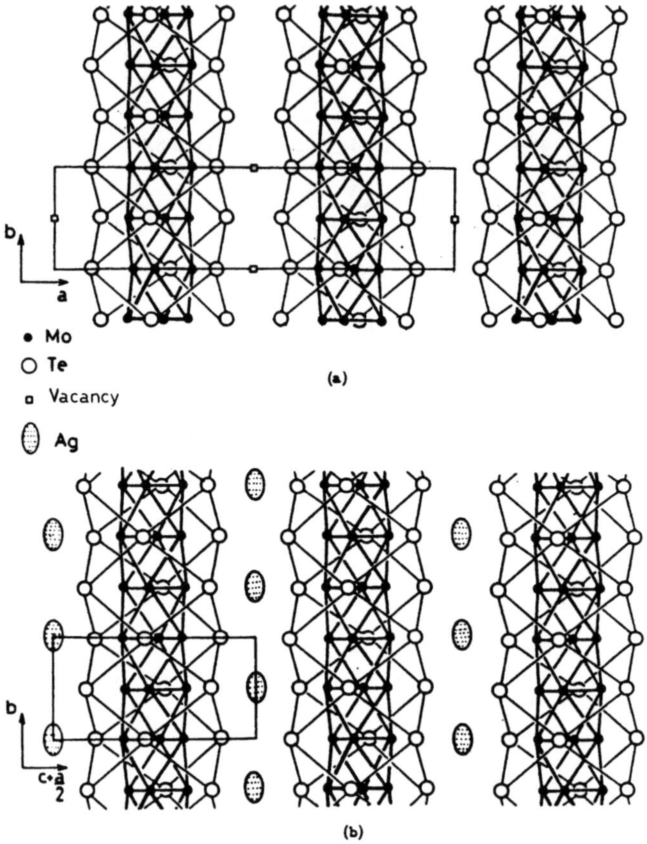

Fig. 36. Projection of the $AgMo_6Te_6$ structure onto (a) the (001) plane and (b) the $(\overline{2}01)$ plane.

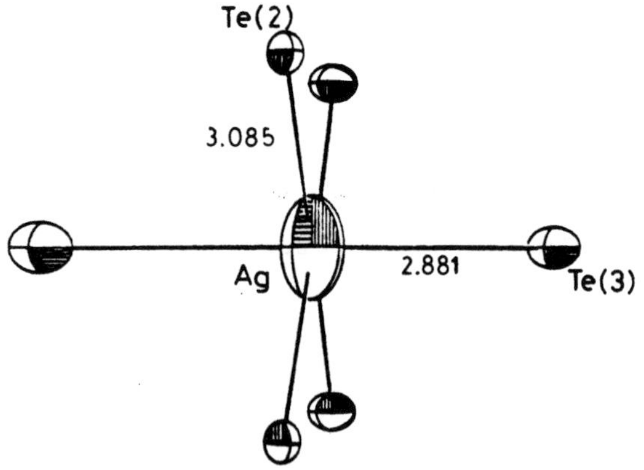

Fig. 37. Octahedral environment of Ag atoms in the $AgMo_6Te_6$ compound.

the c translation, the sequence is exactly analogous to that of the $M_2Mo_6X_6$ compounds. On the other hand, along the a and b directions, the C face-centered lattice translation ($\frac{1}{2}$, $\frac{1}{2}$, 0) implies a shift of equivalent chains, by a $b/2$ translation (Figure 36). This new situation destroys the 9-prismatic sites between chains, and creates four pseudo-octahedral sites, with only one of them being occupied by silver atoms (Figure 37).

6.3. Physical Properties

The transport properties of low-dimensional compounds have been intensively studied in the past decade [65, 66]. One major motivation for many investigations on materials like TTF—TCNQ, HMTSeF—TCNQ, KCP, $(SN)_x$, and $(NbSe_3)$ has been the hope of finding a new type of superconductivity in such materials. For 1D system, different ground states depending on relative strengths of interchain and intrachain couplings are possible at low temperature: charge density wave modulation, associated with a periodic lattice distorsion, spin density wave modulation, or a singlet or triplet superconducting state [67, 68]. Superconductivity and other instabilities are antagonistic and the superconducting state is stabilized when an external parameter, such as pressure, suppresses the charge or spin density wave modulation. So far however, only superconductivity in 1D compounds has been discovered in the family of $(TMTSF)_2X$ compounds, with anions such as NO_3, PF_6, AsF_6 or ClO_4 [69], in $(BEDT{-}TTF)_2ReO_4$ [70] and $(TMTSF)_2FSO_3$ under pressure [71], in $NbSe_3$ under pressure (T_c = 3.5 K) [72] and $TaSe_3$ (T_c = 2 K) [73]. That is one of the reasons why the physical properties of single crystals of new $M_2Mo_6X_6$ compounds are now extensively investigated. The electrical resistivity, the Seebeck coefficient, and magnetic susceptibility have been measured on single crystals (needles growing along the c hexagonal axis) [9, 33, 74, 79]. The band structure calculations of these linear chains, carried on by Hughbanks and Hoffmann [47], as we have seen in Section 5.1, suggest that some of these materials could undergo a 'Peierls distortion' leading to a semiconducting ground state [80]; however, so far, three compounds of this large family do not undergo a semiconducting transition and remain metallic throughout the whole range of temperature.

6.3.1. *Properties of $Tl_2Mo_6Se_6$, $Tl_2Mo_6Te_6$, and $In_2Mo_6Se_6$ Compounds*

Among these compounds, $Tl_2Mo_6Se_6$ has been extensively investigated, because it presents a superconducting behavior. Moreover, the different processes of synthesis of the compound produce single crystals with different physical properties. The measurements of $Tl_2Mo_6Se_6$ samples from different batches show a critical temperature about 2.5—3 K (α-batch); for the other samples, a critical temperature T_c = 5.5 K (β-batch) is obtained; and, finally, some samples show T_c ~ 6.5 K (γ-batch). The transition width is between 0.02 and 0.5 K (Figure 38). The resistivity is measured along the needle axis by means of the usual four-probe technique (Figure 39). The absolute resistivity at room temperature is around 100 μΩ cm for β-batch [76], a value intermediate between 220 μΩ cm for samples A in the α-batch and

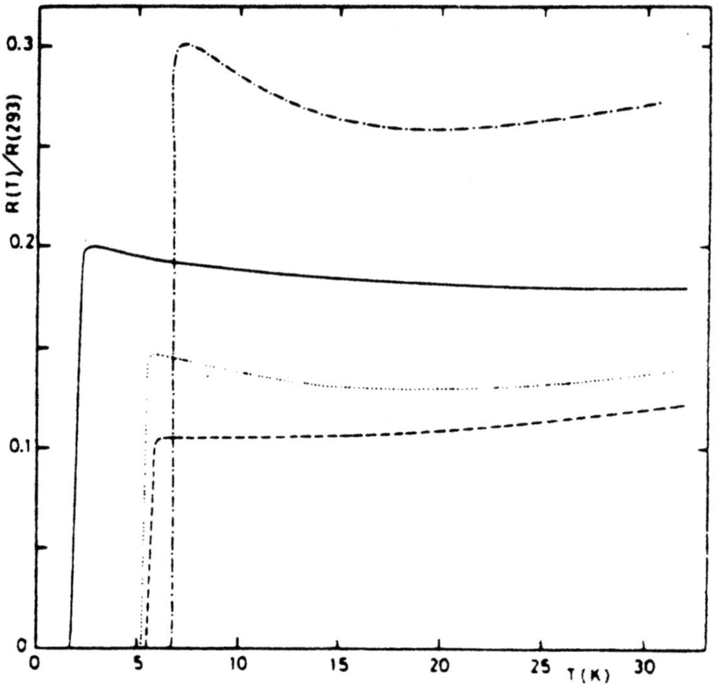

Fig. 38. Variation of the normalized resistance as a function of temperature in the vicinity of their critical superconducting temperature for $Tl_2Mo_6Se_6$ samples from the β-batch (· · ·) and (---), from the γ-batch (-·-·-·), and for the $Eu_{1.5}Tl_{0.5}Mo_6Se_6$ sample (—).

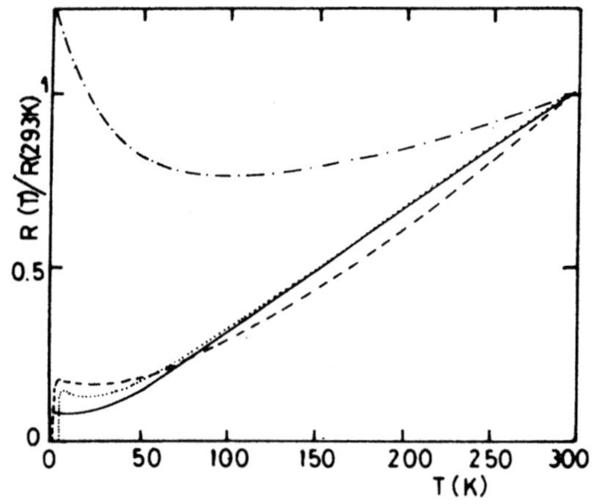

Fig. 39. Variation of the normalized resistance as a function of the temperature for $In_2Mo_6Se_6$ (-·-·-·), $Tl_2Mo_6Te_6$ (—), $Tl_2Mo_6Se_6$ (· · ·), and $Eu_{1.5}Tl_{0.5}Mo_6Se_6$ (---).

30 μΩ cm for one sample, B, found in this same batch [74]; sample B shows an anomalous T^2 dependence on resistivity and a higher T_c around 5.8 K as the samples of β-batch (Figure 40).

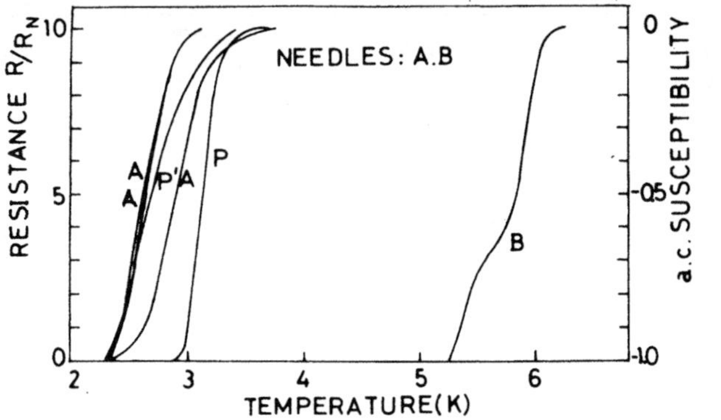

Fig. 40. The superconducting transition temperature of several samples (A, B) of the α-batch as seen along the c axis resistivity. For the polycrystalline sample (P), both resistive (P) and inductive (P') transitions are shown.

Armici *et al.* [74] and Lepetit *et al.* [75, 76] have measured the critical magnetic field Hc_2 with different angular orientations. For these measurements, angles have been defined: two Θ polar angles, when H, the applied field, is rotated between the parallel orientation $\Theta = 0$ ($H \| c$ hexagonal axis) towards the transverse orientation $\Theta = \pi/2$ ($H \perp c$ axis), i.e. towards the a_{hex} (Θ_1) or $|210|$ hexagonal (Θ_2) axes, and one azimuthal angle ϕ between H and the normal to the (010) face in the hexagonal (001) plane. The resistive transition has been measured in a magnetic field as a function of ϕ, and Θ with an angular resolution better than 0.2°.

6.3.1.1. *Azimuthal anisotropy.* $Tl_2Mo_6Se_6$ presents a Hc_2 minimum when H is applied perpendicular to the ribbon plane or to the (010) face, i.e. when $\Theta = \pi/2$. Figure 41 shows resistive transitions for two different azimuthal angles in the (001)

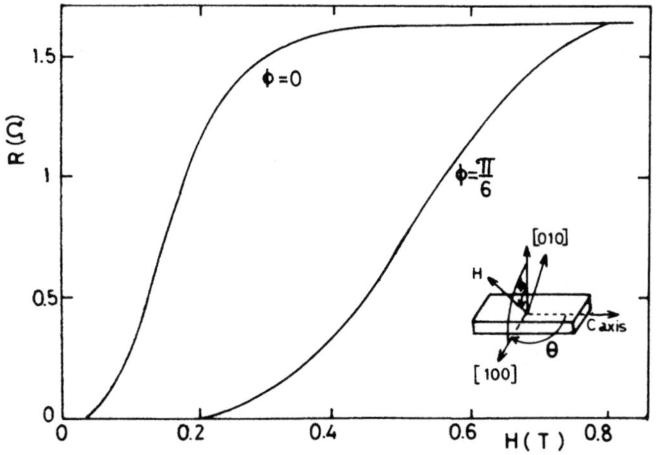

Fig. 41. Resistive transitions of $Tl_2Mo_6Se_6$ (β-batch) at 4.2 K as a function of the magnetic applied field for two values of the ϕ azimuthal angle (as defined in the inset).

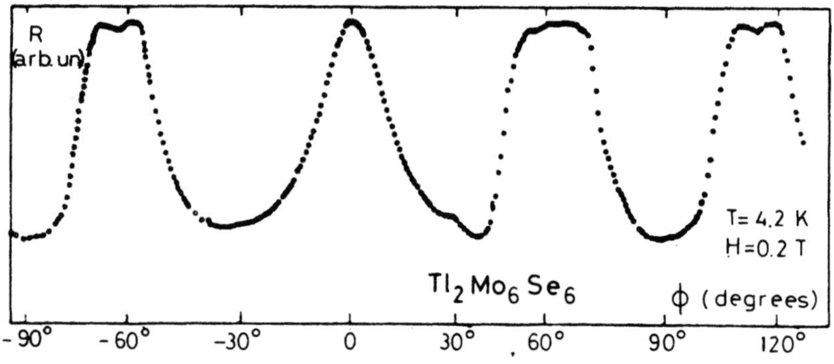

Fig. 42. Variation of the resistance of $Tl_2Mo_6Se_6$ (β-batch) at 4.2 K with a constant magnetic field of 0.2 T as a function of the ϕ azimuthal angle.

plane. The measurements of the resistance as a function of ϕ show a periodic variation of 60° (Figure 42). This anisotropy of Hc_2 follows the crystallographic hexagonal symmetry. As seen in Figure 42, the largest dimension of the cross section of the ribbon, determined by X-rays, is the $|100|$ direction. The ϕ azimuthal angle is defined as $\phi = 0$ when H is parallel to the normal to the (010) face, or to the ribbon plane, i.e. along the $|210|$ direction. In this transverse orientation, Hc_2 is maximum when H is applied along the crystallographic axis $|100|$ or $|010|$, but minimum when H is applied along the $|210|$ direction. The anisotropy ratio is Hc_2 $(\phi = \pi/6)/Hc_2\ (\phi = 0) \sim 2.0$—2.5.

6.3.1.2. *Polar anisotropy.* In Figure 43, the resistive transitions are plotted versus the magnetic field for different values of the Θ_1 polar angle. From these different

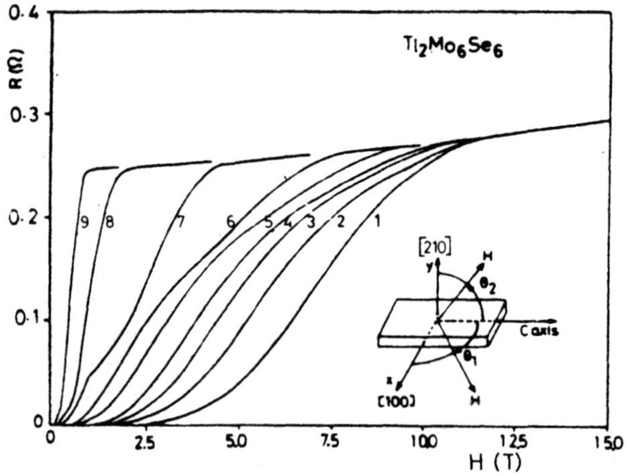

Fig. 43. Resistive transition of $Tl_2Mo_6Se_6$ (β-batch) at 4.2 K as a function of the magnetic field for the different values of the Θ_1 polar angle (as defined in the inset). curves 1—9 are for Θ_1 respectively equal to 0°, 0.29°, 0.67°, 1.15°, 2.10°, 3.75°, 7.6°, 19.6°, 43.6°.

curves, the anisotropy of the Hc_2 critical field is plotted versus the Θ_1 polar angle as well as for the Θ_2 polar angle in Figure 44.

The anisotropy of the critical field is very large:

$$Hc_{2\parallel}/Hc_{2\perp}(\Theta_2 = \pi/2) \sim 26, \qquad Hc_{2\parallel}/Hc_{2\perp}(\Theta_1 = \pi/2) \sim 13.$$

This latter anisotropy is of the same order of magnitude as the one for the low-T_c $Tl_2Mo_6Se_6$ samples of the α-batch [74].

The critical fields of filamentary superconductors have been calculated by Turkevitch and Klemm [81] extending the model of Lawrence and Doniach [82] of Josephson coupled planes. When H is applied perpendicular to the chains, the vortex current follows the chains and tunnels by the Josephson effect from one chain to another. When H is parallel to the chains, the vortex current is entirely a Josephson tunnel current. Near T_c, the coherence lengths (ξ) are much larger than the interchain distances and a generalized effective mass model can be used.

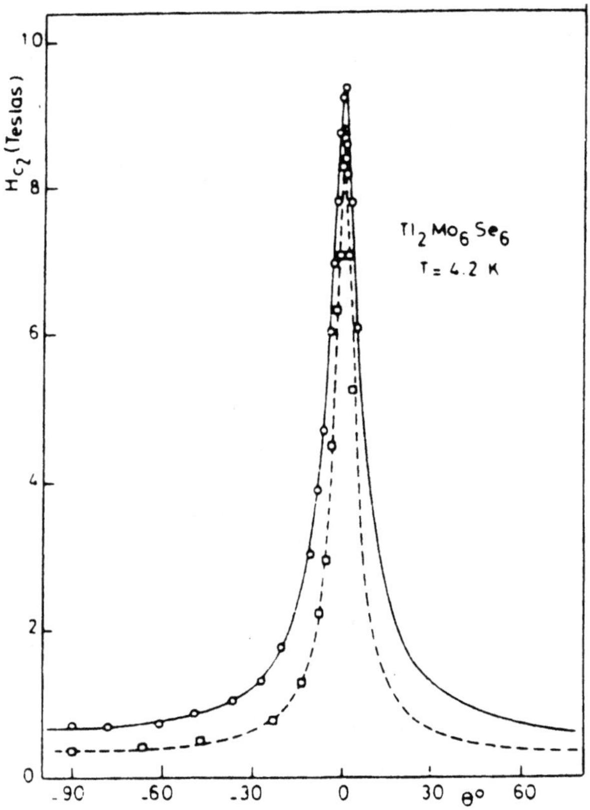

Fig. 44. Variation of the Hc_2 critical field of $Tl_2Mo_6Se_6$ (β-batch) as a function of the Θ_1 (○) and Θ_2 (□) polar angles as defined in the inset of Figure 43. The straight and dashed lines plot the theoretical variations of Hc_2 (○) from the expression seen in Section 6.3.1 with $\varepsilon_x = 14.8$ and $\varepsilon_y = 27.2$, respectively. The Hc_2 field is defined as the field for which the resistance is 90% of its normal resistance.

The angular polar dependence of Hc_2, when H is rotated from parallel to perpendicular orientation to the chains, is:

$$Hc_2(\Theta) = \frac{\Theta_0}{2\pi\xi^2}\,\frac{1}{\varepsilon_y}\,\frac{1}{(\cos^2\Theta + \varepsilon_x^2\sin^2\Theta)^{1/2}}\,,$$

with $\varepsilon_x = (m/mx)^{1/2}$ and $\varepsilon_y = (m/my)^{1/2}$ when mx and my are the effective masses, respectively in the x, |100| direction, and y, |210| direction. The theoretical variations of $Hc_2(\Theta)$ for the two polar dependences fit well the experimental points: in the (010) planes, at 4.2 K, the best fit between the experimental data and the theoretical variation of $Hc_2(\Theta_1)$ corresponds to $\varepsilon_x = 14.8$—15 for $Hc_{2\parallel} = 9.35$ T and $Hc_{2\perp} = 0.63$ T measured, and in the plane defined by the perpendicular to the (010) face and the c axis, the best fit of $Hc_2(\Theta_2)$ corresponds to $\varepsilon_y = 27.2$ for $Hc_{2\parallel} = 8.57$ T and $Hc_{2\perp} = 0.32$ T [76].

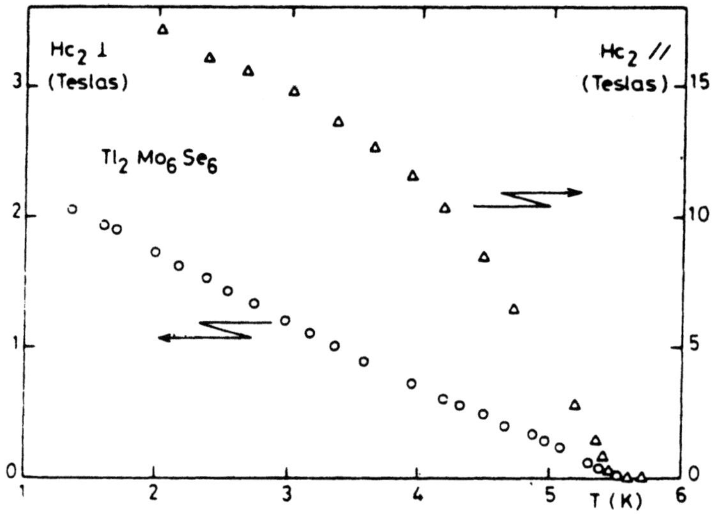

Fig. 45. Variation of the critical field $Hc_{2\parallel}$ and $Hc_{2\perp}$ ($\phi = \pi/6$) of $Tl_2Mo_6Se_6$ crystal (β-batch) as a function of temperature.

TABLE XIII
Critical temperatures, slopes of anisotropic critical field, and anisotropic critical fields at 4.2 K for $Tl_2Mo_6Se_6$

Crystals $Tl_2Mo_6Se_6$	Batch α	Batch β	Batch γ
T_c	2.5—3	5.5—5.8	6.57
$dHc_{2\perp}/dT$ (T K^{-1})	0.1	0.43 ($\phi = \pi/6$) 0.2 ($\phi = 0$)	0.6
$dHc_{2\parallel}/dT$ (T K^{-1})	2.6	6.0	8.0
$Hc_{2\parallel}$ at 4.2 K		9.35—8.57	20
$Hc_{2\perp}$ at 4.2 K		0.63—0.32	1.54

The curve of the Hc_2 critical field versus temperature has been plotted for two preferential directions: $Hc_{2\parallel}$ and $Hc_{2\perp}$ to the c axis (with $\phi = \pi/6$) for $Tl_2Mo_6Se_6$ (Figure 45). Curiously, from the γ-batch, with a high T_c, there is no observed azimuthal anisotropy, the Θ polar anisotropy is equivalent to the previous case but the critical field $Hc_{2\parallel}$ is very high (~ 20 T). In Table XIII are reported the whole superconducting properties of three different batches.

6.3.1.3. *Normal state properties.* The anisotropy of the normal state resistivity of $Tl_2Mo_6Se_6$ crystal (α-batch) has been investigated [74]: the $\rho_\perp/\rho_\parallel$ resistivity ratio is about 10^3. The temperature dependence of the transverse resistivity is typical metallic with a resistance ratio $\rho(300\text{ K})/\rho(4.2\text{ K}) \sim 4$. Similarly, the parallel resistance $\rho(300\text{ K})/\rho(4.2\text{ K})$ is about 5 to 6.

The effective mass approximation can also be obtained from microscopy theory. This is the same effective mass that enters the normal state resistivity and in the dirty limit, from the linearized Ginzburg—Landau equation, the anisotropy Hc_2 is obtained as proportional with the ρ resistivity [83]

$$Hc_{2\parallel} = 0.31\rho_\perp \gamma T_c = 0.31\varepsilon^2\rho_\parallel \gamma T_c,$$
$$Hc_{2\perp} = 0.31(\rho_\perp \rho_\parallel)^{1/2}\gamma T_c = 0.31\varepsilon\rho_\parallel \gamma T_c,$$

and

$$\rho_\perp = \varepsilon^2 \rho_\parallel,$$

with ρ in Ω cm, H in Tesla, γ in J K^{-2} m^{-2}.

Thus,

$$Hc_{2\parallel}/Hc_{2\perp} = (\rho_\perp/\rho_\parallel)^{1/2} = \varepsilon.$$

From this equation, the anisotropic resistivity ratio experimentally found of $\rho_\perp/\rho_\parallel \sim 10^3$ corresponds to an anisotropy of 30 in Hc_2 and is in reasonable agreement with the measured anisotropy of the Hc_2 critical field of ~ 26—27. This observed anisotropy is a result of intrinsic effects related to the anisotropy of the Fermi surface. Actually, the weak electronic interchain coupling which will result from the

Fig. 46. Schematic view of the excellent overlap between a_1'' orbitals of d_{xz} or d_{yz} character, creating a very wide a_2 band: (a) onto the (0001) hexagonal plane, and (b) along the c axis. This view along the chain exhibits the propagation of electrons according to a helical symmetry (results of SCF—SW—Xα calculations).

van der Waals Se—Se interactions and the very strong coupling along the chain (the very large a_2 band (~ 4 eV) coming from helical molybdenum molecular orbitals) of molybdenum bondings (Figure 46), imply a very high anisotropy of the Fermi surface. A remarkable result is the very large coherence length of about 2000 Å in the direction of the chains. This is only found in very pure materials, and for materials with very high Fermi velocity. It supports our above conclusions for a large mobility along the chains. These $Tl_2Mo_6Se_6$ compounds are very anisotropic, more anisotropic than other inorganic chain-like compounds such as $NbSe_3$ [84] or $TaSe_3$ [85]. As seen elsewhere, the calculated Fermi surface consists of 1D sheets perpendicular to the chain direction overlapping with a small pocket of electrons, located at the zone boundary. These electrons protect the structure from a Peierls distortion and are responsible for the superconductivity. The anisotropy is very sensitive to the shape, and the volume of the electrons pockets. The measured azimuthal and polar anisotropies have their origin in the shape of the pockets. The occupancy of these pockets must be connected very subtly with the sample preparation (stoichiometry, defects, . . .), in order to explain, for example, the absence of azimuthal anisotropy in some samples from the γ-batch. The occupancy of the electron pockets might be supposed to change with pressure. Huang *et al.* [86] has shown on $Tl_2Mo_6Se_6$ crystals, similar to those α-batch, that T_c decreases under pressure at a rate of $\Delta T_c/\Delta P$ equal to -7.5×10^{-5} $Kbar^{-1}$, which can indicate a significant band structure change. In addition, the electrical conductivity increases sharply under pressure in the whole temperature range, which leads to the suggestion of a change in the electron—phonon scattering mechanism under pressure. However, preliminary Tc measurements on $Tl_2Mo_6Se_6$ crystals from the β-batch [76] show no pressure dependence or a slight decrease of T_c at least one order of magnitude less than that measured by Huang *et al.*

In conclusion, the $Tl_2Mo_6Se_6$ crystals from several batches show different behaviors opposite to the same property. But all these crystals have identical lattice parameters and their differences might only be due to factors related to their preparation, such as furnace temperature, temperature gradient, weight of the batch and whether or not there was an excess of thallium. Samples from the γ-batch have the highest T_c, and the highest dHc_2/dT (Table XIII) which is in agreement with a smaller mean free path as deduced from the resistance ratio between 293 K and helium temperature ($\rho(298\ K)/\rho(4.2\ K) = 3$) compared to 10 found in the β-batch.

In order to study the substitution of Tl^+ ions by M^{2+} magnetic ions, samples containing Eu^{2+} ions have been prepared [76] with the nominal composition $Tl_{2-x}Eu_xMo_6Se_6$ with $x = 0.25$, 0.5, 0.75, 1, or 1.5. X-ray measurements on powder indicate for $x = 0.25$ a single phase similar to $Tl_2Mo_6Se_6$, but reveal, for $x > 0.5$, the presence of $EuMo_6Se_8$. Table XIV gives the variation of the unit cell parameters as a function of x. The decrease of the a_{hex} parameter essentially is a consequence of the smaller ionic radius of Eu^{2+} compared with the Tl^+ one, and the decrease of c_{hex} corresponds to a larger charge transfer onto the chain. The change of the unit cell parameters, as a function of the divalent cation concentration, can be compared with the one for $Ba^{2+}_{1.8}Mo_6Se_6$ as also given in Table XIV.

TABLE XIV
Hexagonal lattice parameters of $Tl_{2-x}Eu_xMo_6Se_6$ compounds with nominal composition and of $Ba_{1.8}Mo_6Se_6$ compound

	a_H (Å)	c_H (Å)
$Tl_2Mo_6Se_6$	8.94	4.500
$Tl_{1.75}Eu_{0.25}$	8.89	4.487
'$Tl_{1.5}Eu_{0.5}$'	8.88	4.485
'$Tl_{1.25}Eu_{0.75}$'	8.87	4.482
$Ba_{1.8}Mo_6Se_6$	9.05	4.290

Needle-like monocrystals of $Tl_{2-x}Eu_xMo_6Se_6$, with $x = 1$ and 1.5 nominal compositions, have been synthesized after long annealing at 1200 °C. The variation of the resistance below room temperature for the $Eu_{1.5}Tl_{0.5}Mo_6Se_6$ sample is shown in Figure 39 and near superconducting transition in Figures 38 and 47. The large roundness of the resistance indicates the competition between the increase of the resistance in the normal state, for temperatures lower than the resistance minimum and the superconducting transition. This increase of resistance at low temperature is due to Eu^{2+} magnetic ions (Kondo effect). The critical temperature T_c at the mid-value of the resistive transition is 2 K.

The variation of the resistance of needles of $Tl_2Mo_6Te_6$ as a function of temperature exhibits a metallic behavior in Figure 39. Its resistance decreases linearly when the temperature is reduced and saturates at low temperature. The resistance

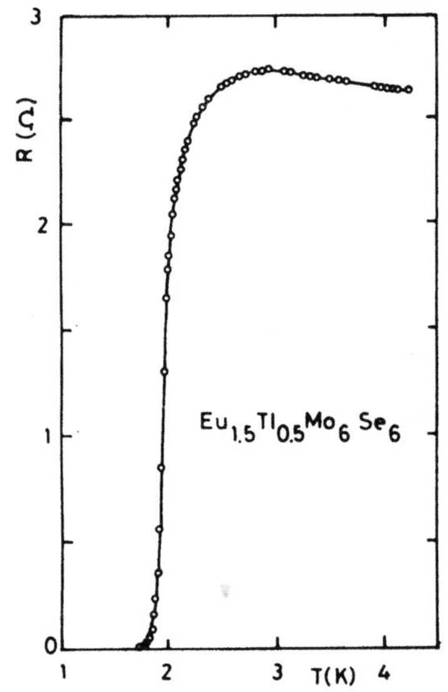

Fig. 47. Variation of the resistance of $Eu_{1.5}Tl_{0.5}Mo_6Se_6$ as a function of temperature in the vicinity of its superconducting transition.

ratio between $T = 293$ and 4.2 K is 14. No superconducting transition is observed down to 2 K [76].

For $In_2Mo_6Se_6$, several type-behaviors have been observed. Mori *et al.* [78] have measured the room-temperature resistivity of $In_2Mo_6Se_6$ about 230 μΩ cm. This value is about the same as that of $Tl_2Mo_6Se_6$ (β-batch) (~ 220 μΩ cm). The resistivity ratio is about 10 from the room temperature value to 4.2 K. At high temperature, the temperature dependence is linear (ρ proportional to T). One of the measured samples showed a sudden decrease of resistivity at 5.8 K (temperature corresponding with the superconducting transition temperature of $Tl_2Mo_6Se_6$). In the other metallic samples, no sign of superconductivity was found down to 1.8 K. For Lepetit *et al.* [76], when T is reduced the resistance of $In_2Mo_6Se_6$ needles crystals decreases, but less than linearly with T. Around $T = 100$ K, a broad minimum occurs, and the resistance increases at low temperature with a value at 2 K around 1.5 times the room-temperature values, as seen in Figure 39.

The Seebeck coefficient has been measured by Mori *et al.* [78] on single crystals of $Tl_2Mo_6Se_6$ and $In_2Mo_6Se_6$ in Figure 48. Both $Tl_2Mo_6Se_6$ and $In_2Mo_6Se_6$ show negative thermoelectricity and similar temperature dependence, but the magnitude of $Tl_2Mo_6Se_6$ (−40 μV K^{-1}) is about eight times larger than that of $In_2Mo_6Se_6$ (−5 μV K^{-1}). At high temperature, the Seebeck coefficients were approximately

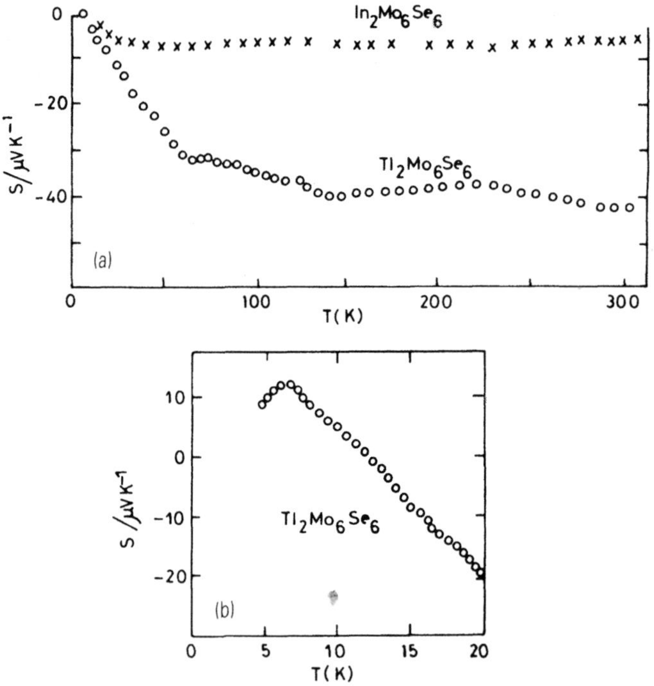

Fig. 48. (a) Seebeck coefficient of $In_2Mo_6Se_6$ and $Tl_2Mo_6Se_6$ compounds along the c-needle axis. (b) Seebeck coefficient of $Tl_2Mo_6Se_6$ at low temperature.

independent of temperature; at low temperature, the Seebeck coefficient for $Tl_2Mo_6Se_6$ becomes positive and makes a maximum around 7 K and again tends to zero.

The Seebeck coefficient due to electron scattering is generally expressed by the equation [87]:

$$S = -\frac{\pi^2 k^2 T}{3\,|e|\,E_F} \left| \frac{\partial \ln \sigma(E)}{\partial \ln E} \right|_{E=E_F}.$$

The difference of S arises only from E_F and can be explained by a two-band model where the main band is represented by a 1D free-electron-like band (a_2 band viewed in Figure 46) contributing mainly to the S Seebeck coefficient and a sub-band making a little contribution to S. The structure of the linear chain, thus of the main band with d-Mo character, varies little between $Tl_2Mo_6Se_6$ and $In_2Mo_6Se_6$ and leads to the same temperature dependence of S. This subband appears come from the p band of the M^+ ion, overlaps differently with the main band, and thus alters the occupancy of the main band. Really, the larger covalency of the Tl ions with respect to the In ions in $M_2Mo_6X_6$ compounds, as seen in Figure 33, can explain the different overlapping of this subband with the main band. In a free-electron-like band, the electron density of states $N(E)$ and other factors show an E dependence as $N(E)$ proportional to $E^{-1/2}$. Using the first equation, S is related to $N(E_F)$ as:

$$\frac{S_{Tl}}{S_{In}} = \left(\frac{N_{Tl}(E_F)}{N_{In}(E_F)} \right)^2$$

(where the subscripts Tl and In represent $Tl_2Mo_6Se_6$ and $In_2Mo_6Se_6$). From this relation, the experimental result found $S_{Tl}/S_{In} \sim 8$ involves $N_{Tl}(E_F)/N_{In}(E_F) \sim 2.8$, i.e. that $N(E_F)$ density of states of $Tl_2Mo_6Se_6$ is qualitatively larger than the $In_2Mo_6Se_6$ one. The superconducting critical temperature is closely related to the $N(E_F)$ density of states at the Fermi level; so, the smaller $N(E_F)$ in $In_2Mo_6Se_6$ corresponds well to the nondiscovery of superconductivity in this compound.

Remark: The one-dimensional $AgMo_6Te_6$ compound with shifted $(Mo_3Te_3)^1_\infty$ chains presents a metallic behavior on the whole interval of temperature (from 4.2 K to room temperature). The $\rho_\parallel$ resistivity is equal to 159 μΩ cm at room temperature and varies linearly with decreasing temperature down to 100 K. The residual resistivity at 4.2 K is 15.3 μΩ cm [9].

6.3.2. *Properties of $M_2Mo_6X_6$ (M = alkaline ions)*

Electrical resistivities have been measured along the needle axis for $Cs_2Mo_6Te_6$, $Rb_2Mo_6Se_6$, and $Rb_2Mo_6Te_6$ by Tarascon *et al.* [79]; and for $Cs_2Mo_6X_6$ (X = S, Se, Te), $Na_2Mo_6Se_6$, $K_2Mo_6Se_6$, and $Rb_2Mo_6Se_6$ by Gougeon *et al.* [9, 58, 77]. At room temperature, the values ranging from 100 to 400 μΩ cm are similar to the value of metallic $M_2Mo_6X_6$ compounds. The anisotropy of resistivity ($\rho_\perp/\rho_\parallel$) measured on a $Cs_2Mo_6Se_6$ single crystal is about 10^4. These compounds reveal a semiconductor-like behavior at low temperature, and a metallic behavior near room temperature.

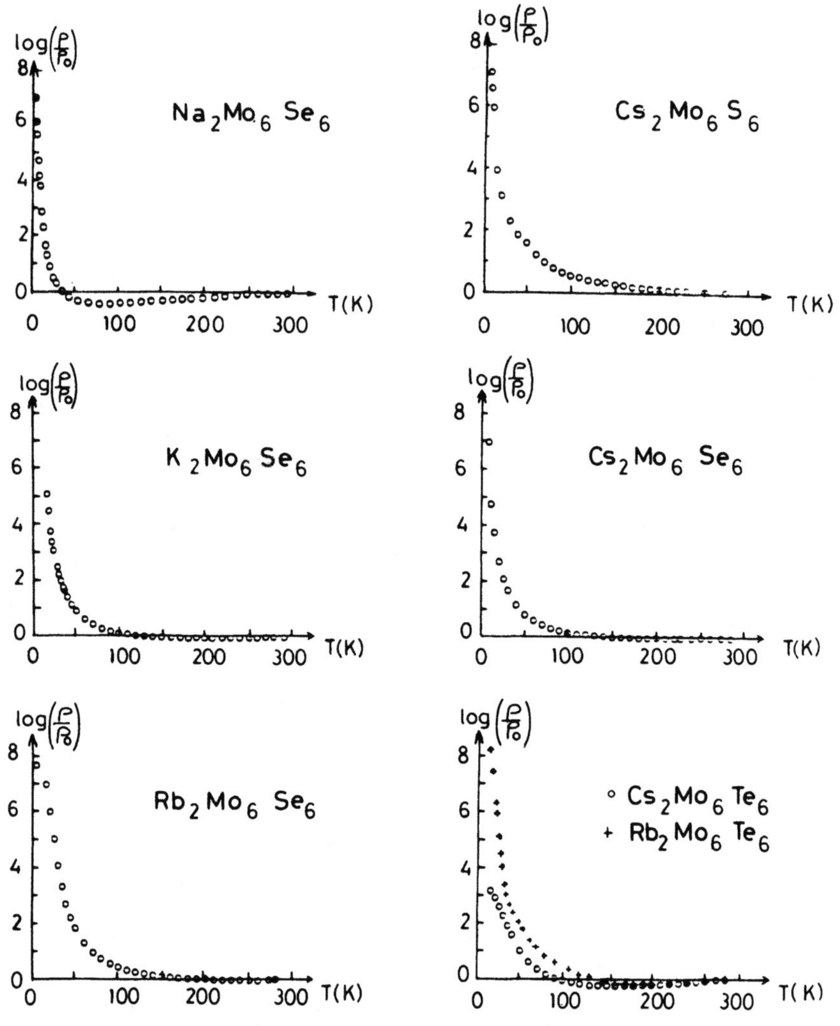

Fig. 49. Variation of the normalized resistivity (log scale) as a function of temperature for $M_2Mo_6X_6$ compounds.

Between both behaviors, the resistance goes through a broad minimum between 70 and 150 K. Below this minimum, the resistance exponentially arises to reach values of 10^3—10^8 higher than that of room temperature, as seen in Figure 49. It is obvious that the nature of the ternary element has a major effect on the physical properties of these materials. As seen above, the band structure calculations for the $(Mo_3Se_3)^1_\infty$ chain indicate that a more or less half-filled a_2 wide band crosses the Fermi level. This is consistent with the small magnitude of the observed Pauli susceptibility in these phases [79], smaller than those of Chevrel phases, corresponding with the low derived $N(E_F)$ densities of states at the Fermi level.

7. Conclusion

In an ideal 1D metal, especially with a half-filled band, the regular chain is unstable with, consequently, periodic lattice distortions (Peierls distortions): a semiconducting behavior at low temperature resulting from these lattice distortions is so predicted from the band structure calculations. In these $M_2Mo_6X_6$ compounds, the presence of the half-filled a_2 band will provide a driving force for a Peierls distortion. The effect of this distortion on the a_2 band will be to open up a band gap by splitting the two branches of the band, stabilizing filled levels at the zone edge, and destabilizing empty levels. That would manifest in alternatively long and short spaces between the Mo_3Se_3 units in the chain, implying a semiconducting ground state. No structural phase transition has yet been observed in the range of temperature corresponding to this metal-insulator electronic transition.

One of the more realistic explanations for this semiconducting-like behavior at low temperature (broad resistive transition) is the competition between the very weak interchain coupling and the very strong Mo—Mo intrachain interaction (a_2 band), which should prevent a sharp phase transition (Peierls transition). Another explanation is that the phase transition is completely suppressed by doping (impurities, defects, stoichiometry, . . .) as has been seen in the pseudo-1D systems (K_xMoO_3, MEM$(TCNQ)_2$, . . .). Now, if the interchain coupling is fairly large, the instability will be completely quenched and the properties will be typical of a somewhat anisotropic metal ($Tl_2Mo_6Te_6$) and even, it appears, a superconducting transition temperature ($Tl_2Mo_6Se_6$). This increase in interchain coupling is linked to the possibility of overlapping of the wave functions of the *p* levels of thallium ions with the *d*-Mo levels with respect to the *s* levels of alkaline ions. An analogous behavior is observed in the MMo_4O_6 compounds reported by Torardi and McCarley [88]. These compounds consist of 1D chains of *trans*-edge-sharing Mo—octahedral clusters. The transport properties are very similar to those of the $M_2Mo_6X_6$ compounds. A metallic behavior has been found in $InMo_4O_6$ and a broad metal-semiconductor transition in $NaMo_4O_6$ [88].

The $M_2Mo_6X_6$ compounds, although no sharp Peierls transition temperature has been seen, are remarkable 1D materials because of the very high anisotropy of their transport properties as mentioned above, and also because of the anisotropy of their polarized reflectance spectra [89]. In Figure 50, the optical reflectance of $Tl_2Mo_6Se_6$ single crystals was investigated between 0.2 and 5 eV under irradiation with polarized light. When the light polarized is parallel to the metallic *c* axis, a pronounced plasma edge is observed at 0.7 eV recalling the quasi-1D organic conductors and the KCP salts. The optical reflectance spectra give clear evidence for a strong electrical anisotropy. Recently, Tarascon *et al.* [90] have succeeded in dissolving the pseudo-1D $M_2Mo_6Se_6$ compounds when exposed to highly polar solvants: acetonitrile, dimethylsulfoxide or N-methylformamide. These solutions contain individual $(Mo_3X_3)^1_\infty$ chains, proving the weak interchain coupling and the quasi-1D character of these materials.

The architecture of chain-like $M_2Mo_6X_6$ compounds is very attractive, because of its extreme sensitivity to the interchain coupling, which opens the way for many

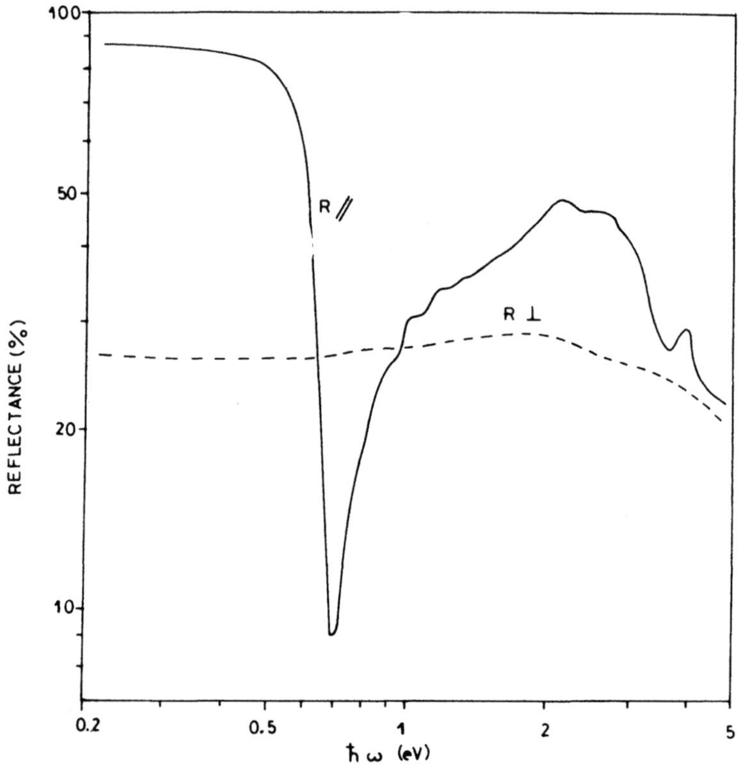

Fig. 50. Polarized reflectance spectra of needle-shaped single crystals of $Tl_2Mo_6Se_6$ (dashed line represents the mean value of the optical perpendicular reflectivity).

further fruitful chemical and physical investigations. However, a large amount of experimental work still remains to be done to understand their properties. These are fascinating materials that constitute the first large family of compounds exhibiting the crossover from tridimensional character to one-dimensional character with a quasi-continuous linear condensation of finite Mo_6—octahedral clusters $Mo_6 \rightarrow Mo_9 \rightarrow Mo_{12} \rightarrow Mo_{18} \rightarrow Mo_{24} \rightarrow Mo_{30} \rightarrow \cdots (Mo_{6/2})^1_\infty$.

Acknowledgements

The authors thank particularly Patrick Gougeon for allowing us to quote many results prior to publication and for his substantial contribution to the work summarized in this article. They are grateful to M. Potel, J. Padiou, O. Peña, A. Le Beuze, R. Lissillour, P. Monceau, and H. P. Geserich, for many stimulating discussions and for communications of unpublished data. They are also grateful to Mrs P. Poulain, Miss V. Boulais, and Mr J. C. Jegaden for their invaluable help in preparing the manuscript and drawings.

References

1. D. L. Kepert, *The Early Transition Metal*, Academic Press, New York (1972).
2. M. C. Baird, *Prog. Inorg. Chem.* **9**, 1 (1968).
3. R. B. King, *Prog. Inorg. Chem.* **15**, 287 (1972).
4. A. Simon, *Angew. Chem.* **20**, 1 (1981); *Ann. Chim. Fr.* **7**, 539 (1982).
5. H. G. von Schnering, *Z. Anorg. allg. Chem.* **385**, 75 (1971).
6. C. Perrin, Thesis, Rennes (1981); L. Leduc, Thesis, Rennes (1983); L. Leduc, A. Perrin, and M. Sergent, *C.R. Acad. Sci.* **296**, 961 (1983).
7. C. Perrin, M. Sergent, F. Le Traon, and A. Le Traon, *J. Solid State Chem.* **25**, 197 (1978).
8. C. Perrin and M. Sergent, *J. Chem. Res.* **5**, 38 (1983).
9. P. Gougeon, Thesis, Rennes (1984).
10. P. Gougeon, J. Padiou, M. Potel, M. Sergent, and M. Couach, *Ann. Chim. Fr.* **6**, 1083 (1984); P. Gougeon, M. Potel, J. Padiou, and M. Sergent, *C.R. Acad. Sci.* **297**, 339 (1983)
11. H. Schafer and H. G. von Schnering, *Angew. Chem.* **76**, 833 (1964).
12. C. C. Torardi and R. E. McCarley, *J. Am. Chem. Soc.* **101**, 3963 (1979).
13. D. A. Lokken and J. D. Corbett, *Inorg. Chem.* **12**, 556 (1973); A. Simon, N. Holzer, and H. J. Mattausch, *Z. Anorg. allg. Chem.* **456**, 207 (1979).
14. K. R. Poeppelmeier and J. D. Corbett, *Inorg. Chem.* **16**, 1107 (1977).
15. D. A. Adolphson and J. D. Corbett, *Inorg. Chem.* **15**, 1820 (1976).
16. R. L. Daake and J. D. Corbett, *Inorg. Chem.* **16**, 2029 (1977).
17. K. R. Poeppelmeier and J. D. Corbett, *Inorg. Chem.* **16**, 294 (1977).
18. J. D. Corbett, *Adv. Chem. Ser.* **186**, 329 (1980).
19. J. D. Corbett, *Acc. Chem. Res.* **14**, 239 (1981).
20. J. D. Corbett, *Pure Appl. Chem.* **56**, 1527 (1984).
21. J. Rouxel and A. Meerschaut (this volume, p. 205).
22. T. Sambongi (this volume, p. 281).
23. R. Chevrel, M. Sergent, and J. Prigent, *J. Solid State Chem.* **3**, 515 (1971). For reviews of the Chevrel phases: Ø. Fischer, *Appl. Phys.* **16**, 1 (1978); K. Yvon, *Curr. Top. Mat. Sci.* **3**, 53 (1979); R. Chevrel, 'Chevrel Phase High Field Superconductors' in *Superconductor Material Science* (eds S. Foner and B. B. Schwartz), chap. 10, p. 685 (Nato Asi Series), Plenum, New York (1981); Ø. Fischer and M. B. Maple (eds), 'Topics in Current Physics' in *Superconductivity in Ternary compounds*, I and II, Springer-Verlag, Berlin, Heidelberg, New York (1982).
24. R. Chevrel and M. Sergent, 'Chemistry and Structure of Ternary Molybdenum Chalcogenides', *Superconductivity in Ternary Compounds*, I, (eds Ø. Fischer and M. B. Maple) Springer-Verlag, Berlin, Heidelberg, New york, p. 25 (1982).
25. R. Chevrel, M. Sergent, and J. Prigent, *Mat. Res. Bull.* **9**, 1487 (1974).
26. R. Schollhorn, M. Kumpers, and J. O. Besenhard, *Mat. Res. Bull.* **12**, 781 (1977); R. Schollhorn, M. Kumpers, and D. Plorin, *J. Less Common Met.* **58**, 55 (1978); R. Schollhorn, M. Kumpers, A. Lerf, E. Umlauf, and W. Schmidt, *Mat. Res. Bull.* **14**, 1039 (1979).
27. M. Potel, P. Gougeon, R. Chevrel, and M. Sergent, *Rev. Chim. Min.* **21**, 509 (1984).
28. C. D. Grossman, D. P. Olsen, and C. H. Duffey, *J. Chem. Phys.* **38**, 73 (1963); F. A. Cotton and T. E. Haas, *Inorg. Chem.* **3**, 10 (1964); S. F. A. Kettle, *Theor. Chim. Acta* **3**, 211 (1965); L. J. Guggenberger and A. W. Sleight, *Inorg. Chem.* **8**, 2041 (1969); F. A. Cotton and G. G. Stanley, *Chem. Phys. Lett.* **58**, 450 (1978).
29. L. F. Mattheiss and C. Y. Fong, *Phys. Rev.* **B15**, 1760 (1977); O. K. Andersen, W. Klose, and H. Nohl, *Phys. Rev.* **B17**, 1209 (1978); H. Nohl, W. Klose, and O. K. Andersen, in *Superconductivity in Ternary Compounds* (eds Ø. Fischer and M. B. Maple) Springer-Verlag, Heidelberg, New York, chap. 6 (1981); D. W. Bullett, *Phys. Rev. Lett.* **39**, 664 (1977); T. Jarlborg and A. J. Freeman, *J. Magn. Magn. Mat.* **15–18**, 1579 (1980) and *Phys. Rev. Lett.* **44**, 178 (1980); A. Le Beuze, M. A. Makhyoun, R. Lissillour, and H. Chermette, *J. Chem. Phys.* **76**, 6060 (1982); D. Certain, A. Le Beuze, and R. Lissillour, *Solid State Commun.* **46**, 7 (1983); D. Certain and R. Lissillour, *Ann. Chim. Fr.* **6**, 953 (1984).

30. J. D. Corbett, *J. Solid State Chem.* **39**, 56 (1981); **37**, 335 (1981).
31. E. Teatum, J. Waber, and K. Gschneidner, quoted by W. B. Pearson, *The Crystal Chemistry and Physics of Metals and Alloys*, Wiley Interscience, New York, p. 151, (1972).
32. W. Honle, H. D. Flack, and K. Yvon, *J. Solid State Chem.* **49**, 157 (1981).
33. M. Potel, R. Chevrel, M. Sergent, M. Decroux, and Ø. Fischer, *C.R. Acad. Sci.* **288**, 429 (1979).
34. R. Chevrel, M. Potel, and M. Sergent, *Ternary Superconductors* (eds Shenoy, Dunlap and Fradin) Elsevier, North Holland, p. 7 (1981).
35. A. Grüttner, K. Yvon, R. Chevrel, M. Potel, M. Sergent, and B. Seeber, *Acta Cryst.* **B35**, 285 (1979).
36. R. Chevrel, M. Sergent, B. Seeber, Ø. Fischer, A. Grüttner, and K. Yvon, *Mat. Res. Bull.* **14**, 567 (1979).
37. R. Chevrel, M. Potel, M. Sergent, M. Decroux, and Ø. Fischer, *Mat. Res. Bull.* **15**, 867 (1980).
38. M. Potel, R. Chevrel, and M. Sergent, *Acta Cryst.* **B37**, 1007 (1981).
39. P. Gougeon, M. Potel, O. Peña, and M. Sergent, *1st National Conference of French Chemical Society*, Nancy, Sept. 1984.
40. K. Yvon, A. Paoli, R. Flukiger, and R. Chevrel, *Acta Cryst.* **B33**, 3066 (1977).
41. M. Potel, R. Chevrel, and M. Sergent, *Acta Cryst.* **B36**, 1319 (1980).
42. R. Chevrel, M. Potel, M. Sergent, M. Decroux, and Ø. Fischer, *J. Solid State Chem.* **34**, 247 (1980).
43. P. Gougeon, M. Potel, J. Padiou, and M. Sergent, *C.R. Acad. Sci.* **296**, 351 (1983).
44. P. Gougeon, J. Padiou, J. Y. Le Marouille, M. Potel, and M. Sergent, *J. Solid State Chem.* **51**, 218 (1984).
45. P. Gougeon, O. Peña, M. Potel, M. Sergent, and R. Brusetti, *Ann. Chim. Fr.* **6**, 1079 (1984).
46. R. Chevrel, P. Gougeon, M. Potel, and M. Sergent, *J. Solid State Chem.* **57**, 25 (1985).
47. T. Hughbanks and R. Hoffmann, *Inorg. Chem.* **21**, 3578 (1982); *J. Am. Chem. Soc.* **105**, 1150 (1983).
48. A. Le Beuze, R. Lissillour, H. Chermette, M. Potel, R. Chevrel, and M. Sergent, *Solid State Commun.* **43**, 19 (1982).
49. A. Perrin, M. Sergent, and Ø. Fischer, *Mat. Res. Bull.* **13**, 259 (1978).
50. A. Perrin, R. Chevrel, M. Sergent, and Ø. Fischer, *J. Solid State Chem.* **34**, 43 (1980).
51. A. Le Beuze, P. Lamande, and R. Lissillour, *Phys. Rev.* **B31** (1985).
52. R. Chevrel, M. Potel, M. Sergent, and J. Prigent, *Ann. Chim. Fr.* **7**, 92 (1982).
53. B. Seeber, M. Decroux, Ø. Fischer, R. Chevrel, M. Sergent, and A. Grüttner, *Solid State Commun.* **29**, 419 (1979).
54. M. Potel, Thesis, Rennes (1981).
55. D. Jerome, *Ann. Chim. Fr.* **6**, 921 (1984).
56. J. C. Armici, M. Decroux, Ø. Fischer, M. Potel, R. Chevrel, and M. Sergent, *Solid State Commun.* **33**, 607 (1980).
57. M. Potel, R. Chevrel, M. Sergent, J. C. Armici, M. Decroux, and Ø. Fischer, *J. Solid State Chem.* **35**, 286 (1980).
58. P. Gougeon, M. Potel, J. Padiou, M. Sergent, and P. Monceau, *Ann. Chim. Fr.* **6**, 1087 (1984).
59. W. Honle, H. G. von Schnering, A. Lipka, and K. Yvon, *J. Less Common Met.* **71**, 135 (1980).
60. J. Huster, G. Schippers, and W. Bronger, *J. Less Common Met.* **91**, 333 (1983).
61. M. Sergent, Thesis, Rennes (1969).
62. J. C. Tarascon, G. W. Hull, and F. J. Di Salvo, *Mat. Res. Bull.* **19**, 915 (1984).
63. K. Yvon, *Ternary Superconductors* (eds Shenoy, Dunlap, and Fradin), Elsevier, North Holland, p. 15 (1981).
64. M. Potel, R. Chevrel, and M. Sergent, *Acta Cryst.* **B36**, 1545 (1980).
65. For a review of recent works on pseudo one-dimensional materials, see *Chemistry and Physics of One-Dimensional Metals* (ed. H. J. Keller), Plenum, New York (1977); T. D. Schultz and R. A. Craven, *Highly Conducting One-Dimensional Solids* (eds J. T. Devreese, R. P. Evrard, and V. E. Van Doren) Plenum, New York, p. 147 (1979).
66. J. Chaussy, P. Haen, J. C. Lasjaunias, P. Monceau, G. Waysand, A. Waintal, A. Meerschaut, P. Molinie, and J. Rouxel, *Solid State Commun.* **20**, 759 (1976).

67. P. A. Lee, T. M. Rice, and P. W. Anderson, *Solid State Commun.* **4**, 703 (1974).
68. R. Horowitz, H. Gutfreund, and M. Weger, *Phys. Rev.* **B12**, 3174 (1975).
69. J. Friedel and D. Jerome, *Contemp. Phys.* **23**, 583 (1982).
70. S. S. Parkin, E. M. Engler, R. R. Schumaker, R. Lagier, V. Y. Lee, J. C. Scott, and R. L. Greene, *Phys. Rev. Lett.* **50**, 270 (1983); S. S. Parkin, E. M. Engler, R. R. Schumaker, R. Lagier, V. Y. Lee, J. Voiron, K. Carneiro, J. C. Scott, and R. L. Greene, *J. Phys. C*, 3, **44**, 791 (1983).
71. R. C. Lacoe, S. A. Wolf, P. M. Chaikin, F. Wudl, and E. Aharon-Shalom, *Phys.Rev.* **B27**, 1947 (1983).
72. P. Monceau, J. Peyrard, J. Richard, and P. Molinie, *Phys. Rev. Lett.* **39**, 161 (1977).
73. K. Yamaya, T. H. Geballe, J. F. Kwak, and R. L. Greene, *Solid State Commun.* **31**, 627 (1979).
74. J. C. Armici, M. Decroux, Ø. Fischer, M. Potel, R. Chevrel, and M. Sergent, *Solid State Commun.* **33**, 607 (1980).
75. R. Lepetit, P. Monceau, M. Potel, R. Chevrel, and M. Sergent, in *Proc. IVth Conf. on Superconductivity in d and f Band Metals, Karlsruhe, June 1982* (eds W. Buckel and W. Weber) p. 133 (1982).
76. R. Lepetit, P. Monceau, M. Potel, P. Gougeon, and M. Sergent, *J. Low Temp. Phys.* **56**, 219 (1984).
77. P. Gougeon, M. Potel, J. Padiou, M. Sergent, and P. Monceau, *Proc. Conf. on Cryogeny and Superconducting Materials, Rennes, June 1984.*
78. T. Mori, Y. Yokogawa, A. Kobayashi, Y. Sasaki, and H. Kobayashi, *Solid State Commun.* **49**, 249 (1984).
79. J. M. Tarascon, F. J. Di Salvo, and J. V. Waszczak, *Solid State Commun.* **52**, 227 (1984).
80. R. E. Peierls, *Quantum Theory of Solids*, Oxford Univ. Press (1964).
81. L. A. Turkevitch and R. A. Klemm, *Phys. Rev.* **B19**, 2520 (1979).
82. W. Lawrence and S. Doniach, *Proc. 12th Int. Conf. on Low Temperature Physics, Kyoto 1970* (ed. Eizo Kanda), Academic Press of Japan, Kyoto, p. 361 (1971).
83. R. R. Hake, *Appl. Phys. Lett.* **10**, 189 (1967).
84. N. P. Ong and J. W. Brill, *Phys. Rev.* **B18**, 5265 (1978).
85. M. Yamamoto, *J. Phys. Soc. Japan* **45**, 431 (1978).
86. S. Z. Huang, J. J. Mayerle, R. L. Greene, M. K. Wu, and C. W. Chu, *Solid State Commun.* **45**, 749 (1983).
87. R. D. Barnard, *Thermoelectricity in Metals and Alloys*, Taylor and Francis, London (1972).
88. K. H. Lee and R. E. McCarley, A.C.S. Meeting, St Louis, April 8—13, 1984.
89. H. P. Geserich (private communication).
90. J. M. Tarascon, F. J. Di Salvo, D. W. Murphy, and G. W. Hull, A.C.S. Meeting, St Louis, April 8—13, 1984; J. M. Tarascon, F. J. Di Salvo, C. H. Chen, P. J. Carroll, M. Walsh and L. Rupp. *J. Solid State Chem.* **58**, 290 (1985).

INDEX OF FORMULAS

INDEX OF NAMES

INDEX OF SUBJECTS

GPSR Compliance

The European Union's (EU) General Product Safety Regulation (GPSR) is a set of rules that requires consumer products to be safe and our obligations to ensure this.

If you have any concerns about our products, you can contact us on ProductSafety@springernature.com

In case Publisher is established outside the EU, the EU authorized representative is:

Springer Nature Customer Service Center GmbH
Europaplatz 3
69115 Heidelberg, Germany

Batch number: 10815974

Printed by Printforce, the Netherlands